W9-AJO-518

Life Histories of
North American Birds of Prey

by

Arthur Cleveland Bent

in two parts

Part 1

Dover Publications, Inc.
New York   New York

This new Dover edition, first published in 1961, is an unabridged and unaltered republication of the work first published in the United States Government Printing Office. Part 1 was originally published in 1937 as Smithsonian Institution United States National Museum *Bulletin 167;* Part 2 was originally published in 1938 as Smithsonian Institution United States National Museum *Bulletin 170.*

Manufactured in the United States of America

Dover Publications, Inc.
180 Varick Street
New York 14, N. Y.

ADVERTISEMENT

The scientific publications of the National Museum include two series, known, respectively, as *Proceedings* and *Bulletin*.

The *Proceedings* series, begun in 1878, is intended primarily as a medium for the publication of original papers, based on the collections of the National Museum, that set forth newly acquired facts in biology, anthropology, and geology, with descriptions of new forms and revisions of limited groups. Copies of each paper, in pamphlet form, are distributed as published to libraries and scientific organizations and to specialists and others interested in the different subjects. The dates at which these separate papers are published are recorded in the table of contents of each of the volumes.

The series of *Bulletins*, the first of which was issued in 1875, contains separate publications comprising monographs of large zoological groups and other general systematic treatises (occasionally in several volumes), faunal works, reports of expeditions, catalogs of type specimens, special collections, and other material of similar nature. The majority of the volumes are octavo in size, but a quarto size has been adopted in a few instances in which large plates were regarded as indispensable. In the *Bulletin* series appear volumes under the heading *Contributions from the United States National Herbarium*, in octavo form, published by the National Museum since 1902, which contain papers relating to the botanical collections of the Museum.

The present work forms No. 167 of the *Bulletin* series.

ALEXANDER WETMORE,
Assistant Secretary, Smithsonian Institution.

WASHINGTON, D. C., *March 12, 1937.*

CONTENTS

INTRODUCTION

This is the tenth in a series of bulletins of the United States National Museum on the life histories of North American birds. Previous numbers have been issued as follows:

107. Life Histories of North American Diving Birds, August 1, 1919.
113. Life Histories of North American Gulls and Terns, August 27, 1921.
121. Life Histories of North American Petrels and Pelicans and their Allies, October 19, 1922.
126. Life Histories of North American Wild Fowl (part), May 25, 1923.
130. Life Histories of North American Wild Fowl (part), June 27, 1925.
135. Life Histories of North American Marsh Birds, March 11, 1927.
142. Life Histories of North American Shore Birds (pt. 1), December 31, 1927.
146. Life Histories of North American Shore Birds (pt. 2), March 24, 1929.
162. Life Histories of North American Gallinaceous Birds, May 25, 1932.

The same general plan has been followed, as explained in previous bulletins, and the same sources of information have been utilized. The nomenclature of the new Check-List of the American Ornithologists' Union has been followed, but it has seemed best to continue in the same order of arrangement of families and species as given in the old (1910) check-list.

An attempt has been made to give as full a life history as possible of the best-known subspecies and to avoid duplication by writing briefly of the others and giving only the characters of the subspecies, its range, and any habits peculiar to it. In many cases certain habits, probably common to the species as a whole, have been recorded for only one subspecies; such habits are mentioned under the subspecies on which the observations were made. The distribution gives the range of the species as a whole, with only rough outlines of the ranges of the subspecies, which cannot be accurately defined in many cases.

The egg dates are the condensed results of a mass of records taken from the data in a large number of the best egg collections in the country, as well as from contributed field notes and from a few published sources. They indicate the dates on which eggs have been actually found in various parts of the country, showing the earliest and latest dates and the limits between which half the dates fall, the height of the season.

The plumages are described only in enough detail to enable the reader to trace the sequence of molts and plumages from birth to maturity and to recognize the birds in the different stages and at the different seasons. No attempt has been made to describe fully the adult plumages; this has been done very well in the many manuals now available. The names of colors, when in quotation marks, are taken from Ridgway's Color Standards and Color Nomenclature (1912), and the terms used to describe the shapes of eggs are taken from his Nomenclature of Colors (1886). The boldface type in the measurements of eggs indicates the **four** extremes of the measurements.

Many of those who contributed material for previous bulletins have continued to cooperate. Receipt of material from over 365 contributors has been acknowledged previously. In addition to these, our thanks are due to the following new contributors: Klauss Abegg, R. G. Bee, Henry Beston, J. C. Braly, J. F. Brenckle, J. V. Coevering, A. A. Cross, C. T. Dalgety, F. R. Decker, H. C. Denslow, J. B. Dixon, W. S. Duncan, C. L. Field, F. H. Fowler, A. F. Ganier, H. K. Gloyd, W. A. Goelitz, W. C. Hanna, H. L. Harllee, Eric Hearle, H. G. Heggeness, John Helton, Jr., F. H. Holmes, J. C. Howell, R. H. Imler, L. B. Kalter, Curtis Kingsbury, W. S. Long, E. D. Lumley, V. L. Marsh, J. H. McNeile, Lotta T. Melcher, D. V. Messer, James Moffitt, J. A. Moore, T. E. Musselman, Margaret M. Nice, W. H. Nicholson, W. P. Owen, Theed Pearse, Mrs. H. R. Peasley, J. S. Rowley, C. D. Scott, A. R. Sharp, Jr., L. O. Shelley, C. F. Smith, F. R. Smith, G. D. Sprot, Lawrence Stevens, Paul Thompson, R. W. Tufts, C. E. Underdown, H. S. Vaughn, L. H. Walkinshaw, and R. S. Woods.

Through the courtesy of the Bureau of Biological Survey, the services of Frederick C. Lincoln were again obtained to compile the distribution paragraphs. With the matchless reference files of the Biological Survey at his disposal, his many hours of careful work have produced results far more satisfactory than could have been attained by the author, who claims no credit and assumes no responsibility for this part of the work.

Dr. Charles W. Townsend and Dr. Winsor M. Tyler rendered valuable assistance in reading and indexing, for this group, the greater part of the periodicals relating to North American birds, which saved the author many hours of tedious work and for which his thanks are due. Dr. Townsend also contributed the entire life histories of two species, Dr. Tyler one, and the Rev. F. C. R. Jourdain one.

Thanks are due to the late Owen Durfee for many hours of careful work in collecting and arranging a great mass of data on egg dates, and to F. Seymour Hersey for figuring egg measurements.

The manuscript for this volume was completed in May 1936. Contributions received since then will be acknowledged later. Only information of great importance could be added. The reader is reminded again that this is a cooperative work; if he fails to find in these volumes anything that he knows about the birds, he can blame himself for not having sent the information to

THE AUTHOR.

LIFE HISTORIES OF NORTH AMERICAN BIRDS OF PREY

ORDER FALCONIFORMES (Part 1)

By Arthur Cleveland Bent
Taunton, Massachusetts

Order FALCONIFORMES

Family CATHARTIDAE: American Vultures

GYMNOGYPS CALIFORNIANUS (Shaw)

CALIFORNIA CONDOR

HABITS

Far from the haunts of man, in the wilder portions of southern California, among the most rugged and rocky gorges and canyons of the less frequented mountain ranges, this magnificent vulture, the largest and grandest of its tribe, still survives. Here in the remote fastnesses of the untamed wilderness it still finds comparative freedom from the dangers of advancing civilization and may long continue to exist. To see one of these great birds in the solitude of its native haunts gives a thrill well worth the time and effort required. Few have enjoyed the experience, and many are not equal to the task. Only once have I had the opportunity, on March 17, 1929, when the Peyton brothers guided us to the home of the condors in the mountains of Ventura County. It was a long hard climb up a steep, brush-covered slope to the top of a ridge and then a long walk down a wooded mountain trail to the head of a deep rocky canyon. From the trail we could look across the canyon to the rocky summits of the mountains, the home of the condors, where we were delighted to see four of the great birds soaring above the summits or sitting on the rocks. Once two of them sailed over us, near enough for us to see their yellow heads and the conspicuous white patches in their wings. As the trail dipped down into the canyon we found ourselves in the bed of a rocky mountain stream, where we separated to visit three former nesting sites of the condors. My party scrambled down the rough bed of the stream and then up a very steep,

1

rocky slope, climbing over the rocks and clawing our way over or around cliffs, for an hour and a half, until we reached a huge, irregular boulder perched on the shoulder of the mountain, near the top. Under this boulder, in more or less remote cavities, were three old nesting sites of the California condor, one of which was quite open and visible from the outside. L. G. Peyton said that the condors had not nested under this rock for several years, but the nests smelled and looked as if they had been occupied more recently. I brought home three large black feathers as trophies. (Pl. 2.)

The California condor has never enjoyed a wide distribution, being confined mainly to the hot interior valleys and mountains of California, west of the Sierra Nevada. It formerly ranged north to the Columbia River and even Vancouver Island, as a straggler; its range also extended south into northern Lower California. With the spread of civilization in California its numbers have been steadily reduced and its range gradually restricted to the few remote localities where it is still found. Several of the earlier writers on California birds noted the alarming decrease and predicted its early extinction. I doubt if it was ever an abundant species, as compared with the other vultures, although "Dr. Canfield informed Dr. Cooper that he has seen as many as one hundred and fifty of these birds at one time and place in the vicinity of antelopes he had killed" (Baird, Brewer, and Ridgway). William R. Flint wrote to Major Bendire (1892): "The largest number I have ever seen at one time during late years was in the summer of 1884, when I saw fourteen together." Major Bendire himself had seen "from six to fifteen on several occasions" in Inyo County. Mrs. Bailey (1902) says that "in 1894 Mr. Stephens actually encountered a flock of twenty-six of these magnificent birds." I was told of 17 being seen at one time recently in Ventura County. The birds seem to be holding their own in certain restricted localities and might survive permanently if rigidly protected and if poisoning were stopped.

Nesting.—For most of our information on the home life of the condor we are indebted to William L. Finley (1906), who found a nest in the mountains of southern California on March 10, 1906; by making many subsequent visits to it, he gave us a very interesting life history study of this species. He describes the nest as follows: "We climbed to the rock above and found it was a huge bowlder set well into the mountain. Against this was leaning a big stone slab about ten feet high. This left a space about two by six feet and open at each end. This cave was lined with leaves and fine rock and in the middle was one big egg. We thought it was not far from hatching by its glossy surface and the tenacity with which the mother stayed on her nest."

The condor is no nest builder but lays its single egg on the bare soil, gravel, or rocky floor of some more or less inaccessible cave or crevice in a cliff, or under rocks or boulders on the side of a mountain canyon. Sometimes the crevice is barely large enough to admit the bird and at other times it is quite open. H. R. Taylor (1895) tells of a nest in a large open cave "about 20 feet wide, 30 feet high, and 16 feet deep" in a cliff 120 feet high on the south side of a mountain. "The nest was on the bare stone. In front was a slight ridge of decomposed stone, which had been raked up by the bird to keep the egg from rolling out, while on the other side was the bare rock." Bendire (1892) mentions, apparently on hearsay evidence, "the eggs having been laid in the hollow of a tall old robles oak, in a steep barranca, near the summit of one of the highest peaks." Again he says that "it is possible that at times they make use of the abandoned nests of the Golden Eagles." Both of these statements seem doubtful and need confirmation.

A nest found in San Luis Obispo County is thus described by W. L. Dawson (1923):

The aperture of the nesting cave was midway of the face of a sloping stretch of sandstone, not too steep, perhaps, for inspection without the aid of a rope, but too steep for comfortable work. The entrance was just twelve inches high in the clear and nineteen inches wide; but the struggles of the emerging birds had broken out fragments of the thin wall on each side, so that three inches of this total width was plainly "artificial." This opening gave access to a lens-shaped cavity some six feet in horizontal depth by ten in length and two or two and a half feet high in the clear. The floor was of fine dry sand several inches in depth, and upon this at the remotest distance a baby Condor hissed and roared.

There are three California condors confined in a large flying cage in the National Zoological Park in Washington. They were received, as birds of the year, in 1901 and 1903. Two of them are supposed to be a mated pair. When about 12 years old one of these birds laid an egg on the bare floor of a large wooden shelter, and she has continued to lay an egg nearly every year since. But the eggs have never hatched, even when placed in an incubator.

Eggs.—The California condor lays only one egg in a season; and apparently it does not lay every year; hence it reproduces very slowly. The egg is quite elongated, varying in shape from elliptical-ovate to elongate-ovate. The shell is finely granulated and without gloss when fresh; after it has been incubated for some time it becomes smoother and glossier. Some specimens have small pimples or wartlike excrescences on the surface. The color is plain greenish white, bluish white, or dull white. The measurements of 46 eggs average 110.2 by 66.7 millimeters; the eggs showing the four extremes measure **120** by 68, 110.5 by **71, 102.4** by 67.4, and 103.6 by **62.9** millimeters.

Young.—The period of incubation is said to be from 29 to 31 days. Whether both sexes incubate does not seem to be known.

Mr. Finley (1906) was fortunate enough to begin his study of the young bird at the very beginning, for on his second visit the condor chick had only recently hatched. He writes:

> When we climbed over where we could look between the rocks and see into the cave, the old bird was on. I went closer and could see her bald head of orange color, and the great black bird still sat on the nest. I climbed up within four feet of her and whistled and yelled till she rose on her feet. She looked so big that I shrank back at the thought of her pitching in to defend her young, for when she rose, I glanced in and saw a youngster not larger than the egg [see pl. 3]. His head was bald like his mother's, but baldness did not signify age in this case, altho his head was fleshy-pink in color. He was weak for he could hardly kick, and he seemed to raise his head with difficulty as he cried out in a wheezing, hissing note. Beside him lay the end of the egg from which he had emerged not many hours before. He was not yet dry. He was not even well clothed, for behind his little wings, the flesh was bare and his belly was bare, while the rest of his coat was down of pure white.
>
> At first the mother arose and her neck feathers ruffled up in anger. Then as her baby began to squirm, she put her head down and covered him partly with her bare neck.

Being unable to scare the old bird from the nest and wishing to photograph the young one, Mr. Finley gently removed the young bird. The little fellow became so chilled during this process that its mother would not accept it. Mr. Finley revived it with the heat from his own body and returned it to the nest again. "For an instant she paid no attention to him, but just then he began to stir and wriggle. Her eyes changed from their vacant stare; she suddenly seemed to recognize her nestling, and putting her bill down she drew him gently near, crouching down at the same time and finally drawing him under her breast."

When Mr. Finley (1908) made his third trip to the nest, on April 11, 1906, the young condor was 19 or 20 days old. He says:

> When we climbed around to the nest, we found the condor nestling had grown from the size of the egg, or from about a double handful, till he filled my hat. The down on his body had changed color from a pure white to a light gray. Instead of the flesh color on his head and neck, it had changed to a dull yellow. He sat with his shoulders humped and his head hung as if in the last stage of dejection. The minute he saw me, he began crying in a note most peculiar for a bird, for it sounded exactly like the hoarse tooting of a small tin horn. However, he only used this note a few times; then he began hissing. He showed his resentment by drawing in his breath and letting it escape as if thru his nose. His feet were short and stubby, the feet of a scavenger. What a deterioration from the eagle! The claws were like those of a chicken rather than a bird of prey. The head, the bill and even the look in the eye were very different from the savage expression of the eagle even in his babyhood.

On April 25, they found the old bird sound asleep in the nest, brooding the nestling; after she left the youngster showed fight.

The young condor was growing steadily, for he was now thirty-five days old and as large as a good-sized chicken [see pl. 3]. His whole body was covered with dark gray down with the outer edgings of lighter gray. When I put down my elbow, he lunged forward and struck it such a hard blow with his bill that it would have drawn blood had he hit my bare hand. The minute I appeared, his neck puffed out with wind and his whole crop filled till it felt just like a rubber ball. He seemed to use his crop as a supply tank for air, which he blew out slowly thru his nose to express his anger. He sat with his head down and mouth open. The front part of his tongue was round and it folded over from each side and met in a little crease down the front. About an inch back, it looked as if it were partly cut in two, for it was narrower and flatter. Such a breath as that youngster had! I could not describe it, and I tried to forget it as soon as possible.

That evening we watched the old condor to see if she would go back to the nest. But at six o'clock she settled down on her perch with her head drawn in, and went to sleep. The young condor had to sleep alone.

Of the later development of the young condor Mr. Finley (1908) says: "The young condor was now fifty-four days old, but he was still clothed in gray down [see pl. 4]. It was over two months before the first black feathers began to show on his wings, and they developed very slowly; for by the first week in July when we had expected to complete our series, the young bird was not half feathered out, altho he was three months and a half old and weighed over fifteen pounds."

By July 6 the young condor was about two-thirds grown and was transported to Mr. Finley's home in Oregon. He was fed twice a day with about a pound of raw meat and given plenty of water. He showed a decided preference for fresh beef and would reject anything else, unless forced by hunger; he especially disliked any stale or tainted meat. He made a most interesting pet and was very tame, affectionate, and playful. By the middle of August he "was well fledged except that his breast was still covered with gray down. By another month this was replaced by brown feathers. With wings extended, he measured over eight feet. He weighed twenty and a half pounds and was forty-six inches in length. The wing feathers were strong, but they could not yet support his heavy body, for as yet he could fly but a few yards." (Pl. 4.)

On September 29, 1906, General, as he was named, went to New York to take up permanent quarters in the New York Zoological Park. His former master was not forgotten, however, for Mr. Finley (1910) writes:

During the month of December, 1906, while I was in New York, I went out to see General and was allowed to enter the cage with him. The minute I got near enough, he began nibbling my buttons and putting his head under my arm.

I did not see the young condor again until December 6, 1908, when I was in New York. I again entered his cage and found him as friendly and affectionate as ever. He nibbled the buttons on my coat and wanted to be petted. I was very much surprised to find that he showed no signs of bright color about his head, as it was covered with short gray down. He had been in good health, but at the age of almost three years he had not acquired the bright coloring of his parents. It is interesting to note that the head of a newly-hatcht condor, as well as that of the old bird, is perfectly bald; yet the head of the immature condor for the first few years is covered with a thick coat of furry down.

Plumages.—The foregoing quotations from Mr. Finley's articles tell us all we know about the development of plumages in the young condor, white down at first, followed by gray down, the first plumage appearing during the fourth month.

Even at three years of age the condor's head was still covered with gray down, showing none of the bright colors of the adult. Mr. Finley (1908) describes the colors of the adult as follows:

Their bills were of dark horn color and the red skin of the head extended down covering the bill about half way. The feet were of similar color, but on each knee was a patch of red. There was a brighter patch of red on the breast of each bird, which could occasionally be seen when they were preening and when they spread their breast feathers. Both had light-colored wing-bars and the primaries were well worn. The skin on the throat hung loose and the lower mandible fitted in close under the upper, giving the bird a peculiar expression. The chin was orange and below this on the neck was a strip of greenish-yellow merging into brighter orange on the sides and back of the neck. The top and front of the head were bright red, but between the eyes was a small patch of black feathers, and these extended down in front of the eye till they faded into the orange red of the neck. The pupil of the eye was black, but the iris was deep red and conspicuous. The top of the head was wrinkled as if with age. The ruff, or long shiny black feathers about the neck, was often ruffled up, giving the bird a savage appearance. Behind the ruff on the back the feathers were edged with dark brown.

I have not seen enough material to work out the molts of adults, but, as the parents referred to above were in worn plumage in July and as Mr. Finley (1908) saw a third bird with feathers missing from wings and tail, a complete molt probably occurs late in summer. Experience with the birds in the National Zoological Park suggests that even in a wild state the young birds require a long time to reach the breeding age. Young birds show little or no white in the under wing coverts.

Food.—Mr. Finley's young condor was a very clean feeder, rejecting any meat that was not fresh or the bodies of dead game birds and mammals. When the flesh of squirrels or birds was mixed with fresh beef, he would always pick out the beef and leave the other things. But the California condor is a vulture and naturally has food habits similar to those of other vultures, though it probably prefers to feed on a freshly killed animal. Baird, Brewer, and Ridgway (1905) have covered the subject very well, as follows:

Often when hunting in the Tejon Valley, if unsuccessful, they would be several hours without seeing one of this species; but as soon as they succeeded in bringing down any large game, these birds would be seen rising above the horizon before the body had grown cold, and slowly sweeping towards them, intent upon their share of the game. In the absence of the hunter, unless well protected, these marauders will be sure to drag out from its concealment the slain animal, even though carefully covered with branches. Dr. Heermann states that he has known them to drag out and devour a deer within an hour. This vulture possesses immense muscular power. Dr. Heermann has known four of them to drag the body of a young grizzly bear, that weighed over a hundred pounds, the distance of two hundred yards. Dr. Cooper states that it visits the Columbia River in autumn, when its shores are lined with great numbers of dead salmon, on which, in company with other birds and various animals, it feasts for a couple of months.

Behavior.—The flight of the California condor is a superb exhibition of graceful ease and grandeur as it floats steadily along on its great wings, a powerful and skillful master of the air. On account of its great size its flight seems slow, but it really travels very fast; a mere speck in the distant sky rapidly develops into a great black bird, sweeping overhead with seven or eight strokes of its white-lined wings, curved upward at the tips, followed by prolonged periods of graceful sailing, until, all too soon, it disappears in the distance. From its perch on a tree or rock the bird launches itself with a few great flaps into a glorious sailing flight; but when rising from the ground it must run, hop, and flap along for 50 or 60 feet before taking the air, much like the take-off of an airplane. Then it soars in wide circles, mounting higher and higher on the ascending currents of warm air, until it is almost lost to sight in the ethereal blue. Illustrating its mastery of the air, Mr. Dawson (1923) relates the following incident, as witnessed by Claude C. L. Brown:

Just because the sails of this bird are so accurately trimmed for the utilization of light breezes, the craft itself is unable to make headway against a strong wind. Not even by flapping can the Condor negotiate a breeze above a certain intensity. What the bird does in such an emergency is best told by Brown, who was once present on a quite critical occasion. * * * Presently he descried four Condors approaching from the far northeast, but before they came up a smart breeze sprang up from the southwest, and presently it whistled over the peaks with increasing fury. The birds were baffled on the very last mile of their approach. They tacked back and forth, down wind, or struggled valiantly in the teeth of the gale, only to be swept away again and again. The cold sea breeze had it in for them, and though it was only mid-afternoon, it began to look to the observer like a case of sleeping out that night. But off to the southeastward some twenty or thirty miles, the Carisso plains lay baking in the sun. The focal point of this great oven was sending up a huge column of heated air, as evidenced by clouds slowly revolving at the height of a mile or so above the plain. What followed can best be given in Mr.

Brown's own words: "Presently one of the Condors gave up the fight, sailed a mile or so to the eastward, and, after circling to gain elevation, made away in a bee-line for the southeast. In a short time the other three went through the same manoeuver and followed after their companion. I now brought my telescope into action and I never took the glass off the birds although they became mere specks in the sky. The Condors did not swerve from their course until they entered the spiral cloud. Upon striking that ascending column of air they rose rapidly, apparently without effort, as a balloon might rise, being now and again lost to view in the fleecy folds of ascending vapor, until within an incredibly short space of time they emerged above the clouds, into a higher region of absolute clearness, say three miles above the earth. Here they must have found themselves well above and quite free from the lower currents of air which had plagued them, for now they sailed straight to the westward, descended and—glided triumphantly homeward on the wings of their ancient enemy, the southwest gale!

"I do not think that more than thirty minutes had elapsed from the time the Condors gave up the fight till they were safely at roost in their rookery; yet these birds must have traveled somewhere from fifty to seventy miles to accomplish their purpose, and the whole performance took place without the flap of a wing."

Audubon (1840) quotes J. K. Townsend as saying: "In walking they resemble a Turkey, strutting over the ground with great dignity; but this dignity is occasionally lost sight of, especially when two are striving to reach a dead fish, which has just been cast on the shore; the stately walk then degenerates into a clumsy sort of hopping canter, which is any thing but graceful."

California condors are generally considered to be very shy birds; most observers have been unable to approach them near enough for an effective shot with a gun or even a rifle; but there are exceptions to the rule. W. R. Flint in 1884, was able to approach to within 30 yards of a flock of 14, according to Bendire (1892); and Dr. Cooper (1890) walked right up to an apparently healthy adult bird and could have killed it with a hammer. Mr. Finley (1908) won the confidence of the pair of condors that he studied, as is amply illustrated in the marvelous series of pictures he and Mr. Bohlman took within a few feet of them. These birds were very gentle and affectionate with each other and with their offspring. Mr. Finley (1908) writes:

While ascending the steep slope to the nest, a large bowlder was accidentally loosened and narrowly missed taking the camera man along as it dropped into the canyon with a loud report. The next moment, the old condor, aroused from her nest, flapped to her perch in the dead tree directly over our heads. We watched and waited, hoping she would return to the nest. But after about fifteen minutes, she raised her wings, hooked her bill about the stump, parrot fashion, and climbed to a higher perch. We crawled on up behind a cover of rocks to get a picture. While fixing the camera, I looked up and the old male was just alighting beside his mate on the dead tree. We crouched down to watch. If the birds saw us, they paid no attention to our presence. The mother edged along the limb and put her head under his neck. Then she nosed him as if asking to be fed, but he responded rather coldly by moving away and she followed. This crowded him out where the limb was too small, and he jumped across back of her. He seemed to get more friendly and the two sat there side by side, nibbling and caressing each other.

He says also that "they were almost devoid of fear, for several times they stood within five or six feet of us in perfect unconcern." But they were not so friendly to a third condor that twice appeared on the scene; once the old male condor gave chase and eventually drove the intruder away.

Mr. Finley's condors were very clean about their nest, and the young captive bird seemed to be very fond of clear running water. A. M. Shields (1895) says: "The California Condor is a much cleaner bird than is generally accredited, as one of its favorite habits is to assemble on the bank of some secluded mountain pool and spend hours at a time in bathing and standing around the margin of the clear, cold water. Hunters on coming upon a far removed body of water in localities frequented by the birds, often find numbers of immense feathers around the edge of the stream, discarded by the birds during some of their fresh-water baths."

Carroll Dewilton Scott contributes the following notes:

As in the case of most birds and animals with strong individualities, condors appear to be fond of play. In the wild state this most often takes the form of swooping down at another condor. The other bird never seems to resent it and parries the pretended stroke with a deft turn of the body. After swooping at each other several times both birds will presently be sailing about in intersecting circles. One day while I was watching a pair about their nesting cave, both birds lit on neighboring rocks. After a few minutes one of them swung toward the other, and both navigated for a quarter of a mile, first one, then the other, making the dashes. At length they turned and calmly glided back to their respective lookouts. On another occasion, in winter, I was watching a group from a mountain ridge. Presently three immature condors came gliding overhead, their wings partly bowed, making a rushing sound like a stormwind through pine trees. As indicated by their future movements they were not going anywhere in particular. They were just playing. One of them evidently was surprised to see me, for he tried to turn so suddenly that he almost turned a somersault. But he recovered his balance and sailed back over me, then bowed his wings again and shot away in pursuit of his companions. In captivity, condors are fond of toying with bones, ribbons, or pieces of paper, no doubt to relieve the tedium of imprisonment. One day a friend of mine and I played for half an hour with a condor at the San Diego Zoo. His kittenish antics were laughable as he thrust his head along the sand or stuck his beak through the wire meshes of his cage coaxing us to give him attention.

Field marks.—The immense size of the California condor, larger than even the golden eagle, and the white under wing coverts are the most conspicuous characters. Its wings measure 9 or 10 feet in extent, and when the bird is soaring the tips of the primaries are curved upward and slightly forward. If near enough the brightly

colored head may be seen. Young birds have darker under wing
coverts and dusky heads and necks.

Enemies.—The condor has no enemies of consequence except man;
and man has gone a long way toward the extermination of this grand
species. It seems to be a common trait in human nature to want to
kill any large creature, and in the early days when these birds were
unsophisticated many were wantonly killed. Poisoned carcasses set
out to destroy predatory animals killed a great many of them. Many
condors were killed for their quills, which were useful for carrying
gold dust. In the early days it was easy to kill them with a rifle or
even with a shotgun loaded with buckshot. Sometimes, when gorged
with food, they could be lassoed or even killed with some missile
thrown at them. Mr. Shields (1895) says:

Among the latter contrivances for their destruction, one of the most frequently employed was "penning." This consisted of four-sided portable pens
about six feet square and five in height. These were placed in convenient
localities with the carcass of a sheep or goat temptingly displayed within; the
voracious bird would soon spy the tempting morsel, and settle down for a feast,
but when he came to rise it was different, as the small diameter of the pen
absolutely prevented the full stretching of his wings, and, being unable to
make the upright leap of four or five feet, he was a secure prisoner and an
easy prey to the herder and his club, when making the rounds of his traps. It
was strange that this bird, so conspicuously wary at the present time, should in
those days have manifested so little of that quality, as certain it is that the
traps would constantly claim their victims practically as long as the birds
held out.

There is a modern menace in the high-tension power lines, on which
many large birds are electrocuted by making a contact across the
wires when they spread their wings. As I have not seen such casual-
ties mentioned, perhaps the condors have not learned to alight on
such dangerous perches. Public sentiment now seems to favor the
condor, and, as it is protected by law, we hope it will long continue to
survive in the wilder portions of California, as one of the many glories
of the Golden State.

Carroll Dewilton Scott has sent me the following interesting notes
on the condor-killing ceremony practiced in primitive times by the
southern California Indians, as one of several mourning festivals:

Three birds were used as convenience dictated, the bald and golden eagles
and the California condor. One idea back of the ceremony was that the spirits
of the dead, especially the spirits of children, could mount to the Indian's
heaven on the wings of great fliers like the eagle and the condor. Another was
that the bird was a messenger from the living to the dead. Though an authentic
and picturesque incident in the life history of the condor, there is no evidence
that it played the least part in the destruction of condors that took place mainly
between 1875 and 1895, when Americans were rapidly settling the State.

The essential part of the ceremony was as follows: The Indians gathered
around a campfire in the evening. Groups of 10 or 20 Indians, under leaders

skilled in singing ancient songs and executing dances relating to the mythology of the ceremony, would perform. The dancing and singing continued until 2 or 3 o'clock in the morning. About this time the bird was brought in and danced with around the fire and passed from one performer to another. Finally the chief would seize the bird, now nearly dead with rough handling and suffocation. It was supposed to be killed by magic, without the shedding of blood, but was practically put to death by twisting its backbone or by pressure on its heart. It was then skinned, the feathers being saved for future dancing skirts, and the body placed in a hole within the circle of performance. Old women then gathered about the place of interment, threw seeds and food on the carcass, asked enigmatic questions of the dead bird, and indulged in weeping and lamentations. After the usual exchange of presents, wearing apparel and food, the party dispersed.

I examined two condor-feather dancing skirts that contained 48 and 70 plumes, respectively. The plumes were fastened at the quill end to cord belts made of twisted strands of milkweed fibers. The primaries were placed at ends of the belt, the secondaries and tail feathers in the middle. One of the skirts was made from a condor that was shot at the reservation in 1926.

Winter.—Mr. Scott has this to say about the concentration of condors in winter:

After the condor chick is half a year old and able to perch about the ancestral cave, the condors of a region congregate in companies as do the turkey vultures. The buzzards go south or gather along the California coast in "roosts", but the condors are nonmigratory. In earlier times, however, they were accustomed, no doubt, to move over wide areas. Pioneers testify that in the sixties and seventies within their chosen range flocks of condors were fully as large as those of buzzards. Where food was plentiful they often gathered in enormous numbers. One such concentration was witnessed by Hector Angel, of Mesa Grande. In March 1886, just after the buzzards had returned, a late snow killed 3,000 lambs on the famous Warner Ranch. Angel rode for a mile through acres of turkey vultures and condors. "There may have been 1,000 condors and 5,000 buzzards for all I can tell", he declared.

Nowadays it is rare for an individual condor to leave the protection of the Santa Barbara National Forest at any season. But the remnant gather as of old for social or feeding purposes. I had the good fortune to witness the activities of a band of 15 on the ranch of Eugene Percy, 8 miles northeast of Fillmore, on January 17, 1936. There were several dead cattle on the ranch and condors were in the sky all day. Twice the company of 15 sailed overhead. They glided and spiraled and shot through the air like rockets. They wheeled about old carcasses, sometimes alighting in trees or on the ground. In the evening they went to roost in groups of two to five in dead trees on the mountainsides, where it was possible to ride a horse around a tree without making them take wing. Only three days before, my host had counted 30 in the sky at once and had surprised a flock of 12 at the carcass of a dead calf.

DISTRIBUTION

Range.—Western and Southern United States and northern Lower California. Nonmigratory.

The California condor is now greatly reduced in numbers and is confined to the south-central coast ranges of California from Monterey, Bear Valley, and San Benito south to the Cuyamaca Mountains, Santiago Canyon, and Ventura County.

Its former range extended **north** to northern Oregon (mouth of the Columbia River and Multnomah). **East** to Nevada (cave remains near Las Vegas) and New Mexico (cave remains in Rocky Arroyo, northwest of Carlsbad). In the **south** the range extended into northern Lower California (San Fernando, Colorado Delta, Laguna Mountains, and San Pedro Martir Mountains).

Casual records.—Fleming (1924) records a specimen from Fort Vancouver, Wash., in the spring of 1827. In British Columbia, Fannin (fide Kermode, 1904) reported seeing two at Burrard Inlet in September 1880, while Rhoads (1893) states that condors were reported on Lulu Island as late as "three or four years ago."

This species undoubtedly was more widely distributed in geologic times, as Wetmore (1931) has identified condor bones in Pleistocene deposits of fossils from the Seminole area, near St. Petersburg, Fla. Abundant remains of California condors also have been obtained from the Pleistocene asphalt beds of Rancho La Brea, Los Angeles, Calif.

Egg dates.—California: 38 records, February 17 to May 28; 19 records, March 23 to April 25.

CATHARTES AURA SEPTENTRIONALIS Wied

TURKEY VULTURE

HABITS

CONTRIBUTED BY WINSOR MARRETT TYLER

When we travel southward along the Atlantic Seaboard, soon after we cross the invisible line that separates the Transition Zone from the Upper Austral (the map shows us to be in the State of New Jersey), we begin to see from time to time dark spots high up in the sky. They seem stationary at first sight; then, as we watch, we see that they are moving, swinging often in wide circles, and when one of them comes near, we see it is a great dark bird sailing through the air. We have entered the domain of the turkey vulture, the chief avian scavenger of the United States; a big bird with long broad wings, with a keen sense of sight and of smell and, utilitarian as well as aesthetic, a plumage that does not show the dirt, and a naked head and neck like the bare arms of a butcher; a bird of prey, one of the Raptores, but one that does not inflict death, but searches and watches and waits until it comes upon the dead. Then the feast begins.

Spring.—Toward the northern limit of the bird's breeding range an increase in its numbers is noted at the approach of spring. Thus

Thomas H. Jackson (1903), writing of southeastern Pennsylvania, says, "Early in April, with the advent of settled weather, they become quite numerous, and at once show an attachment for the old nesting sites, to which they seem to return for many years."

Louis B. Kalter reports (MS.) a flock of 30, apparently migrants, going westward over Yellow Springs, Ohio, on March 20, 1933, at 4.30 p. m. Dr. O. L. Inman reported to him that "during part of the time they were in sight, they seemed to hold rather a loose formation, when most of them would be going in the same general direction. Then they would break and wheel about for a short time, only to reform their loose formation."

Dr. F. M. Chapman (1933) made observations at Barro Colorado Island that point to an extensive northbound migration of vultures late in February and March. Large numbers of the birds, several hundred together on one occasion, passed over the island, following the course adopted by kingbirds and barn swallows on their route northward (at this point southwest). He says: "Usually they sailed straight ahead without stopping but at times they circled, though still drifting southward."

One day they flew over the island at a height of 4,000 to 5,000 feet, and on another day, in the morning, Dr. Chapman found a number of vultures collected in trees. "These", he says, "were apparently migrating birds which had roosted over night on the island and, becalmed, were waiting for enough wind to resume their flight."

Wherever these birds were bound—a question impossible to answer definitely—Dr. Chapman's observations indicate that the turkey vulture is, to a certain extent, highly migratory and that many individuals, gathered in flocks of considerable size, make a very long journey between their winter quarters and their breeding grounds.

Nesting.—In any region, no matter how widely it may range, there is a limited number of places in which a bird as large as a turkey vulture can hide its nest. The vulture is a big bird; it must have room for its nest somewhere either inaccessible to predatory animals or where they cannot easily reach its eggs or young. There is the added danger that the odor of the food may proclaim the whereabouts of the nest after the eggs are hatched and the young birds have to be fed.

Many situations meet these requirements in some degree, notably on precipitous cliffs, of access only through the air, or in caves or hollow stumps, or in the midst of dense shrubbery where a narrow entrance limits the attack by enemies to one direction. In such locations the vulture lays its eggs on the ground, or on the bare stone of a cliff, or on the rotten chips in a hollow log with little or no attempt to make a nest.

The literature contains descriptions of many nest sites of the vulture. The following citations show a variety of ways in which the bird has solved the problem of protecting its nest. The nest site is almost always on or near the ground, but in one case Isaac E. Hess (1910) found a nest "twenty feet up in a dead stump, and [another] six feet below the surface of the ground in the hollow of a rotten stump." Manley B. Townsend (1914) notes another tree nest. He says: "At the very top [of a gigantic elm tree] there was the hollow, dead shell of the main trunk, and, in this, upon the bare decayed wood, two eggs as large as Turkey eggs." William Lloyd (1887) speaks of the birds in western Texas as "breeding in caves, but frequently on the bare edge of a bluff", and in Texas also James J. Carroll (1900) mentions them as "selecting brush-heaps, clumps of chaparral, caves in arroya banks, and hollow trees." Dr. T. Gilbert Pearson (1919) says: "I have found the eggs of chese birds on a level with the ground in the hollow snag of an old tree, the entrance to which was at the top, 14 feet above."

In the two following quotations water plays the part of the ancient moat in defending the vulture's castle. In a letter to Mr. Bent, W. A. and George M. Smith describe a nest in Orleans County, N. Y. They write: "The nest we found was located in an old decayed hollow log which had fallen from its stump many years ago, and lay rotting amid a luxuriant growth of ferns and other swamp plants. There was nearly one foot of water all around, but the two eggs were placed on a bed of dry leaves and decayed wood." (See pl. 6.)

Russell M. Kempton (1927) describes a similar nest in detail thus:

The nest is in a live soft maple tree, whose trunk slants on a sixty degree angle east by north and has a southern exposure; inside dimensions of the cavity are diameter twenty-eight inches; height, forty-two inches and its bottom is about forty inches from the ground. The top of the cavity is closed by dry decayed wood. The surrounding ground is swampy and during wet seasons water stands thirty inches deep around the base of the tree.

The nest is unlined, and the eggs were deposited on clean broken up punk * * *. It was always clean (also the ground around the tree), from the time the eggs were laid until the nestlings left the nest. No offensive odors were noted during the five years of observation, (except when the nestlings would regurgitate for me).

Continuing, Kempton shows the vulture's simple method of making her nest with materials near at hand.

The parent birds arrived March 18, and used the same nest to roost in during the cold wet spring. On several occasions during daylight in April, I found them in the nest standing with heads together, and they did not fly when I approached within ten feet of the tree. Visiting the nest on April 28, I watched them preparing the nest, by pulling at the dry rotten wood on the side walls of the cavity with their beaks. When a large piece came loose the female would hold it down with one foot and tear it into small bits, which she spread about on the floor, where the eggs were to be deposited. The interested male bird, was a hindrance in nest making, and every now and then the female placed her head under his breast and pushed him out of the way. Once he tumbled out of the tree. However, undaunted, he clambered back keeping his head down, so that his mate could not repeat her attack.

Mr. Kempton observed that "both birds alternately covered the eggs during incubation."

Paul G. Howes (1926), in an account of a vulture's nest in a cave in the State of New York, says: "Another point of interest to me was that there was absolutely no odor about the nest." At this time the nest contained eggs, but on a later visit he remarks: "A very offensive odor emanated from the rocky shelter now for the first time, and as we approached very quietly, the old bird bounded clumsily to the rear of the cave. * * * The young one had hatched safely, had had its first taste of carrion, as its vile odor attested."

Thomas H. Jackson (1903) reports an unusual nest site. He says: "I found a pair that had taken possession of an abandoned pig-sty in the woods, which furnished them an admirable place to set up housekeeping. Unfortunately, the smooth board floor had allowed one of their two eggs to roll away, and only one was hatched. Here they were safe from the attack of foxes, raccoons or other night prowlers."

A. L. Pickens (1927), reporting some "out-of-the-ordinary" records, includes the following on the turkey vultures.

On May 1, 1927, I was at the home of Mr. Elihu Wigington in Anderson County, S. C., and he took me to an old and neglected barn in a wood near his home to see a nest of this bird. I found the eggs, two in number, on the refuse of the stable floor, close up in a corner. About ten feet away a domestic hen was brooding on her nest in a pile of forage, the two being separated, however, by a low partition. The vulture could gain access to its nest through a small window in the stable, or through a door at some greater distance. Mr. Wigington told me that this was the third year this place had been used by the Vulture for a breeding spot.

Of the time of nesting Bendire (1892) speaks as follows: "In most of the Southern States nidification begins usually about the latter part of March, occasionally even in February; in the Middle States generally about the last week in April or the beginning of May, and in the more northern portions of its range it may be protracted till June, according to the season."

Charles E. Doe writes to us of a set of eggs he took from an old caracara's nest, 20 feet up in a lone palmetto on a Florida prairie.

Eggs.—[AUTHOR'S NOTE: The turkey vulture lays almost invariably two eggs, occasionally only one and very rarely three. I have a photograph (pl. 7) of a nest containing four, but they were in pairs a short distance apart and were probably the product of two females. The eggs are usually elliptical-ovate or elongate-ovate in shape, but a few are ovate and rarely one is cylindrical-ovate. The shell is smooth or very finely granulated, with little or no gloss. The ground color is dull white or creamy white. The eggs are prettily marked and sometimes nearly covered with spots, blotches, and splashes of bright browns. Generally they are boldly and irregularly blotched and spotted, sometimes sparingly or finely spotted, with dark browns, such as "chestnut", "liver brown", or "chocolate" and more rarely with "russet" or "cinnamon-brown"; they are often washed with one of the above browns and many have underlying spots in shades of "Quaker drab." Very rarely one is nearly or quite immaculate. The measurements of 52 eggs average 71.3 by 48.6 millimeters; the eggs showing the four extremes measure 83.5 by 50, 76 by 53, 62.5 by 50.5 and 71.3 by 43.7 millimeters.]

Young.—When the nature of the vulture's food is considered, it seems almost inevitable that the young birds, in their earlier days, be fed by the process of regurgitation. Thus, one of the first associations that the nestlings learn is that of the odor of decomposing animal matter with appetite and good digestion.

A. G. Lawrence, writing from Winnipeg, Manitoba, to Mr. Bent, describes the process, which he watched from a blind on the side of a cliff. "Both young rushed toward the female parent with widespread wings. The first to reach her thrust its bill well into the parent's gullet, the old bird stretching out low over the rock to facilitate the exchange of regurgitated food. The feeding process was carried on so vigorously that it resembled a tussle, both birds swaying their heads up and down and from side to side and balancing themselves by raising their wings."

As these young birds "were fully grown, but unable to fly", this method of feeding may continue through the major portion of their protracted life as nestlings. Lawrence continues: "The young spent much of their time sunning themselves on the rocks outside the nesting cavity. They continually exercised their wings, spreading them out to their full extent whenever the sun shone and closing them when a cloud cut off its rays. They stood with backs to the sun, and their wings immediately responded to its warmth."

Thomas H. Jackson (1903) estimates the period of incubation as "very close to thirty days, possibly a day more or less" and "the period between hatching and flight eight or nine weeks." In the case of a nest under the observation of C. J. Pennock (MS.), these periods were somewhat longer; the incubation lasted about 41 days, and at the end of 74 days the young birds "had not been away from the near proximity of the site and had not flown at all."

Dr. Pearson (1919) says: "From eight to ten weeks are passed in and about the nest before the young are able to fly." Jackson (1903) continues:

Young Turkey Vultures at a very early age display more intelligence than the young of any other raptores with which I am familiar. Their eyes are open from the first, and in less than a week they move about in their home, hiss vigorously, and show considerable alertness, but do not seem to have any fear at that age. At two weeks they show a great increase in size and weight, but otherwise have changed but little in appearance. They now resent being disturbed and snap at the intruder, and as they get older become quite pugnacious, rushing at one with extended wings, uttering continually their loud hissing sound, which comes the nearest to any vocal performance I have ever heard from these birds. Their beaks are quite sharp and capable of injuring an unprotected hand.

On being approached they retire to the farthest corner of their den and there disgorge the contents of the stomach or crop.

H. Justin Roddy (1888), speaking of the nestlings as pets, says:

The young birds kept in captivity drank water freely from any vessel as a fowl drinks, but were fonder of drinking from some vessel, as a bottle, with a narrow opening partially inverted, that the liquid might flow out. This must be because it is similar to the opened bill of the parents.

They are very fond of thrusting the bill into the opening formed by the partially closed hand. I inferred from this fact the manner of feeding before I had an opportunity of observing it.

They are very fond of being caressed, or at least handled, especially so while feeding. In a few days after being placed in captivity they become fond of being handled, and soon follow persons about like dogs. They express pleasure by a low hiss; displeasure by a more forcible hiss.

Leon L. Gardner (1930), in a study of the body temperature of nestlings, states that "A young Turkey Buzzard early in its development attained the temperature normal to the species * * *. At one week of age it was 102.5 and at two weeks of age it had risen to and above the normal for the species. Thereafter there was a remarkable constancy of temperature at about 103.6 except when influenced by other factors [struggling]. The Buzzard therefore apparently stabilizes its temperature long before the appearance of feathers and while the general development of the body is still immature."

Plumages.—[AUTHOR'S NOTE: The young vulture is clad in a coat of long, cottony, white down, covering the whole body; a thinner coat of shorter white down extends onto the crown.

W. Bryant Tyrrell has sent us some elaborate notes based on his studies of a brood of young turkey vultures, from soon after hatching up to the flight time, and has contributed an interesting series of photographs illustrating the development of the juvenal plumage (pls. 8 and 9).

The nest contained eggs on April 15 and April 30, and the young were hatched but still very small and helpless when the nest was next visited, on May 21; they were probably not over three or four days old; they were unable to hold up their heads and were completely covered with white down, except on the black face. On June 4, the young were still covered with down, except that the primary quills were beginning to show, the sheaths protruding about three-quarters of an inch and tipped with down; they were now about 17 days old and about three times as large as they were on May 21. When about 37 days old, on June 24, the young were still covered with down, but "the primaries and secondaries and their coverts were about 4 inches long, and the tail feathers were about 3 inches long." On July 4, when about 47 days old, "the birds were about two-thirds grown. The wing feathers were well developed, though still growing, and the tail feathers 3½ inches long. The rest of the body was still covered with down, which came off easily." When about 60 days old, on July 17, the wings appeared to be quite fully developed, the back was well feathered, and the plumage was coming in on the sides of the breast, but the neck and the remainder of the under parts were still downy. Mr. Tyrrell made his last visit to the nest on July 25, when the young birds were about 68 days old. "Both were well feathered, with down only about the breast and belly, and, in one, some still clinging to some of the feathers of the neck, looking like a ruffled collar."

Apparently the juvenal plumage is not fully assumed, or the flight stage reached, until the young vulture is well over 10 weeks old.

The juvenal plumage is much like that of the adult, but with lighter edgings on the feathers of the mantle; the plumage is said to be darker when fresh, but it fades out to dull brown, "Verona brown" to "warm sepia." The naked skin of the head and neck is blackish or livid brown, not red as in the adult; and the neck and crown of the head are scantily covered with short, dark brown, hairy down. This plumage is worn through the first winter. The annual molt of both young birds and adults begins late in winter or early in spring and continues gradually through the summer and early fall, the wings and tail being molted in September and October. The new body plumage, which appears first on the breast, neck, and back, is glossy, bluish black at first, but fades out later to dull, dark brown.]

Food.—The inability of the vulture to kill its prey has forced it to play the part of a scavenger, and the struggle for existence has driven it further. Where the bird is abundant, it cannot, like other Raptores, select its victim; it has to accept what chance presents.

When death comes to any animal, its body becomes food for the vultures. As soon as the animal can no longer move, the meal is ready, and if a vulture finds a dead body, although it be warm from the life just flown, the bird begins at once to feed. But a large animal—a horse or a cow—cannot be finished, even by a company of voracious vultures, while the body is fresh. Putrefaction works fast and overtakes the birds, and the end of the meal becomes far advanced in decomposition. Also it often happens, owing to the position of the body, or because it is submerged, or because the hide is too tough for the vulture's beak to tear, that little or none of it is accessible to the birds. Then the vultures gather about the carcass, in large numbers if it be a big one, and wait patiently near at hand until time and decay, making it soft and ripe, shall fit it to their needs. Then they descend and strip it to the bone.

Thus evolution has led the vulture in its search for food away from the other Raptores and has compelled it to develop feeding habits that it shares with few companions among birds and mammals.

The vulture shows apparently little preference in its choice of food. It is a useful bird in the Southern States, where it disposes of the dead animals about the farms, and, as Dr. Pearson (1919) says, "in many a southern city the Vultures constitute a most effective street-cleaning department, and the garbage piles on the city's dump-heaps are swept and purified by them. When the rancher of the West dresses cattle for home consumption or the market, his dusky friends in feathers gladly save him the trouble of burying the offal."

Wright and Harper (1913), writing of the Okefinokee Swamp, remark that "it is astonishing how soon the buzzards appear over a spot where an alligator has been shot, and how quickly they transform its carcass into a bare skeleton."

Florence A. Merriam (1896) reports from California that "Mr. W. W. Merriam watched two of the buzzards eating skunks. They began by pulling the skin from the head and ate till they came to the scent gland, which they left on the ground."

Snakes appear to be a favorite food. Ivan R. Tomkins, in a letter to Mr. Bent, says: "I flushed two turkey buzzards from a clump of willows. On looking into the bushes to see what they were discussing, I found the partly eaten remains of a cottonmouth moccasin that had been dead some time. The head had dried instead of decomposing, and I picked it up with a forceps, and was able positively to identify it. The buzzards had eaten the meat off the back bone rather than swallowing the snake whole, as I would have expected."

R. M. Kempton (1927) says "a reptile was evidently a choice relish, because one dead snake will call fifty vultures, more or less, to the vicinity of its demise."

Dr. Pearson (1919) writes: "To a limited extent, our southern Vultures feed on living animals. Newly-born pigs are killed by them, and, in some of the bird-colonies, * * * young Herons and Ibises are often eaten." W. E. D. Scott (1892) corroborates this statement. He says that the vulture in Jamaica "is certainly not a carrion eater from choice, fresh meat being eagerly taken whenever an opportunity offers, and when sore pressed young and weakly chickens, etc., are taken up."

The birds have been known to feed on grasshoppers, and they readily eat fish. Dr. Alexander Wetmore (1920) mentions that in New Mexico "they clambered over the piles of Potamogeton and algae cast up last year and left on the shore, picking at it experimentally, pulling off the surface and digging into the interior with their bills as they would into carrion."

James Green (1927) reports a remarkable observation of finding a flock of 62 vultures, hard pressed for food, feeding on pumpkins. He says: "A few had been touched by frost, making them soft, and these had been all but pecked to shreds. But there were the marks of the buzzard's powerful beaks on the sides of the big yellow pumpkins that otherwise were sound as a dollar."

The young pigs left dead on the road by automobiles in the Southern States afford opportunities to observe the dissecting habits of the vulture. A bird sails along, doubles back, alights, and, folding its great wings, slowly approaches the pig. With head high and tail held well above the ground, it sidles about, wary and watchful lest the pig move, it seems; then reassured, it steps upon the body and, with a deft hook of its beak, extracts the eyeball and swallows it.

The vulture next nips through the skin and by tearing or pulling it back lays bare the muscles beneath. Three times I have seen vultures make their first incision over the upper part of the shoulder blade and pick out and devour the supraspinatus muscle before they touched any other part of the body, except the eye.

The vulture at its meal moves deliberately, but, like a skilled workman, surely. It is watchful of intrusion and will not tolerate the approach of other vultures while eating as small an animal as a young pig. It turns upon any vulture that comes near, but more with a remonstrance than a threat. Indeed, as we watch, we see that a solemn but strict etiquette governs the bird at its meal.

Mr. Bent once surprised a party of turkey vultures that were feeding on a lot of dead tadpoles that had become stranded by the drying up of a small pool in Florida. He has also often seen them feeding on the main highways there, where snakes, turtles, small birds, or small mammals have been killed by passing automobiles, or where fish have been thrown away by fishermen. Often, when disturbed by an approaching automobile, the vulture will pick up some such small object in its bill and fly away with it.

Behavior.—On the ground the vulture is an awkward bird, hopping clumsily, sometimes with a hitch sideways; it has a gawky walk. To get into the air it leans forward, stumbles onward with a few steps or hops, gives a push with its legs, and, with a visible effort, flops its wings, until at last it is under way and sails off.

In the air the vulture wins our admiration. Its great wings, long and broad, hold the bird aloft like a kite. Adjusting its wings to the wind, it progresses for miles with never a wing beat, or rises very high in the air, nearly out of sight from the ground. In soaring, the vulture raises its wings to a slight angle above the line of the back, making a shallow **V** in the sky, and often the wind pushes upward the separated tips of the primary feathers. As it moves along it sways a little from side to side, not rolling like a ship at sea, but teetering, balancing like a tight-rope walker, but slower. When the bird sweeps past us just above the treetops, we see the flight as a steady rush through the air; we see the head turn as the bird studies the ground.

Usually we see the turkey buzzards flying alone at no great height, but sometimes they collect in the sky, dozens together, and wheel about. The habit of gathering into flocks is much less marked than that of the black vulture, and they do not go in packs during the day as the latter birds do.

M. P. Skinner, writing to Mr. Bent of the buzzards' habits, says of their roosting: "At night they gather in a roost, usually located on high trees in a low or swampy area in the depths of the forests. At other times I have seen single buzzards in pines comparatively near the ground as well as on the very tops."

Ludlow Griscom tells me that the vulture is a late riser, seldom being seen on the wing until an hour after sunrise. The ground mists, which often obscure the southern lowlands early in the morning, probably delay the vulture's search for food until long after the time when most birds are stirring.

Mr. Tyrrell, in his notes, thus describes a flock of turkey vultures going to roost:

Today (February 22, 1932) at about 4.30, we were seated beneath a large white oak, whose upper limbs were white with excreta, while on the sombre floor of the forest beneath there was a whitish ring of the same material. As we sat there, the great birds would sail noiselessly over, sometimes their small, naked, red heads gleamed in the last rays of the sun, their dark, silver-lined wings moving only to catch movements of the air currents, as they glided by. Some, after alighting, would shake themselves until every feather was ruffled, giving them a most unkempt appearance. Others would alight on a branch where one or more were already roosting, and the impact of the landing bird would throw them off their balance and result in many awkward and ludicrous balancing movements. Often they would sit and preen, and some were always watching the movements of the neighborhood, cocking their heads first on one side and then on the other to see each newcomer. A few seemed to be resting, oblivious of what was going on. There were 71 in three trees at 5 p. m., 113 at 5.25, and 147 at 5.55.

While brooding small young, the turkey vulture sticks to the nest tenaciously, appearing very tame or stupid, allowing itself to be handled or even feigning death. Mr. Tyrrell got close enough to one "to grab her, and lifted her off the young, she not showing any resistance. As I lifted her, she disgorged a mass of half-digested, decayed flesh that was plenty odorous. I held her over the nest by her wings, and every time we let her go, she would put her head under the log with only her back showing. We thought she must be sick or wounded, for she acted so queerly, always with her head hanging and not showing the least inclination to get away. It was suggested that we put her on top of the log and possibly get a picture of her, so we did; but no sooner had she touched the log than off she flew, soon to be joined by her mate."

Speaking of their relation to other birds, Skinner notes that "small birds had no fear of the buzzards and vultures flying over, although they quickly took alarm if a hawk appeared. Buzzards often swept over within a hundred feet of doves, meadowlarks * * * and many others without alarming them in the least."

Since the days of Audubon naturalists have speculated on whether the vulture finds its food by sight or by scent. They have sought to find the answer by experiments on the bird and have published the results of many of these. After going carefully over the literature on this fascinating subject—too large a field to do more than summarize here—a reader cannot feel convinced that the problem has been definitely solved, even now. The evidence shows him that the vulture has keen eyesight and that it has an acute sense of smell. The reader finds running through the controversy, however, a great deal of contradiction and refutation; no one article stands out as indisputable proof on either side to the exclusion of the other side, and many experiments present to the vulture problems that it would never meet under natural conditions.

Experiments in which food is concealed in boxes, covered by canvas, or wrapped up in paper parcels make trial only of the bird's ingenuity; they do not call for the employment of the faculties with which nature has equipped the bird to use in finding its food. On the other hand, the experiment of exposing to the vulture's view the stuffed skin of an animal arranged to simulate a carcass does call into play the food associations of the bird. However, when the bird is not lured to the bait, it may be either because its nostrils do not inform it of the presence of food or because its eyes do inform it of the deception. Another difficulty in interpreting the vulture's behavior arises from its habit of reconnoitering before it begins a meal.

The vulture does not have to move quickly to catch its prey; it has only to find out where it is, and to make sure that the body is ready to be eaten—that it will not move. There is never need to hurry, so the bird reconnoiters, examining from a distance with a deliberation that allows time for the use of all its senses. Therefore, experiments conducted on birds in the field, presumably in possession of all their senses, do not suffice to show whether the bird is seeking its food by one sense or another, or by a combination of senses, but merely test the bird's general intelligence.

P. J. Darlington, Jr. (1930) has made, from the viewpoint of an entomologist, some very interesting and novel observations on this subject, noting "a possible factor in the bird's behavior which seems to have been overlooked." Here is his story:

The first incident took place at the Harvard Tropical Laboratory on the Soledad sugar estate near Cienfuegos, Cuba. In November, 1926, some dead fish were put out near Harvard House to attract beetles, but were stolen by Turkey Buzzards the first day. The bait had been hidden under fairly large stones, and since it was placed beside a garden where people were frequently moving about, there is no reason to suppose that the birds were attracted by my actions. They may, indeed, have smelled the fish, but it seems just as likely that they saw the insects which collected and which would have given the set away to any intelligent human being. Near Santa Marta, Colombia, in 1928, the same sort of thing happened, for when dead iguanas were put out they were invariably discovered by Vultures, even when the baiting was done in scrubby woods. The most rational explanation in this case seemed to be that the birds had heard the carrion-drawn flies.

The literature to date leaves the reader with the belief that the vulture is a bird not very intelligent from the human standpoint, but alert and keen to detect the presence of food by every sense at its command. The problem is discussed further herein under the black vulture.

Voice.—For the most part, the vulture is a silent bird. Dr. Pearson (1919) says: "Over the coveted carcass they flop and hiss and even fight in a bloodless sort of way. Aside from this hissing and an occasional low grunt, the birds appear to be voiceless." The grunt he speaks of is a raucous growl or snarl, suggesting a note of some of the larger herons.

Field marks.—The turkey buzzard and the black vulture, large, dark-colored birds with a soaring flight, resemble each other some-what in the air. The buzzard, however, is dingy brown; its tail is rounded at the tip and, carried nearly closed, projects beyond the wings in flight, whereas the vulture is black with a square-tipped tail, which fits snugly between the wings. The posterior half of the buzzard's wing, seen from below, is gray, the color extending to the end of the wing. The vulture's wing is black with a gray tip, and the bird flaps its wings much more frequently than the buzzard does. At close range the buzzard's head and neck are seen to be dull red. These parts in the vulture are black.

Coues (1874) brings out the difference in the shape of the wings when he says of the buzzard that "the fore-border of the wing is bent at a salient angle, and there is a corresponding reëntrance in its hind outline", and of the vulture that "the front edge of the wing is almost straight, and the back border sweeps around in a regular curve to meet it at an obtuse point, where the ends of the quills are neither spread apart nor bent upward."

The bald eagle, a much larger, sturdier bird than the buzzard, is at once distinguished by its more conspicuous head, proportionally longer secondaries, and powerful, driving wingbeats.

The California condor, compared to the buzzard, is a giant.

Enemies.—W. E. D. Scott (1892) reports that in Jamaica the vulture is "said to have decreased greatly in numbers in the past few years, being preyed upon, like all other ground, and many low tree builders, by the mongoose."

The nestlings are subject to attack from predatory animals, but the adults have few enemies.

It was feared at one time that the vulture should be held responsible for the spreading of hog cholera, but the bird has been cleared recently of the suspicion. Howell (1932), quoting from the 26th Annual Report of the State Board of Health of Florida, 1914, says that "the virus of hog cholera is digested in the intestinal tract of buzzards and the droppings of buzzards fed on the flesh of hogs dead from cholera do not produce cholera when mixed in the feed of hogs."

Game.—The turkey vulture plays a negligible role as a game bird, although G. B. Benners (1887) reports that in Texas the Negroes eat the young birds.

Frank L. Burns (1906) recounts that when a number of birds, among which was a vulture, were presented to an Italian workman, "the vulture, being the largest, was naturally considered the prize, so it was cleaned, and stuffed with plenty of garlic, and the entire household proceeded to make a meal of it; with the result that all were made deathly sick."

Winter.—Winter, with its frost and snow, drives the bird from the northern part of its summer range, for, as Thomas H. Jackson (1903) says, "to obtain food here [Pennsylvania] in zero weather, with deep snow covering everything, would seem for them an impossibility."

Range.—The Western Hemisphere from Patagonia north to southern Canada.

Breeding range.—The turkey vulture breeds **north** to northwestern Washington (Bellingham Bay); southern British Columbia (Okanagan); central Alberta (Lake Astotin); Saskatchewan (Muscow and Indian Head); Manitoba (Duck Mountain); Minnesota (Elk River and Lanesboro); southern Michigan (Three Rivers and Ann Arbor); southern Ontario (probably Harrow, Kerrwood, and Coldstream); and New York (near Canandaigua Lake and Westchester County). **East** to New York (Westchester County); New Jersey (Spring Lake and Newfield); Maryland (Easton and Cambridge); Virginia (Hog Island and Hampton); North Carolina (Fort Macon); South Carolina (Summerville and Sea Islands); Georgia (Savannah and Blackbeard Island); Florida (Gainesville, Micanopy, Seven Oaks, Passage Key, and Miami); Bahama Islands (Abaco, Little Abaco, and Andros Island); Puerto Rico (Guanica); Guiana (Georgetown and Cayenne); Brazil (Para, Parahyba, Goyanna, Rio de Janeiro, and Santa Catharina); Uruguay (probably Montevideo and probably Santa Elena); Argentina (Rio Negro); and the Falkland Islands. **South** to the Falkland Islands and southern Chile (Straits of Magellan and Tierra del Fuego). **West** to Chile (Tierra del Fuego, Chiloe Island, probably near Santiago, Corral, Temuco, Copiapo, and Iquique); Peru (Arequipa, Talara, and Callao); Ecuador (Jambeli Island, Las Pinas, Babahoyo, Gualaquiza, and La Plata); Colombia (Santa Elena and Ocana); Costa Rica (La Palma, San Jose, and Miravalles); Nicaragua (San Juan del Sur); Guatemala (Duenas); Oaxaca (Tehuantepec); Colima (Colima); Nayarit (Tres Marias Island); Lower California (San Jose del Cabo, Victoria Mountains, Santa Margarita Island, Magdalena Bay, Cerros Island, and San Martin Island); California (San Diego, Escondido, Fullerton, Whittier, Big Creek, Mount Hamilton, San Francisco, and probably Eureka); Oregon (Newport and Portland); and Washington (Camas, Kalama, Steilacoom, Tacoma, Everett, and Bellingham Bay).

The range as outlined is for the entire species, which has been separated into several subspecies. The North American form, *C. a. septentrionalis*, extends south only to northern Mexico and Lower California.

Winter range.—During mild winters some turkey vultures may be found over a large part of the breeding range. Mostly, however, they retire southward. The following appears to represent the northern limits of their normal range at this season: California

(Marin County, Vaca Valley, Gridley, and Chico); southern Arizona (Fort Whipple, Salt River Bird Reservation, Catalina Mountains, and Tombstone); rarely southern New Mexico (Fort Fillmore, Fort Thorn, Rio Grande Bird Reservation, Cooney, and Carlsbad); rarely Kansas (Manhattan, Winfield, and Ellis); rarely Missouri (Jonesburg, Cardwell, and Warrensburg); southern Illinois (Anna and Mount Carmel); Indiana (Bicknell, Worthington, Greenville, and Bloomington); Ohio (Sidney, Wooster, and Hillsboro); Pennsylvania (Parkesburg, Carlisle, Lenape, Lititz, Concordville, and Darling); and New Jersey (Princeton, Moorestown, and Trenton).

Migration.—In parts of the country, particularly the Atlantic and Pacific coastal regions, the migrations of the turkey vulture are frequently unnoticed, as in large areas the species is present in more or less numbers throughout the year. The southward movement in autumn is usually more conspicuous, and in the vicinity of Washington, D. C., flocks of 30 or more may be regularly seen in October and November as they circle steadily toward the south.

Spring migration.—Early dates on which the turkey vulture has been observed to arrive in spring are: New Jersey—Vineland, February 2; Camden, February 8; New Providence, February 15; Hackettstown, February 22; Morristown, March 2; and Princeton, March 6. Pennsylvania—Berwyn, February 2; Jeffersonville, February 12; and Lionville, February 22. Ohio—Columbus, February 1; Hillsboro, February 11; New Richmond, February 21; New Paris, February 22; and Circleville, March 1. Michigan—Three Rivers, February 17; Ann Arbor, March 20; Brant, March 22; Petersburg, March 25; and Sault Ste. Marie, April 5. Ontario—Harrow, April 11; Point Pelee, April 23; and London, April 26. Indiana—Bicknell, February 2; Bloomington, February 6; Brookville, February 9; and Rushville, February 10. Illinois—Carlinville, February 3; Odin, February 8; Martinsville, February 15; Hillsboro, February 20; and Rantoul, March 7. Wisconsin—St. Croix Falls, March 29; Elroy, March 31; Jefferson, April 4; and Viroqua, April 6. Iowa—Keokuk, February 7; Iowa City, February 28; La Porte City, March 7; and Hillsboro, March 9. Minnesota—Fosston, March 22; Wilder, March 25; Dassel, March 27; White Earth, April 1; and Minneapolis, April 8. Nebraska—Omega, February 12; Belvidere, March 7; and Falls City, March 31. South Dakota—Custer, March 27; Springfield, April 4; and Huron, April 9. North Dakota—Larimore, April 4; and Tepee Buttes, April 10. Manitoba—Treesbank, March 30; Margaret, April 9; and Aweme, April 10. Saskatchewan—Eastend, March 29; Indian Head, April 28; and Qu'Appelle, May 4. Colorado—Mesa County, March 18; Pueblo, March 28; Lay, April 9; Loveland, April 10; and Grand Junction, April 20. Wyoming—Fort Sanders, May 13. Idaho—Rathdrum, March 13; Meridian, March 23, and Grangeville, April 5. Montana—Corvallis, March 4; Billings, April 4; and Terry, April 10. Alberta—Mundare, April 24. Nevada—Carson City, April 2; and Nixon, April 10. Oregon—Mercer, March 4; Monmouth, March 7; Rickreall, March 10; Klamath Lake, March 19; and Corvallis, March 21. Washington—Camas, March 14; Seattle, March 25; and Tacoma, March 29. British Columbia—Courtenay, March 24 (once on February 24, 1920); Chilliwack, April 4; Okanagan Landing, April 10; and Osoyoos Lake, April 11.

Fall migration.—Late dates of departure in the autumn are: British Columbia—Okanagan Landing, September 28. Oregon—Newport, September 28; Klamath Lake, October 4; and Camp Harney, November 27. Alberta—Red Deer River, September 22; and Islay, October 3. Montana—Anaconda, November 1; and Ravalli County, November 29. Idaho—City of Rocks, October 3. Wyoming—Sundance, October 19. Colorado—Yuma, September 14; Littleton, September 20; Mesa County, September 27; and Lyons, October 2. Manitoba—Margaret, September 20; Treesbank, October 13; and Aweme, November 2. North Dakota—Harrisburg, September 3. South Dakota—Douglas County, September 8; Harrison, September 17; and Forestburg, November 6. Nebraska—Lincoln, October 7; Arago, October 10; and Badger, October 24. Minnesota—St. Vincent, September 27; Lake Andrews, October 8; and Minneapolis, October 12. Wisconsin—Lake Koshkonong, November 2. Iowa—Keokuk, October 29; Sioux City, November 7; and Grinnell, November 23. Ontario—Point Pelee, October 26. Michigan—Ann Arbor, October 13; Hillsdale, October 23; Manchester, October 26; and Schoolcraft, November 13. Ohio—Campbellstown, November 10; Bowling Green, November 12; Scio, November 24; and Columbus, November 27. Indiana—Greensburg, November 12; Bicknell, November 25; Lyons, November 27; and Richmond, December 4. Illinois—Johnsonville, November 13; Odin, November 23; Port Byron, November 27; and Rantoul, November 31. Pennsylvania—Doylestown, November 4; Columbia, November 4; Jeffersonville, November 24; and Berwyn, December 1. New Jersey—New Brunswick, November 3; Princeton, November 12; Morristown, November 12; and Hackettstown, November 14.

Casual records.—The turkey vulture has a penchant for extensive wandering and so has been frequently recorded outside (usually north) of its normal summer and winter ranges. There are many records for the New England States (cf. Forbush, 1927). Among others the following may be cited as examples: One seen near Hatley, Quebec, July 31, 1917; New Brunswick, several records, one of which is for a bird taken about January 10, 1884, at Nequac, Miramichi Bay; one was killed at Renews, Newfoundland, in 1905; one was re-

corded from Calhoun County, Mich., on December 22, 1929; in
Ontario one was taken at Moose Factory in June 1898, one was
seen at Ojibway, May 15, 1925, one was taken at Scarborough, November 17, 1908, and another was taken at Warsaw in the summer of
1895; one was killed at Dawson Bay, Manitoba, on September 15,
1913; a specimen was obtained at Camrose, Alberta, in January
1911; one was recorded from Comox, Vancouver Island, British
Columbia, February 24, 1920; and one was killed in Bermuda in
1853.

Egg dates.—Colorado, Oregon, and British Columbia to Saskatchewan: 115 records, March 14 to June 25; 57 records, April 4
to 23.

Ohio, Illinois, Missouri, and Kansas: 24 records, April 16 to June
13; 12 records, April 26 to May 19.

New York to Georgia: 118 records, April 3 to June 2; 59 records,
April 15 to May 5.

Oklahoma and Texas to Florida: 59 records, February 25 to June
3; 29 records, March 31 to May 1.

CORAGYPS ATRATUS ATRATUS (Bechstein)

BLACK VULTURE

HABITS

CONTRIBUTED BY CHARLES WENDELL TOWNSEND

If one could forget the unsavory feeding habits of the black
vulture and remember only the pleasing attributes of its flight, one
would place this bird among the most distinguished and interesting
of avian friends. As a feature of the landscape in its flight and
soarings on high—and after all this is the feature most evident—
the black vulture appeals to our aesthetic feelings, while the mental
effort needed in distinguishing it from the turkey vulture and from
larger hawks and eagles adds greatly to its interest. It is a bird
well worth seeing and watching.

Spring.—As the black vulture is a resident throughout its breeding range except in the extreme northern parts, a marked spring
migration does not occur. It seems to be fond of the neighborhood
of the sea and generally outnumbers the turkey vulture in these
regions, while it is outnumbered by the latter farther inland. As a
straggler or wanderer it has been recorded as far north as Quebec
and New Brunswick, while its breeding range extends north only as
far as Maryland and Virginia.

Courtship.—Aretas A. Saunders (1906) thus describes the courtship of this vulture, which "took place on the ground in the shade
of a small lime tree": "In a circle in front of the female were three

admirers, who, with their wings partly spread, were rapidly ducking their heads to her like well-trained servants. She paid little
attention, and soon turned her back on them. They persisted in
their attentions till she finally got disgusted and flew away, with her
suitors in close pursuit."

Audubon (1840) gives a more graphic account as follows:

At the commencement of the love season, which is about the beginning of
February, the gesticulations and parade of the males are extremely ludicrous.
They first strut somewhat in the manner of the Turkey Cock, then open their
wings, and, as they approach the female, lower their head, its wrinkled skin
becoming loosened, so as entirely to cover the bill, and emit a puffing sound,
which is by no means musical. When these actions have been repeated five or
six times, and the conjugal compact sealed, the "happy pair" fly off and remain
together until their young come abroad.

Simmons (1925) describes the courtship of the black vulture as
observed in Texas:

During February and to the middle of March, the love-flight or courtship
flight of the two birds may often be seen at the breeding grounds, lasting
from two to ten minutes, in rapid, prolonged, wide-spreading circles. In the
air over a thickly-populated nesting area, such as a honey-combed cliff or
canyon wall in the hills, as many as 25 or 50 pairs may be seen going through
these nuptial ceremonies during early March, presenting a slowly-moving,
gyrating maelstrom, circling and sailing in close spirals, one of a pair continually following the other; out of this maelstrom a female occasionally
drops, the male a few feet behind, and then a chase ensues, dropping, darting,
wheeling with incredible speed, wing tips of one touching the wing tips of
the other in the twists and turns of the play. A male performing before a
female perched high on a dead tree overlooking the chasm often circles high
in front of her, half folds his wings and dives straight for the earth, his
wings shrilling and whistling until he zooms upward again to resume his
circling.

C. J. Pennock writes: "What I take for a mating-time flight I
have noted frequently in February and early March [in Wakulla
County, Fla.], namely two birds in rapid flight in close company
through the tree tops and open country, sometimes lasting three to
five minutes."

Wayne (1910) says of South Carolina: "The birds mate in February, and when engaged in this pleasure utter a hissing sound which
can be heard at a distance of several hundred yards."

S. A. Grimes sends us the following account:

I was returning home from a short trip to Baldwin Bay and noticed two
vultures in a tall dead cypress in a swamp about 300 yards off the highway.
This aroused my curiosity, and I turned into a road that put me within 100
yards of the birds. Without getting out of the car, I focused my glasses on
the birds and presently saw one hop over to the branch on which the other
was perched. This bird, which was undoubtedly the male, alighted with his
wings outstretched above his back and, holding them in this position, sidled
up so close to the other that she was forced to back away on the limb. Losing

her balance, she flopped to another branch and was followed by the male, who continued to hold his wings above his back in such a manner that the tips almost touched. The two birds "necked" a little, and the female pecked feebly at her mate's head and breast when he pressed her too hard out on the branch.

This performance was repeated four times, and each time I looked for copulation to take place, but the female was not agreeable. The male finally folded his wings and perched quietly beside his mate. They remained thus for about 15 minutes, each occasionally pecking gently at the other; and once I noticed that they grasped each other's bill, as doves are wont to do. I could plainly see, too, that the birds parted their mandibles repeatedly, as if making some sound, but none was audible at my distance. The birds suddenly sprang into the air and flapped away, after I had watched them 25 minutes. This was on January 31.

Nesting.—The heading of this section is in a literal sense incorrect, for no nest is made by the black vulture, and the eggs are laid without this preparation. As bits of stick and weed stalks as well as dead leaves strew the ground in many places, the eggs may be deposited on these, but only incidentally, for these are not collected to form even the semblance of a nest, and there is no hollowing out of the ground as a receptacle for the eggs. Wayne (1910) called attention to an aesthetic habit of the bird that may have been peculiar to his region in South Carolina and that does not appear to have been noticed by other observers. He says: "It is a peculiar habit of this bird, which I have found to be almost constant, to have pearl, bone, and china buttons, as well as pieces of glass and figured china, around and under the eggs."

As there is no nest to hold the eggs, these cannot be placed on branches of trees but must necessarily rest on the firm foundation of the ground or at the bottom of hollow stumps, sometimes as much as 8 feet above the level of the ground.

Hollow stumps, access to which is only from the top, are commonly chosen for nesting by the black vulture. In some instances there may be an opening at the ground by which the bird may enter and leave. When the stump is 6 or 8 feet high and the nest is at ground level, the entrance and exit of the bird from this chimneylike structure must require the use of both the wings and feet, when the bird scrambles up and out. I was told by an ornithologist that once when a boy he climbed down for the eggs into one of these nesting sites, and was unable to get out until a companion came to his rescue. Edward J. Court (1924) reports a nesting "in a large white oak stump in a cavity about two feet below the level of the ground."

C. J. Pennock describes the nesting two years in succession of a black vulture "in a large decayed hollow tree, the entrance five feet above ground * * * the eggs being placed on a level with the outside ground * * * In every instance when the nest was visited, the brooding bird became alarmed at our approach, and we could hear her flapping to scramble up and escape at the elevated entrance."

J. J. Carroll, of Houston, Tex., says in a letter that he has seen many nestings in standing trees hollow at the base; "sometimes the eggs were at a level not far below the entrance, but I have known the eggs to be placed on the ground in the hollow, with the entrance six or eight feet up. Usually these entrance holes are not higher than that from the ground, but I have seen them as high as fifteen feet." A hollow in a standing tree sufficiently large even if at a considerable height above the ground might be used by this bird, and I was able to find one such record. Charles R. Stockard (1904) found the eggs of the black vulture "about sixty feet up in a huge poplar tree which stood in a cotton field that had been cleared for five years. In the crotch of this tree there was a large hollow running down about three feet and slightly sheltered above by the inclination of one of the limbs that formed the crotch. The eggs were deposited on the floor of this hollow. This was the only nest of this species that was observed more than a few feet from the ground. It is probable that the birds occupied this tree while it stood in the woods and when the land was cleared in 1897 the tree, being a large one, was deadened and left standing and the birds continued to use it as a nesting site." This is, of course, a very exceptional case. A still more unusual site is recorded by O. E. Baynard (1910), who in Florida found a black vulture incubating its eggs in a Ward's heron's nest in a cypress tree some 90 feet above the ground. As he collected the eggs, there is no doubt about the identity.

Where hollow stumps and standing trees occur, they seem to be favorite nesting sites for this bird, but elsewhere the eggs are laid on the ground, often in dense thickets of palmetto, yucca, tall sawgrass, or small trees, although sometimes exposed to the full light of day in the open. The shade of a partly fallen tree trunk is another favorite site, as well as the shade of a rock or under boulders, and, especially in limestone country where caves abound, the eggs are often laid in a shallow cave on a cliff side.

In its nesting habits the black vulture is often gregarious, as shown in the following description by Walter Hoxie (1886) of the nesting on Buzzard Island, 3 miles from Beaufort, S. C., where a dozen or more pairs nested yearly:

There is never the slightest attempt at forming a nest, or even excavating a hollow. The eggs are laid far in under the intertwining stems of the yucca, and in the semi-shadows are quite hard to be seen. The parent birds, however, have a habit of always following the same path in leaving and approaching their precious charge, and after a little experience I learned to distinguish these traces so well that I seldom failed to follow them up and secure the coveted specimens. This track is seldom, if ever, straight. It winds under and around

the armed stems, and, the difference in bulk between a man and a Buzzard
being considerable, the pointed leaves find a good many of a fellow's weak
points before he reaches his prize * * *

Quite rarely I have found eggs on the other parts of the island, and once or
twice in completely exposed situations, with not even an attempt to get under
the protection of an overhanging bush. Possibly these belonged to young birds
which had still much to learn in regard to ways of housekeeping.

Charles R. Stockard (1904) says: "The black vulture was found
depositing her eggs in more widely different situations than any
other bird observed. The favorite site was a large hollow log, or a
tree having a huge hollow base with an opening only a few feet
up, so that the female might be able to jump out of the nest." He
notes the following nestings of this bird: "One pair for three seasons nested in a large hollow sycamore log that lay across a small
stream and served as a 'foot log' for a little-used path in a swampy
wood. At least three people a day must have walked over the log
as the vulture sat calmly on her eggs." In another case "a set of
two eggs was found lying on the bare ground under a large tree that
had been uprooted and had fallen so that its trunk made an angle
of about fifteen degrees to the earth. The eggs were placed below
this trunk, which was four and a half feet above them, and thus
slanting sun rays could have fallen upon the spot but for the heavy
foliage of the wood." Two sets of eggs were found on the ground
in a dense cane thicket. Another set was found in a cave in a steep
clay bank bordering a creek. The entrance of the cave was 7 feet
wide, it was 2½ feet high, and ran back 6 feet. The eggs lay in the
back of the cave.

James A. Lyon, Jr. (1893), writing of the limestone bluffs on the
Cumberland River in Tennessee says: "The most of these bluffs have
'caves' or holes running back into them only two or three feet deep,
others deeper. It is in these 'caves' that the black vulture usually
deposits its eggs, though sometimes they are found under an overhanging ledge of rock. As a general rule they do not go far into the
bluff, but lay near the entrance to the hole."

Eggs.—[AUTHOR'S NOTE: The black vulture lays normally two
eggs, occasionally only one and very rarely three. They vary in
shape from ovate to elliptical-ovate or elongate-ovate, very rarely
fusiform. The shell is smooth but not glossy. They can usually
be distinguished from turkey-vulture eggs by being somewhat larger,
having a peculiar ground color, and being much less heavily marked.
The usual ground color is pale gray-green, sometimes pale bluish
white or dull white and rarely creamy white. There are usually
a few large blotches or spots, mostly near the large end or in a
ring around it; some eggs are more evenly spotted and some are
nearly immaculate. The markings are mostly in dark browns,

"chestnut", "liver brown", or "chocolate", but sometimes in lighter browns, "russet" or "tawny", with occasionally a few "Quaker drab" spots. One very pretty egg is heavily blotched with "pale purple drab", with a few spots of "bay"; another is heavily blotched and finely spotted with "burnt sienna"; but such eggs are exceptional. The measurements of 51 eggs average 75.6 by 50.9 millimeters; the eggs showing the four extremes measure **90.5** by **55.9**, 75 by **56, 66.5** by 51, and 67.3 by 47 millimeters.]

Young.—The incubation period is variously stated to be anywhere from 28 to 39 days; and both parents assist in the incubation. Baynard (1909) watching 21 nests found the incubation was usually 28 to 29 days, in one case 30 days. Edward S. Thomas (1928) reports it as about 39 days in one case. The young, helpless at first, may stray a little from the nest on the ground at a comparatively early age, but, according to Baynard (1913), they are about 14 weeks old before they are able to fly. Simmons (1925) quotes H. J. Kofahl's statement that the young remain at the nesting site for about 60 days. Howard Lacey (1911) says "the young feign death when disturbed."

The Rev. James J. Murray, of Lexington, Va., gives the following interesting account of an experience with young birds on the summit of House Mountain in Virginia, an elevation of about 3,000 feet:

The nest cavity was under a pile of huge boulders. The cave had an opening above large enough for a man to crawl into, and tunnels from two sides at the ground level * * *. One of the parent birds flew out of the upper opening as we approached. There were two young birds, one somewhat larger than the other. They appeared to be three or four weeks old and to weigh about three pounds. They had no feathers and were covered with a thick down of cream buff color, almost reddish above. As we went into the hole they began to vomit large pieces of meat, almost choking in the effort, and continued to do so at intervals as long as we were there. They constantly made a loud blowing noise through slightly opened mouths. It was not a hiss but more like the noise of a bellows. At every effort to get them out into the open they scrambled back into the darkness, jamming themselves under the overhanging rocks and burying their heads in the cracks. When we finally pulled them out to the end of the tunnel in a vain effort to get a good picture in the dim light, they fought each other fiercely and pecked at our hands.

Edward S. Thomas (1928) describes the feeding of the young as follows:

The adult bird lowers its bill to the young, which immediately inserts its beak between the opened mandibles of the adult. The adult, with or without a perceptible gulping movement, regurgitates the food, which is eaten by the young with a nibbling movement of its mandibles. We were certain that at times the adult extruded broth-like drops of liquid which the young secured from the scarcely opened mandibles of the old bird. At other times the young birds obtained the food from the middle part of the adult's beak, but the preferred source of supply, without question, was far up in the corner of the old one's mouth, where the young birds thrust their bills whenever they were able to do so.

The adults fed either from a standing position, or while brooding the young. The day was cold and the old birds brooded almost continually. The young were fed repeatedly. Between the hours of 6 : 03 a. m. and 5 : 48 p. m., there were 17 distinct feeding periods, some of which continued over an interval of seven minutes or more.

The young were very matter-of-fact about taking their food, and at no time showed the eagerness which characterizes the young of most birds. This perhaps may be accounted for by the fact that the young were kept gorged with food continually, the distended stomachs being plainly visible from the blind. On this occasion, the young apparently were fed liquid food only, the liquid being described by Geist as having a milky appearance. On several occasions, solid food, having the appearance of flesh or connective tissue, was regurgitated, which the young attempted to seize, but on each occasion the parent re-swallowed the material.

Plumages.—[AUTHOR'S NOTE: The young black vulture is warmly covered, except on the head, with long, thick, heavy down of a rich buffy color. Mr. Thomas (1928) says:

At 17 days the pin-feathers of the wing begin to show. At 39 days, the young were almost full-grown, but the wing quills were only five inches in length, and the tail feathers of the larger of the two birds, two inches. At this age, there were no other feathers. On June 12, when the young were about 52 days old, the scapulars, tertials and practically all of the wing coverts were feathered, and quills were appearing on the breast. A week later, the upper parts were practically covered, although there was still a great deal of down showing, but while feathers were appearing on the breast and underparts, they were concealed by the down. On June 26, at 66 days, one of the young was able to fly up to the top of the box blind. By July 4, they had left the cave, having a period of from 67 to 74 days in the nest.

Immature birds during their first year are much like adults, but the plumage is duller black and less glossy, and the naked skin of the neck and head is partially concealed by a scanty growth of short, black, hairy down. I have been unable to trace subsequent molts.]

Food.—The principal food of the black vulture gives it its common name of carrion crow, for carrion is the chief article of its diet. This food is to be found in the sewers and dump heaps and about butchers' shops in southern cities, as well as in the fields and forests where animals have come to untimely ends. The methods used in searching for and disposing of this food will be described farther on under "Behavior." As scavengers, especially in cities where these functions are not attended to by man, the black vulture is considered a valuable servant. Black vultures will also eat fresh meat, and butchers must watch their stalls carefully when these birds are about.

J. D. Figgins (1923) found that black vultures in the neighborhood of Bird Island, La., were very destructive in some of the heron rookeries and stated that "it is a frequent occurrence to observe a vulture with a struggling young heron dangling from its beak * * *. In regions where cattle raising has replaced the cultivation of rice, the Black Vulture is credited with considerable damage to the herds by tearing the eyes from calves at the time of birth and instances are cited of a like treatment accorded cows while in a weakened condition. I personally saw one of these tear the tail from a small pig, and was informed that the practice was of too common occurrence to excite comment." O. E. Baynard (1909) reported that these birds were very destructive to young pigs and lambs in Florida, and he has known them to take young chickens. Young herons are frequently devoured. Audubon (1840) says of his experience with the bird in Florida: "I observed them many times devouring young cormorants and herons in the nest."

The United States Biological Survey recommends local control where "through their predatory habits and concentrated numbers, both turkey buzzards and black vultures have become a menace to new-born pigs, calves, lambs, and kids" (Redington, 1932).

Although it is common knowledge that black vultures eagerly devour fresh meat at butchers' stalls, C. J. Maynard (1896) says of this vulture that they "are more emphatically carrion feeders than the latter described species [turkey vulture] and will seldom eat fresh meat but prefer to wait until decomposition has set in before beginning their feast. Thus I have frequently seen the Turkey Buzzards gather around the freshly skinned carcass of an alligator, and eagerly devour the flesh, while the Black-heads would wait until it had lain for a day or two in the broiling sun before they would attack it; then, when the odor from the decaying mass became insufferable to human nostrils, they would eat to repletion. * * * They not only eat decomposed meat but feed upon animal excrement and various kinds of offal."

Behavior.—When a carcass of an animal is discovered, black vultures gather at the feast, which in many cases they must share and fight for, not only among themselves, but with turkey vultures and sometimes with eagles and dogs. Alexander Wilson's (1832) classic description of one of these feasts on a dead horse near Charleston, S. C., is well worth quoting:

The ground, for a hundred yards around it, was black with carrion crows; many sat on the tops of sheds, fences, and houses within sight; sixty or eighty on the opposite side of a small run. I counted at one time two hundred and thirty-seven, but I believe there were more, besides several in the air over my head, and at a distance. I ventured cautiously within thirty yards of the carcass, where three or four dogs and twenty or thirty vultures, were busy tearing and devouring. Seeing them take no notice, I ventured nearer, till I was within ten yards, and sat down on the bank. Still they paid little attention to me. The dogs being sometimes accidentally flapped with the wings of the vultures, would growl and snap at them, which would occasion them to spring up for a moment, but they immediately gathered in again. I remarked the vultures frequently attack each other, fighting with their claws or heels, striking like a cock, with open wings, and fixing their claws in each other's head. The females, and, I believe, the males likewise, made a hissing sound, with open mouth, exactly resembling that produced by thrusting a red hot poker into water; and frequently a snuffling, like a dog clearing his nostrils, as I suppose they were theirs. On observing that they did not heed me, I stole so close that my feet were within one yard of the horse's legs, and again sat down. They all slid aloof a few feet; but, seeing me quiet, they soon returned as before. As they were often disturbed by the dogs, I ordered the latter home: my voice gave no alarm to the vultures. As soon as the dogs departed, the vultures crowded in such numbers, that I counted at one time thirty-seven on and around the carcass, with several within; so that scarcely an inch of it was visible. Sometimes one would come out with a large piece of the entrails, which in a moment was surrounded by several others, who tore it in fragments, and it soon disappeared. They kept up the hissing occasionally. Some of them having their whole legs and head covered with blood, presented a most savage aspect. Still as the dogs advanced, I would order them away, which seemed to gratify the vultures; and one would pursue another to within a foot or two of the spot where I was sitting. Sometimes I observed them stretching their necks along the ground, as if to press the food downwards.

The black vultures are often obliged to share their feasts with turkey vultures, and, according to Golsan and Holt (1914), they always get the better of the latter in a quarrel. On the other hand, according to Audubon (1840), "should eagles make their appearance at such a juncture, the Carrion Crows retire, and patiently wait until their betters are satisfied, but they pay little regard to the dogs." In tearing off choice morsels from the carcass with their bills the vultures brace their feet firmly on the ground and flap violently with their wings to aid them in pulling away.

Their movements on the ground are not graceful. Aretas A. Saunders (1906) graphically describes them as follows: "When the vulture is taking his time about getting around, he moves with a very solemn, sedate walk, carefully placing one foot in front of the other. When he is in a hurry, however, he slightly spreads his wings and indulges in what looks like hopping but is really a very one-sided run. At first sight he seems to put both feet on the ground at once, but in reality he puts down the left foot first and takes his long step with the right foot."

In the air, on the other hand, the black vulture is much more at his ease, but he is far inferior in flight to the turkey vulture, owing to his shorter wings and tail and to his greater weight. While the turkey vulture sails in majestic circles on motionless wings, borne up by the air currents, the black vulture on the same up-currents is obliged to flap his wings from time to time. If the up-currents are strong, his need for flapping is reduced, but he never equals the grace of the turkey vulture. I once compared the flight of the two birds on a calm warm day in Georgia, as they were soaring over a sparse pine forest. They were both about 60 yards above my head as I reclined on the ground and about 40 yards over the forest. The turkey vulture soared in small circles, neither rising nor falling and without once flapping its wings, which with the tail were merely adjusted from time to time to the air currents. The black vulture, on the other hand, flapped its wings quickly at frequent intervals. The contrast was very marked. After a while they both sailed off. Whether they were inspecting me as possible carrion I do not know. On another occasion, when lying outstretched on a sandy Florida beach, I was startled by the shadow of a vulture passing over me and at once sat up. I have been told that this is a habit of vultures to determine whether a body is alive or dead. That they fly near for this purpose is not improbable but one cannot believe that they are able to plan to have their shadow fall on the body.

When a black vulture flying and circling at a great height becomes aware of a carcass lying far below it, the bird at first circles down but soon drops with great swiftness, with legs hanging and, at times, wings flapping furiously. Such actions of descent from a height immediately attract the attention of other vultures on the ground or roosting in trees and they at once follow up the clue. One such action, even a mistaken one, can quickly collect a flock of vultures.

The question that has been much discussed then arises, as to how these birds find the carrion. It is evident that sight is of great importance, and the way in which vultures turn their heads in flight suggests that they are all the time on the lookout for their food. As carrion is so evident to our own sense of smell, even from a great distance, it is natural for us to suppose that these birds also are guided by the sense of smell, especially when trees or bushes partly conceal the carrion. In fact, this belief in the use of the smelling powers has always been a popular one, but since the experiments of Audubon and Bachman (1835) it has generally been accepted that sight alone guides the birds to their food. These experiments, made chiefly on the black vulture, are summarized briefly as follows:

(1) A carcass securely hid in a brier and cane brake was not detected by the birds, although the odor was very marked and attracted dogs.

(2) Carrion on ground covered by a frame of brushwood 12 inches above it was not detected by vultures who passed over it during the 25 days of the experiment.

(3) Fresh meat, placed on canvas covering carrion, was devoured by vultures standing on the canvas, but they did not detect the carrion.

(4) A blinded black vulture did not notice carrion placed within an inch of its nostrils.

A few observers since Audubon occasionally have tested the sense of smell in black vultures, but their findings are generally not conclusive, are not free from the possibility of error, and are often contradictory. Thus, C. J. Pennock writes to me that in Florida he placed "the offal from a large green turtle on the ground 15 or 20 yards within a grove of closely growing pine trees, averaging perhaps 50 feet in height and with tops thickly interlocked but with no side limbs for 30 feet up. At 8:15 a. m. the meal was ready; at 9 o'clock a single black vulture was atop the fence nearby; at 9:40 there were 40 birds, all black vultures sitting on the ground, perched in trees or regaling themselves. No vultures were in sight when the table was spread, and it was thought the repast could not be seen by a flying bird at the nearest open side of the grove, but of this there is a possible doubt."

Dr. Frank M. Chapman (1929) at Barro Colorado Island has made the latest and most careful experiments. Most of them were on the turkey buzzard, and he says that "some of my results leave no room for doubt that the turkey buzzard has a highly developed sense of smell. From others, exactly the opposite conclusion may be drawn." On one occasion two black vultures perched on a tree about 125 feet to leeward of a small house where carrion was concealed. These were the first black vultures he had seen alight on the island.

There is one source of error that so far as I know has not been considered in these experiments and may account for some of the contradictory results. This was brought out by Darlington (1930), who in collecting beetles by the use of carrion bait in tropical regions also attracted vultures, and was led to the following conclusions:

Soon after the death of an animal, except in unusual cases or during cold weather, the body attracts numbers of flies and beetles, some of which may continue to circle about it for several hours or days. The resulting congregation of insects is noisy and conspicuous, and of a sort which does not occur except about decaying material, so that it may be considered more or less characteristic of the latter. Since Vultures can undoubtedly see and perhaps hear such insect swarms at a distance, they have probably learned to recognize their significance, just as we recognize the significance of gatherings of the Cathartidae.

Aretas A. Saunders (1906) found that the lives of black vultures on a rubber plantation in Nicaragua followed a regular routine, influenced only by hatching and the character of the weather. Early

in the morning they sat on fence posts or walked about the plantation in search of bits of food. At noon in fine weather they circled
high in the air, coming down toward evening for another walk. At
sunset they flew one after another to fence posts, thence to the top
of a large tree, where they waited until all were congregated. All
at once they flew to another tree and thence to another, until they
found one to suit their fancy. They seldom slept in the same tree
two nights in succession, though they always commenced operations
from the same tree. Saunders continues:

Butchering day, which occurs at irregular intervals, is the important day in
the life of the Vulture. As soon as the men go down to the potrero to drive
up the cattle, they know what is coming. They gather together on the fence-
posts and shed-roofs, watching the movements of the men with an air of expectancy. Sometimes they wait for three or four hours before the butchering
is finished and the remains thrown out to them. Then there is an instantaneous
scramble. Each Vulture takes hold with his beak and begins to pull and hiss
and flap until the piece he holds breaks off, when it is swallowed as quickly
as possible and a fresh hold taken. At this rate the whole feast is consumed
in an hour or two, when the Vultures go back to the fence-posts and sit in
silence for the remainder of the day.

Black vultures are very social in their habits and often resort to
regular roosts. One such I visited at Buzzard Isle, Lake Iamonia, in
northern Florida. The roost was in big live oaks, mostly dead, and
at about 11 o'clock in the morning contained some 200 black vultures
and half a dozen turkey vultures. The birds did not leave when I
walked beneath them on a ground devoid of vegetation and covered
with their droppings and many bones. The odor was strong of a
chicken yard, but not of carrion. Toward sunset I saw from a distance a number of flocks of about 20 vultures each, sailing and
flapping high up toward the roost. On another occasion on the Vermilion River, La., I passed at sunset about a hundred of these black
creatures sitting on the limbs of moss-draped cypresses, many more
in a nearby field and six or eight on the roof of a deserted house.
A short distance away several were perched on the floating body of
a dead cow. It was a mournful sight.

Audubon (1840) describes a visit by John Bachman and himself
to a roost of black vultures that attended to the offal of Charleston,
S. C. This roost was in a swampy wood of about two acres, across
the Ashley River, two miles from the city. "When nearly under the
trees on which the birds were roosted, we found the ground destitute
of vegetation, and covered with ordure and feathers, mixed with
the broken branches of the trees. The stench was horrible. The
trees were completely covered with birds, from the trunk to the very
tips of the branches." They estimated the number of vultures at
several thousands.

Simmons (1925) states: "Just before daybreak, when a reddish glow begins to show in the eastern sky, black vultures begin to leave their roosts in the mountainous country, passing over in a continuous long string by ones and twos, or as many as half a dozen at a time, moving eastward towards the slaughter pens or to spread out over the open country and begin their tireless vigil for carcasses." They return just after sunset.

B. J. Blincoe (1922) observed an unusual flight of these birds in March in Nelson County, Ky., where the black vulture is generally scarce at this season. The flock of 92 individuals "presented a beautiful appearance as the birds soared in a spiral column, each bird taking, intermittently, a few short wing strokes. At times the whole flock in a long train coursed across country on set wings in an orderly manner suggesting the movement of a flock of water fowl, but not a bird moved a wing until they again maneuvered into a spiral column." There was not a single turkey vulture among them.

J. J. Murray (1928), at Lexington, Va., found about 60 vultures at a slaughter pen, and at least 40 of them, he says, were black vultures. "As we disturbed them, they began walking in single file in a long procession up a steep hillside for 200 or 300 yards, and then near the top took flight." In a letter he says: "This procession was not in order to reach a high place from which to take off, for many of them had jumped to the ground from the top of the slaughter house as the procession started." To rise from the ground in calm weather it is sometimes necessary for the black vulture to hop or run along for 20 or 30 feet, beating its wings violently until it is able to take off.

In cold weather these vultures often sit around chimney pots and on chimneys to obtain some of the warmth. In wet weather they present a most dejected appearance, with wings drawn in close to the body and with back and tail in an almost vertical position. They have a habit of spreading their wings and tail to dry and air when the sun is shining. When alarmed or caught they eject the contents of their stomachs with great quickness and power.

In southern regions it is unnecessary to bury a dead animal to prevent long pollution of the air, as in the North; the farmer merely drags the carcass to a secluded spot and the vultures soon strip off and consume the flesh and entrails. Around butchers' stalls and in cities where offal is thrown into the street, the birds are semidomesticated and walk around almost underfoot. Owing to these habits of the black vultures in consuming carrion and offal of all sorts, the danger of their spreading disease by pathogenic bacteria dropped directly from the vultures' feet and plumage, or by their dejecta, has been given serious consideration. If, for example, anthrax may be spread in this way from the carcass of a horse dead of that disease, it may be better economy to burn or bury the body than to leave it to the vultures.

Dr. Casey A. Wood (1922) relates an experience with black vultures in Georgetown, British Guiana, where until 1921 large numbers of them frequented the city, especially in the region of the slaughterhouses, and were to be seen daily perched on the roofs of the houses. The prejudice among the inhabitants in their favor as scavengers was strong, but it was found that the birds polluted the drinking water, which was largely supplied by roof drainage. It was proved that serious pollution of the drinking water was brought about by the birds' habit of bringing filth to the roofs and also by the pathogenic bacteria in their feces. Analysis of the cistern water of houses protected by wires stretched above the ridge pole to prevent roosting showed it to be free from pathogenic bacteria, while cisterns filled from unprotected roofs, especially those known to be patronized by black vultures, were often shown to be infected by morbific germs. Since 1921 the black vultures have been nearly banished from the city by shooting and systematic frightening away.

Herbert W. Brandt communicates the following about the turkey and black vultures in Kleberg County, Tex., where both are abundant: "It is claimed by Mr. Kleberg that they spread the deadly anthrax to the cattle, and also other cow diseases. He trapped 3,500 buzzards on the Laureles Ranch alone during the winter of 1918–19. The trap is simply a wire-enclosed yard with some loud-smelling carcasses of cows, hogs, etc., as bait and an entrance that closes behind the bird and keeps it in. A Mexican then enters the trap with a club and kills the birds and burns the bodies."

In a publication of the Biological Survey (Redington, 1932), it is stated that "the Biological Survey has discouraged the general destruction of turkey buzzards and black vultures. These birds have been accused of being important carriers of livestock diseases, but skilled investigators have shown that the virus of charbon, or anthrax, is destroyed in passing through the digestive tract of the turkey buzzard. There also are on record similar data regarding the virus of hog cholera. Experimental work of the Bureau of Animal Industry has indicated that the transmission of hog cholera on the feet or feathers of birds is by no means so likely to occur as is generally supposed."

Voice.—The black vulture is a very silent bird. Hissing, grunting, and blowing compose its entire vocabulary, and these sounds are rarely to be heard except when the birds are feeding or fighting. Aretas A. Saunders (1906) describes its voice as consisting of "a

hiss and low guff, guff, guff, like a dog barking in the distance."
Pennock describes a cry as sounding like *watt* or *waugh*. The
blowing sound resembles that made by bellows. Donald J. Nicholson (1928b) says that the young hiss at an intruder and utter a
blowing note very similar to that of a rattlesnake. Edward S.
Thomas (1928) writes:

The birds were heard to give a variety of notes. Adults and young, when
cornered or annoyed, give a rasping, hissing snarl, also described as a "snore",
and "half-way between a wheeze and a squeel." The young give this frequently in the presence of the parents. The young also frequently make a
sound which, when they were very young, was described as "Phuh!" or
"Whuh!" Later this note became in the older birds, "Woof!" or "Wooft!"
This note apparently denotes suspicion, and may be the counterpart of a grunting sound which the adults frequently emit. In addition, I heard the adult
give a low, croaking "Coo," very much like a one-syllabled coo of the domestic
pigeon.

Field marks.—When this bird is seen at close range, alighted on
the ground or on a tree, it is unmistakable. Its black head and neck
bare of feathers proclaim it to be a black vulture, although it must
be remembered that the head of the immature turkey vulture is also
dark and not red as in the adult. The other characteristics of the
black vulture are best seen in flight. Here its short, nearly square-
ended tail, as contrasted with the longer rounded tail of the turkey
vulture, is evident. The feet may sometimes be seen against the tail
as they reach nearly to the end and even project a little, but it is
more difficult to see them in the turkey vulture. The wings seen
from above and below both show a light-colored space at the outer
end of the primaries, while in the turkey vulture all the primaries
and secondaries are light colored, giving the effect of a light posterior
border to the wings. While the wings of the turkey vulture are
held up at an angle in soaring, those of the black species are as a
rule more nearly horizontal, and the ends of the primaries are more
distinct and spread out like fingers. The heavier, clumsier flight of
the black vulture, with frequent flappings of the wings, easily dis-
tinguishes the two birds, although in very favorable airs the black
vulture may soar nearly as well as the turkey vulture.

Enemies.—The black vulture is fortunate in having few if any
enemies. Eagles and wolves may chase it away from a carcass, and
ospreys may wrathfully pursue it if it appropriates a fish from the
osprey's nest. Even man treats it with consideration in return for
its services in cleaning up carrion and offal, although in time most
southern cities may adopt the more sanitary but more expensive
methods needed in northern cities in order to escape the defilements
of these scavenger birds. In some regions, as has already been mentioned under "Food", it may be necessary for man to control these
birds when they kill young domestic animals.

While smaller birds take alarm quickly at the sight of a hawk, they are not disturbed by the presence of these vultures. M. P. Skinner thus writes of a black vulture in a roost among the sandhills of North Carolina: "They never bothered small birds—wood ducks, blackbirds, meadowlarks and myrtle warblers among others—who seemed to know this and to be able to recognize the vulture readily. They showed no alarm at the vultures sailing over them, although quick to dive out of sight when even a small hawk appeared."

DISTRIBUTION

Range.—The Southeastern United States, Central and South America; casual in the West Indies, the Northern and Western States, and southeastern Canada. Not regularly migratory.

The normal range of the black vulture extends **north** to southeastern Kansas (Chetopa); Missouri (Ozark Mountains and the vicinity of St. Louis); southern Illinois (Anna and Mount Carmel); Indiana (Annapolis and Brookville); Ohio (Hocking County); eastern Kentucky (Lexington); and eastern Virginia (opposite Plummers Island, Md.). **East** to Virginia (opposite Plummers Island, Md., probably Newport News, and Suffolk); North Carolina (Raleigh and probably Fort Macon); South Carolina (Oakley Depot); Georgia (Savannah, Blackbeard Island, McIntosh, and St. Marys); Florida (Gainesville, Orange Lake, Fruitland Park, Titusville, Kissimmee Prairie, Big Cypress Swamp, and Royal Palm Park); southeastern Mexico (Chichen-Itza, Yucatan, and the territory of Quintana Roo); British Guiana (Georgetown); French Guiana (Cayenne); eastern Brazil (Counani River, Para, Capim River, Cantagallo, Rio de Janeiro, and Taquara); Paraguay (Puerto Pinasco, Concepcion, and Asuncion); and Argentina (Formosa, Las Palmas, Resistencia, Santa Fe, Buenos Aires, and Rio Negro). **South** to Argentina (Rio Negro, Cordoba, Mendoza, and Tunuyan); and Chile (Ancud). **West** to Chile (Ancud, Valdivia, Concepcion, Santiago, Valparaiso, and Coquimbo); Peru (Ica and Callao); Ecuador (Babahoyo, Riobamba, and Quito); Colombia (Honda and Antioquia); Panama (Ancon and Culebra); Costa Rica (La Palma, San Jose, Juan Vinas, and Miravalles); Nicaragua (Escondido River); Guatemala; Tepic (Acaponeta River); Sinaloa (Mazatlan); Sonora (Guaymas and Tonichi); western Texas (San Angelo, Fort Worth, Decatur, and Gainesville); Oklahoma (Caddo, Limestone Gap, and Tulsa); and southeastern Kansas (Chetopa).

The range as above outlined is for the entire species, the typical race, *C. a. atratus*, occupying the northern regions south to Mexico and Central America, while the South American bird is known as *Coragyps a. foetens.*

Casual records.—The black vulture has a penchant for extensive wandering, which many times takes it far north of its regular range. At such times it has been recorded north to Colorado (one taken on October 8 or 9, 1921, near Boulder) ; Nebraska (one taken on Wolf Creek sometime prior to 1905) ; Michigan (three seen near Tecumseh, October 4, 1924) ; New York (one seen at West Seneca in June 1884; one at Medina on May 28, 1892; one taken on Plum Island, May 19 or 20, 1896; and one shot near Pulteney, July 11, 1909) ; Connecticut (one seen at Bolton reservoir, October 10, 1879, and one shot at East Lyme on July 6, 1901) ; Massachusetts (several records, among them being one on November 16, 1889, at Essex; one on July 2, 1890, at Plymouth; one on October 5, 1902, at Taunton; one shot September 15, 1905, at Waltham; one taken on May 12, 1916, at Pigeon Cove; and one seen on November 2, 1924, at Ipswich) ; New Hampshire (one shot at Randolph, April 17, 1926; one seen at East Westmoreland, May 9, 1933; and one seen about May 1, 1926, at Whitefield) ; Vermont (one taken July 11, 1884, near Montpelier, and one shot at Pawlet, July 7, 1912) ; Maine (nine records, the more recent being one captured alive near Dover, on August 20, 1901; one shot at Lubec, August 26, 1904; one shot on September 25, 1907, at Whitefield; one taken on July 6, 1909, at Monhegan Island; and one seen on July 11, 1915, at Scarboro) ; Quebec (one taken on October 28, 1897, at Beauport) ; New Brunswick (one about August 1879 at Campobello Island, and one seen August 9, 1924, at Grand Manan) ; and Nova Scotia (one taken January 12, 1896, at Pugwash).

Several were seen during May 1890 in the Tonto Basin, Ariz., and about a dozen were observed a few miles south of Tucson on May 7, 1922. No specimen is recorded for Cuba, but Danforth (1928) records four seen late in June or early in July 1926, near El Cobre. The species has been said to occur in Jamaica (Sclater, 1910), while in the Lesser Antilles it has been reported from Grenada (Clark, 1905) and Trinidad, where it is said to be common (Chapman, 1894). It is probable, however, that the black vultures of Trinidad are the South American race.

Egg dates.—Texas to Florida and North Carolina: 198 records, January 20 to July 7; 99 records, March 12 to April 17.

Family ACCIPITRIIDAE: Kites, Hawks, and Allies

ELANOÏDES FORFICATUS FORFICATUS (Linnaeus)

SWALLOW-TAILED KITE

HABITS

This elegant bird seems to have largely withdrawn from its former wide range in North America and is now confined, in this country, mainly, if not wholly, to Florida and perhaps the other Gulf

States. I have never seen it anywhere but in southern Florida, where it is still fairly common. Here we may look for its arrival early in March; Harold H. Bailey's (1925) earliest date is March 3; but Charles J. Pennock tells me that he has seen it at St. Marks as early as February 28. Audubon (1840) says: "In the States of Louisiana and Mississippi, where these birds are abundant, they arrive in large companies, in the beginning of April, and are heard uttering a sharp plaintive note. At this period I generally remarked that they came from the westward, and have counted upwards of a hundred in the space of an hour, passing over me in a direct easterly course."

I first made my acquaintance with this beautiful species in the Cape Sable region of extreme southern Florida. While crossing the narrow strip of prairie between Flamingo and Alligator Lake, we saw seven of these lovely birds sailing about over the prairie, soaring in circles high overhead, or scaling along close to the ground, like glorified swallows. They seemed to be quartering the ground systematically in the search for prey, for, as they circled, they gradually moved along over new ground. It was a joy to watch their graceful movements and a pity to disturb them, but my companion, the late Louis A. Fuertes, and I both wanted specimens. We concealed ourselves in the long grass and had not long to wait before we had two of the birds down on the ground and five others hovering over them, after the manner of terns, uttering their weak squealing or whistling notes. We shot no more; they were too beautiful; and we were rapt in admiration of their graceful lines, the purity of their contrasting colors, and the beautiful grapelike bloom on their backs and wings, which so soon disappears in museum specimens. I shall never forget the loving reverence with which the noted bird artist admired his specimen, as he began at once to sketch its charms.

Courtship.—I have never seen what I was sure was a courtship performance, but apparently this consists of spectacular aerial evolutions. Major Bendire (1892) quotes J. W. Preston as follows:

Of all aërial performances I have ever witnessed, the mating of the Swallow-tailed Kite excels. Ever charming and elegant, they outdo themselves at this season. In the spring of 1886 they chose as their mating ground an open space over the mouth of an ice-cold brook that made its way out from a dark tangled larch swamp. From my boat on the lake I had an excellent view of them. All the afternoon seven of these matchless objects sported, chasing each other here and there, far and near, sailing along in easy curves, floating, falling, and rising, then darting with meteor-like swiftness, commingling and separating with an abandon and airy ease that is difficult to imagine.

Col. N. S. Goss (1891) says: "I once saw a pair of these birds in the act of copulation. They were sitting on a small, horizontal limb, close together and facing each other, when, quick as a flash, the female turned or backed under the limb, the male meeting her from the top."

Nesting.—Much has been written about the nesting habits of the swallow-tailed kite in various parts of the country. In Florida its favorite nesting sites are in the tall, slender, Cuban pines near cypress swamps. The nests are seldom found very far from the cypresses and are sometimes placed in the tops of these trees. The kites are quite dependent on the long Spanish moss for nest building, and H. H. Bailey told me that recent hurricanes in extreme southern Florida have destroyed so much of this moss that the kites have largely moved away from certain sections. Dr. W. L. Ralph, who had considerable experience with these kites in Florida, sent the following notes to Major Bendire (1892): "They usually commence laying about the middle of April, and I have found them sitting on their nests from that time until the 1st of June, the latter being the latest date I have ever remained in Florida. Most of them have their eggs laid by the middle of May. * * * As nearly as I could judge, about three-fourths of the nests of this species found by me were about the same distance above the ground, i. e., they were 90 feet, and the remainder from a little above that height to 125 or 130 feet."

He describes a typical nest as follows:

It was situated 90 feet above the ground in, or rather on, the top of a very slender pine tree growing on the edge of a cypress swamp. The trunk of this tree at a height of 5 feet above the ground was not more than 15 inches in diameter, and at the place where my climber stood, as he took the eggs, it was less than 3 inches, while the limbs he stood on were only about an inch thick. The nest was composed of large twigs thickly covered with Spanish moss (*Tillandsia usneoides*) and long moss (*Usnea barbata*), lined with the same materials, with the addition of a few feathers from the birds. It measured 20 inches in length, 15 inches in width, and 12 inches in depth on the outside, and 6 inches in diameter by 4 inches deep on the inside. * * *

The Swallow-tailed Kite has a peculiar way of leaving its nest, for instead of flying directly from one side, as other birds do, it nearly always rises straight up for a short distance first, as if it were pushed up with a spring, and, when about to alight on its nest, it will poise itself a short distance above its eggs and then gradually lower itself down on to them. When they are thus poised above their nests there is scarcely a perceptible movement of their wings, and they often lower themselves so gradually that one can hardly tell when they have reached their eggs.

Bendire (1892) quotes J. W. Preston as follows:

* * * Nesting materials (twigs and moss) are carried by the female in her talons, the male following close, and going on the nest to arrange them. Days, and sometimes even weeks, are required to suitably complete the structure. Durii g this time they work in the morning and fly over the lakes and

woods in the afternoon. The nest is usually built on the foundation of an old one of a previous year. The female does not alight to secure nesting materials, but snatches them while in full flight. Once, while standing in a larch swamp, a Kite dashed by me and took up a small twig, heavily draped with *usnea*, and proudly soared out over the woods with it.

Colonel Goss (1891) watched a pair building a nest in the top of a large hickory tree, and says: "When either came to the nest alone with a stick, it would place it hurriedly upon the nest, but when both met at the nest they would at once commence fussing about, pulling at the sticks and trying to arrange the material, first one getting upon the nest and then the other, turning around as if trying to fit a place for the bodies. I think at one time they must have worked at least ten minutes trying to weave in or place in a satisfactory manner a stripping from the inner bark of the cotton-wood. As builders they are not a success."

In Texas these kites sometimes nest in tall pines, but oftener in the tops of the largest and loftiest deciduous trees, such as cotton-woods, elms, sycamores, pin oaks, cypresses, or pecans, along the banks of streams or in the river bottoms. The nests are often 100 to 150 feet above the ground, seldom less than 60, and placed among the slender topmost branches, concealed in the thick foliage; occasionally a nest is placed far out on a horizontal limb.

G. B. Benners (1889) mentions a nest that was over 200 feet from the ground in a giant cottonwood. He describes another nest as follows:

It is about one foot wide by two feet long, and four inches deep (or high), perfectly flat on top, with just the least depression in the middle to hold the eggs. Composed of a harsh green moss with a little Spanish moss among it, and with a mass of small twigs mixed in among the moss. These twigs must have the moss growing on them, for I saw several Kites carrying twigs with moss hanging from them, during our trip. The nest is just a platform, and what keeps the eggs from rolling out during the high wind, when the bird is not on, I cannot see. All the other nests we saw were of the same description, with the exception of one, which was composed wholly of Spanish moss. As the trees were all covered with this moss it was very hard indeed to see the nests.

J. W. Preston (1886) records the nesting of this kite in the wilderness of Becker County, Minn., and says of the locality: "Somewhere back from the shores of one of these lakes, where the rich flat land had sent up a heavy growth of basswood, elm and balsam, and the higher ground was covered with poplar, sugar tree and birch, a pair of Swallow-tailed Kites (*Elanoides forficatus*) had chosen a nesting place." The nest was finally found, after much watching, in the extreme top of a tall white birch, "whose greatest diameter was less than twelve inches, with scarcely a dozen branches, and these close to the nest, which was borne fifty feet upwards, and swayed by the slightest breeze." He says of the nest:

The nest consisted of small, dead larch branches, thickly interwoven with a long, fine moss, or lichen, found in great abundance on the larch everywhere in that region. This substance also formed a soft lining to the deep, well-shaped structure. In the nest were over two hundred separate pieces, which had been carried, one at a time, from a marsh a mile distant. It therefore required the travelling of four hundred miles to do the work; and there were certainly as many pieces strewn upon the ground as appeared in the nest. The birds also made long circuits while about the nest and at the swamp, where the material was gathered, so that no less than eight hundred miles must have been traversed while constructing the nest.

The swallow-tailed kite seems to have disappeared entirely from the northern portions of its breeding range during recent years. Dr. Thomas S. Roberts (1919) says: "The seemingly almost complete disappearance of this beautiful and once frequent bird is difficult to understand."

All observers seem to agree that the swallow-tailed kite is a very bold and aggressive bird in the defense of its nest. In many cases the birds have attacked the climber, diving at him repeatedly, dashing through the branches above him, and threatening to strike him, all of which is quite disconcerting while he is clinging to the slender, swaying treetop. Evidently the collector of a set of eggs earns his prize.

Eggs.—The swallow-tailed kite lays usually two eggs, sometimes three; four eggs have been reported and may occasionally occur, but the larger numbers reported were doubtless errors. They are rounded-ovate or nearly oval in shape; the shell is smooth and not glossy. The ground color is white or creamy white. They are usually boldly and irregularly, sometimes heavily, blotched or spotted, the markings often concentrated at one end; sometimes they are more evenly spotted and rarely finely or sparingly marked with fine dots. The usual colors of the markings are dark browns, "bone brown" to "liver brown"; but they often are brighter browns, "chestnut" or "Kaiser brown", or "ochraceous-tawny."

Occasionally a few small shell markings of light lavender are seen. The measurements of 50 eggs average 46.7 by 37.4 millimeters; the eggs showing the four extremes measure 50 by 39, 49.3 by 39.5, and 41.9 by 34.5 millimeters.

Young.—The incubation period for this species does not seem to be definitely known, but for other kites it is said to be from 21 to 24 days. Both parents share the duties of incubation and care of the young. Beyond the fact that they are very devoted and will fiercely defend their offspring, very little seems to be known about their home life.

Plumages.—I have seen but two rather large nestlings of the swallow-tailed kite. The smaller one, largely downy, was covered with short, thick, white down, faintly tinged with yellowish, and glossy black feathers were sprouting in the wings and tail. In the larger bird the back was well covered with black feathers, narrowly edged with white; "cinnamon-buff" or buffy-white feathers were appearing on the breast, belly, crown, and hind neck.

I have not seen a fully grown young bird in fresh juvenal plumage, but older birds in summer have lost the white edgings on the mantle and the buff colors on the under parts, probably by wear and fading; but August birds still have the dusky shaft streaks on the crown and breast, which gradually fade and probably disappear at the fall molt; in this plumage the mantle is browner than in adults, with greenish rather than purplish reflections, and the grapelike bloom is lacking; the wing and tail feathers and the primary coverts are narrowly tipped with white.

I have been unable to find any molt of the flight feathers in August birds (5 examined) and infer that this molt is accomplished after the birds leave for the south.

Food.—The food of the swallow-tailed kite consists mainly of small reptiles, amphibians, and insects. Dr. A. K. Fisher (1893) says that "it never molests small mammals and birds"; and some other observers agree with him. But George Finlay Simmons (1925) includes in its food "field mice, young Western Mockingbirds and Texas Painted Buntings which it takes on the wing from nests in mesquite growth." On the whole its food habits are neither beneficial nor particularly harmful. Its food includes small snakes, for which it is often called "snake hawk", lizards, frogs, and tree toads. It feeds very largely on grasshoppers, locusts, crickets, cicadas, beetles of various kinds, bees, wasp grubs, dragonflies, cotton worms, and various other insects. Practically all its food is procured on the wing and eaten while flying. Audubon (1840) says:

They dive in rapid succession amongst the branches, glancing along the trunks, and seizing in their course the insects and small lizards of which they are in quest. Their motions are astonishingly rapid, and the deep curves which they describe, their sudden doublings and crossings, and the extreme ease with which they seem to cleave the air, excite the admiration of him who views them while thus employed in searching for food. * * * In calm and warm weather, they soar to an immense height, pursuing the large insects called Musquito Hawks, and performing the most singular evolutions that can be conceived, using their tail with an elegance of motion peculiar to themselves. Their principal food, however, is large grasshoppers, grass-caterpillars, small snakes, lizards, and frogs. They sweep close over the fields, sometimes seeming to alight for a moment to secure a snake, and holding it fast by the neck, carry it off, and devour it in the air. * * *

The Fork-tailed Hawks are also very fond of frequenting the creeks, which, in that country, are much encumbered with drifted logs and accumulations of sand, in order to pick up some of the numerous water-snakes which lie basking in the sun. At other times, they dash along the trunks of trees, and snap off the pupae of the locust, or that insect itself. Although when on wing they move with a grace and ease which it is impossible to describe, yet on the ground they are scarcely able to walk.

Baird, Brewer, and Ridgway (1905) write:

One was noticed as it was hunting after grasshoppers. It went over the ground as carefully as a well-trained pointer, every now and then stopping to pick up a grasshopper, the feet and bill seeming to touch the insect simultaneously. They were very fond of wasp grubs, and would carry a nest to a high perch, hold it in one claw, and sit there picking out the grubs. * * *

Mr. R. Owen, while travelling from Coban to San Geronimo, in Guatemala, among the mountains, came suddenly upon a large flock of two or three hundred of these Hawks, which were pursuing and preying upon a swarm of bees. At times they passed within four or five yards of him. Every now and then the neck was observed to be bent slowly and gracefully, bringing the head quite under the body. At the same time the foot, with the talons contracted as if grasping some object, would be brought forward to meet the beak. The beak was then seen to open and to close again, and then the head was again raised and the foot thrown back. This movement was repeatedly observed, and it was quite clear to him that the birds were preying upon the bees.

Behavior.—The flight of the condor or the eagle may be grand, majestic, but the flight of the swallow-tailed kite is beautiful in the extreme, unsurpassed in grace and elegance. Coues (1874), in his usual matchless style, describes it as follows:

Marked among its kind by no ordinary beauty of form and brilliancy of color, the Kite courses through the air with a grace and buoyancy it would be vain to rival. By a stroke of the thin-bladed wings and a lashing of the cleft tail, its flight is swayed to this or that side in a moment, or instantly arrested. Now it swoops with incredible swiftness, seizes without a pause, and bears its struggling captive aloft, feeding from its talons as it flies; now it mounts in airy circles till it is a speck in the blue ether and disappears. All its actions, in wantonness or in severity of the chase, display the dash of the athletic bird, which, if lacking the brute strength and brutal ferocity of some, becomes their peer in prowess—like the trained gymnast, whose tight-strung thews, supple joints, and swelling muscles, under marvellous control, enable him to execute feats that to the more massive or not so well conditioned frame would be impossible. One cannot watch the flight of the Kite without comparing it with the thorough-bred racer.

Holt and Sutton (1926) write: "That this kite is playful, or mischievous, was obvious. Once a pelican flew slowly along under a soaring kite. The kite swooped down at the pelican and nagged the big clumsy creature for half a mile, crying loudly the while in a high voice, *kii-ki-ki*. Again, when a Barred Owl was flushed from a thicket, two kites slashed furiously down at the owl, crying loudly, and clearly intent on driving the creature away."

Donald J. Nicholson (1928a) relates the following:

On several occasions I had the privilege of witnessing at close range the bird taking a bath and a cooling drink from a deep pool hidden in a big cypress swamp. I was sitting under the shade of an oak, eating my lunch, when I saw a Kite come sailing around over the lake, finally coming down lower and lower. Satisfying itself that no harm was near, it swooped down to the surface and merely brushed its belly in the water for several yards, as if wishing to cool off; it was probably a setting bird. It then rose, circled about, and again swooped down, this time trailing its entire underparts and long tail in the water, taking a drink by dipping its bill in the lake. This was repeated six or eight more times with variations; sometimes merely trailing its body and tail feathers and not drinking, or doing both at the same time. After about ten minutes, the bird circled high, shook itself, folding its wings as it did so, dropped several feet, and then sailed from sight.

Voice.—I recorded the cries of distress or anxiety over fallen companions as weak, squealing, or whistling notes. Bendire (1892) says: "Their call notes are a shrill keen '*e-e-e*,' or '*we-we-we*,' uttered in a high key, which is very piercing and may be heard at a great distance." When several are flying together they have been heard to give soft twittering notes. Mr. Nicholson refers to their notes as "shrill, sweet cries, sounding like *peat, peat, peat.*"

Field marks.—The white head, neck, and underparts, the black wings and back, and the long, forked, black tail are unmistakable. But, above all, the graceful, swallowlike flight makes the bird recognizable as far as it can be seen. It need never be mistaken for anything else.

Fall.—Most observers record the swallow-tailed kite as a summer resident in the United States, departing in August or September for its winter home in Central or South America. It often occurs in large flocks while migrating. There are, however, some late fall and winter records for even the northern portion of its former range. D. H. Talbot (1882) saw a flock of 50 or more near Bismarck, N. Dak., on November 17, 1881. And Dr. Elliott Coues (1878) was informed by Dr. C. E. McChesney of the presence of this kite at Fort Sisseton, Dakota, during nearly the whole of the previous winter.

DISTRIBUTION

Range.—The United States east of the Rocky Mountains, south to Argentina. Casual in the Northern States and in southern Canada; accidental in Great Britain. Now practically extirpated from the northern part of its range.

Breeding range.—The breeding range of the swallow-tailed kite has extended **north** to probably formerly Nebraska (Doss and London); formerly Minnesota (Lake Minnetonka); formerly Wisconsin (Fort Atkinson and Racine); probably formerly Ohio (Portage and Stark Counties); and North Carolina (Lake Ellis). **East** to North Carolina (Lake Ellis); South Carolina (Chester and Charleston); Georgia (Marshallville and probably St. Marys); Florida (Palatka, San Mateo, Orlando, Lake Gentry, St. Johns

Prairie, probably Lake Worth, and Miami); probably Cuba
(Habana, Bahia, and Cienega Oriental de Zapata); probably
occasionally the Lesser Antilles (St. Bartholomew and Trinidad);
British Guiana (Waini River, Georgetown, and Aremu River);
Brazil (Para, Capim River, Bahia, Cantagallo, and Pirahy); and
Argentina (Territory of Misiones and Buenos Aires). **South to**
Argentina (Buenos Aires) and southern Bolivia (Chiquitos).
West to Bolivia (Chiquitos); Peru (Chamicuros and Huallaga
River); Ecuador (Santo Domingo de los Colorados and Bucay);
Colombia (Nechi, Bucaramanga, and Cali); Costa Rica (San Jose
and Naranjo); Nicaragua (Escondido River and Chontales); Guate-
mala (Coban); Jalapa; Nuevo Leon (Saltillo); Texas (San Antonio,
Austin, Waco, formerly Decatur, and Gainesville); Oklahoma
(Caddo); Kansas (Neosho Falls, Topeka, and probably Manhat-
tan); and probably formerly Nebraska (Doss).

The range as above outlined is for the entire species. The South
American form has been separated as *Elanoïdes f. yetapa*, but the
area of demarcation or intergradation between the two races, though
believed to be in Costa Rica, is at present imperfectly known.

Winter range.—During the winter season the swallow-tailed kite
withdraws almost entirely from the United States, although a few
are reported to winter in southern Florida (Harney River). The
distance that the northern form goes southward at this season is
not yet known, but a specimen from Bucay, Ecuador, taken in December, is referable to this race.

Coues (1878) quoted a report to him that some were seen almost
all the winter of 1877–78 at Fort Sisseton, Dakota.

Spring migration.—Early dates of arrival are: Florida—Titus-
ville, March 1; Pensacola, March 8; St. Marks, March 11; and Royal
Palm Hammock, March 13. Georgia—Cumberland, April 4. South
Carolina—Mount Pleasant, March 19. Mississippi—Biloxi, March
18. Louisiana—New Orleans, March 23; and Holden, April 5. Mis-
souri—Bolton, April 10; and Warrensburg, April 15. Texas—
Nunnsville, February 1; Giddings, February 13; Corpus Christi,
March 12; and Gainesville, March 21. Oklahoma—Caddo, April 1.
Kansas—Richmond, April 15; and Neosho Falls, April 27.
Nebraska—Vesta, April 3.

Fall migration.—Late dates of fall departure are: Iowa—Grinnell,
September 16; and Hillsboro, September 24. Missouri—Courtney,
September 4; and St. Louis, September 15. Texas—Corpus Christi,
September 1; and Tivoli, September 2. Mississippi—Bay St. Louis,
September 7.

Casual records.—The swallow-tailed kite has been recorded outside of its normal range on numerous occasions. Among these are
the following: Virginia, one at Aylett, on August 31, 1895; Maryland,

one taken at Ellicott City, on August 7, 1879, and one taken in Montgomery County, August 3, 1895; District of Columbia, one seen at the Virginia end of the Aqueduct Bridge, on April 11, 1897 (Bartsch); Pennsylvania, one taken near Philadelphia, April 4, 1791, and another in 1857, one captured at Olney in the spring of 1888, and one taken at Jerseytown, August 8, 1894; New Jersey, one taken about 1872 at Chatham (Herrick) and one seen at Morristown, September 18, 1887; New York, one at Raynor South in 1837, one about 1845 on the south shore of Long Island, one shot at Pittstown, on July 17, 1886, one seen at Stephentown on April 10, 1895, and another recorded on August 22, 1900, from Piermont; Connecticut, one seen July 2, 1877, at Lyme, another noted near Portland during the summer of 1861, while a third was recorded from Saybrook on June 16, 1889; Massachusetts, one taken at West Newbury about September 25, 1882, one seen near Northampton in 1880, while sometime prior to 1870 one was seen at Whately; Vermont, one seen at Waitsfield on April 26, 1913; New Hampshire, one recorded from Franklin in 1875; Ontario, one seen at Port Sydney on July 15, 1897, and Macoun (1903) records one seen at Ottawa prior to 1881, while Fleming (1907) records one from London "said to have been taken [there] many years ago"; Michigan, one taken near Detroit in the summer of 1881, one killed at Saline, on September 15, 1880, two obtained at Petersburg, on June 19, 1882, one taken at Ann Arbor on October 4, 1924, and another the same day near Ypsilanti; South Dakota, one shot several years ago near Vermillion, according to S. S. Visher (letter, 1912); North Dakota, in addition to the winter record of Coues, about 50 were reported near Jamestown between November 14 and 17, 1881 (Talbot, 1882); Manitoba, Seton (1908) reports that two were taken near Winnipeg in 1889 and 1892; Saskatchewan, while the species has been reported from this province, Mitchell (1924) considers the records as doubtful; New Mexico, one reported from the Capitan Mountains on July 10, 1903, one taken at Carlsbad about 1907, while a third was obtained at Cantonment Burgwyn about August 5, 1859; and Colorado, one shot in August 1877 in Manitou Park.

Swallow-tailed kites have been on a few occasions recorded from Great Britain as follows: One in 1772 at Balachulish, Argyllshire; one on September 6, 1805, at Shawgill, Cumberland; one in the summer of 1833 at Farnham, Surrey; one shot in June 1853 on the Mersey River; and probably another taken in April 1853 at Eskdale, Cumberland (Dalgleish, 1880).

Egg dates.—Texas to Florida: 81 records, March 10 to May 18; 41 records, April 7 to 26.

Iowa: June 3.

ELANUS LEUCURUS MAJUSCULUS Bangs and Penard

NORTH AMERICAN WHITE-TAILED KITE

HABITS

The above name was applied to the North American bird by
Bangs and Penard (1920) to distinguish it from the smaller South
American race, to which the name *leucurus* was originally applied.
The northern bird is larger, with longer wing and tail and relatively
wider tail feathers.   They say of the two ranges: "The small south-
ern form ranges from Argentina and Chile, northward to Vene-
zuela; the large northern form from California, Texas, Oklahoma,
South Carolina, and Florida, southward through Mexico to British
Honduras and Guatemala.   There is thus a wide area in southern
Central America and northern South America between the ranges
of the two forms as outlined above, where the species apparently
does not occur at all."

This gentle and attractive bird seems to have become exceedingly
rare, or to have been entirely extirpated, in the eastern portions of its
North American range.   During my six seasons, or parts of seasons,
spent in various portions of Florida I have never seen this kite;
once a special trip was made to a section where our guide said they
had recently nested, but no sign of them was found.   Donald J.
Nicholson tells me that he has not seen one there since 1910.   We
could not find it in southern Texas, and I have no recent records
of it there.   In certain sections of California it seems to be holding
its own, though exceedingly local in its distribution, and nowhere
universally abundant.   I doubt if it ever was very abundant, although Cooper (1870) referred to it as "quite abundant in the middle
districts of California, remaining in large numbers during winter
among the extensive tule marshes of the Sacramento and other
valleys", and Belding (1890) considered it "still a common resident"
about these marshes "in the centre of the State."   But Belding quotes
Dr. B. W. Evermann, as calling it "a rare resident" in Ventura
County, as early as 1886; and he quotes W. E. Bryant as saying
that "it is still a very rare resident" in Alameda County.   It seemed
to be the general opinion, at that time, that the white-tailed kite
was a disappearing species.   As a result, it has since been rigidly
protected by law and exempted from collecting permits.

Now comes more recent light on the subject, which is more encouraging.   Dr. Gayle B. Pickwell (1930) has published the results
of his exhaustive study of the literature and his field work in the
Santa Clara Valley.   Referring to past and present conditions in
that region, he says:

In spite of the fact that Taylor, in 1889, wrote of the Kite, "I venture to assert that there are not more than four pairs this year breeding within a radius of seven miles of that city [San Jose]", today, forty-one years later, there are still that many or more. * * *

Let us estimate that an average of four pairs of Kites (too high an estimate for some, too low, perhaps, for others) frequents each. We have then sixteen pairs of Kites in this entire valley. Twenty pairs, forty birds, I feel convinced, account for every Kite from Gilroy to the Bay and from Mount Hamilton to the summit of the Santa Cruz Mountains. * * *

The Kite was certainly more numerous in San Joaquin and Sacramento counties forty to sixty years ago than it is now. In other regions where it was present, especially in marsh districts, undoubtedly it has been seriously reduced in numbers. The condition in hill sections inhabited by it can be but guessed at. Here it probably has suffered least. * * *

This Kite is probably a dying species, never within historical times having predominated as such raptorial birds as the Desert Sparrow Hawk or Red-tailed Hawk for instance.

Since the above was written Dr. Pickwell (1932) has published a "requiem" for the kites in this valley; whereas he estimated that there were possibly 16 to 20 of these kites in the Santa Clara Valley in 1928, he now says: "This day (October 30, 1931) there cannot be more than two or three, and all too possibly none." We hope that this is a mere local condition.

His observations on the home life of these kites were made in the foothills of the Mount Hamilton Range in Santa Clara County:

The Slatore ranch lies in the foothills whose summits are grass-covered with wild oats and bromes, with scattered valley oaks and live oaks, and here and there a cluster of California coffee berry (*Rhamnus californica*) and gnarled *Sambucus*. Rocky outcrops, where more moisture may be trapped, have curious copses of scrubby growths of toyon, holly-leaved cherry, sages and sage brush; and the gullies lined with buckeye, California laurel, and poison oak run down to Silver Creek where the laurels and willows predominate. But the hills are mostly smooth as velvet, golden velvet most of the year, and green oaks are scattered over the velvet, like buttons on a buxom vest. In three buttons on this velvet vest were occupied nests of the White-tailed Kite. [See pl. 18.]

That such a habitat is not an unusual Kite home is shown by the fact that all the Kites of Santa Clara Valley today are, excepting one or two pairs, restricted to the lower foothills of the Mount Hamilton Range and Santa Cruz Mountains, on either side of the north end of the Valley. The exception is of not more than two pairs that occur to the north of San Jose between that city and the Alviso salt marshes. These frequent the cottonwoods and eucalyptus trees of the Coyote Creek and, not infrequently, are seen hunting over the treeless marshes at the foot of the Bay in common with Marsh Hawks, native there, and Turkey Vultures and Red-tailed Hawks from the hills.

Bendire (1892) says of their haunts: "Their usual resorts during the breeding season are the banks of streams or the fresh water marshes, especially if a few scattered live oaks or willow groves are close by, and their favorite nesting sites are the tops of live oaks, although other trees are also made use of whose foliage securely conceals the nest during incubation."

The impression I gained from men I talked with in California and from my own limited experience there was that this kite shows a decided preference for the vicinity of water, fresh-water marshes and streams; in such places it finds its food readily available all through the year, and it probably does not wander far away even in winter. According to Audubon (1840) it was found in similar haunts in Texas and Florida.

Nesting.—The white-tailed kite nests in a variety of situations. Usually the nesting pairs are widely separated, but sometimes several pairs may be located near each other in favorable situations. Two of the nests studied by Dr. Pickwell (1930) were in "valley oaks (*Quercus lobata*), and the third a coast live oak (*Quercus agrifolia*). The three formed an oblique or scalene triangle on the rolling hills with the longest side 320 yards and the others 200 and 175 yards respectively. To anyone conversant with the wide spacing of most raptorial birds this juxtaposition of the Kite nest territories seems unusual—indeed, so much in contrast with their near-relatives, semi-communal." The data, which he compiled from the literature cited, show that 11 nests were in live oaks, 3 or more in unspecified oaks, 2 or more in sycamores, and 1 in a maple. The heights from the ground varied from 18 to 50 feet; another that he measured was 59 feet. The nests were made of sticks and twigs of oaks in most cases, one being made of willow twigs. They were lined with grasses, dry stubble, barley straw, weed stems, rootlets, or Spanish moss. Some were described as flat, flimsy structures, and others were large, well-made, substantial, and deeply hollowed. Of five references that describe nesting sites, "two describe foothills (with oaks), two stream banks (or marshes with live oaks and willow groves nearby), and one a willow swamp."

Dr. B. W. Evermann wrote to Major Bendire that his first nest "was near the end of one of the topmost limbs of a cottonwood." Chester Barlow (1897), for one season at least, indulged in the bad practice of robbing the kites of their second sets. He found that they required about three weeks, or from 19 to 23 days, to lay a second set after the first set had been taken. These birds will almost always make a second attempt to raise a brood, in which they should not be discouraged, for whether they will make a third attempt or not is an open question.

I can add a little from my limited personal experience with the nesting habits of the white-tailed kite, as two of the three nests I saw were in situations different from any mentioned above. I was told that there were about six pairs of these kites nesting on an island in the Suisun Bay marshes. On April 15, 1929, my informant,

James Moffitt, took me there to investigate it. It was a low flat island a mile or more square, mostly covered with long, thick grass, quite marshy in places, but largely dry. It was partially surrounded by a canal, which we' navigated in a power boat. Extending along the banks of this canal in a curving line was a row of tall eucalyptus trees over a mile long. It was in these trees that the kites were nesting. As we approached we saw a kite sitting in the top of a dead tree, so we landed; and, after a short search, we saw what looked like a nest about 40 feet up in the thick top of a eucalyptus. After we had rapped the tree several times the kite flew off. It was a very uncomfortable tree to climb, but I managed to reach the nest, which was firmly lodged in the topmost crotch. I was surprised to find in it four small young, recently hatched. The nest was well made of small fine twigs, deeply hollowed, and profusely lined with dry grass; it was rather bulky and filled the crotch quite deeply. It had probably been used in previous years, as these kites have often been known to repair and use their old nests. Wishing to find a nest more conveniently located for photography, we spent considerable time hunting through the long row of eucalyptus trees; but, although we located at least three other pairs of kites, we could not find another nest. Although well hidden from below, the nests are open from above and give the birds a good lookout; the birds probably left the nests as they saw us coming.

Another nest was shown to me by M. C. Badger on April 27, 1929. It was located in an extensive tract of small willows and cottonwoods, mixed with a dense tangle of underbrush and vines, growing over many dead or fallen trees and branches, all of which covered a broad sandy plain along a river in Ventura County. The nest was not over 15 feet from the ground, yet well hidden in a thick mass of tangled vines in the top of a small dead willow. It was a well-made nest of coarse sticks and fine twigs, deeply hollowed and lined, in the bottom of the hollow only, with strips of inner bark. It measured 21 inches over all, and the inner cavity was about 7 inches in diameter; it held three eggs. One of the birds was seen in the vicinity, but it did not come near the nest. As the eggs were warm, she had probably slipped off when she heard us coming through the thick brush. Another nest (pl. 17) that he showed me was about 30 feet up in the topmost twigs of a small willow in the middle of another extensive tract of willows, cotton-woods, and thick underbrush.

Eggs.—The eggs of the white-tailed kite are among the most beautiful and richly colored of any of the hawks' eggs; consequently they are greatly in demand among oologists. The set usually consists of four or five eggs, sometimes only three, and I have one record of six eggs. In shape they vary from ovate to oval, and the shell is smooth but not glossy. The white, or creamy-white, ground color is usually largely, and often wholly, concealed by the profuse markings of rich browns, large blotches of dark ".bone brown" or "liver brown", over washes or splashes of brighter browns, such as "burnt sienna", "amber brown", "hazel", "tawny", or "ochraceous-tawny"; some eggs are finely spotted with the darker browns over the lighter washes, or more rarely over the whitish ground color; in some eggs the heaviest markings are concentrated at one end and very rarely the rest of the egg or the entire egg is mainly white; the splashes and blotches have a longitudinal trend. The measurements of 50 eggs average 42.5 by 32.8 millimeters; the eggs showing the four extremes measure **45.3** by 33.3, 42.4 by **35.6** and **38.1** by 30 millimeters.

Young.—Dr. Pickwell's (1930) evidence "indicates that the incubation period is not less than 30 days. Young are in the nest about 30 days." Probably both sexes incubate; the sexes are so much alike that this is difficult to determine unless the act of nest relief is seen; such an observation does not seem to have been made. But both parents are known to share in the care of the young and sometimes an exceptionally aggressive pair will swoop down at the intruder. Chester Barlow (1895) relates the following:

After leaving the female flew over and around me a few times and was presently joined by the male, both flying near and uttering a raspy, clacking note which I had never heard before. This no doubt was giving vent to their anger. Now and then the short, sharp whistle characteristic of the bird was uttered. Soon the female flew to an oak a short distance away and the male took up the battle in earnest. Soaring away perhaps 100 yards he came swiftly toward me almost on a level with my head until within about ten feet when he would switch upwards. Then he would soar up and swoop down at lightning speed, always changing his course before reaching me. The rush of his wings was plainly audible. Again he was joined by the female but after a few attacks both flew to near-by trees where they remained till I had departed.

The young, according to Dr. Pickwell (1930), show the usual reactions, common to all raptorial birds, when too closely approached. "At first approach the young Kite spreads wide the wings and backs off with mouth agape, emitting a rasping note If the tormentor persists, the bird thrusts its feet forward with a resultant dropping back upon the tail. The third and last stage is to drop completely on the back and to present the most impressive weapons a Kite has, the talons."

Plumages.—The smallest young, such as I found in the nest, are sparsely covered with short, dull-white down, tinged with "pinkish buff" on the crown and dorsal tracts. At a later downy stage Dr. Pickwell (1930) found the young bird clothed in "heavy bluish down." A nearly full-grown juvenal is a beautiful bird; the fore-

head is white and the crown mostly "cinnamon", heavily streaked
with dusky; the back and scapulars are "hair brown" to "drab-gray",
broadly edged with "cinnamon", or white and "cinnamon"; the tail
is "pale to pallid mouse gray", with a darker subterminal band and
white tips; the lesser and median wing coverts are brownish black,
the latter tipped with white; the remiges are "light to pale mouse
gray", mostly white-tipped, the primaries darker near the tips; the
under parts are white, heavily suffused with "cinnamon" on the breast
and less so on the belly; the lores are dusky. Dr. Pickwell (1930)
adds: "Toes and tarsus, yellow; beak and claws, black; eyelids, blue;
iris, brown."

This plumage is worn but a short time, and the bright colors soon
disappear by wear and fading. A postjuvenal molt begins in July
and continues through the fall; it involves all the contour plumage
and the lesser and median wing coverts. Some November birds have
nearly completed the molt but are still largely brown on the back.
A January bird shows the last of this molt and is renewing the
scapulars and tail feathers. Except for the wing quills, which are
probably not shed until later, the young bird is practically adult by
spring.

Adults apparently have a prolonged molt late in summer and in
fall; a December bird has not yet completed the molt of the wings
and tail but is otherwise in fresh plumage. I have seen South American birds molting their flight feathers in July and October, their
winter and spring.

Food.—The food of this kite includes field mice, wood rats, pocket
gophers, ground squirrels, shrews, small birds, small snakes, lizards,
frogs, grasshoppers, crickets, beetles, and other insects. Probably
very few birds and few of the larger mammals are taken, but mainly
the smaller vertebrates and the insects named. It is evidently a
highly beneficial species. Dr. Loye Miller (1926) noted, from the
examination of a well-filled stomach—

* * * that both its appetite and its table manners are far from dainty.
Remains of four meadow mice (*Microtus*) and an entire shrew (*Sorex ornatus*)
were identified in the contents of stomach and crop. The shrew was absolutely
entire. The largest mouse had been torn apart in the lower thoracic region
and the hinder portion bolted entire with skin and fur in place. Two mouse
heads had been swallowed hair and all. The fore quarters of the mice seemed
to have been stripped of skin, but great masses of skin and fur had been
swallowed after stripping them off. Viscera and small bones indicated that
most of both mice had been eaten, and there is no reason to believe that any
part had been discarded. Well cleaned bones from two other *Microtus* skulls
were still retained in the stomach.

Dr. Pickwell (1930) writes:

The Kite hunts, not by soaring and searching from a lofty position as do
Buteos, nor by the low harrier method of the Marsh Hawk, but by a rather

erratic scouting from a position intermediate between these two. When prey is seen the bird "stands" with wings quiet if the air is moving sufficiently to permit it to "kite", as its name would intimate its habit to be, or beats the wings slowly from an angle well above the back. During such a stand it drops its legs. If it stoops it makes no falcon drop of lightning speed with wings drawn into a thin wedge along the sides of the body, but keeps them up in a V angle above and slips down with legs hanging and at a speed one would never guess was more than fast enough to catch a snail. But that they do catch prey, some of it very agile, there is no doubt. And that this method is used to catch it there is no doubt either, for they have been observed to do so.

Laurence G. Peyton (1915) says: "One morning, while working near the nest, my brother saw one of the Kites returning from the direction of the river with something in its claws. While still some distance from the nest it began calling and was quickly joined by the other bird. The first bird remained hovering in the air like a Sparrow Hawk, while the other darted up underneath it, took the food from its claws and returned to the nest while the other sailed away."

Behavior.—The flight of the white-tailed kite is light, airy, and graceful; often it is a pretty fluttering flight with quick wing beats, or a stationary hovering flight like a sparrow hawk; and at times it is quite swift. I noticed that when the bird is soaring or scaling there is a bend in the wing, as in the osprey. Dr. Pickwell (1930) describes it as "with wings slightly raised and down-curving at the tips." Also he says: "The leg-dangling habit of the Kites is one of their most conspicuous oddities. On the nesting territory the protesting birds flew here and there nearly constantly, uttering their cries, beating the air slowly with short strokes, the wings held up at a sharp angle above the back, the legs dangling from a point about the center of the body."

W. H. Hudson (1920) says of the South American form:

Its wing-power is indeed marvellous. It delights to soar, like the Martins, during a high wind, and will spend hours in this sport, rising and falling alternately, and at times, seeming to abandon itself to the fury of the gale, is blown away like thistle-down, until, suddenly recovering itself, it shoots back to its original position. Where there are tall Lombardy poplar-trees these birds amuse themselves by perching on the topmost slender twigs, balancing themselves with outspread wings, each bird on a separate tree, until the tree-tops are swept by the wind from under them, when they often remain poised almost motionless in the air until the twigs return to their feet.

Although ordinarily gentle birds, these kites are often very pugnacious toward certain large birds, crows and hawks, that invade their territory. Several observers have seen them persistently drive away crows and the various Buteos. Dr. Pickwell (1930) writes:

In fact many of our records of Kites have come about because our attention has been drawn first to a large hurried Buteo in the distance and glasses

showed there not only Buteo but Kites above swooping down, one, then the
other (Kites are nearly always in pairs), in huge parabolas reaching a
hundred feet or more above the harried giant. Down one comes with a rush
and swings up again. Immediately after, the other one drops, then up, and
so around and around they alternate until the distance and blue swallows
up Buteo and tormentors. This game is played the year around, in the breeding
season and out. Perhaps, as with the excitement that small birds display
over the discovery of an owl, there may be a meaning in the Kites' pugnacity.
It may well be that the contents of the Kite nest, in the very top of its oak,
concealed from below but completely exposed from above, are a temptation to
these big hawks the Kites so persistently annoy. If so, then there is something
of significance in the fact that Turkey Vultures, though they have always
been, in the Kite territory, more numerous than all other large birds, are
never molested.

Voice.—Dr. Pickwell (1930) also gives the best description of this
bird's notes, as follows:

The notes are several in number and no one word or term describes them
all. The most frequently uttered is a spasmodic short whistle: *kēēp, kēēp, kēēp.*
At a distance it sounds like *chĭp, chĭp, chĭp,* or *kĭp, kĭp, kĭp, kĭp,* or even more
chicken-like, *chēēp, chēēp, chēēp.* This is the note that is given as the birds
beat slowly here and there with legs dangling, and it expresses the mildest
solicitude. Undoubtedly Dawson (1923) means this note with his *"clewk".*
The next is more highly pitched and longer, a "plaintive whistle" in truth.
It may be transcribed as *krēēk* or *krēē-ēēk.* It may be as repeatedly and rapidly
uttered as the former and expresses greater solicitude. The last and most
solicitous, uttered usually only when an intruder is climbing the tree to a
nest, is a prolonged *kēē-răk* or *kēē-rĕk.* This note comes at the end of a series
of *kēēp* notes. Its terminus is lower and almost guttural, reminding me
much of the whang of a focal-plane shutter. The notes of the young are two.
They have a mild, high-pitched *krēē-ēēk* like the adults, and when at the
height of their intimidation display they have a harsh scream uttered with
the mouth enormously agape. This reminds one much of the rasping scream
of the Barn Owl.

Field marks.—The most striking field mark of this kite is its
whiteness; in the distance it seems to be wholly white; it might easily
be mistaken for a white domestic pigeon, except for its peculiar flight.
But it can be recognized by its flight, described above, as far as
its outline can be seen. If near enough its black shoulders and, at
times, its dangling legs are diagnostic. As seen from below, it
appears wholly white with a dark crescent at the bend of the wing
and gray at the extreme tip; its tail is decidedly rounded.

Enemies.—Milton S. Ray has sent me some extensive notes on his
experiences with these kites in several of the central counties of California, from the late nineties up to 1932. He says that jays,
magpies, or crows will sometimes puncture or destroy the eggs in an
incomplete set. Once he saw a raccoon leaving a nest, and the eggs,
which it had contained previously, had entirely vanished. He mentions a very loosely built nest, "so frail and open that one of the four
eggs partially fell through the nest." Another nest "was so com-

pactly built that it held water" and, after a storm, the eggs were "almost submerged"; the nest was subsequently deserted.

He agrees with other observers as to the recent disappearance of these kites, saying: "Occasional birds were recorded in the last decade but at the present writing (1932) the birds seem to have disappeared from almost every point simultaneously." As to the cause of its decline, he says:

This Kite is peculiarly friendly and unsuspicious and therefore exceptionally easy to shoot. This is particularly true during the nesting period. Through a mistaken belief that the bird preys on quail, ducks, and other game birds the kites have been widely shot by hunters, gamekeepers, and ranchers. The "hunts" of gun clubs instituted by the various cartridge companies to exterminate owls, hawks, jays, and crows (these hunts are a curse of the present generation) have been largely responsible for the extermination of these beautiful birds. In a number of cases I have actually been able to prevent the birds being shot. In some instances I have found that the rather close resemblances this kite bears to the smaller gulls, as Bonaparte's and the kittiwake, has also prevented it from being killed.

DISTRIBUTION

Range.—The Southern United States south to central South America; accidental in central and northern States. Not considered migratory and now apparently almost extinct in North America.

Although the white-tailed kite is a transcontinental species, its range (in the United States) is more or less discontinuous, there being great areas from which it is practically or entirely unknown. The range extends **north** to central California (Geyserville, St. Helena, and Stockton); Oklahoma (Fort Arbuckle); and Florida (near Lake Kissimmee). **East** to Florida (near Lake Kissimmee and Fort Myers); eastern British Guiana (Demerara River); eastern Brazil (Porto Real, Bahia, and Itarare); and eastern Argentina (Concepcion, Baradero, and Buenos Aires). **South** to Argentina (Buenos Aires); and Chile (Arauco). **West** to Chile (Arauco and Santiago); northwestern Argentina (Tucuman); northern Brazil (Forte de San Joaquim); western British Guiana (Mount Roraima); Lower California (San Carlos and Cape Colnett); and California (Alamitos, Saticoy, Santa Barbara, Hollister, San Jose, Santa Clara, Lake Merced, Nicasio, and Geyserville).

The range as outlined is for the entire species, but the United States form, *E. l. majusculus*, is not known south of Lower California.

Casual records.—Audubon recorded the white-tailed kite as breeding on the Santee River, S. C., but Wayne (1910) believes this to be an error. A specimen was recorded from Marthas Vineyard, Mass., on May 30, 1910; one was shot near Kenner, La., on October

11, 1890; Ridgway reported a pair seen at Mount Carmel, Ill., during the summer of 1863 or 1864; one was said to have been taken near Ann Arbor, Mich., in September 1878, and one in Livingston County on April 21, 1879 (Barrows, 1912); while it also has been reported from northern California, as a specimen was obtained about August 6, 1924, at Miranda, and there is also a record from Red Bluff (C. H. Townsend, 1887).

Egg dates.—California to Texas: 120 records, February 12 to June 21; 60 records, April 2 to 29.

ICTINIA MISISIPPIENSIS (Wilson)

MISSISSIPPI KITE

HABITS

As I have never seen this kite in life, I shall have to rely wholly on the observations of others. It is a bird of the Lower Austral Zone, being seen chiefly in the Southern States from South Carolina and northern Florida to Texas, Oklahoma, and Kansas. Walter Colvin writes to me that he found this kite quite common in Barber County, Kans. "A bend of the Medicine Lodge River, where the timber consisted of elm, cottonwood, walnut, white locust, black locust, redwood, mulberry, boxelder, and cedar, which grew in parklike fashion, seemed to be a favorable location. Here more than a dozen kites were seen in the air at once."

Although rather widely distributed within the region outlined above, it seems to be localized in breeding communities, rather thickly populated, and to be entirely absent from apparently similar intervening territory. It also seems to gather in very large numbers, at other times, on particularly favorable feeding grounds.

Spring.—The Mississippi kite is a summer resident in the United States, arriving from the south in March or April. Dr. Frank M. Chapman (1891) witnessed a heavy migration near Corpus Christi, Tex., of which he writes: "This species was first observed April 24, when nine individuals were seen flying northward. The following day we crossed a great flight of these birds. They could be seen to the limit of vision both to the north and south, and about twenty-five were in sight at one time. They flew northward at varying heights; some were within gunshot, while others were so far above the earth that they looked no larger than swallows."

Audubon (1840), in his usual flowery style, describes the coming of spring in southern Louisiana, where he says that this kite arrives "about the middle of April, in small parties of five or six, and confines itself to the borders of deep woods, or to those near plantations, not far from the shores of rivers, lakes, or bayous. It never moves into the interior of the country, and in this respect resembles *Falco furcatus*. Plantations lately cleared, and yet covered with tall dying girted trees, placed near a creek or bayou, seem to suit it best."

G. W. Stevens tells me that it arrives in northern Oklahoma from May 1 to 15. And Charles J. Pennock gives me his earliest date for northern Florida as March 1. He says that during the spring this kite frequents "the neighborhood of the more dense, low hammocks, while later in the season it might be found in the vicinity of the rivers and ponds."

Nesting.—Although the Mississippi kite often builds its nest in the top of some tall tree, Mr. Colvin has sent me some notes on several nests that he found in the valley of the Medicine River, Kans., which were at rather low elevations. He refers to one nest that "was 50 feet up in the outer branches of a cottonwood"; but the others, ten or more, found on two or more days spent in the kite country, were in low elms or walnut trees. The timber in which the kites were nesting on May 31 and June 7, 1931, "was made up largely of elm, walnut, chinaberry, and elder. Most of the trees were stunted by the wind and storms and most of the elms were blighted." One nest in an elm was "situated on a limb about 12 feet from the ground, small and compactly built of sticks of trees, 6 to 8 inches in length. The usual sticks were one-fourth to three-eighths of an inch in diameter and broken clean at both ends. The nest was lined with green walnut leaves" (pl. 22). Another nest was "in the upper branches of a small walnut tree some 18 feet from the ground." Two other nests mentioned were on horizontal limbs of dwarf elms, 14 and 18 feet up, one of these measured 10 inches in diameter and 8 inches in height; it had "a small quantity of dried plants in the center" and was lined with green walnut leaves.

Mr. Stevens tells me that in northern Oklahoma it nests in scattering trees, 12 to 40 feet up, usually in the larger forks but sometimes in the smaller forks and occasionally on horizontal limbs. Elms are most commonly chosen, but also black jack oaks and occasionally cottonwoods, hackberries, and soapberries. He says the nests are always lined with green leaves, often with twigs attached; these may come from the nesting tree or another, commonly the sumac (*Rhus glabra*).

Albert F. Ganier has sent me excellent photographs of three nests taken near Vicksburg, Miss. (pl. 21). One of these was 80 feet up in a sweetgum tree, "located at the crest of a ridge in a wooded pasture"; it was a well-built nest, containing much Spanish moss; it had been used the previous year and was occupied the following year. Another nest was 60 feet up in a red oak on a ridge in thin woods; this was the "only nest of 18 examined that was built in an oak; they usually select the sweetgum because of its dense foliage and the tall erect form of the tree."

In his excellent article (1902) on this kite he describes the nest-building activities as follows:

On looking up I was surprised to find them soaring high in the air, apparently with nothing more upon their mind than to satisfy their appetites. Suddenly, however, one of them remained stationary for a second, then with half-closed wings came swift as an arrow down through the trees and reappeared above my head with an oak twig in his talons; wheeling, he sailed swiftly upward to a crotch in a gum tree, which showed a bunch of sticks, the beginning of a nest. Only for a moment did he remain; then, dropping over one side of the nest, he sailed upward and rejoined his mate.

For over an hour and a half I lay there and watched them slowly constructing their nest; both birds worked, darting in among the trees as on the first occasion, and reappearing with either a twig or spray of green leaves. At last, as the midday hour began to cast short shadows, one of the birds perched on the edge of the nest, while its mate lit on the topmost branch of a cottonwood tree some two hundred yards away.

He says that the nests are very difficult to see as the birds "show a great preference for the tip-top branches of gum and cottonwood trees whose dense foliage is almost impenetrable to the eye." One big cottonwood tree that he felled and measured was over 2½ feet in diameter and 131 feet high; the nest in it had been 119 feet above the ground.

Another nest that he examined was "composed of sticks and twigs with a thick lining of locust, gum, thorn and other green leaves"; it measured "25 inches from tip to tip of the longest twigs, while the width of the nest proper was 14 inches, the area covered with green leaves being 6 inches square. The nest as usual was almost flat on top."

In certain parts of Texas this kite nests in mesquite trees at such extremely low elevations as 4, 5, or 6 feet above the ground, making small nests lined with mesquite leaves. In Louisiana, according to George E. Beyer (Bendire, 1892), "the nests are placed in the tops of loblolly pines (*Pinus taeda*) or white oaks (*Quercus alba*), at a height of from 50 to 60 feet. Pine woods are the favorite localities."

The highest nests of which I can find any record are reported by Arthur T. Wayne (1910) in South Carolina; one was 111 feet and another 135 feet from the ground in the tops of gigantic short-leaf pines. He says that a pair nested within a mile of his house for ten years and for five years used the same nest. Other observers have noted that these kites often use the same nest for several years in succession. They also often return to their own nest after a lapse of a few years and sometimes appropriate an old crow's nest. The great variation in the height of the nest indicates that the kites select their nesting site where they can find the best food supply regardless of timber conditions, and then build their nest in the highest tree available. If their nest is robbed they will lay a second set, either in the old nest or a new one, about two weeks later.

Eggs.—The Mississippi kite lays only one or two eggs, rarely three, with some variation in different parts of its range. Mr. Ganier tells me that in Mississippi he has found two eggs or two young in only two out of some 13 or more cases; all the other nests contained only one egg or young. Mr. Stevens, referring to Oklahoma, says in his notes, "one occasionally, two usually, and three very rarely"; in some 500 nests examined during seven years, he has found only three sets of three. Of 40 nests under observation by Dr. George M. Sutton, 38 held two eggs and two held one egg each. Most of the sets in collections consist of two eggs, but there are very few sets of three. The eggs vary in shape from ovate to rounded-ovate or nearly oval. The shell is smooth but without gloss when fresh. The color is white or pale bluish white. They are normally unmarked and are often more or less nest stained, and some may appear to be faintly spotted, but such markings are, I believe, wholly adventitious; true pigment markings must be exceedingly rare. The measurements of 50 eggs average 41.3 by 34 millimeters; the eggs showing the four extremes measure 45.7 by 35.2, 44.5 by 36.5, 37.7 by 33.8, and 41 by 31.2 millimeters.

Young.—Dr. George M. Sutton writes to me that "the period of incubation is 31 or 32 days. An egg laid on May 18 hatched June 18. * * * In an attitude of repose the young bird rests on the *outer* part of its feet only. The cry is a thin, feeble squeal, a hair-thin sound." Both sexes assist in incubation and in the care of the young. Mr. Ganier (1902) writes:

The nest could plainly be seen from several points and I soon made out the form of a young bird on the edge of it, looking out among the trees and occasionally spreading its wings as though impatient to be free.

While still looking, a shadow glided through the trees and an old bird lit on the edge of the nest with something in her beak; slowly the young bird turned around to receive its food and then assumed its old position. The parent bird lingered but a minute, then glided away as silently as she had come.

I sat on a log and watched them for an hour, the parent birds taking turns at feeding the young one, whose restless wings seemed to trouble him much more than his appetite.

Plumages.—I have seen no very small young of this kite, but Dr. Sutton describes it for me as follows: "The natal down is pure white, with a small faint spot of buffy brown on the nape and a wash of the same pale brown over the back and upper surface of the wings. The area in front of and about the eyes is dull gray, the marking occupying almost precisely the same position as the black facial marking of the adult. Bill dull blue-gray. Cere dull brownish orange.

Corners of mouth light orange. Feet pale, clear yellow-orange, with gray claws. Eyes dull gray-brown, with bluish pupils. Eyelids dull gray." The juvenal plumage appears first on the scapulars, then on the wings and tail, and then on the back and the sides of the breast; the last of the down is seen on the head and belly.

In fresh juvenal plumage the head is white, streaked with black; the back and wing coverts are sooty black, almost clear black, with narrow edgings of "russet" or buffy white; the scapulars are broadly banded with white; the greater wing coverts, all the rectrices, and all the remiges are jet black, tipped with white, most broadly on the tertials and scapulars and most narrowly on the tail; the under wing coverts are "pale ochraceous-buff" spotted with rusty brown; the tail feathers are deeply notched or barred with white on the inner webs; the under parts are from "cinnamon-buff" to buffy white, heavily spotted with browns, the breast feathers being centrally "hazel" surrounded by blackish brown and broadly edged with "cinnamon-buff."

This plumage, with considerable fading of the browns and buffs, is worn only through the summer and fall. During the first winter and spring progress is made toward maturity by a gradual molt of the contour plumage; but considerable white still shows on the under parts owing to basally white breast feathers, the white increasing on the belly and under tail coverts. One-year-old birds in May, July, and August still retain the juvenal wings and tail and show the last of the first winter plumage on the under parts. Apparently the adult plumage is assumed at this first postnuptial molt, which is complete and much prolonged; I believe that the wings and tail are not molted until after the birds go south. Mr. Stevens has seen birds breeding in this immature plumage. Adults probably have a similar, prolonged, annual molt.

Food.—Mr. Stevens says in his notes that these kites feed on the wing, snatching locusts from plants and seizing cicadas in flight. A flock of from 3 to 20 will sail about a person, a horseman, or a team, traveling through grassy flats and bushy places, and seize the cicadas as they are scared up. The insect is grasped in the claws and eaten in the air. Usually only the abdomen of the cicada is eaten and the remainder is dropped; the wings and legs of locusts are often picked off and the remainder swallowed. He has found the remains of toads, mice, and young rabbits in the nests with young.

Audubon (1840) graphically describes its feeding as follows:

He glances towards the earth with his fiery eye; sweeps along, now with the gentle breeze, now against it; seizes here and there the high-flying giddy bug, and allays his hunger without fatigue to wing or talon. Suddenly he spies some creeping thing, that changes, like the cameleon from vivid green to dull brown, to escape his notice. It is the red-throated panting lizard that has made its way to the highest branch of a tree in quest of food. Casting upwards a sidelong look of fear, it remains motionless, so well does it know the prowess of the bird of prey; but its caution is vain; it has been perceived, its fate is sealed, and the next moment it is swept away.

All writers seem to agree that the Mississippi kite feeds almost exclusively on the larger insects, such as cicadas, locusts, grasshoppers, crickets, katydids, dragonflies, and large beetles, but small snakes, lizards, and frogs are sometimes taken. Birds apparently are never molested, and small birds show no fear of it.

Behavior.—Audubon (1840) writes:

Its flight is graceful, vigorous, protracted, and often extended to a great height, the Fork-tailed Hawk being the only species that can compete with it. At times it floats in the air as if motionless, or sails in broad regular circles, when, suddenly closing its wings, it slides along to some distance, and renews its curves. Now it sweeps in deep and long undulations, with the swiftness of an arrow, passing almost within touching distance of a branch on which it has observed a small lizard, or an insect it longs for, but from which it again ascends disappointed. Now it is seen to move in hurried zig-zags, as if pursued by a dangerous enemy, sometimes seeming to turn over and over like a Tumbling Pigeon. Again it is observed flying around the trunk of a tree to secure large insects, sweeping with astonishing velocity. While travelling it moves in the desultory manner followed by Swallows; but at other times it is seen soaring at a great elevation among the large flocks of Carrion Crows and Turkey Buzzards, joined by the Fork-tailed Hawk, dashing at the former, and giving them chase, as if in play, until these cowardly-scavengers sweep downwards, abandoning this to them disagreeable sport to the Hawks, who now continue to gambol undisturbed. When in pursuit of a large insect or a small reptile, it turns its body sidewise, throws out its legs, expands its talons, and generally seizes its prey in an instant. It feeds while on the wing, apparently with as much ease and comfort, as when alighted on the branch of a tall tree. It never alights on the earth, at least I have never seen it do so, except when wounded, and then it appears extremely awkward. It never attacks birds or quadrupeds of any kind, with the view of destroying them for food, although it will chase a fox to a considerable distance, screaming loudly all the while, and soon forces a Crow to retreat to the woods.

Dr. E. W. Nelson (1877b) says:

Their power of sight is truly wonderful. I saw them repeatedly dart with unerring aim upon some luckless grasshopper, from an elevation of at least one hundred yards.

No less remarkable is their power of flight. * * * I repeatedly saw them dart down from a great height with such velocity that it would seem an impossibility for them to escape being dashed to pieces on the ground, but instead, when within a few feet of the earth, they would suddenly spread their wings and the reaction would lift them with almost equal rapidity to about one-half their former elevation. They were so shy that it was impossible to get within gunshot of them.

Although a gentle, inoffensive bird at ordinary times, it can put up a stiff fight when wounded. Wilson (1832) tells of one that fastened its claws so firmly in his hand that he had to cut the tendons in its leg to release its grip. It is brave too in the defense of its nest, driving away such predatory birds as crows and jays. It will even occasionally attack a man that is climbing to its nest, as Mr. Ganier (1902) relates:

I had scarcely made half the distance when three or four Kites began to circle about on the level with the tree-top, and as I seated myself to rest on a branch, twelve feet below the nest, one of the birds began to dart at me. It was a very pugnacious fellow and would circle around within twenty feet of me until it would catch my eye; then, pausing for a moment, it would dart directly at me, to within six or eight feet of my face, when it would swoop suddenly upward, emitting at the same time a sharp shrieking cry. This performance was kept up until I descended, the birds darting closer as I reached the nest.

Voice.—The Mississippi kite is usually a rather silent bird except when the vicinity of its nest is invaded. Mr. Stevens refers to the alarm note at such times as a whistling cry of three or four syllables, the first and last on a lower key and the middle on a higher key, "longer, more forceful and tremulous"; it is the only note he has heard. C. J. Pennock describes it as a "clear but not loud call, *kee-e-e*, repeated sometimes two or three times in succession." Dr. Sutton tells me: "The usual cry of the kite I should write down as *phee-phew*. I heard this cry hundreds of times. I did not hear a three-syllabled cry. In mating the birds sometimes chipper at each other, a cry similar to one of the marsh hawk's calls."

Fall.—Mr. Stevens says that these kites leave Oklahoma in rather large flocks in September, usually by the fifteenth. Mr. Ganier (1902) writes:

Near the middle of August the birds seem to be very active at feeding; evidently they are then preparing for their southward journey. A specimen shot in the last days of August was so fat that I found it impossible to make a first-class skin of it; the breastbone sank far below the level of the breast meat.

As the first days of September approach the last individuals may be seen slowly flying southward; then the woods lose their charm to me for the sky has lost its gem, the Mississippi Kite.

DISTRIBUTION

Range.—Southeastern United States; accidental north to Pennsylvania and New Jersey and south to Guatemala. Only slightly migratory.

The Mississippi kite breeds **north** to northern Texas (Tascosa and Lipscomb); Kansas (Sun City, Medicine Lodge, and Baldwin City); Missouri (Webster and Howell Counties); probably formerly southern Illinois (Mount Carmel); Georgia (probably Marshallville and Augusta); and South Carolina (Columbia and

Charleston). **East** to South Carolina (Charleston, Yemassee, and probably Bluffton); and Florida (Waukeenah and Gainesville). **South** to Florida (Gainesville, St. Marks, Whitfield, and Pensacola); southern Mississippi (Bay St. Louis); Louisiana (probably New Orleans and Avery Island); and southern Texas (Beaumont, Sour Lake, Giddings, Austin, San Antonio, and Sheffield). **West** to western Texas (Sheffield, San Angelo, and Tascosa).

Winter range.—In winter the species has been detected north at least to Texas (Eagle Pass and Nunnsville) and Florida (Panasofkee Lake and Fort Myers).

Migration.—It is not apparent that this species makes a regular migration, but probably in winter it withdraws to some extent from the northern part of its breeding range. It has been noted to arrive at Copan, Okla., on March 16; at Neosho Falls, Kans., on May 5; and at Huger, S. C., on May 9.

Casual records.—Among casual or accidental records of occurrence are the following: Woodhouse stated he obtained two in New Mexico, probably in the Canadian River section (Cassin, 1860); Aiken reported seeing one near Colorado Springs, Colo., during the summer of 1873; a specimen was taken near Omaha, Nebr., in November 1912; one was recorded from Grinnell, Iowa, on October 4, 1886; Pindar (1925) records it as a rare summer visitant in Fulton County, Ky., but gives no details; one was reported from Benns Creek, Knox County, Ind., on September 18, 1911; one was seen on October 20, 1852, in Chester County, Pa.; several observers, including Dr. Witmer Stone, reported seeing one at Cape May, N. J., on May 30, 1924; and one was taken near Andrews, N. C., on May 26, 1893. Specimens have been reported from Wisconsin, but in one way or another they are considered doubtful (Kumlien and Hollister, 1903), and it is probable that others of the sight records listed above are open to question.

The Sennett collection is reported to contain a specimen collected at Tampico, Mexico, on May 17, 1888, while Salvin (1861) recorded a specimen received by him from Coban, Guatemala.

Egg dates.—Texas, Oklahoma, and Kansas: 95 records, March 15 to June 25; 48 records, June 3 to 12.

ROSTRHAMUS SOCIABILIS PLUMBEUS Ridgway

EVERGLADE KITE

HABITS

The Tamiami Trail runs due west from Miami for 35 miles straight through the southern half of the Everglades. For most of this distance one sees only a broad expanse of marsh, mainly covered with long grasses as far as one can see, but dotted here and there with little clumps of small trees or bushes, with an occasional island of larger trees and bushes, amphibian willow, alligator apple, waxmyrtle, swamp bay, silver-leaved magnolia, and cocoa plum. Toward the western border the grassy glades are dotted with very small isolated cypresses, 8 to 10 feet high, so scattered that the broad view is not obstructed. Farther west the cypress becomes thicker and taller, often forming dense cypress clumps. During wet seasons these glades are covered with clear, fresh water 1 or 2 feet deep; but since the drainage operations in dry seasons the glades are mainly dry and the abundant bird life disappears. The winter of 1929 and 1930 was unusually rainy, the glades were full of water, and we were favored with many interesting views of water birds. Little blue herons, in both blue and white plumages, were the most abundant birds, feeding in the shallow water or flying away in immense flocks to their evening roosts; with them were many American and snowy egrets and Louisiana herons; and frequently a stately Ward's heron stood and calmly watched us as we drove by. This was the former home of the everglade kite, and here it was that we saw a lone individual in March 1930. After many days of careful scrutiny of every hawk, crow, or other suspicious bird, we finally discovered one sitting on a tiny cypress in the western part of the glades. Its slaty-gray appearance attracted our attention, but when it spread its broad wings and circled over the marsh, showing the white base of its tail, both above and below, and we caught a glimpse of its reddish legs as it wheeled, all doubt was dispelled. Twice we saw it dart down into the grassy marsh, pick up something, and alight on a small cypress to eat it. This was probably a snail and we thought we could see it extract the meat with its long, hooked beak.

When I first visited southern Florida, in 1904, everglade kites were breeding commonly all through the southern Everglades, west of Palm Beach and back of Miami and Homestead; there was even said to be a breeding colony of them near Paradise Key, now Royal Palm State Park. But the draining of the Everglades has changed all this; most of their former haunts are so dry, except during especially wet seasons, that the great marsh snails (*Ampullaria depressa*), their principal food, have died and their pearl-like egg clusters are no longer seen on the marsh vegetation; during temporary wet spells the snails do not become established again and the kites must look elsewhere for their food supply. What few everglade kites still remain to breed in Florida may be found only where there are permanently wet marshes and where the snails still survive, such as still exist in some portions of the upper St. Johns River region. Further drainage operations may dry up these marshes, and these interesting birds will disappear permanently from the North American fauna. On March 20, 1930, we visited one of the localities in Brevard County, where the everglade kites were still breeding in some vast marshes near the river. These marshes were very difficult to explore, as the water was from knee deep to waist deep; the vegetation was so thick and high that in many places a man disappeared from sight while wading; and it was infested with plenty of deadly moccasins. The deeper and more open spaces were full of floating "lettuce" and "bonnets", with yellow flowers in bloom; and among them were a few large white pond lilies. There were many large and small clumps of sawgrass and large areas of lower growth of *Sagittaria* and *Pontederia;* in some places were patches of blue *Iris* in bloom and some extensive tracts of cattails (*Typha*). There were large islands, small clumps, and isolated bushes of myrtle and willow, a few scattered small cypresses, and occasional tangles of morning-glory vines. Here also we saw the usual Florida marsh birds, both gallinules, limpkins, herons, white ibises, bitterns, grackles, and blackbirds. Such was the setting in which we located five or six pairs of kites, but we found only one empty nest.

Courtship.—I did not see anything at the above locality that I thought was a courtship performance, but one of my companions on that trip, John H. Baker, told me that he saw a group of three kites soaring at a height of about 500 feet above the marsh; they "were seen repeatedly folding their wings for sudden dips of short duration, much as do kingfishers and terns when plunging." After some 5 minutes spent in these evolutions they set their wings and sailed away out of sight. Such behavior in March looked like part of a courtship display, but it may have been caused by the presence of two men in the marsh hunting for nests. Dr. Charles W. Townsend (1927) witnessed a somewhat similar performance, which he describes as follows:

In the marshes of the upper waters of the St. John River, Florida, on March 4, 1926, I watched three of these birds flying together. Presently one departed and the other two circled about, darting at each other from time to time. Occasionally one would turn on its side and stretch out its legs as if to grapple. After playing in this way for a short time, one of the Kites circled upwards and, reaching a considerable elevation, dove swiftly downwards with wings curved back, and then turned completely over, end to end. This maneuver was repeated several times, the bird crying out at the same moment in a bleating fashion very much like a sheep.

Nesting.—While I was collecting near Miami, Fla., in 1903, a guide brought me a set of three everglade kite eggs, together with the parent bird, taken near there on April 28. He described the nest as located 7 feet up in a solitary "custard apple tree" about 9 feet high, in a sawgrass slough; it was made of sticks and leaf-bearing twigs, with both dry and green leaves, and was lined with fine twigs and bay leaves.

C. J. Maynard (1896) seems to have been the first to discover the nest of this species. His first nest "was small, flat in form, composed of sticks somewhat carelessly arranged, and was placed on the top of the grass [sawgrass] which supported it and which grew so luxuriantly at this point that it bore" him up as he "was endeavoring to reach the nest." About three weeks later, on March 24, he found another nest in a magnolia bush; "it was placed about four feet from the water, was quite flat, about a foot in diameter, was composed of sticks quite carelessly arranged, lined with a few dry heads of sawgrass and contained one egg."

Bendire (1892) quotes J. F. Menge as writing to him: "According to my observations the female does not assist in the building of the nest. I have watched these birds for hours. She sits in the immediate vicinity of the nest and watches while the male builds it. The male will bring a few twigs and alternate this work at the same time by supplying his mate with snails, until the structure is completed."

Bendire continues:

A nest of this species now before me, taken by Mr. Menge, and kindly forwarded, measures 16 by 13 inches in diameter, and is about 8 inches thick. It is not an artistic looking structure, but rather carelessly put together. The base consists of dry willow twigs, some of them half an inch in diameter; the greater portion are, however, smaller. The inner cavity is about 7 inches wide by 1½ inches deep. This is lined with small stems of a vine and a few willow leaves. The latter look as if the twigs, to which some of them are still attached, might have been broken off by the birds while green; the first mentioned material predominates in the lining. * * *

Donald J. Nicholson (1926), who has had considerable experience with this kite, has found as many as 10 nests in one day; these were in two separate colonies about 150 yards apart. Three of the nests were in sawgrass clumps, but all the others were built in dead or partly dead myrtles, 3 to 7½ feet above water. He says of one nest: "The nest was a fairly compact structure, about one foot deep, and fifteen inches across, with a hollow for the eggs, three and one-half inches deep. Upon nearing the nest the female flew towards me with a cackling note similar to that of an Osprey, but finer in tone, and not so loud. Soon the male appeared, scolding with notes exactly like those of the female. At times they both circled around together, again only one flew around while the other sat perched on a myrtle nearby."

A set of four eggs in my collection is said to have been taken from a "nest of sticks and grasses on the ground in a dense marshy growth." In Argentina the South American subspecies often nests in colonies. Major Bendire (1892) quotes Mr. Gibson, as saying: "In the year 1873, I was so fortunate as to find a breeding colony in one of our largest and deepest swamps. There were probably twenty or thirty nests placed a few yards apart in the deepest and most lonely part of the whole 'cañadon.' They were slightly built platforms, supported on the rushes and 2 or 3 feet above the water, with the cup-shaped hollow lined with pieces of grass and water rush."

Eggs.—The everglade kite usually lays three or four eggs, but sometimes only two. These are mostly oval in shape, with an elliptical tendency in some; the shell is smooth but without gloss. The ground color is dull white or rarely creamy white, but is usually mostly concealed by profuse markings. Some eggs are heavily and boldly blotched, some irregularly spotted or blotched, some finely sprinkled with minute dots, and some washed with light browns, "hazel" or "ochraceous-tawny", so completely as to conceal the ground color. The markings are usually in shades of "chestnut", "auburn", or "chocolate", but sometimes lighter browns, "hazel" or "tawny"; rarely the browns are combined with "fawn color" or "cinnamon-drab" in a pretty pattern. An occasional egg is largely white with only a few scrawls or small spots of dull light browns. The measurements of 65 eggs average 44.2 by 36.2 millimeters; the eggs showing the four extremes measure 59.4 by 37.9, 47.4 by 38.3, 40.1 by 34.8, and 43.9 by 33 millimeters.

Young.—The period of incubation seems to be unknown. Both sexes incubate and assist in the care of the young. Mr. Nicholson (1926) says: "When a nest was found with young, the little fellows would remain perfectly quiet and still; sometimes one would squat in the nest as if to hide. The note of the young birds is hard to describe but is much different from that of the adult. * * * On April 27, young kites six days old were in the downy stage, and upon our return May 12 they were practically fully feathered and would have likely been able to fly by May 20. They showed a remarkable growth in sixteen days' time."

Mr. Menge wrote to Major Bendire (1892): "They feed and care for their young longer than any other birds I know of, until you can scarcely distinguish them from adults."

Plumages.—A small downy young everglade kite, recently hatched, is sparsely covered with short down, "cartridge buff" in color, tinged with "cinnamon" on the crown, with "cinnamon" and "snuff brown" on the rump, and with "warm sepia" on the wings. On an older downy young this first buff down is being replaced by short thick down of a much darker color, "dark grayish brown." The bill in both cases is long and decidedly hooked.

In fresh juvenal plumage the young bird is quite richly colored. The crown and occiput vary from "sayal brown" to "ochraceous-tawny", heavily streaked, especially on the occiput, with "mummy brown"; the under parts are "ochraceous-tawny" to "tawny", heavily marked with "mummy brown", in the form of narrow streaks on the throat and involving large central portions of the breast feathers; the flank feathers are "mummy brown", notched with "tawny"; the tibiae are unmarked "tawny"; the primaries are nearly black, tipped with "cinnamon"; the tail above is "mummy brown" to nearly black, broadly tipped with "cinnamon-buff"; the upper tail coverts are "warm buff" to "light buff"; the rest of the upper parts are "mummy brown", broadly tipped with "cinnamon" or "cinnamon-buff" on the back and wings, except that the lesser coverts are very broadly edged with "tawny" or "russet." The sexes are about alike in this plumage, but they can be distinguished by the tails, which show the same differences as in adults.

The juvenal plumage is worn through the first winter, subject to much wear and fading, the lighter edgings disappearing by wear and the bright colors fading to pale buff or nearly white; I have seen this faded plumage in March, April, and May birds. But usually an extensive molt takes place in spring, at which the sexes begin to differentiate. This molt involves much of the body plumage, the wing coverts, and the tail. Young males acquire much slate-colored plumage on the upper parts and some on the breast; but, in both sexes, much of the new plumage of the under parts is broadly edged or notched with "tawny" or "cinnamon." Whether this plumage is worn throughout the second year the material does not show. An adult male that we collected, and another that we saw closely, in March, were molting the primaries, so it may be that the complete annual molt begins in spring and that the young birds referred to above were undergoing a molt into a second-year plumage. Summer and fall material is needed to settle the question.

Some of the manuals imply that the sexes are alike, in adult plumage, or fail to make the difference clear. They are easily recognized in life. The female is somewhat larger than the male; her general color is "mummy brown" or "bister", instead of bluish slate-color, mixed with whitish or pale buff below, with much whitish streaking on forehead and throat and with duller colors on the soft parts; the under side of the tail is different, the dark portion being browner and more restricted and the subterminal light portion more extensive.

Dr. Alexander Wetmore (1926) has described the colors of the soft parts as follows: "The immature female taken October 28, when fresh, had the bill, anterior to the cere, black; base of bill, including the mandibular rami, the skin back as far as the eye and a narrow external rim on the eyelids zinc orange; iris liver brown; tarsus and toes dull yellow ocher; claws black. The male in adult plumage secured on October 31 had the bill mainly black; cere, bare skin in front of eye, gape, and mandibular rami flame scarlet; iris carmine; tarsus and toes apricot orange; claws black. The adult thus was much brighter in color."

Food.—The everglade kite has been well named "snail hawk", for it feeds exclusively on the meat of a large fresh-water snail (*Ampullaria depressa*), which formerly abounded all over the Everglades and is still abundant in some other fresh-water marshes and sluggish streams in Florida and in many places in South America. It is useless to look for this kite where these snails have been killed off by drainage or drought, as in southern Florida. Their presence can be detected by their pearly egg clusters on the sawgrass or reeds. The kites search for the snails in the open places in the marshes or in shallow ponds, beating slowly back and forth, low over the ground, after the manner of marsh hawks, or hovering over the water like a gull. When the snail is located the kite plunges down to secure it and flies with it in its claws to some favorite perch on a stump, post, low tree, or bush; often an old deserted nest is used as a feeding station. Here the snail is neatly extracted with the aid of the kite's long hooked beak, admirably suited for the purpose, and the shell is dropped unbroken. That the birds use the same perch regularly is shown by the large number of empty shells often found in such places, sometimes as many as 200 or 300. There is no evidence to indicate that this kite ever eats anything but these mollusks.

Dr. John B. May (1935) quotes Herbert Lang (1924) as follows, regarding its methods of feeding, as observed at Georgetown, British Guiana:

The snails remain in the water during the hotter part of the day, but in the early morning and late afternoon are found at the surface or creeping about on the marsh vegetation. The kite quarters back and forth low over the water, suggesting a sea gull at a distance. Often it hovers over one spot for a considerable interval, then dives down to pick up a snail which it carries in its talons to some favorite perching place in a bush or low tree. Here it stands for several seconds motionless, on one leg, holding the snail in the long claws of the other foot. Soon the snail, which had withdrawn into its shell when picked up, closing tightly its operculum, begins slowly to extrude its slimy body. Suddenly, like a flash the Kite grasps the body of the snail, between the operculum and the shell, in its blunt-edged but deeply hooked beak. The muscular contraction of the snail's body apparently detaches it from its attachment within the shell, and a moment later, with a shake of the Kite's head, the shell is tossed aside and the body swallowed, including the operculum.

Behavior.—Although it has a broad expanse of wing, this kite flies with a slow, desultory flight; it seemed to me rather floppy and heronlike, as if lacking the muscular power to move its great wings vigorously. Its flight has been compared to that of the marsh hawk, as it flies low over the marshes while hunting. But it often soars to great heights, gliding along easily and gracefully; its slender body is easily supported on its broad wings and tail. Bendire (1892) quotes Sclater and Hudson: "When soaring, which is their favorite pastime, the flight is singularly slow, the bird frequently remaining motionless for long intervals in one place, but the expanded tail is all the time twisted about in the most singular manner, moved from side to side, and turned up, until its edge is nearly at a right angle with the plane of the body."

It is a gentle, harmless species and lives so peacefully with its neighbors that even the small song birds do not seem to fear it at all. It is not particularly shy and sometimes even shows some curiosity; one that we were watching from the Tamiami Trail twice flew out over the road near us, as if to look us over. About its nest it is mildly solicitous, but not bold enough to attack the intruder.

Voice.—Mr. Nicholson (1926) heard, upon nearing the nest, "a cackling note similar to that of an osprey, but finer in tone and not so loud." Dr. Wetmore (1926) found them rather noisy; they "emitted a rasping chattering call that was audible at no great distance." Bendire (1892) gives it as "a peculiar cry, resembling the shrill neighing of a horse."

Field marks.—The everglade kite can be easily recognized at a great distance by its dark color, its broad rounded wings and square tail, and by its slow flight. When nearer, the white upper and under tail coverts are quite distinctive and the brilliant orange-colored feet and cere are very conspicuous, especially in the male.

DISTRIBUTION

Range.—Florida, Cuba, eastern Central America, and South America.

The everglade kite breeds **north** to northwestern Florida (probably rarely Waukeenah and near Crescent City). **East** to Florida (near Crescent City, Lake Norris, St. Johns Marsh, probably Micco, Fellsmere Marsh, Loxahatchee Marsh, Lake Hicpochee, Miami, and probably Cuthbert Lake); Cuba (Isle of Pines); British Guiana (Demerara River, Mahaica River, Abary River); southeastern Brazil (Iguape); and Argentina (Buenos Aires, Barracas al Sud, and Cape San Antonio). **South** to Argentina (Cape San Antonio, Espartilla, Conchitas, and Tucuman). **West** to northwestern Argentina (Tucuman and Jujuy); Ecuador (Babahoyo); Colombia (Remedios, Bonda, and Barranquilla); Nicaragua (Los Sabalos and Ometepe Island in Lake Nicaragua); Guatemala (probably Lake Peten); Veracruz (Catemaco, Cosamaloapan, and Mirador); and northwestern Florida (Wakulla Springs, and probably rarely Waukeenah).

The range as outlined is for the entire species. The North American form, *R. s. plumbeus*, is probably confined to the peninsula of Florida, Cuba, eastern Mexico, and Central America.

Migration.—It appears likely that the everglade kite withdraws slightly from the northern and southern limits of its range during the winter seasons, but the extent of the movement is not known. The species has been observed to arrive in the vicinity of Waukeenah, Fla., on May 9, while in the Province of Buenos Aires, Argentina, Dr. Wetmore found them fairly common on October 28 and judged that they had only recently returned from the north.

Egg dates.—Florida: 68 records, February 15 to July 20; 34 records, March 13 to April 28.

CIRCUS HUDSONIUS (Linnaeus)

MARSH HAWK

HABITS

The above name recalls to mind those delightful days, now long past, when we sat for hours in a flimsy blind on the Cape Cod marshes, listening for the startling whistle of the yellowlegs or the mellow notes of the plover. The day is one of those lovely Indian summer days; only a gentle breeze is stirring, and the autumn haze softens the brilliant colors with which the waning summer has painted the marsh vegetation and the distant woods. As we sit there in the soft sunshine, dreamily drinking in the beauties of the scene, our eyes are alert to what is going on around us. Off on the distant mud flats are flocks of gray and white gulls, with scattered groups of shorebirds; over the extensive salt marshes black terns are winnowing the air, or plunging down into the grass for grasshoppers, and numerous swallows, now nearly ready to migrate, are skimming low over the meadows or the little pools; on a nearby sand flat some turnstones are digging holes in the sand; occasionally a great blue heron or a bittern flaps lazily over the marsh. There is always something moving; and, whether the yellowlegs and plover come to our decoys or not, we are sure to see, sooner or later, a dark speck in the distance that soon develops into a large, long-tailed, long-winged bird. On it comes with an easy gliding flight, its long wings slanting upward; as it turns we see its brownish breast and then its white rump, a young marsh hawk. A lazy, loafing, desultory flight it seems, but really it is full of purpose, as it quarters low over the ground in a systematic search for its prey. Often during the day it circles near us, but not too near, for all hawks have learned to avoid gunners. A peaceful day on the marshes would hardly be complete without an occasional glimpse of this industrious harrier, to add its touch of life to the picture.

But the marsh hawk's haunts are not limited to marshes. It is very common on the prairies and plains of the Middle West, though it shows a preference for the vicinity of sloughs and wet meadows. M. P. Skinner tells me that in Yellowstone National Park he sees "more of these hawks hunting over the rolling upland prairies than anywhere else." Here they "choose both the grassy meadows and the sage- and brush-covered hills to hunt over." He has even seen them "hunting across the open lands high up on the mountains", between 5,300 and 10,300 feet. Anywhere in open country, where its prey may be found, the marsh hawk is likely to be seen.

Spring.—The marsh hawk is a migratory species. Most individuals spend the winter in the Southern States or in the milder sections of the country. But, even as far north as Manitoba, C. L. Broley tells me he has seen the species every month but January. There the light-colored males are the first to arrive, around the middle of March, and the brown females come about three weeks later. The season is about the same in southern New England, where some birds remain all winter near the coast.

Courtship.—Many accounts of the spectacular nuptial flight have appeared in print, but I prefer to use the following description of it, one of the best, in some notes sent to me by Mr. Broley:

This is a vigorous and pleasing series of nose dives, mostly done by the male, although the female frequently takes part in them. This takes place sometimes at an altitude of 500 feet, but the usual flight averages 60 feet up, swooping down to 10 feet from the ground. It might be illustrated by placing a number of capital U's together as UUUUUUUUUUUUUUUUUU, as the turn at the bottom is well rounded out, but at the apex the bird almost stalls, tipping downward again to continue the movement. Some observers claim it makes a somersault as it turns, but only on one occasion have I seen any indication of this. The wings are kept fully extended during the whole period, and they appear to be working easily all the time. I have seen a male make 71 of these dips in succession fly on for a short distance and commence anew. The average number of dips would be perhaps 25. The flight is frequently made while the female is flying along near the ground hunting for mice, below the male, or again he may swoop continually in one location while she is standing on the ground. The movement is extremely graceful and is a welcome sight each spring.

Other observers have described a similar performance, which seems to be characteristic of the species, but most of them have noted a complete somersault, or a sidewise turn, at the top of the rise. E. H. Forbush (1927) says: "As it bounds up and down in the air, it seems to move more like a rubber ball than a bird. * * * When two of these birds are mated or mating they keep together much of the time, either on the ground or in the air. When the female alights the male follows her and walks or flies around her. On the ground he bows to her and swells with amorous ordor. Sometimes the male flies alone across the marsh rising and falling alternately and with each fall turning a complete somersault, as if to show his larger mate what a clever and wonderful bird he really is. Again he 'carries on' in the same way while flying in her company."

Nesting.—In southeastern Massachusetts, at least in the region I hunt over, the march hawk is a rare breeder. My first nest was found in a sphagnum bog, overgrown with low huckleberries, pitcherplants, and scattered small larches, and surrounded with thickets of alder and swamp honeysuckle, a secluded spot. The nest was a flimsy structure of light, dry sticks and straws, loosely placed on the flattened tops of the low huckleberry bushes, only a few inches above the water and the thick growth of sphagnum moss and pitcherplants. It held five eggs on April 30, the last two having been laid during the past three days, indicating that the eggs may be laid on successive days (pl. 26).

Another and better nest was found in a somewhat different swamp; it was densely overgrown with alders, swamp azaleas, huckleberries, and other bushes, in some places higher than my head and difficult to penetrate, but in the center was a more open space, where the bushes were lower and more scattered, with a few brakes and flags growing up among them. Here the nest was placed on slightly elevated ground among some small bushes and brakes. It was a handsome and well-made nest of dry straws, weed stems, and sticks and lined with finer straws, brake stems, and thistle tops; it measured about 23 by 20 inches in outside and 9 by 8 inches in inside diameter; the material in the center of the nest was about 2 inches deep. It held five spotted eggs on May 26.

But our local birds do not always nest in swamps. We have found them nesting on high and dry ground in what we call sprout lands, where woods have been cut off and where sprouts are growing on the stumps, but usually near a swamp or meadow. In such a place a similar nest to those described above is built on the dry ground and the larger stumps are used as perches or feeding stations. Other observers have described similar nesting sites at various eastern points.

Charles A. Urner (1925) made a careful study of three nests on the salt marshes of New Jersey, of which he says:

One nest found was in the center of a large clump of High-tide Bush (*Iva oraria*), and two were even more securely hidden in large beds of thick reeds (*Phragmites communis*). One was on dry, sandy ground, the other two on the wet marsh, occasionally flooded by tide.

Here I found an interesting difference indicating that the Harrier varies the height of its nest with the danger of floods in its chosen location. A nest found on dry ground, above all tide levels, constructed of weed stalks and grasses, nicely lined, was only an inch or two thick. A nest located on the marsh over a mile inland from the shore of Newark Bay, but more or less exposed to floods and unusual tides, was similarly constructed, but was about 5 or 6 inches thick. A third nest, found nearer the Bay shore and in a location frequently flooded, was remarkable for its greater size and bulk. It was built of weed stalks and finer material to a height of fifteen to eighteen inches, and it measured over three feet long and two feet wide. It was of uniform construction from the ground up with no indication of a "foreign" foundation.

In the more western States the marsh hawk sometimes nests in bushy swamps or in brush-covered slopes, or even hillsides, but more commonly it selects more open grassy situations, the margins of sloughs, wet grassy hollows, or even extremely wet situations among reeds, flags, or tules. In Nelson County, N. Dak., we found five nests in one day, June 3, 1901. One was well made of sticks and straws and lined with soft grasses; it was built up 14 inches above the water in a patch of dead flags on the edge of a slough; it contained two young hawks, three normal eggs, one runt egg, and a dead spermophile. Another still finer nest, made of sticks, reeds, and coarse weeds, was built up 18 inches above the water in a wet meadow and measured 30 inches across the top. Other nests were similarly located (pl. 26).

Dr. John W. Sugden writes to me that in Salt Lake County, Utah, on July 18, 1928, he found a nest, containing five half-incubated eggs, "near the center of a 30-acre wheatfield on a dry farm, at least 4 miles from the nearest water. The nest was a shallow depression in the ground lined with a few sticks and straws." Bendire (1892) mentions a nest found by George G. Cantwell on a haycock.

Both birds assist in building the nest, the male bringing some of the material and dropping it for his mate to arrange, but most of the gathering and arranging of material is done by the female. E. L. Sumner, Jr., watched a female building her nest and has sent me his notes on it. He saw her make seven trips to the nest within 10 minutes. He says:

In carrying the sticks, if they are small, she nearly always uses her bill alone; if they are large she uses her feet; in one case of a particularly large branching stalk she carried it in beak plus both claws; in another case she transferred a piece from her beak to her claws while sailing toward the nest. Once she carried a particularly large weed in her feet, but all the other times she used her bill instead. Once she picked up a piece, started to fly with it, but stopped and picked up another piece in addition, but in flying away with them, dropped first one and then the other so that she had to continue on across the rush patch to the other side and pick up another load. Once I saw her tug violently at a weed that was still rooted, but it did not give way, and so she walked a few steps farther on and picked up a loose piece instead.

W. H. Laine (1928) reports finding a marsh hawk incubating on a nest of 12 prairie-chicken eggs; the experiment was not a success, as only one chick hatched and it promptly ran away. Perhaps the hawk's nest had been destroyed and she adopted the nearest available nest.

Eggs.—Perhaps the commonest number of eggs is five, but four or six are frequently found and occasionally as many as seven or eight, or even nine, are seen in a marsh hawk's nest. In shape they are ovate, short-ovate, or nearly oval. The shell is smooth, with little or no gloss. The color is dull white or very pale bluish white. They are generally unmarked, but about 10 percent of the sets show, more or less, scattered spots of very pale browns, "cinnamon-buff" or paler, dull buff. The measurements of 84 eggs average 46.6 by 36.4 millimeters; the eggs showing the four extremes measure 53 by 38, 48 by 39.5, 41.4 by 35.6, and 43 by 34 millimeters.

Young.—The period of incubation has been variously estimated as 21 to 31 days; the latter figure was definitely noted by Aretas A. Saunders (1913). It is difficult to determine, as it often begins when the first egg is laid; an egg is usually laid each day, but often a day or two may intervene between layings. Both sexes share the duties of incubation and care of the young, and they are very devoted parents. Mrs. Irene G. Wheelock (1904) writes:

In eighteen to twenty days the young Hawks break their hard shells, *one each day*, and cuddle down among the feathers and straw of the crude nest. From the day the first little ball of down appears, one or the other of the adults may be seen constantly on the wing over that meadow. The same tactics are pursued as before, for the food is dropped to the parent on the nest, who, after the first few days, holds it fast in her beak while the nestlings tear off bits from it for themselves. In this way the muscles of the bill and neck are developed. Later on the food is simply dropped to them, both parents being off on the hunt, and the little fellows grasp it in their sharp claws and tear from it with a right good-will.

Aretas A. Saunders (1913) noted that three eggs in a set of five hatched between the evening of June 30 and the morning of July 1. The fourth bird hatched before the morning of July 2, the fifth on the afternoon of July 4, and the sixth on July 7. He noted that they were born with their eyes closed, but that they opened within a few hours. Following is his account of their development:

For the first six or seven days the young showed no change in appearance except that they grew larger and became somewhat more active. On July 8, just after the youngest bird had hatched, I noted that the oldest birds were about three times the size of the youngest. About July 10 the two youngest birds disappeared, probably having died. I believed that this was because they were so much smaller and weaker than the four older birds that they were unable to get their proper share of food.

Sheathed feathers began to appear in the oldest birds at the tips of the wings on July 8, when they were seven days old. On July 14, when twelve and thirteen days old, the birds began showing fear and crawled back into the cinquefoil bushes when I approached. When I attempted to handle them, they sat up and threatened me with their beaks, and called in a high, squeaky, baby voice. On July 17 the feathers at the tips of the wings began to break the sheaths, and sheathed feathers were appearing thickly on back, shoulders, breast and tail. At this time the feet and cere were beginning to turn from a light pinkish color to yellow. On July 22 the feathers were breaking the sheaths in many places, those at the tips of the wings being broken for about two inches of their length. The feet and cere were now bright yellow. The birds stood with outstretched wings and open beak, turning to face me no matter to which side of the nest I went. They were in about the same condition on July 24, so that I found it almost impossible to handle them. When I attempted a photograph of them they crawled off into the bushes, so that I could only get two at a time in the picture.

During the week following this the birds changed rapidly. Feathers unsheathed all over them, and much of the white down came off. On August 4, when the birds were thirty-three and thirty-four days old, I approached the nest and found three of them able to fly a little. One rose at my approach and flapped away for about 150 feet before it sank in the grass.

Mr. Urner (1925) found that the time from hatching to flight was about 30 to 35 days. He refers to them as "sturdy, fearless, wide awake, active, noisy and hungry youngsters. * * * The readiness with which the young imitate their parents is worthy of note. On July 7, I visited a brood which had left the nest and learned to fly, though still in the vicinity of the nesting site. They flew in all directions as I approached, uttering an immature peeping call. The adult male turned immediately to attack and I was surprised to see two of the young, probably males, follow suit, flying in very close and making a more or less unsuccessful effort to imitate the long rolling call."

As to the food of the young he says:

As far as I can judge from remains picked up in the general vicinity of the nests, mice and small birds, supplemented with insects, constitute the principal fare during early life. But as the birds grow, rats assume a more important role, and in or near two different nests I found remains, picked clean, of practically full-grown American Bitterns (*Botaurus lentiginosus*). Now the young American Bittern is no mean antagonist, and the fact that such large birds are actually killed and carried to the nest indicates the calibre of the Harrier as a hunter. * * * During the fourth week of the young Harrier's life pellets of fur and feathers, containing some bone, begin to appear about the nest. These pellets are often as large, as compact and as well formed as those of the Short-eared Owls, constituting an interesting similarity between the two species. It is probable that the failure to find pellets about the nests earlier in the young brood's growth is due to the thorough removal of waste by the adults, rather than any change in feeding habits.

The main reason why pellets are not found about the nest during the early life of the young is that the old bird feeds the young, at that age, with small pieces of pure flesh. Dr. Frank N. Wilson (1927) saw, at close range, a marsh hawk feed a field mouse to her small young. "Holding it in her beak, she walked to the edge of the nest and, placing both feet upon it, tore off small pieces of the raw flesh and fed the young in turn. The coarser parts she ate herself."

After the young are able to fly they are often fed by their parents while on the wing. Dr. Charles W. Townsend (1905) writes: "Three weeks later near the same place, the female flew over my head, and whistled as she approached the nesting site. Upon this, four full grown young Hawks flew up to meet her and she dropped from her talons a mouse, which after falling about five feet was skillfully caught in the air by one of the youngsters. How it was done, whether in the bill or in the talons, I could not make out in the confusion. It certainly did not get by the birds, who at once retired to the ground, the successful one to eat its prize."

For a long time after the young are able to fly, the family group hangs together, hunting over the familiar grounds near their former home, the young learning from their parents and practicing the serious business of earning a living.

When the time comes for migrating, young birds are apt to wander widely in different directions. Young birds banded as nestlings by William I. Lyon, at Waukegan, Ill., were recovered that season, one at 50 and one at 300 miles northwest, and another at 500 miles southwest.

Plumages.—When first hatched the chick is covered with short down, very scanty on the under parts; it is pure white with only a slight tinge of buffy on the upper parts. As the chick grows, the down increases in length and becomes darker, "pinkish buff", on the upper parts; the lores and a space around the eyes are naked. The development of the juvenal plumage is described by Mr. Saunders (1913) above. In fresh juvenal plumage, in August, the upper parts are "mummy brown", many feathers narrowly tipped, or broadly margined, or deeply notched, with "tawny" or "cinnamon"; the white upper tail coverts are tinged with "cinnamon"; the tail has four dark "mummy brown" bands, the four intervening bands being dark gray on the central pair of feathers and much mixed with "tawny", "cinnamon", gray, and white on the other feathers; the primaries are brownish black above, glaucous on the outer webs; the entire under parts are rich yellowish brown, "amber brown" to "ochraceous-tawny", broadly streaked on the chest and narrowly on the flanks with "bister", but otherwise immaculate. The sexes are alike in plumage, but there is a marked difference in size.

The juvenal plumage is worn for about a year but becomes much faded by spring; young males fade out to almost white below. Molting sometimes begins in April but usually not until summer, when a complete molt takes place from July to October or later. This produces a second winter plumage in which the sexes are different.

Young males are quite dark above, "bister" to "mummy brown"; the under parts are largely white, with considerable drab and buffy mottling, especially on the chest, which is heavily clouded with drab; the wings and tail are much like those of the adult. Young females show similar progress toward maturity, but they still show many rufous edgings above; they can be distinguished from first-year females by their spotted breasts. At the next complete molt, the following summer, the young become practically adult in plumage, though probably males continue to grow whiter as they grow older. Adults have their complete annual molt during July, August, and September.

Food.—The marsh hawk is regarded by many as a highly bene- ficial species, mainly because of the large numbers of mice, rats, and other injurious small mammals that it destroys. It certainly is a great mouser; it lives largely on frogs and small snakes and de- vours many injurious insects, but the records show that many small birds and some larger ones are killed by it. Dr. A. K. Fisher (1893) gives the following summary of its food:

Of 124 stomachs examined, 7 contained poultry or game birds; 34, other birds; 57, mice; 22, other mammals; 7, reptiles; 2, frogs; 14, insects; 1 indeter- minate matter, and 8 were empty.

Although this hawk occasionally carries off poultry and game birds, its eco- nomic value as a destroyer of mammal pests is so great that its slight irregu- larities should be pardoned. Unfortunately, however, the farmer and sports- man shoot it down at sight, regardless or ignorant of the fact that it preserves an immense quantity of grain, thousands of fruit trees, and innumerable nests of game birds by destroying the vermin which eat the grain, girdle the trees, and devour the eggs and young of the birds.

Maj. Allan Brooks (1928) condemns the marsh hawk, as "the most destructive hawk in all America to our marsh loving waterfowl for at least three months in the year." He accuses it of killing large numbers of young ducks and says that it does not kill its victim out- right "but slowly wears the wretched captive out and literally eats it alive commencing at the breast muscles." He cites another case where a family of marsh hawks killed over two dozen old and young blue and ruffed grouse during one nesting season. These cases are probably exceptional, or extremely local in effect, for most of the evidence is in favor of the marsh hawk. Herbert L. Stoddard (1931) found remains of cotton rats, which destroy the eggs of quail, in 925 out of 1,100 pellets of this hawk. Several observers have men- tioned the great service that marsh hawks perform in the southern ricefields by driving away bobolinks and blackbirds more effectively than hired men with guns, thus saving considerable expense.

Meadow mice seem to constitute the bulk of the food, according to nearly all observers. Judge John N. Clark wrote to Major Bendire (1892): "One I examined contained not less than eleven, another nine, and nothing else." Among other mammals taken are young rabbits, young skunks, pocket gophers, rats, spermophiles, squirrels, shrews, and moles. The long list of birds includes bittern, green heron, teal and other ducks, coot, rails, grouse, quail, partridges, pheasants, plovers, sandpipers, woodcock, snipe, sparrow hawk, screech owl, flicker, doves, starling, meadowlark, blackbirds, grackles, numerous sparrows, cardinal, towhees, warblers, wrens, mockingbird, catbird, thrashers, robin, bluebird, and thrushes. Frogs form a large item; and small snakes and lizards are eaten. It also feeds on large numbers of grasshoppers, locusts, crickets, and other insects. Ivan R. Tomkins tells me that in the salt-water marshes of South Carolina and Georgia "its winter food is mostly marsh rabbits (*Sylvilagus palustris*)." At times it is quite destructive to poultry and game. E. S. Cameron (1907) writes: "This bird is the common 'Henhawk' of eastern Montana and is the most pertinacious of any in attacks on the poultry yard. Young marsh hawks weighing about ten ounces will endeavor to disable a chicken weighing a pound, by pecking it on the head and striking on the back at the same time with the feet, their strong wings enabling them to keep directly above it no matter where the prey may run. Birds of the year, through inexperience, are the most daring, and my wife has taken a screaming pullet from the claws of one of them which found the prize too heavy to lift."

Henry K. Coale (1925) reported that a marsh hawk killed 7 of a flock of 14 Hungarian partridges within two weeks, before it was caught in a trap. "It would tear the back open and rip the flesh and skin off in strips."

The well-known habit of quartering the ground over fields or marshes, barely high enough to clear the tallest vegetation, is the common method employed to hunt its principal prey, small mammals and small birds. Its keen eyes are quick to detect its quarry, and its flight is under such perfect control that it can stop suddenly and drop quickly down upon the victim. Usually it is devoured right there on the ground, but often it is carried to some convenient stump or post, or carried away to feed its mate or young. A mouse or small bird may be almost wholly eaten, but a larger animal or bird will be skinned or plucked and the flesh torn off. When the victim is too large to be eaten at one meal, the hawk may return later to finish the feast. Dead game or even carrion is often welcome. A. G. Lawrence writes to me: "E. Robinson informs me that he has seen marsh hawks hovering in front of a prairie fire, picking up the mice as they fled before the flames. I have seen a marsh hawk hover for more than 5 minutes over a bush in which a small bird had taken refuge, darting rapidly from side to side when the bird ventured to fly out, but mainly hovering over the bush about 10 feet up. Eventually it swept down beyond the bush and secured its victim as it tried to escape."

Several observers have noted the interesting way in which the male feeds his mate. C. L. Broley has sent me the following note on it: "The male flies with the mouse near where the female may be nesting and calls to her; upon which she takes to the air; and, flying 12 to 20 feet over his mate, the male drops the mouse. The female either turns partly on her back and catches the mouse with her claws or, as on one occasion, just swings her feet out to the side and catches the mouse neatly. I have seen the male carry a mouse 15 minutes awaiting the return of his mate to present it to her. Another time the male became tired of waiting for her and ate half the mouse but kept the other half till she returned."

Eugene S. Rolfe (1897) noted the following interesting attempt to secure a meal: "Many times I have watched the Marsh Hawk sailing low and keenly scanning the ground on the open prairie, and suddenly pouncing down and quickly ascending again with an empty mouse nest in its talons, and on one occasion I followed behind for fully 2 miles and in that distance it picked up and dropped seven of these empty nests. On examination they proved to be simply wads of fine dried grasses, and it was easy to see that if these had all chanced to be occupied by families of young mice, the foray of that particular Hawk would have been most fruitful in the destruction of these small pests."

E. L. Sumner, Jr. (1931) witnessed a playful reaction of a marsh hawk with a horned lark that it had captured:

All at once the hawk dropped the lark, whereupon the latter, still alive, flew weakly to the ground about seven feet away, its captor with outstretched talons hovering meanwhile about two and one-half feet above it but not pouncing upon it. When the lark reached the ground, the hawk lit beside it, then gave a little jump into the air and landed with spread talons upon its prey. It seemed not to bite the lark, but after examining it with many twistings and turnings of the head rose about three feet into the air with it, and then dropped it again, the lark still fluttering, and pounced upon it just as before. This the marsh hawk did seven or eight times, and I marveled at the clumsiness of the bird until I realized what was going on—it was playing.

At length the lark fluttered into a tangle of shrubby weeds, which circumstance seemed to furnish even more interest for the hawk. It would prance about in the weeds, taking great high steps, and now and again bend down to peer intently in at the lark. I do not think the hawk at any time really lost its prey. This continued for about ten minutes from the time when I had started to watch, after which the bird settled in a little depression with its victim and was then out of sight.

Behavior.—Much of this subject has already been covered under other headings. The characteristic low flight, as it quarters over the wide open spaces in search of food, is light, buoyant, graceful, and easy, as well as long protracted and apparently tireless. William Brewster (1925) has described it perfectly, as follows: "Flying ever in the buoyant, unhurried manner so characteristic of their race, now renewing waning impetus by a few deliberate wing-strokes, next gliding for several rods on wings set with the tips held well upwards, much as those of a gliding Turkey Vulture are held, tilting their bodies more or less perceptibly from side to side and rarely pursuing a perfectly straight course for more than a few yards at a time, they may skirt the shore for miles, following all its windings closely, and keeping just outside the outer ranks of living trees, but taking no especial pains to thus avoid outstanding dead ones."

While migrating it flies at a higher elevation with steadier wing beats. Its nuptial flight is spectacular and shows its ability as an aviator and a stunt flier, for which the long wings and tail, combined with a light body, are well adapted. Its lofty evolutions are not so well known, but it compares favorably with other hawks in its soaring ability. Mrs. Bailey (1915) says: "When flying high enough to be exposed to the strong prairie wind, her maneuvers, and those of the male when he joined her, were fascinating and beautiful to watch. After flapping low over the ground, they would set their wings and, perfected monoplanes, rise with the wind, tilting and turning, changing their angles with enviable skill to meet the vagaries of the air-currents. They would sail with set wings, buffeted by the wind, and then, as if their sailing muscles were tired, turn tail in midair and sweep back with a beautiful downward curve."

Marsh hawks occasionally perch on trees or bushes, but only rarely; they normally stand on the ground or perch on stumps, fence posts, or telegraph poles. They even roost on the ground at night. They have favorite perching, feeding, and roosting stations, which are well marked with pellets, excrement, and feathers. Mr. Stoddard (1931) says: "This species has the un-hawklike habit of roosting on the ground, frequenting the same spot night after night. If numerous, the hawks form a loose roosting group numbering from two or three, to as many as thirty. A large field grown up to heavy broomsedge and preferably upon a hilltop is chosen as a roosting site. Each bird has a beaten-down spot in the sedge, well 'limed' with the droppings."

J. D. Smith shot a male marsh hawk just after daybreak of a very frosty morning; its back and tail feathers were covered with frost.

I have no brief for the marsh hawk as a gentle, harmless bird; on the other hand, I consider it a decidedly intolerant, aggressive, and pugnacious defender of its home territory, as everyone knows who has ever attempted to invade its precincts. Especially when there are young in the nest, or even after the young are on the wing, one or both parents are sure to attack the intruder. Some say that the male is the more aggressive, but I have seen very little difference. I have had them dash at my head repeatedly, and keep it up as long as I was anywhere near the nest; flying off for a short distance, the hawk would turn and come like a flash straight for my face, as if it would surely strike me; but it always just missed me by a few inches. A. D. DuBois writes me: "While I stood near a nest, trying to arrange a tripod and camera, the parent marsh hawk repeatedly struck me on the head. In one of these onslaughts she lifted my hat and dropped it on the ground. Her claws penetrated the hat sufficiently to scratch the scalp."

Mr. Saunders (1913) had a marsh hawk attack him frequently when he was a long way from the nest and often not headed in that direction, once when he was a mile away from it. Mrs. Bailey (1915) had similar experiences. Elon H. Eaton (1910) had the bellows of his camera, which he had concealed near the nest, torn to pieces by the attacking hawk. Paul L. Errington (1930) gives an interesting account of the territory disputes of three pairs that nested within 400 yards of each other; each pair had its definitely outlined territory, on which none of the others were allowed to trespass.

No less intolerant is their behavior toward other species. They have been seen repeatedly attacking red-tailed and red-shouldered hawks that were peacefully soaring over their domains. They always drive away crows and have been known to attack and drive away eagles. They often drive away sparrow hawks, blackbirds, and other small birds without attempting to catch them. Walter B. Savary writes to me that he "saw a marsh hawk with a mouse in its claws trying to escape from three crows that were pursuing it in an endeavor to get the mouse. So close at last were the crows that the hawk let its prey drop; without checking its flight, the leading crow snatched up the mouse and continued on, to be at once followed by a caracara, which, in turn, forced the crow to drop its prize. This happened so near me that the hawk dared not to pick up the mouse, but perched on a nearby stub and waited."

Even the bold and dashing duck hawk is sometimes robbed of its prey, but sometimes the tables are turned. Forbush (1927) relates a story, told him by William G. Means, of a duck hawk that knocked a marsh hawk off a fallen duck it was eating. On the other hand, C. J. Maynard (1896) writes: "The Marsh Hawks are, as a rule, not very bold but I once knew an exception to this and, while in Florida, some years ago, repeatedly saw one of these birds rob a Peregrine Falcon of Ducks which it had captured. This appears almost incredible, but I was once quite near when the Marsh Hawk took possession of the booty of the Falcon that was sitting on the ground, and I distinctly saw the latter give up his prey, almost without a struggle, to the venturesome Hawk which coolly began to eat it, utterly disregarding the screams of the Falcon that was darting about a few yards above him."

I have often noticed, in a large colony of breeding terns, that as soon as a marsh hawk appears on the scene their otherwise cease-less din suddenly stops, every voice is still; the silence is so strik-ing that we look up to see the cause, as thousands of white wings are diving after him in an angry mob, and he is forced to beat a hasty retreat. I have no evidence that the hawk ever molests the terns. I have seen the same phenomenon in a densely populated colony of yellow-headed blackbirds in a western slough.

Lewis O. Shelley (1930) enjoyed an unusual experience in taming some young marsh hawks that he raised from the nest, of which he writes:

They flew anywhere they wished and were always called by a whistling note. They flew all about the village and to points a mile or more distant at least, without harm by humans befalling them. Their maneuvers were at once interesting and unbelievable at the same time to everybody, including myself. That they became perfectly tame and came to me when called, was a reaction considered remarkable in a wild raptorial bird. * * *

During the fall migration, vireos, warblers, sparrows—many species—would feed contentedly in the same tree, on the same limb, with one of the hawks. I never saw an attempt of the hawks to molest them. Our own and the neighboring hens became used to the hawks and did not become frightened when they alighted in the henyards. * * *

At any time when I wanted them to exhibit to visitors or for other reasons, if within hearing distance they always came. If I merely whistled to answer their common "contented" call they took it for what it meant and remained where they were, often shifting their positions to be able to watch me. A sharp whistle served as "mess-call" and was responded to promptly—quite so. Perhaps the greatest thrill was in having them alight on my person, any-where, at any time; to be able to handle them to my utmost content without fear of injury; to call them when I left work at the store and have them fly home with me for the evening meal. * * *

As to sight and hearing, their instincts were unsurpassed. Any noise, and a good many too slight to be detected by human ears, was noted instantly with whatever reaction suited the case at hand. To illustrate the eyesight: I once held an inch cube of meat in my finger tips over my head, uttering no sound. A hawk perched in a tree about one hundred and fifty yards dis-tant immediately rushed to me, eyes upon the tid-bit, and without slacking speed perceptibly, grasped it with a downward lunge of one foot and wheeled back to its perch triumphant. If a piece of meat about an inch square was accidentally dropped in the tall herd's grass when flying to the woods, where I, searching keenly, could not find it, the bird poised in mid-air above the spot would see it instantly, alight and eat it. They did this on several occasions. I believe this well illustrates the power of the eyesight when a foraging Marsh Hawk sails low over a meadow searching for field mice. Small chance a moving body has of escaping the keen eye!

Voice.—The several notes of the marsh hawk have been variously interpreted by different writers. Mr. Forbush (1927) has made a choice collection of these, which I quote with authorities:

Alarm call of male, "a shrill screaming *cha-cha-cha-cha-cha-cha*" (Florence M. Bailey); female, "a prolonged shriek—*kee, kee, kee, kee, kee, kee, kee, kee*"; or "*check-eck, check-eck, check-eck, check-eck, check-eck, check-eck*" (Bailey); "a series of syllables like *kuh! kuh! kuh!* repeated very fast and quite a number of times without pause" (H. O. Green); female when disturbed at nest, a flicker-like call sounding like *pé-ter pé-ter pé-ter;* another call *stee-whit-a-whit-a-whit,* also *pee pee pee* repeated fifteen to twenty times and *swit, wat, wat,* the notes sometimes run together like a whinny (C. W. Townsend); rather weak nasal whistle, also a sort of chuckle; at nest with eggs *quip-quip-quip-quip-quip;* male at times has a complaining, scolding note like *chu-chu-chu* or *choo-choo-choo,* quite unlike the usual short, weak but sharp whistle of the bird—this when nesting-area is invaded. The male's voice is deeper, fuller, and heavier than the female's higher-keyed note (J. A. Farley).

Field marks.—The adult male is the whitest of any of our common hawks, with black wing tips. In all plumages, the white rump is conspicuous. The everglade kite, Harris's hawk, and the rough-legged hawk all have similar white patches; the first two have comparatively restricted habitats in the South and have other field marks; the rough-legged hawk is a more heavily built bird and has the white mainly on the tail instead of on the rump (upper tail coverts). At any reasonable distance the marsh hawk can be recognized by its slender form, its long slim tail, and its long wings, held at an upward angle except when soaring. Its manner of flight, described above, is distinctive.

Fall.—Late in August or early in September the fall migration begins in New England. Mr. Forbush (1927) says: "The principal migration here seems to move along the coastal plain. Many marsh hawks coming south through the region below Boston follow down the west side of Buzzards Bay and then turn westward across Narragansett Bay and along the coasts of Rhode Island and Connecticut."

By the middle of November most of these hawks have left the northern parts of their range, though they linger on the way as long as they can find enough mice and small birds to hunt. Audubon (1840) writes: "I have observed it in our western prairies in autumn moving in flocks of twenty, thirty, or even as many as forty individuals, and appearing to be migrating, as they passed along at a height of fifty or sixty yards, without paying any attention to the objects below; but on all these occasions I could never find that they

were bent on any general course more than another; as some days
a flock would be proceeding southward, on the next to the northward
or eastward."

Maurice Broun's (1935) records for 1934 at Kittatinny Ridge in
Pennsylvania "extend from September 24 to November 24. The
majority of the 105 individuals recorded passed through between
October 10 and November 10. The greatest number seen on one day
was 11 on October 18, and 11 on November 3. The females precede
the males, apparently, as most of the 51 birds that occurred up to
October 19 were of the former sex. Of 38 Marsh Hawks observed
from November 1 to 12, 28 were males."

Winter.—A few individuals remain, during mild winters, on the
coastal marshes of southern New England, or in other suitable localities throughout the Northern States; but the great majority follow
the migrations of the small birds southward, and spend the winter
in the Southern States, the land of plenty.

DISTRIBUTION

Range.—North America, Central America, and (rarely) the West
Indies; accidental in northern South America.

Breeding range.—The marsh hawk breeds **north** to Alaska (Kobuk
River); Mackenzie (Lower Anderson River, Fort Rae, Fort Resolu-
tion, and probably Fort Smith); Manitoba (probably Fort Churchill,
probably Cape Churchill, and York Factory); Ontario (fork of the
Albany River and Moose Factory); and Quebec (Quebec, Kamou-
raska, Pointe des Monts, and mouth of the Natashquan River).
East to Quebec (mouth of the Natashquan River); New Brunswick
(Chatham); Prince Edward Island; Nova Scotia (Pictou and Hali-
fax); Maine (Portland); Massachusetts (Chatham and Marthas
Vineyard); Connecticut (New London); New York (New York
City); New Jersey (Summit, Princeton, Lawrenceville, Long
Beach, Great Egg Harbor, and Cape May); Virginia (Wallops
Island and Cobbs Island); and rarely Florida (Gainesville and Mican-
opy). **South** to rarely Florida (Micanopy); Ohio (Toboso and
Circleville); Indiana (Marco); Illinois (Philo); Missouri (Mount
Carmel and Stotesbury); Kansas (Neosho Falls, Wichita, Fort Hays,
and Ellis); New Mexico (Salt Creek); Arizona (Tucson); and
Lower California (El Rosario). **West** to Lower California (El
Rosario and Cape Colnett); California (San Diego, Riverside, San
Luis Obispo, probably Seaside, Santa Cruz, probably Point Reyes,
and Humboldt Bay); Oregon (Fort Klamath and Salem); Wash-
ington (Seattle and Port Angeles); British Columbia (Chilliwack,
Lac La Hache, Cariboo District, and probably the Kispiox Valley);
and Alaska (probably Homer, probably Nushagek, St. Michael,
Nulato, and Kobuk River).

Winter range.—During the winter season the species may be found **north** to southern British Columbia (Ladner, Okanagan Landing, and Chilliwack); rarely Alberta (Belvedere); rarely Saskatchewan (McLean); probably rarely North Dakota (Argusville); Wisconsin (Cottage Grove, Elkhorn, and Madison); Michigan (Ann Arbor); casually southern Ontario (London, Toronto, and Ottawa); New York (West Point); and Massachusetts (Danvers). **East** to Massachusetts (Danvers and Dennis); Connecticut (Branford and Saybrook); eastern New York (Gardiners Island); New Jersey (Plainfield, Moorestown, and Cape May); Delaware (Lewes); Virginia (Fort Union and Bowers); North Carolina (Louisburg and Raleigh); South Carolina (Mount Pleasant and Sea Islands); Georgia (Savannah, Blackbeard Island, and Darien); Florida (Ponce de Leon Inlet, St. Lucie, and Royal Palm Hammock); the Bahama Islands (New Providence and Great Inagua); eastern Cuba (Guantanamo); rarely Haiti (Tortue Island, St. Michel, and Trou Caiman); and rarely Puerto Rico (Cartagena Lagoon). **South** to rarely Puerto Rico (Cartagena Lagoon); Panama (Chiriqui and Gatun); Costa Rica (Boruca and San Jose); Guatemala (Duenas and Quezaltenango); Puebla (Llano de Chapulco and Puebla); and southern Lower California (Sierra de la Laguna). **West** to Lower California (Sierra de la Laguna and Don Lorenzo); California (Salton Sea, Escondido, Santa Barbara, San Mateo, Marin County, Marysville, and Chico); Oregon (Klamath Lake, Malheur Lake, Harney, and The Dalles); and southern British Columbia (Ladner).

Migration.—While the species is resident over a very large portion of its range, there are many areas where it is not of regular occurrence throughout the year, and in other regions it may be temporarily absent (or unnoticed) for a month or more, usually during the winter season. This subject might be discussed at length, but it seems desirable to limit the treatment of the migration of the marsh hawk to that portion of its breeding range that is definitely north of the northern limits of the winter range.

Spring migration.—Early dates of spring arrival are: Vermont—Bennington, March 14; Rutland, March 16; Bethel, March 19; and St. Johnsbury, March 20. New Hampshire—Monadnock, March 18; Manchester, March 25; Tilton, April 1; and Charlestown, April 5. Maine—North Livermore, March 9; Auburn, March 12; Portland, March 14; and Lewiston, March 16. Quebec—Quebec, March 4; Montreal, March 13; Kamouraska, April 5; and East Sherbrooke, April 6. New Brunswick—Scotch Lake, March 26; St. John, March 29; and Grand Manan, April 4. Nova Scotia—Pictou, April 1; Wolfville, April 7; and Yarmouth, April 20. Prince Edward Island—North River, April 10. Manitoba—Margaret, March 10;

Aweme, March 11; and Winnipeg, March 25. Saskatchewan—
McLean, March 12 (observed once on January 19); Dinsmore,
March 18; Muscow, March 22; and Indian Head, March 24. Mackenzie—Fort Simpson, April 23; and Hay River, May 1. Alberta—
Edmonton, March 14; Flagstaff, March 19 (once on February 23);
Camrose, March 21; Alliance, March 22; Veteran, March 27; Eastend, March 29; and Carvel, April 2. Alaska—Kuiu Island, April
29; Fort Yukon, May 4; Kigluaik Mountains, May 8; Mount McKinley, May 12; Portage Bay, May 14; and Nulato, May 18.

Fall migration.—Late dates of departure are: Alaska—St. Michael,
September 2; Sitka, September 3; Kowak River, September 3; and
Kenai River, October 6. Alberta—Edmonton, September 22; Nan-
ton, October 12; and Veteran, October 15. Saskatchewan—South
Qu'Appelle, November 1; and Indian Head, November 20. Mani-
toba—Margaret, November 2; Treesbank, November 4; Aweme,
November 8; and Reaburn, November 12. Ontario—Toronto (oc-
casionally winters), November 8; Ottawa, November 11; and Point
Pelee (occasionally winters), November 23. Michigan—Sault Ste.
Marie, October 23; Vicksburg, November 3; Detroit, November 5;
and Ann Arbor (occasionally winters), November 24. Prince
Edward Island—North River, October 21. Nova Scotia—Yarmouth,
September 30; and Pictou, October 20. New Brunswick—Grand
Manan, September 12; St. John, September 16; and Scotch Lake,
October 30. Quebec—Montreal, October 31. Maine—Portland,
October 13; Pittsfield, October 23; Lewiston, October 31; and Win-
throp, November 9. New Hampshire—Dublin, October 11; Jaffrey,
October 16; and Durham, October 17. Vermont—St. Johnsbury,
September 30; Clarendon, October 10; Woodstock, October 21; and
Wells River, October 23.

Marsh hawks have been observed to arrive at San Jose, Costa Rica,
on October 1, and to leave in spring on February 2; at Cape San
Lucas, Lower California, an early arrival date is October 21; and in
Cuba early fall arrival dates are Guantanamo, October 4; Isle of
Pines, October 12; and Santiago de las Vegas, October 13; while
spring departure dates for Cuba are Trinidad, April 14; Guantanamo, April 24; and Isle of Pines, April 25.

Casual records.—The marsh hawk has on several occasions been
recorded outside its regular range. Among these occurrences are
the following: An immature male was collected on the Atrato River,
Colombia, on November 23, 1909, and there is another record from
Medellin, Colombia (Chapman, 1917); Peters saw one on several
occasions late in the winter of 1916 at Sosua, Dominican Republic;
one was taken at Barbados, Lesser Antilles, in September 1886; in
Bermuda one was shot in 1845, one in December 1851, one at War-

wick in November 1874, and there are probably one or two other records (Reid, 1884); one was seen at Cape Blossom, Alaska, on July 26 and again on August 6, 1898; and Henshaw (1901) reports that several specimens have been taken on the island of Oahu, Hawaii. Macoun (1909), quoting Audubon and others, states that it is rare in Newfoundland, but no definite record for the colony has been found.

Egg dates.—New England and New York: 30 records, May 5 to June 14; 15 records, May 23 to June 4.

New Jersey to Maryland: 33 records, April 18 to June 23; 17 records, May 9 to 25.

Indiana to Iowa and North Dakota: 58 records, April 6 to June 30; 29 records, May 21 to June 7.

Alaska and Canada: 41 records, May 7 to July 15; 21 records, May 20 to June 2.

Colorado, Utah, and Washington: 34 records, March 16 to July 18; 17 records, April 10 to May 17.

Florida and Louisiana: 3 records, April 16 to 29.

ACCIPITER VELOX VELOX (Wilson)

SHARP-SHINNED HAWK

HABITS

This bold and dashing little hawk, the terror of all small birds and the audacious murderer of young chickens, is widely distributed in North America, very common at some season in practically all the United States and Canada. Although it breeds throughout most of its range, more or less rarely, its center of abundance during the nesting season is in the eastern Provinces of Canada.

It is best known to most of us as a migrant, coming along with the migrations of small birds and frequenting the open country, the edges of the woods, clearings, hedgerows, bushy pastures, and shore lines, where migrating birds may be found. It is not a forest-loving species and is seldom seen in heavily wooded regions. It has been well called a bushwhacker from its habit of beating stealthily about the shrubbery to the fatal surprise of many a little songster.

Spring.—W. J. Brown, of Montreal, Quebec, says in his notes, sent to me:

The sharp-shinned hawk reaches this Province during the first week in April. Some pairs evidently start nesting shortly afterward, as I have found nests all completed and ready for eggs on April 24 while snow still remains in the evergreen woods. * * *

The sharp-shinned hawk is a common summer resident in the Province of Quebec. I know of 50 different localities where at least one nest could be found if time permitted, and I have no doubt that the bird is equally abundant throughout the Province. While exploring new timber late in fall, I can always find nests of this species where young have been raised during the year. At this time of the year the down and droppings still remain glued to the nest of fine twigs. The following spring one is sure to find a hawk's nest in the same neighborhood.

Courtship.—Lewis O. Shelley has sent me the following note:

At this time both hawks were seen to come from the woodland and flap along beside an old roadway, dashing and circling at and about each other over a nearby mowing. Inside of 5 minutes they returned to the maple tree, alighted, the male on a dead branch some 5 feet directly below the female; both facing east, standing crosswise of their perches with heads turned to the right. The female moved first, was heard to call several times in a modulated key, and the male answering once, both notes the same and similar to the *peep* of young chicks. Suddenly the female crouched along the limb and, as though this were a signal, the male launched forth on set wings, banked and alighted about 4 feet from the female, then sidled toward her until their wings touched. The male then settled on his perch immovable, looking away and uttering a feeble whine. With this whine, the throat could be seen in agitation, I believe due to the vocal efforts while having a full crop. It was fully 3 minutes before mating took place, the female remaining crouched the while, and, with mating, both went in for much wing-flapping for 40 seconds. The male then returned to his perch beside the female and both sat still for nearly half an hour of utter silence. Then the low whining on the male's part was repeated and mating immediately followed. The birds did more fluttering, but the display lasted less than 30 seconds. After another interval mating again took place. And this time, losing their balance, both birds actually tumbled head over heels to the ground and not until then did one fly. The male was seen to be gripping the feathers of the female's back, but this alone could not have buoyed their descent together. At the first of their fall, however, the female was seen to spread her wings and beat them several times as when rising in air, and thus probably hindered a more abrupt fall than was the case.

Nesting.—In southeastern Massachusetts the sharp-shinned hawk was formerly a fairly common breeding bird, though we always considered the nest a desirable find. We used to find the nest practically every year that we hunted for it, and one season we found five nests. We could generally count on finding the nest in the same vicinity for several years in succession. But in recent years, with the growing scarcity of small birds in this section this hawk has been steadily decreasing in numbers, until now we seldom find a nest. With us the standard nesting site has always been in a dense grove of medium-sized white pines (*Pinus strobus*), one of our commonest forest trees; 11 out of 18 nests definitely recorded in my notes were in such dense places; 17 in all were in these pines. Occasionally we have found the nest in more open groups of these pines or in mixed woods of pines and oaks. Once I found a nest on Cape Cod, where the white pine does not grow, in a slender little pitch pine (*Pinus rigida*) in oak woods; it was only 14 feet above the ground and contained six eggs; and in one of the small oaks near it, at about the same height, was an old nest that was shown to me as their nest of the previous year. The height from the ground, of my other 17 nests, varied from 20 to 55 feet, and about half of them were between 30 and 35 feet. The nests were all made of small sticks or twigs, and about half of them had no lining at all, except a smooth layer of finer twigs in the hollow of the nest; in others a few chips of outer bark of the pine had been added. Most of the nests were freshly built, but some of them were evidently old nests, to which new material had been added. The presence of many old nests, in a grove occupied by these hawks, indicates that they prefer to build a new nest each year. This hawk often builds a very large nest in proportion to its size, so that the incubating bird is invisible from below; but often, on the smaller nests, the bird's tail may be seen projecting over the edge. A typical large nest, which was in use for its second consecutive year, had outside measurements of 26 by 25 inches; it completely encircled the trunk of the tree and from the trunk to the outer edge it was 16 inches wide; it was 7 inches in height; it measured 6 inches across the inner cavity, which was 3 inches deep, very deeply hollowed for this species. There is much individual variation in the behavior of different birds; sometimes the incubating bird will sneak quietly off the nest, as the intruder approaches, and not show herself again; in such cases it is easy to pass by a nest and not notice it; another may not leave the nest until the tree is rapped; still another may stick to the nest until the climber is part way up the tree; and once I saw the climber within 3 feet of the nest before the sitting hawk left. Even if the hawks are not seen or heard, there are other signs to guide the collector to the nest. During the courtship season in April, the shrill plaintive call notes of the male may be heard in some likely spot, and the chances are that a nest will be built near there later. After incubation begins one may see a small bit of white down on or near an occupied nest; but there is never so much down to be seen on an Accipiter's nest as is usually seen on a Buteo's, and oftener there is none. But almost always a patch of woods occupied by a breeding pair of sharpshins shows ample signs of their bird-killing habits, wings and feathers of domestic pigeons, robins, blue jays, and other small birds; often cast-off flight feathers of the hawks are seen, as they begin to molt in May. Where such signs are abundant it pays to climb to every likely looking nest. A sharp-shinned hawk's nest is usually recognizable as a broad, rather flat platform of clean sticks, built on horizontal branches against the trunk, quite unlike a crow's nest.

I have seen a pair of these hawks acting as if they had a nest in a dense cedar swamp, but I have never found a nest in such a situation. Others have found them in other parts of New England nesting in cedars, hemlocks, spruces, and firs, but very seldom in a deciduous tree. Out of eight Massachusetts nests recorded in Col.

John E. Thayer's notes, seven were in white pines, one 90 feet above the ground, and one was in a hemlock, only 25 feet up. I have a Massachusetts set in my collection taken from a nest in a beech.

W. J. Brown, who has examined over 200 nests of this hawk in the vicinity of Montreal, has sent me some elaborate notes. He says of its nesting habits:

The majority of nests have been found in black spruce trees, a few in balsam, and an occasional one in hemlock, cedar, tamarack, and pine. The height varies from 10 to 60 feet from the ground against the trunk on horizontal branches. The nest does not resemble the bulky structure of the crow, as some authorities aver, but it is easily distinguishable from the latter by the shallow platform of interlaced spruce twigs. The usual nest of this hawk is an affair of twigs, sometimes lined with flakes of bark, and it cannot be mistaken for that of a crow or any other species of hawk, but can be recognized at a glance at any season of the year. A number of nests have been built over old foundations, but as a general rule the bird builds a new nest each season. The tree chosen is on the outskirts of the woods or at the edge of any clearing or opening in the middle of the woods. A favorite location is in a thick clump of spruce near a clearing or on the border of a path. Any large area of coniferous timber usually contains a pair of sharpshins.

Mr. Brown once found a sharp-shinned hawk sitting on a set of five eggs in an old blue jay's nest, 6 feet up in a hemlock sapling, with its "long tail and a portion of its body showing conspicuously over the edge of the nest." In the Thayer collection is a set from a nest 25 feet up "in a crotch in a white poplar", taken in Manitoba, and also one from Utah, taken from a nest lined with grass, leaves, and pine needles, only about 6 feet up in a "native birch, near a creek, in the bottom of a canyon." I have a Utah set taken from a cottonwood, about 25 feet up. While collecting in the Huachuca Mountains, Ariz., we found a typical nest, containing four eggs, on May 28, 1922; it was built on horizontal branches against the trunk of a fir, about 30 feet from the ground, in a clump of tall thick firs, about halfway up the mountain (pl. 32).

Audubon (1840) reports finding two very unusual nests; one was "in a hole of the well-known 'Rock-in-Cave', on the Ohio River"; the other was in "the hollow prong of a broken branch of a syca-more." John Krider (1879) says he has "found its nest built on high rocks in the mountains of Pennsylvania." John Macoun (1909) mentions a nest in Saskatchewan "in the crotch of a willow, less than 10 feet from the ground" in a willow thicket. A nest found by P. M. Silloway (1903) in a Montana thicket was "in a crotch of a haw tree", only 9 feet from the ground. Charles F. Morrison (1887), in Colorado, took a set of three eggs on June 22, 1886, "deposited in a dilapidated magpie's nest, the arched roof of which had fallen upon the main nest, forming a hollow which had been lined with a few feathers upon some dead leaves which had partially filled it the fall before." From the above data, and from many other records not referred to, it is quite evident that the sharp-shinned hawk prefers to nest in thick coniferous trees; but where conifers are not available in the vicinity of good hunting grounds it will nest in almost any other convenient site.

Eggs.—The eggs of the sharp-shinned hawk are highly prized by collectors, as they are among the handsomest of American hawk's eggs and show almost endless variations in color and pattern. The set usually consists of four or five eggs, often only three, and rarely six or even seven or eight. If some of the eggs are taken during the laying period the hawk will keep on laying. C. L. Rawson, "J. M. W." (1882), took 18 eggs from a single pair of birds in one season:

From the nest in a pine grove four eggs were taken the week ending May 23d. The next morning boys Crow-hunting tore down the nest. Before night a new nest resembling a Night Heron's was constructed in the same grove and three eggs taken the second week. By the middle of the third week two more eggs were taken, and a Pigeon's egg substituted, from which were taken successively as laid nine more eggs. The early morning of every alternate day was the rule for a fresh egg. The longest break in the series was from June 2d to June 6th. The seventeenth and last egg in the direct line was laid on June 21st, and when taken the nest was deserted, neither bird being seen for several days. On the 25th, the female ventured back, and apparently as an afterthought or a "positively the last" trial-egg, laid just one more.

The eggs are well rounded, ovate to short-ovate or nearly oval in shape; the shell is smooth but not glossy. The ground color is dull white or very pale bluish white. Some eggs have great blotches or splashes of dark, light, or bright rich browns, such as "burnt umber", "chocolate", "liver brown", "amber brown", or "hazel"; some of the handsomest eggs have underlying washes or great splashes of the lighter browns, or of shades of "vinaceous-fawn", overlaid with the darker markings; and some are largely covered with pale vinaceous tints and spotted with the lighter browns, producing a very pretty effect. The heavy markings may be concentrated, or confluent, at either end, or they may form a ring midway. Some eggs are finely and evenly sprinkled with small spots or dots of any of the browns named above, or with vinaceous shades, or both. Occasional eggs are sparingly marked or nearly immaculate, one or two such eggs occurring in sets otherwise heavily marked. The measurements of 58 eggs average 37.5 by 30.4 millimeters; the eggs showing the four extremes measure **40.6** by 30.5, 39 by **32, 35** by 29, and 36.6 by **28.9** millimeters.

Young.—Incubation lasts about three weeks, perhaps 21 to 24 days, and is shared by both sexes. Henry J. Rust (1914) made a careful study of a brood of young sharp-shinned hawks and published an interesting, illustrated article on it. On the morning of

July 2 he found four of the five eggs pipped and that afternoon one young bird hatched, showing that incubation had not begun until the set was complete, or nearly so. "The eyes were open and very dark in color." The next day three more eggs hatched, and the fifth egg was pipped. On July 7 the young birds "seemed to have increased one-third in size" since July 3. Three days later the young were able to hold up their heads and show some resentment; "the sheathed feathers at the wing tips were about one-half inch long", when the young were about a week old. On July 12 he saw the female feed the young by tearing off strips of meat from a young bird. On July 16 the wing quills were bursting their sheaths, and on July 26, when about 23 days old, the young all left the nest as Mr. Rust climbed the tree. He says: "When I was about half-way up the mother gave what seemed to be a warning cry, and hawks were flying in all directions. They must have all left at once." Their wings were well developed, but their bodies were still largely downy. He caught two of the young birds with considerable difficulty and took them home to study further development. On August 1 the last of the down had disappeared, and on August 9 he liberated the captives near the nesting site where he found the other young and the mother still in the vicinity of the nest.

My one and only experience with a nest of young sharpshins was similar to Mr. Rust's. On July 16 they were all downy except for a few feathers on the scapulars and for wing quills about an inch long; but when I climbed to the nest on July 23, I was surprised to see them all fly away, although one was quite feeble. Two that I kept in captivity made very unsatisfactory pets, always timid, wild, and untamable, but with fierce appetites for raw meat. The old birds must kill large numbers of small birds to keep them satisfied. Mr. Forbush (1927) thinks the young must require three or four birds each every day; he says that J. A. Farley found the twigs of a nest "littered with thrushes' legs." Ralph J. Donahue (1923) gives a different picture; he made seven trips to a nest of young sharpshins, and says: "I am glad to say that I found no evidence of a single bird killed. Locusts, large beetles, and cicadas, with a mouse or two for dessert, was the main type of food."

Plumages.—When first hatched the nestling is scantily covered with short white down, with a faint creamy tinge. This is soon replaced, or covered, with thick, woolly, longer down, covering the whole bird, "pale pinkish buff" in color, but whiter on the belly. The wing quills are the first to sprout, when the nestling is still very small. The plumage then appears on the scapulars, back, and tail, then on the flanks and breast, and finally the head. The young leave the nest before the down is entirely replaced by feathers.

The chronology of the development is given in Mr. Rust's (1914) observations, above.

In full juvenal plumage the upper parts are "sepia" or "bister", edged on the crown and tipped on the back, scapulars, upper tail coverts, wing coverts, and tertails with "tawny"; the under parts are white, or buffy white, with large tear-shaped spots, or streaks, of "snuff brown" or "sayal brown", lighter on the tibiae; in some birds the tibiae are uniform, clear "tawny"; the throat is white, narrowly streaked with dusky. The plumage is worn without much change during the first winter; but it becomes much faded by spring, and the molt begins in May. Both sexes are alike in this plumage, but the male is much smaller. They breed in this plumage. The first postnuptial molt is complete, but much prolonged, from April or May to September or October. It produces a second winter plumage which is nearly adult, but browner above with some tawny edgings, especially in the female; the feathers of the breast and flanks are patterned, much as in the adult, giving a transversely barred effect, but in darker browns, with less white. The full perfection of the adult plumage is acquired at the second, postnuptial complete molt, from July to October, the regular annual molting time for adults. There is considerable individual variation in adults, which is perhaps due to age; a male, which is mostly clear "pinkish cinnamon" on the breast and clear "orange-cinnamon" on the tibiae, is perhaps a very old bird. In all adult females the upper parts are less bluish, more brownish, and the under parts are lighter than in males.

Food.—Dr. A. K. Fisher (1893) gives a long list of the food of the sharp-shinned hawk and then summarizes it, as follows: "Of 159 stomachs examined, 6 contained poultry or game birds; 99, other birds; 6, mice; 5, insects; and 52 were empty." It is especially fond of young chickens and domestic pigeons, and will make frequent raids on the poultry yard, as long as the supply lasts, or until a charge of shot puts an end to it. The larger females are strong enough to carry off a half-grown chicken or an adult pigeon. Herbert L. Stoddard (1931) has seen one carry off a full-grown bobwhite; and other quails are easy prey for it. R. B. Simpson (1911) has seen it pick a red squirrel off a limb and "fly heavily away with its struggling victim, holding it down as far away from its body as possible." He also saw one attack a pileated woodpecker, which was dodging around a tree trunk and screaming; the hawk's career was promptly ended by a charge of shot. C. J. Maynard (1896) relates the following:

These small Hawks are very bold and will not hesitate to attack birds which are larger than themselves, and I once saw one strike down a fully

grown Night Heron that chanced to be abroad by day. The Heron was flying
from one island to another across some marshes, when the Hawk darted out
of a neighboring wood and pounced upon him. The force of the shock was
so great that the slowly moving Heron fell to the ground at once but, fortunately for him, in falling, he gave vent to one of those discordant squarks
which only a bird of this species is capable of uttering, and which so astonished and frightened the Hawk, that it completely forgot to take advantage
of its prostrate prey, but darted away; while the Heron regained its feet,
shook itself, and mounting in air, flew wildly into the nearest thicket.

The bill of fare of this hawk also includes a few mice, young
rabbits, shrews, bats, frogs, lizards, locusts, grasshoppers, crickets,
caterpillars, large moths, butterflies, and beetles. But birds are its
principal food, among which the following have been recorded:
Doves, woodpeckers, swifts, flycatchers, horned larks, sandpipers,
cowbirds, orioles, blackbirds, grackles, jays, meadowlarks, many
sparrows, towhees, vireos, many warblers, mockingbirds, thrashers,
catbirds, wrens, nuthatches, chickadees, creepers, kinglets, robins,
thrushes, and bluebirds.

Its ordinary method of hunting has been very well described by
William Brewster (1925) as follows:

Its invariable method of attack is to pounce unexpectedly on its victims,
after watching for their appearance from an inconspicuous, near by perch, or
seeking them by successive gliding flights of no great length, performed low
over the ground beneath branches that overspread secluded wood-paths, or
across little forest glades, or through brush-encumbered fields or meadows. In-
terrupting such level, skimming flight merely by an abrupt turn or drop, and
then pausing but for an instant, the hawk may continue on its way bearing
in its talons some luckless, fear-stupefied Warbler or Sparrow which has been
plucked from twig or turf with truly admirable dexterity. Or it may achieve
similar success almost as quickly, but with greater effort, at the end of a short,
spirited dash, made at top speed, and perhaps with reckless disregard of stiff
intervening branches.

It is often quite crafty in its approach to a poultry yard, flying
low and keeping out of sight behind buildings or fences until it can
dash over and down into the yard, seize a small chicken before anyone is aware of its presence, and make off with it in a hurry; the
sudden surprise attack is most successful. Col. A. J. Grayson, in
some notes published by George N. Lawrence (1874), says:

One day I witnessed an act of this hawk which goes far to illustrate its
habits of perseverance in hunting out the game it may be in quest of; a brood
of half grown chickens was attacked by it, one of which had taken shelter
beneath the bottom rail of a fence; there was barely room between the rail
and the ground to admit the fowl; the little hawk, after perching for a few
moments on the top of the fence, lit upon the ground, and actually reached its
slender claws under the rail, dragged the unfortunate chicken from its hiding
place, carried it off a hundred yards to the bottom of a dry creek, where I
followed it up and recovered the chicken, with which he was unable to rise
above the bank of the creek.

This persistent little hawk often pursues its quarry on the ground. Sitting on some convenient fence post, rock, or low tree, it scans the ground until it detects some sparrow or other small bird moving about in the grass or herbage; it then makes a dash for it, chasing the little bird with a series of long jumps, aided by its wings, until it catches the victim on the ground as it crouches paralyzed with fear; or, if the bird tries to escape by flight, the hawk dashes after it and catches it on the wing. Mrs. Richard B. Harding told me that while watching, from a blind, a veery brooding a nest full of young, she saw a sharp-shinned hawk alight on the ground and walk toward the nest in a menacing attitude; the veery made a show of defense, but the hawk kept on until Mrs. Harding rushed out of her blind and drove it away. As young birds form a large portion of the food of the young hawks, I have no doubt that the hawks systematically hunt for small birds' nests to rob them.

Lewis O. Shelley has sent me the following note on an interesting feeding habit:

Several times in August and September a pair of sharpshins grew into the habit of using a large meadow as a feeding ground, near which they probably nested, and where woodchucks were often killed by the State patrolmen and left as they lay. Of course, flies, beetles, and other carrion-seeking insects gathered. After repeatedly flushing not only the pair of sharpshins from such carcasses but an occasional marsh hawk as well, I determined that the two Accipiters were quick to recognize the presence of food and make use of such a man-made accessory. Later on the sharpshins became in the habit of appearing at the report of a rifle, playing above the lofty elm trees, nonchalantly watchful of the doings below. What instinct is this that told them man was not there to molest them but the woodchucks, and that later these same spoils would offer up to them a booty?

W. J. Brown contributes the following note:

I have sat for hours in a pile of brush near the nest waiting for the return of the male with food for the sitting female. The male, flying through the trees, approaches the nest very quietly, with the exception of a few soft call notes meant only for the ears of the female, who, equally silent, glides from the nest to the "feeding block." The moment has arrived when we can grasp some idea of the wildness and ferocity of these small hawks as they squeal and tear their victim to pieces. The male is soon off far afield, while the female returns to the nest—the greatest secret of all.

Behavior.—The characteristic hunting flight of this hawk has been described above. The lightning speed with which it selects and seizes the luckless victim in a terrified flock of small birds is astonishing and often too quick for the human eye to follow. I have seen one dash at my feeding station and scatter a little group of birds so quickly that I could hardly see what happened. It is not always successful, however, as the little birds are very quick to dash into cover. It often attacks birds in a playful spirit, perhaps for the pure sport of frightening them, as it fails to catch one when it might easily do so. When attacked by crows or jays it sometimes retaliates and sends its tormenters away screaming, perhaps minus one of their number. I once saw a sharp-shinned hawk chasing some small sparrows in an open field, until some barn swallows came along and began attacking the hawk; they drove him away, and, as he mounted in the air, they followed and kept swooping down at him from above; higher and higher he mounted, soaring at times like a *Buteo;* they did not desert him until he was almost out of sight, way up in the sky; the hawk made no attempt to attack the swallows. This high soaring flight is unusual, except during migrations, when it is regularly practiced. Its usual method of procedure, when not hunting, is to fly at a moderate height, with a series of steady, quick flappings, followed by short periods of rapid sailing, the whole process being swift and graceful. Dr. Charles W. Townsend (1920) writes: "A pair soaring and playing together high in the air gave me a beautiful exhibition. The smaller one, the male, would dart at the larger one, the female, who would shake or tip the wings to spill the air and fall down only to glide up again without movement of the wings to a great height. Again they would dart down with great speed, and turn and glide up again."

Naturally this little villain is greatly dreaded by all the smaller birds, and they have learned to keep out of sight and silent when one of these hawks is near. By the larger birds it is not only feared but is cordially hated and sometimes attacked. Many of the hawk's apparent attacks on birds of its own size or larger are playful feints for its own amusement; and sometimes the game is played on both sides. M. P. Skinner (1928) tells an interesting story of a kingfisher escaping from a sharp-shinned hawk by diving and swimming. Mr. and Mrs. T. T. McCabe (1928a), who have seen many such events, evidently think that the kingfisher enjoys the game, for they say: "Not only is the pursuit and escape a matter of daily occurrence over the grassy, many-channelled creek which flows under our windows, but it is hardly less common to see the Kingfishers approach and circle the seated Hawk. Once, when the latter refused to be 'drawn,' the Kingfisher lit on a limb forty feet away and fifty yards from water, and, vibrating with excitement and hatred, rattled his loud defiance."

Mr. Skinner says, in his notes, that he has seen sharp-shinned hawks chased by a nutcracker, which was always careful to keep above the hawk, by robins that came to the rallying cry of one of them, and by tree swallows; the last seem to be immune from the attacks of this hawk. He has seen the hawk scoop at gulls on a garbage pile and seen one persecute a flying red-tailed hawk. A. G. Lawrence says in his notes:

C. L. Broley and I witnessed a sharp-shinned hawk attack a prairie falcon at West Shoal Lake, Manitoba. The prairie falcon had just left off amusing itself by swooping at a juvenile marsh hawk, pretending to attack it, and was flying high over a field near the lake when the sharpshin quickly mounted into the sky and attacked the prairie falcon as a kingbird does a crow, swooping down in fierce plunges until the falcon turned and fled the way it had come, giving us an excellent view of the little battler. The sharpshin completely outmaneuvered the falcon, mounting above it time after time, and dashing down on its back, apparently delivering blows which were at least irritating, as the prairie falcon repeatedly tried to strike sideways at its spunky tormentor.

Dr. J. M. Wheaton (1882) tells the following interesting story:

I once saw an adult bird of this species pounce upon a Meadowlark, quietly feeding upon the ground. By some means the attack was only partly successful, and the Lark hopped about for a few moments with the Hawk upon his back. The ridiculousness of his position seemed to disconcert the Hawk, who relaxed his grip, only to find himself attacked by bill and claws of his victim. Then followed a fierce fight with claws, bills and wings, in which both contestants appeared equally active and determined. Finally the combatants separated, the Hawk flying in one direction disappointed, dejected and disgusted, the Lark in another, recovering his breath by extraordinary cries of alarm and distress.

Hostility toward the human invader at its nest is also well marked. The individual variation in the behavior at the nest has been referred to above, based on the author's experience. W. J. Brown relates, in his notes, his experience with a pair of these hawks, whose nests he found for four successive years; the male was never seen or heard; and the female always slipped off the nest in silence and made no demonstration. Usually these hawks are quite demonstrative; both birds often start their shrill, cackling notes as soon as the intruder approaches the tree; and when he starts to climb to the nest they become very aggressive, darting down at him, dashing through the branches of the tree and threatening to strike him, all the while keeping up a constant cackling. H. J. Rust (1914) describes the actions of a particularly aggressive pair, the parents of a brood of young, as follows:

The old birds were very ferocious, more so than before. The male struck one hard rap between my shoulders while I was examining the young, and the female kept striking so close to my head as to make it very uncomfortable. After descending to the ground I hid near a small fir tree to watch the old birds. The female flew to the nest and kept up a constant call; the male followed close to where I was standing and swooped at my head; shortly afterwards the female made a swoop direct from the nest and just grazed my head. I moved out of the thicket and both birds followed, perching eight or ten feet from me, uttering their shrill cries, and darting at my head at short intervals. I finally started back down hill and stopping fifty yards or more from the thicket looked up just in time to see the male hawk coming straight for me. I waved my hat and he circled and made for a tall tree near the nest, seeming satisfied that he had finally driven me away.

Illustrating the boldness and reckless audacity of this little feathered bandit, the time-honored statement by Nuttall, that one in an impetuous dash broke through two thicknesses of greenhouse glass and was brought up only by the third, has been quoted many times. It does not hesitate to dash fearlessly through dense tangles of trees, underbrush, and thickets in pursuit of its prey. F. A. E. Starr writes to me that he saw one dash through the rusty wire of a pheasant pen while chasing a sparrow. Even trapping does not dampen its courage. Harold Michener (1930) says that they are much troubled by sharp-shinned hawks at their bird-banding traps; they are now capturing the hawks in traps, baited with birds that the hawks have killed, and banding the hawks. One hawk was trapped three times within a few hours. "Usually the hawks are back and into the traps in a very few minutes, sometimes before the one who has set the trap is out of sight." They have no fear of human beings, or have considerable confidence in their own speed, for they often seize a chicken or a sparrow almost under our noses. C. W. Nash (Thompson, 1890) writes:

On one occasion an impudent villain of this species glanced past my head and snatched up a plover I had shot, carrying it off in front of my dog's nose, and this he did before the report of my gun had died away, and through the smoke from the charge. The act so astonished me that I forgot to shoot at him until he was too far off; when I did remember, I sent the other charge after him, but without effect; he did not even drop his ill-gotten spoil. On another occasion one followed a redpoll almost into my buggy. On the 22d of August I saw one strike at a Bronzed Grackle and carry it off from where it was feeding in a public street, at Portage la Prairie, although there were many people about.

Mr. Shelley relates the following in his notes:

The first seasonal sharp-shinned hawk was seen April 3. On the eighth, at the same place, a pair were seen. This was at the edge of a sugar-maple woods. They were first seen circling about a tree standing away from the other trees, diving at it as if pursuing some intended prey. They did no sailing but flapped in flight. As I drew nearer a gray squirrel was seen part way down the tree, and the two Accipiters constantly lunged at it, driving it to the top of the tree. I had noticed earlier that it commonly fed here on maple buds. Watching the hawks, I decided they were merely playing with the squirrel, as, surely as the squirrel got down so low as 30 or 40 feet, it would be driven back to the treetop, where it clung for a space before again attempting to get away from its tormentors. For fully 20 minutes this farce went on, until the hawks tired of their play. Perhaps 40 minutes elapsed before the gray gained the ground; the hawks merely sitting on a convenient dead limb, not even watchful of its escape. At no time were the rushes and pursuits on the hawk's part of a serious nature but were leisurely, easy, and noiseless.

Voice.—The ordinary alarm note as the nest is approached, uttered also during the attack on the intruder, sounds to me like *kek, kek,*

kek, or *kik*, *kik*, *kik*, a vehement cackling note of anger, similar to that of Cooper's hawk, but shriller and not so loud. Mr. Rust (1914) writes it *"cha, cha, cha"*, and says the male gives "similar, but less shrill cries" than the female. I have also heard a peculiar, plaintive, squealing note uttered by a bird perched on the topmost branch of a tall dead tree, its favorite perch; this was evidently a call note, similar to the courtship call. Mr. Brown says in his notes: "The sharp-shinned hawk has two distinct alarm notes when the nest is approached, the usual cackling call in the earlier stages of the nesting season and a series of squealing notes, not unlike those of the grouse, after the young are hatched, alternating from one call to the other when the young are well grown."

W. L. Dawson (1923) records the alarm note as *yip, yip, yip, yip, yip* and says that a bird in pursuit of a horned owl "shouted *Ricky, ticky, ticky, ticky, ticky*, with an animation which was both thrilling and terrifying."

Field marks.—The sharp-shinned hawk may be recognized as an *Accipiter* by its rather short, rounded wings and long tail, or by its manner of flight, usually rather low, with a series of flappings alternating with rapid sailings. It is much smaller than a Cooper's hawk, but a large female sharpshin is nearly as large as a small male Cooper's. The sharpshin's tail is square, or nearly so, whereas the Cooper's is decidedly rounded; Cooper's hawk also has a black cap, which is not pronounced in the sharpshin. It is quite different in shape and in flight from the small falcons.

Enemies.—Hawks have no enemies of consequence except man, mainly the poultry farmer and the sportsman. The former destroys the nests and kills the birds whenever possible; the latter conducts hawk-shooting campaigns with deadly effect. Once I found a sharp-shinned hawk's nest, which I had been watching, knocked down, the eggs broken, and the decapitated bodies of both parents lying on the ground; their heads had been used to collect the bounty. Hundreds are shot on their fall migrations for sport or because they are considered to be harmful vermin. Dr. George M. Sutton (1928) says that in Pennsylvania on one day in October, "several gunners" killed "in a remarkably short time" 90 sharpshins, 16 goshawks, 11 Cooper's hawks, 32 redtails, and 2 duck hawks. Dr. Witmer Stone (1922) says that sharp-shinned hawks are regarded as game birds at Cape May, N. J.; "in one week in September 1920 no less than 1,400 were known to have been killed, one man securing sixty."

Fall.—Sharp-shinned hawks begin to drift southward through New England during the latter half of August, the heaviest flight coming in September. According to F. S. Hersey's notes the migration was still in progress at Cape Ray, Newfoundland, on September

15, 1913. The course is generally southward to the shores of Long Island Sound, thence turning westward along the coast, and then southward along the New Jersey coast. On Fishers Island, at the eastern entrance of Long Island Sound, according to A. L. and H. L. Ferguson (1922), they get three flights, as a rule, each fall:

The first about September 13; the second about September 20, which has always been the main flight; and the last flight, which is much smaller, near the end of September, or early in October.

* * * On any date after September 5, if a decided change of weather occurs, and is followed by a clear, bright day with a northwest wind and large white clouds, we invariably get a flight. That the wind plays the most important part we know from our records. On some days we have had the flight commence early in the morning, only to have it stop completely when the wind changed from north-west to north or north-east. For the last six years we have made notes of the hawks passing over Fishers Island, and have found that with only a few exceptions the flight has come when the wind was from the northwest. The days when these exceptions occurred the surface wind was northeast, and the hawks were flying at a great height, and at a level where we believe the winds were moving from the northwest, though this could not be determined, as there were no clouds.

* * * The young birds are the first to come, and late in the flight season the adults are met with. It is most interesting to watch a good flight. Some birds will be high up, sailing straight along, keeping up their momentum with occassional beats of their wings. Others will be flying close to the ground, taking advantage of hollows and hillsides, to get the most favorable wind currents, while others may be seen darting through the patches of woods, hunting for small birds.

Most wonderful flights have been seen at Point Pelee, Ontario, during September, where these hawks came along in such enormous numbers that it seemed as if all the hawks in Ontario had gathered at this point to cross Lake Erie. The flight begins about the first of September, but the heaviest flight lasts for only three or four days around the middle of the month, after which the numbers of hawks gradually decrease. Taverner and Swales (1907) have given a full account of it, from which I quote as follows:

After the coming of the first in the fall their numbers steadily increased until from six to a dozen can be noted in a day, which in most localities would be accounted common. Then there came a day, Sept. 11, 1905, and Sept. 15, 1906, when the morning's tramp found Sharp-shins everywhere. As we walked through the woods their dark forms darted away between the tree trunks at every few steps. Just over the tree tops, a steady stream of them was beating up and down the length of the Point, while in the air they could often be discerned at every height until the highest looked like a mote floating in the light. As concrete illustrations of the number present:—In 1905 we stood in a little open glade and at various times of the day counted from twenty-five to thirty in sight at one time and Saunders writes, "When I saw the flight in 1882 it was probably even greater than in 1905. There were more Sharp-shins than one would suppose were in Ontario, and one day my brother and I stood thirty paces apart, facing each other, with double-barrel, breech-loaders, and for a short time the hawks passed so thick that we had to let some go by unmolested because we could not load fast enough to fire at each as it came." A farmer told us of sitting in his front yard one afternoon and shooting fifty-six without leaving his chair. * * * Near the extreme end of the Point is a wooden observatory tower built by the U. S. Lake Survey for the purpose of making observations on the changes of the shore contour. It is about fifty feet high, and stands with its base in the red cedar thicket whilst the platform rises well above all surrounding foliage. On this vantage point Saunders and Taverner took their stand the 18th, and with watch in hand counted the Sharp-shins that passed, nearly all within gunshot. From 11:24 to 11:54, 281 passed us, 207 making for the end of the Point and 74 returning, making 133 that started across the lake within half an hour. As far as we could make out without remaining on the spot the whole time this rate was kept up all day and every day of the greatest abundance of the species.

DISTRIBUTION

Range.—North and Central America, casual or accidental in the Bahama Islands and in Bermuda.

Breeding range.—The breeding range of the sharp-shinned hawk extends **north** to Alaska (Nulato, Salcha Slough, and the Kandik River): Yukon (probably La Pierre House); Mackenzie (probably Fort Good Hope, Lake Hardisty, and Lake Mackay); northern Saskatchewan (probably Otter Lake); Manitoba (probably Norway House, Oxford House, and York Factory); probably northern Ontario (Long Portage); Quebec (Godbout and Anticosti Island); and Newfoundland (Raleigh). **East** to Newfoundland (Raleigh); Nova Scotia (Kentville and Halifax); Maine (Bangor, Auburn, and Portland); New Hampshire (Franklin Falls and Webster); Massachusetts (Taunton, Dennis, and Marthas Vineyard); New York (Lake Grove); New Jersey (Pensauken Creek and Sea Isle City); Delaware (Lincoln); Virginia (Variety Mills and Lynchburg); North Carolina (Raleigh); and probably northern Florida (Alachua County and Hernando County). **South** to probably northern Florida (Hernando County and St. Marks); Alabama (Greensboro); Arkansas (Newport and Clinton); Texas (Texarkana and Edinburgh); New Mexico (Santa Fe Canyon, Lake Burford, and Mount Sedgwick); Arizona (Tombstone and Huachuca Mountains); and southern California (Inyo Mountains). **West** to California (Inyo Mountains, Star Lake, Mount Sanhedrin, Mineral, and Redding); Oregon (Klamath Falls, Bandon, Newport, Salem, Dayton, and Beaverton); Washington (Mount Rainier, Olympia, and Neah Bay); British Columbia (Langley, Alta Lake, Cariboo, and Masset); and Alaska (Kupreanof Island, Admiralty Island, Bethel, and Nulato).

Sharp-shinned hawks were seen in August on the Kowak River, Alaska (Grinnell, 1900), and one was taken at Moose Factory, northern Ontario, in September 1862.

Winter range.—In winter the species is found **north** with fair regularity to southern British Columbia (Victoria and Chilliwack);

central Utah (Manti); Colorado (Clear Creek, Boulder, and Fort Lyon); Nebraska (Neligh); rarely Minnesota (Parkers Prairie, Minneapolis, and Lanesboro); rarely southern Michigan (Grand Rapids, Ann Abor, Ypsilanti, and Wayne County); New York (Rochester and Auburn); and occasionally Maine (Portland). **East** to occasionally Maine (Portland); New Hampshire (Dover); eastern Massachusetts (Taunton); eastern Pennsylvania (Philadelphia); Virginia (Accotink and Fort Union); North Carolina (Raleigh); South Carolina (Oakley Depot); Georgia (Savannah, Blackbeard Island, and Okefinokee); Florida (St. Augustine, Orlando, St. Lucie, and Fort Myers); Yucatan (Chable); Nicaragua (Escondido River); Costa Rica (San Jose and Volcano de Irazu); and Panama (Chiriqui). **South** to Panama (Chiriqui); western Guatemala (San Geronimo); Oaxaca (Tehuantepec); Guerrero (Acapulco); southern Sinaloa (Mazatlan); and southern Lower California (San Jose del Cabo). **West** to Lower California (San Jose del Cabo); California (San Clemente Island, Santa Barbara, Pacific Grove, San Francisco, Berkeley, Marysville, East Park, and Paynes Creek); western Oregon (Salem); probably western Washington (Fort Steilacoom and Port Orchard); and southern British Columbia (Victoria).

Winter occurrences at more northern points are not infrequent. In Alaska, Willett (1927) saw it at Wrangell on January 2, 1921, at Craig on November 22, 1922, and took a female near Ketchikan on February 9, 1926. One was noted at Ottawa, Ontario, on January 9, 1900 (Macoun), while Eifrig (1907b) saw another at this place during the winter of 1903–04. Mousley (1918) reported that it wintered at Hatley, Quebec, in 1916.

Migration.—Despite the fact that this hawk may be found in winter over a large portion of its breeding range, it nevertheless performs a very definite migration. In some years the autumn flight at certain favored points is most conspicuous. Since the usual method of presenting dates of arrival and departure north of the winter range would not portray an adequate picture of the movement of this species, it seems preferable instead to include comment relative to the migration as observed at certain places. Among the more or less famous observation points are: The hills of northern New Jersey; the eastern end of Long Island and smaller adjacent islands, particularly Fishers Island, N. Y.; the eastern end of Lake Ontario, N. Y.; Point Pelee, Ontario; and Charity Islands and Whitefish Point, Mich.

Spring migration.—In northern New Jersey the heavy spring flights of sharp-shinned hawks take place usually during the latter part of April and the first 10 days of May. At Shelter Island, N. Y., the movement has sometimes started the last of March and continued through April.

On the south shore of Lake Ontario, east of Rochester, N. Y., sharpshins are frequently abundant in spring, the height of the flight occurring during the latter part of April and the first week of May. The birds pass along the south shore of the lake and around the eastern end en route to breeding grounds in Canada (Eaton, 1910).

N. A. Wood reports (1911) that for several days during the spring season this species is common on the Charity Islands, in Saginaw Bay, Mich.

At Whitefish Point in the eastern end of Lake Superior, hundreds of these hawks have been observed during the period from May 13 to June 3.

Fall migration.—It sometimes seems that the autumnal movement of the sharpshin is timed to coincide with the southward flight of sparrows and warblers, which at this season furnish a large part of its food.

At Point Pelee, Ontario, the flight begins late in August and continues through September or occasionally to the latter part of October. During the height of the migration it is not unusual for 25 or more of these hawks to be in sight at one time.

Fishers Island, N. Y., is one of the best known "hawk observatories." At that point several flights usually occur, the first generally about September 13, the second and main flight about September 20, while the third or last flight is usually about the end of September or the first of October (Ferguson, 1922). These times agree with autumnal observations at the eastern end of Long Island, and in New Jersey.

Casual records.—In Bermuda, a sharp-shinned hawk was taken near Pennistons Pond on February 23, 1853, and another was shot near Stocks Point sometime prior to 1884. Sometimes this species is fairly common in the Bahama Islands, as Bryant (1861) reported that at Nassau it appeared to be the most common hawk and that a number were seen in different places. Todd and Worthington (1911) reported seeing sharp-shinned hawks at Mathew Town, Great Inagua, on February 22, 1909, and also at Acklin Island.

Egg dates.—Alaska to Quebec: 216 records, May 6 to July 25; 108 records, May 25 to July 8.

New England and New York: 149 records, March 30 to June 29; 74 records, May 22 to June 3.

New Jersey to Georgia: 29 records, May 5 to July 30; 14 records, May 18 to 29.

Ohio, Iowa, Minnesota, and Colorado: 8 records, April 11 to June 22; 4 records, May 8 to June 2.

Washington to California and Utah: 25 records, May 8 to July 3; 13 records, May 22 to June 11.

ACCIPITER COOPERI (Bonaparte)

COOPER'S HAWK

HABITS

If the sharp-shinned hawk is a blood-thirsty villain, this larger edition of feathered ferocity is a worse villain, for its greater size and strength enable it to do more damage. Furthermore, it is much more widely common during the breeding season, being one of our commonest hawks in nearly all parts of the United States. It is essentially *the* chicken hawk, so cordially hated by poultry farmers, and is the principal cause of the widespread antipathy toward hawks in general.

In my early bird-nesting days, 30 and 40 years ago, this was one of our commonest nesting hawks; but for the past 30 years it has been steadily decreasing in numbers. This is perhaps due to constant persecution, but it is largely due also to the marked decrease in the numbers of small birds. I have always considered this and the sharp-skinned hawk as competitive species, each intolerant of the other. I have frequently found one of these Accipiters nesting in the same tract of woods with one of the Buteos, but I have never found the two species of *Accipiter* nesting in the same tract; and several times I have known *cooperi* to replace *velox* in a tract where the latter had repeatedly nested. It is a curious fact that the solitary vireo (*Lanivireo solitarius*) has so often been found nesting in pine woods occupied by a pair of Cooper's hawks as to suggest some significance in the ecology; I find six such cases recorded in my notes, and once the vireo's nest was within 50 feet of the hawk's nest; we have also noted that we never find the vireo in similar woods occupied by sharpshins; the reason seems obvious.

Nesting.—Cooper's hawk is still a fairly common breeder in south-eastern Massachusetts, though my records show that we found three times as many nests during the 20 years previous to 1910 as we have since then. My earliest date for a full set of eggs is April 22, and my latest date for heavily incubated eggs of the first laying is June 3. But I have only six records for April and only three for June, leaving 39 records for the month of May. According to our experience here, Cooper's hawk shows a decided preference for white-pine groves as nesting sites; but it is not nearly so closely confined to this type of woods as is the sharp-shinned hawk. Our notes record 27 nests in white-pine woods, 16 in deciduous woods, mostly oaks and chestnuts, 4 in mixed woods, oaks, chestnuts, and pine, and 1 in a pine on an open knoll among a few scattered oaks; this last was occupied by a pair that had nested the previous year in a tract of oak woods nearby that had since been cut off. Of these 48 nests, 28 were in white pines (*Pinus strobus*), 11 were in oaks (red, scarlet, and swamp white), 6 were in chestnuts, and 3 in maples. The heights from the ground varied from 20 to 60 feet, but approximately half of them ranged between 35 and 45 feet. I believe that the Cooper's hawk prefers to, and generally does, build a new nest each year, but the old nest is sometimes repaired and sometimes a nest is built on an old squirrel's or crow's nest. A new nest is a clean, substantial structure of sticks and twigs, varying in shape and dimensions according to its location. Nests in white pines are rather broad and flat, built on two or three horizontal branches and against the trunk. A typical new nest of this type will measure about 28 by 24 inches in length and breadth and 7 or 8 inches in height. In one case, where a new nest had been built on top of an old one, the combined structure was 27 inches high. Nests in deciduous trees are usually built in upright crotches, and are higher but not so broad. An extreme nest of this type, built in the 4-prong main crotch of a chestnut, measured 18 inches in diameter and 30 inches in height; it was probably built on an old nest of some sort. The inner cavity is usually about 7 inches wide and 2 to 4 inches deep, depending on the condition of the outer rim. The inner cavity is always, according to my experience, lined with chips, or flakes, of the outer bark of pines or oaks; I have never seen a nest containing a full set that lacked this bark; but I believe it is usually added after some, or all, of the eggs are laid. Occasionally a few sprigs of green pine needles are added, and once I found a nest that was profusely lined with this material, together with the usual bark flakes.

A Cooper's hawk's nest can usually be recognized by its size, shape, and location. Bits of down are oftener seen on this than on the sharpshin's nest, but it is never so heavily decorated with down as a Buteo's. Often the long tail of the incubating bird may be seen projecting over the edge of the nest. Usually the bird darts away with great speed and is not seen again; sometimes one or both birds fly about and cackle; but I have never found them so bold and aggressive as sharpshins. Often the location of a nest is betrayed by the cackling. Generally one finds in the nesting woods the feathers of ruffed grouse, poultry, or smaller birds. Sometimes these hawks use an old nest as a feeding station; these nests are well decorated with feathers of the hawks and their victims. In one such nest I once found the feathers of a screech owl. Cooper's hawks often return to the same patch of woods to nest for several years in succession, but we have never found them so constant in their attachment to a locality as the red-shouldered hawk.

Bendire (1892) says: "On the plains where, from scarcity of suitable timber elsewhere, they are confined to the shrubbery of the creek bottoms, consisting mainly of cottonwoods and willows, they sometimes nest as low as 10 feet from the ground, and I have here found some of their nests fairly well lined with the dry inner bark of the cottonwood and with weed stalks; while in the vicinity of Grand Forks, North Dakota, according to information furnished me by Mr. G. G. Cantwell, they are said to nest occasionally directly on the ground."

He quotes from Denis Gale, regarding a peculiar Colorado nest, as follows: "On June 25 I found a nest of a Cooper's hawk containing four unmarked bluish white eggs, resting upon some thin flakes or scales of spruce bark, which alone constituted the lining of the nest, the available contrivance for which was a large bunch of matted scrub, an excrescence upon a horizontal limb, about 18 inches from the trunk and about 20 feet from the ground. This bunch consisted of a wonderful growth of very densely interlaced twigs, the surface of which offered a commodious nesting site, having not only an ample flat area, but a sufficient depression in its center to meet every requirement for a nest."

In Arizona we found several nests of Cooper's hawks generally high up in the tops of the giant cottonwoods or sycamores in the mountain canyons; only one was climbed to, as it was only 40 feet up in a blackjack oak. In Texas these hawks nest in the lofty tops of the heavily timbered deciduous forests in the river bottoms.

John H. Flanagan (1901), in Rhode Island, robbed a red-shouldered hawk's nest on April 20 and found a Cooper's hawk's nest within 20 feet of it; 10 days later he took the eggs of the latter also; and on May 12 he was surprised to see the Cooper's hawk fly from the redshoulder's nest, in which it had already laid three eggs. A climb to the nest showed that every vestige of the inner bark (which the redshoulder always uses) had been removed. A few small sticks had been added and the nest relined with outer bark.

Clarence F. Stone (1899) had an unusual opportunity to watch a pair of Cooper's hawks building their nest, which he describes as follows:

I spied the half completed nest just as one of the hawks left it and thought I had been discovered, but an instant later the mate lit upon the nest and arranged a stick.

Their manner of approaching the nest was a very interesting and curious sight. They came through the low woods flying just above the ground three or four feet, with the speed of an arrow, and when within fifteen or twenty feet of the nest-tree they closed their wings with a quick flip and "slid up" to the nest in a graceful curve.

They did not visit the nest together and apparently the one that was away from the nest could see its mate, for no sooner would one of them drop a few feet below and fly away, than the other was on the upward curve. As if to avoid collision they left the nest from the north side and approached from the west, in which direction—and only a few rods away—all the material seemed to be obtained.

While at the nest their actions were quick, nervous; and they placed the sticks in several places before satisfied, but they did not remain at the nest more than half a minute.

If a set of eggs is taken from a nest the hawk will lay a second set, about ten days or two weeks later, sometimes in the same nest, but oftener in another nest hastily repaired. C. J. Pennock tells me that, after taking four eggs from a nest on April 26, he took two more eggs on May 5 and two additional eggs on May 11, all from the same nest.

If either one of a pair is shot during the nesting season, the survivor usually secures a new mate quite promptly. J. Eugene Law (1919) mentions the following incident, related by Maj. Allan Brooks: "A female Cooper Hawk had been shot from her nest of eggs. Some days later another female, in adult plumage, was found incubating the same eggs, and was likewise shot. What was his surprise later to find a third female occupying the nest, this time a bird in the streaked plumage of a sub-adult. And as a matter of curiosity she was allowed to, and did, raise the brood."

Eggs.—Four or five eggs form the usual set for Cooper's hawk; sometimes only three are laid; I have taken one set of six and heard of two or three others. Just half of the sets recorded in my notes have been of four, and one-quarter of them of five. They are deposited at intervals of one or two days. The eggs are rounded-ovate to ovate in shape, and the shell is smooth but not glossy. The color, when fresh, is bluish white or greenish white, which fades out to dirty white. They are generally nest stained but otherwise immaculate and not attractive in appearance. From 25 to 50 percent of the eggs show more or less scattered spotting in pale browns or buffs; rarely some of them are as heavily marked as some of the paler types of red-shouldered hawks' eggs. Major Bendire (1892) says: "Mr. C. J. Pennock has a set of five eggs in his collection, in which the ground color is a rich bright green, and four of these eggs are distinctly and handsomely marked. They were collected by himself near Kennett Square, Pennsylvania, May 2, 1887."

The measurements of 62 eggs average 49 by 38.5 millimeters; the eggs showing the four extremities measured 54 by 40, 51.5 by 42, and 43 by 34 millimeters.

Young.—Incubation is shared by both sexes, does not usually begin until all, or nearly all, the eggs are laid, and is said to last for 24 days. The young hatch at intervals of one or two days, perhaps less, for there is very little difference in size noticeable among small young. A brood that I watched and photographed were still in the eggs on June 4; on June 18, when probably 10 to 12 days old, their plumage had not started to grow, but they were very bright, active, and playful; only two of the four eggs had hatched; one egg was evidently addled and buried in the lining of the nest, but the other had disappeared. On June 30 they were about two-thirds grown and partly feathered; one, probably the female, was considerably larger than the other. I made my last visit on July 11, when one of the young had disappeared and the other was perched on a limb above the nest. I climbed the tree with my camera, but before I got within range, he climbed out to the end of the limb and then flew clumsily over to the next tree; he repeated the operation when I climbed the next tree and I saw him fly from tree to tree with increasing confidence, until I gave up the chase. His wings and tail were nearly grown, but he was still partially downy; he was then about five weeks old. I had noticed on previous visits that while there were only eggs in the nest it was lined with the usual flakes of outer bark; but after the young had hatched it was relined with fresh green sprigs of pine; this lining, however, was not renewed after the young were half grown.

My attempts to rear young Cooper's hawks in captivity have not succeeded; they have always been wild and untamable. But Dr. H. Justin Roddy (1888) has been more successful; he writes of one he took from a nest when not more than two weeks old:

It was a great eater. When six weeks old it ate nine English Sparrows (*Passer domesticus*) and a common mouse (*Mus musculus*) in one day; and ate on an average eight Sparrows a day from that time until it was ten weeks old. * * * In eating the bird tore its food to pieces with the bill, nearly always beginning at the entrails. It almost always seemed to relish the intestines more than any other part of the bird or animal, sometimes eating only this part and leaving the rest. When the bird or animal was still warm and the blood therefore uncoagulated, it tore it open and apparently bathed the bill in the blood and the visceral juices. It apparently sucked up these fluids in order to allay thirst. But I invariably found it refuse water,—in this respect acting quite differently from the *Cathartes aura*, which drank water freely. * * *

The bird became very much attached to me, and even when it could fly and was allowed its liberty did not leave, but returned every few hours for its food, which I always liberally provided. How long it would have continued to do this I do not know, as the experiment ended with its death. It was shot by one who did not know it was my pet.

William Brewster (1925) writes:

While skirting the edge of a deep and heavily-wooded glen on the north side of Upton Hill, half a mile or more from the Lake, I heard on August 4, 1874, a succession of shrill, squealing whistles repeated at frequent intervals. Cautiously approaching the place whence these sounds came, I presently discovered four young Cooper's Hawks not quite fully grown or feathered, and still tufted here and there with fluffy, whitish down, standing close together in a row on a prostrate log. Every now and then one would unfold and raise its wings, flapping them to preserve its balance, as it took a few unsteady steps along the log, at the same time uttering the whistling cries above mentioned. One and all stood very erect when not in motion, and young as they were lacked little if anything of that stern and dignified bearing so characteristic of adult Hawks at most, although by no means all, times. After watching them awhile I shot one, when the three survivors flew heavily up into a spruce where another was promptly killed, the remaining two being permitted to escape.

Plumages.—In the first downy stage the young Cooper's hawk is thickly covered above, and more thinly below, with short white down, faintly tinged with cream color above. This short down is replaced later, before the plumage starts to grow, with long, woolly, pure-white down, with which the young bird is thickly covered above and below. The juvenal plumage starts to grow before the bird is three weeks old, beginning on the wings and tail, closely followed by the scapulars, back, and sides of the breast. Two young birds, about 22 to 24 days old, had the remiges about one-third grown and the rectrices less than one-quarter grown, all partly in sheaths. The young leave the nest when about five weeks old, before the flight feathers are fully developed and while still partially downy on the head, center of the breast, flanks, and legs.

In fresh juvenal plumage the upper parts are "bone brown" to "clove brown", edged on the crown and tipped on the back, wing coverts, and upper tail coverts with "tawny" or "ochraceous-tawny", lightest on the tail coverts; there is a white line over the eye; the chest is washed with "pinkish cinnamon" and heavily marked with broad, hastate, dusky streaks; the flanks and breast are white, with narrow dusky streaks; the belly is immaculate white; and the legs (tibiae) are marked with "buffy brown" cordate spots. This plumage is worn throughout the first winter and spring, the colors fading somewhat and the edgings wearing away. Young birds begin to breed in this plumage. A complete molt begins in June, starting with the wings and tail and ending with the body molt in summer. This produces a second-year plumage that is practicaly adult, but the full perfection of the adult plumage is not acquired for at least another year.

Adults have a complete annual molt from July to October. The sexes are alike in juvenal plumage; the adult male is more brightly and heavily marked below and more bluish above, the female being duller below and more brownish above. The female is much larger in all plumages.

Food.—Cooper's hawk does more damage in the poultry yard than all other hawks put together. It is very destructive to domestic pigeons, of which it is very fond, and, if not killed, will soon clean out a colony. It soon learns also where it can find a convenient supply of half-grown chickens or young ducks, to which it makes frequent visits, until its career is ended. It is not easily killed, however, as it is usually crafty enough and quick enough to avoid the farmer's gun.   Dr. A. K. Fisher (1893) says: "Dr. William C. Avery, of Greensboro, Ala., informs us that during one year he killed and wounded at least a dozen of these Hawks before the inroads among his doves ceased. * * * Dr. Warren states that a pair of these Hawks destroyed some fifty chickens from one farm, twelve of which were taken in a single day." He summarizes the food of this hawk as follows: "Of 133 stomachs examined, 34 contained poultry or game birds; 52, other birds; 11, mammals; 1, frog; 3, lizards; 2, insects; and 39 were empty."

Among the wild birds mentioned in the food of Cooper's hawk are teal and young of other ducks, least bittern, snipe, screech owl, quails, partridges, grouse, doves, small hawks, flickers, blackbirds, jays, meadowlarks, woodpeckers, grackles, numerous sparrows, towhees, a few warblers, vireos, nuthatches, thrashers, catbird, robin, and thrushes. The list of mammals includes hares, rabbits, opossum, various squirrels, skunks, rats, and mice. It sometimes eats snakes, lizards, frogs, grasshoppers, crickets, beetles, and even butterflies and moths. Austin Paul Smith (1915) says that during a drought in Arkansas, when most of the streams were dried up, he "flushed, on several occasions, this hawk in the act of feeding upon minnows in the desiccating pools."

Behavior.—The flight of Cooper's hawk is very similar to that of the sharp-shinned hawk, a low, swift, dashing flight. It surprises its prey by a sudden, swift dash, pouncing upon it before it has a chance to escape. Its short wings and long tail give it such control of its movements that it can dart in and out among the branches of the forest trees with impunity, or dodge through the intricacies of thickets where its victims are hiding. Dr. Daniel S. Gage has sent me some notes illustrating its crafty methods of approach. He was watching a robin at the base of a tree in some thick woods when he saw a hawk come, flying swiftly and keeping the trunk of the tree between him and the robin; when close to the tree the hawk swerved around it and barely missed catching the robin. Again he saw a hawk approach a field of tall weeds, in which some small birds were feeding, flying close to the ground behind a fence, dash over the fence, and pounce on one of the birds. On another occasion, a hawk had seen, while perched on a distant tree, some chickens in a yard close to a house; he flew low, behind the house until close to it, rose over the house and pounced down on one of the chickens, which had no chance to escape until he was right upon them.

M. P. Skinner relates the following incident in his Yellowstone notes:

As I rode up the loop road through the aspens above Mammoth, I heard quick, frightened bird cries on either side, and I even seemed to sense the excitement in the air. I turned about to see what was happening and a Cooper hawk came shooting up the road past me, four feet above the road and going at great velocity. A Kennicott ground squirrel that no doubt had been attracted to the road by spilled oats, tried to cross the road only to be struck amid a cloud of dust. After striking the squirrel the hawk went on for six feet more before it could turn. Meanwhile the squirrel was stretched out in the road lifeless. The hawk came back and attempted to carry off its booty. But I dashed up at a gallop, and as the prey was too heavy to carry off quickly, the hawk had to drop it. I picked it up and found that only one claw had pierced the skin, and only a drop or two of blood had come out. So I believe that the squirrel was killed by the force of the blow itself.

The following remarkable performance is described by William Savage (1900):

I was standing in a thick brush at the time, when suddenly I heard a loud rushing of wings, rather behind me, and, on looking around, saw a quail flying past at its utmost powers of flight and about twenty feet behind was a Cooper's Hawk, but pursuing with such rapidity that I could plainly see it was gaining on the quail. I discovered in a moment that the quail was endeavoring to reach a clump of hazel bushes nearby, though apparently intending to pass them, but when about six feet above and directly over the desired covert, it suddenly dropped like a dead bird for the refuge. The hawk, however, was not to be eluded by this bit of strategy, for with an extra effort, it shot beneath its prey, at the same time turning with its back next the ground and spreading its murderous claws wide open, the quail actually falling into them; then righting up, sailed away with its prize. This I think was done in about ten seconds.

But the hawk is not always successful in capturing its prey; sometimes it meets more than its match; and sometimes it may attack for the mere sport of it. At least two observers have noted its fruitless attempts to capture a kingfisher in flight over water. As the hawk gained on the slower bird, the latter, at the most critical moment, suddenly dove into the water and the hawk's momentum carried it far beyond. As the kingfisher rose the hawk returned to the attack, with the same result, which was repeated six or eight times. After the last fruitless attempt the hawk gave it up, the kingfisher, as Charles E. Johnson (1925) says, "alighting on a perch at the water's edge, with bristling crest and many a hitch and jerk, as if to reassure itself of its own personal solidarity, burst forth in a rattle, loud and ringing with triumph if not actually vibrant with inexpressible scorn."

Charles W. Michael (1921) relates the following incident:

Looking up we saw the two rather large birds dashing through the treetops. The dark bird with the white wing-patches we recognized at once as a Pileated [woodpecker]; the lighter colored bird turned out to be a Cooper Hawk. A pursuit was apparently in progress, but as the birds dashed through the branches of the tall trees it was impossible to be sure which of the birds was the pursuer and which the pursued. Both birds quickly left our range of vision, but a little farther on we heard gentle tappings and soon located the woodpecker. The hawk was there too, perched on a limb a few feet away.

The woodpecker was drilling and prying off chips with apparent unconcern, while the hawk looked on with seemingly hungry eyes. While we were watching, the hawk flew to a branch a few feet above the woodpecker. Pileated tilted his head and gave the hawk a sidelong glance and then deliberately flew toward him and drove him from the tree. With the hawk gone, the woodpecker went on with his drilling as though nothing had happened.

Lewis O. Shelley has sent me the following note:

In May 1929, while I was watching a wave of migrating warblers from the partly sheltered location beneath an old-growth white pine, in a wood not more than a hundred yards wide, my attention was called to a Cooper's hawk erratically flapping back and forth overhead, diving now and then at something I could not see and accompanying each dive and rise with its screamed cry. Finally I detected a flicker dodging among the pines in what seemed to me perfect safety if it had only alighted instead of dodging. But it was bewildered for it continued flying from one side to the other of the woods, the hawk following each movement just under the treetops. Then, when the flicker made a headlong flight toward the open and a scrub apple tree, the hawk gained speed and lunged, checking its speed with spread wings and spread tail thrown downward and forward as it struck its prey, turned, and carried its booty in under a pine, commencing to pluck the feathers before the flicker's cries had died out. I examined the remains of the feast and found only the bill and a carpet of feathers!

Crows often attack Cooper's hawks, as they do owls, sometimes singly, but more often in mobs, chasing the hawk about until it is forced to seek refuge in the woods. Sometimes the tables are turned and the hawk chases the crows, though I doubt if this often results fatally for the crow. Joseph Mailliard (1908) once saw a pair of these hawks attacking a flock of crows. The crows were quietly perched in some dead trees. "One hawk would perch on top of a tree above the crows, while the other would go off a little way and then swoop down on the flock, repeating the operation with variations."

H. W. Henshaw (1875) writes:

While sitting in my tent one day at Camp Apache, I noticed one of these hawks making repeated attacks upon a raven. It would force the raven to take refuge in a tree, and then fly to some neighboring perch and take its stand. The moment the persecuted raven essayed to move away, the hawk flew out and swooping down upon it struck it and again forced it to cover. This was repeated several times, and apparently for no other reason than for the amusement of the hawk; though, judging from the discontented squawks and cries which the abused raven gave vent to, the pleasure was by no means mutual. So engrossed was the falcon in this sport that it allowed me unnoticed to walk up within a few feet, when my gun settled the dispute.

About its nest Cooper's hawk is usually shy, flying swiftly away and generally not returning to protest, although it sometimes flies about cackling. When there are young in the nest it is much bolder; I have had one dash at me while I was near the nest tree or alight in the tree near me while I was at the nest. I once saw one attack a great horned owl that came too near its nest.

The following account by Dr. Paul L. Errington (1932) illustrates the ferocity and bravery of a Cooper's hawk in defense of its young:

On July 7, 1931, I was visiting a juvenile Great Horned Owl that had previously been tethered on the ground for a study of its food habits in a woodlot west of Pine Bluff, Wisconsin. The adult owl that was taking care of the youngster appeared at my approach, alighted in a tree near by, and started the usual hostile demonstration of hooting and bill snapping. As if in answer to the hoots, the cry of a Cooper's Hawk came from deeper in the woods, and an instant later a female hawk dashed at the adult owl with terrific speed. Like a skilled boxer, the owl ducked, barely evading the hawk's talons. Several times in very short order the owl had to dodge as the raging hawk struck from all sides.

During the first part of this performance, the owl had been nearly as much concerned on account of my proximity to the juvenile as it had been with the attacks of the Cooper's Hawk. Finally, things became sufficiently hot that the owl left the branch upon which it had perched, and launched forth in direct and purposeful chase of the Cooper's Hawk, which kept just ahead of her larger pursuer for several yards before doubling back, to wheel and strike again. The hawk behaved as though utterly maddened, but she never let herself get quite within reach of the owl's talons. Her safety was plainly dependent upon her superior agility and precision of movement. For a brief space the action became so fast that I could not see exactly what was happening, especially at close quarters when it seemed that neither bird could avoid being hit. However, it is improbable that damage was done, for not even a feather was noted to fall. The hawk soon went her way, cackling as she flew, and the owl was free once more to center upon me its earnest attention. The hawk gave no evidence of having seen me.

A search of a few minutes revealed the hawk's nest 110 yards away. Two juveniles, ready to fly, were perched on the rim.

Many demonstrations of its impudent boldness in pursuit of its quarry have been recorded. Dr. Charles W. Townsend (1930) says: "The Cooper's Hawk, like the Sharp-shin, is bold in the presence of man. I have known one, in pursuit of a chicken, to fly into a barn where it was killed with a whip by a farmer. Another had trussed and nearly eaten a Robin on the ground near my house and did not fly until I had approached within ten feet."

Dr. Fisher (1893) tells of one of these hawks that attacked Dr. C. D. Walcott, while he was collecting fossils, apparently with no provocation; the hawk was repelled but renewed the attack and was killed with a geological hammer. Mr. Forbush (1927) quotes Mr. Farley, as follows:

This Sunday morning, May 2, 1909, soon after 9 (apparently his usual hour), the Cooper's Hawk (or another just as bad) which is getting so many chickens from poultry-raisers here on Chiltonville Hill, Plymouth (we have lost 25), appeared, coming for the coops. Mr. Graves fired at him, but the hawk, not stopped by the report, circled within a few rods and came in again. But the second barrel sent him away, apparently hit. During this entire episode there were five people standing close to the coop. A few mornings ago also, as Mr. Graves was pounding away making another coop, the hawk caught and carried off a chicken within a few yards of him. A Cooper's Hawk two years ago in East Bridgewater behaved similarly. Four times this daring bird (with people standing near) tried to get a chicken out of a hen-yard that adjoined the mixed woods where it had its nest. The people "shooed" the hawk away three times, but at the fourth attempt, despite their cries, it carried off a pullet.

Enemies.—The Cooper's hawk has few enemies, except the farmer or the sportsman with a gun, but the following account, sent to me by Clyde L. Field, shows that the spunky little kingbird can make life miserable for this hawk and defend its territory successfully against it:

Once while traveling along some foothills in Arizona, where bird life was very scarce, I came to a small grove of sycamores. Here birds were very abundant. Nesting had not yet started, except for a pair of Cooper's hawks. On account of these hawks, I little hoped that many of these birds would live through the summer, but, much to my surprise a month later, they were still there and as many as before.

The answer was soon forthcoming, for up the creek came a badly scared Cooper's hawk with a flock of kingbirds in hot pursuit. The kingbirds were striking at him from all angles and, at each hit scored, the hawk would let out a squark. This was repeated several times during the day.

I came upon one of the hawks sitting on a limb with two kingbirds diving at it. One of the kingbirds struck him, causing him to lose his balance. Taking advantage of this, the other kingbird hit him again, knocking him off. Down the creek they went, with more kingbirds joining in the chase. The kingbirds made living possible here for the other small birds in the grove.

Voice.—The ordinary alarm note, heard about its nest, is a loud, metallic, cackling note, similar in form to the corresponding note of the sharp-shinned hawk, but louder and on a lower key. I have always written it *kak, kak, kak, kak*. Others have called it *cac, cac, cac,* or *cuck, cuck, cuck*. It is rapidly uttered with considerable emphasis. As the female leaves the nest she sometimes gives a loud scream of fright or anger.

Field marks.—The Cooper's is a larger edition of the sharpshinned hawk and easily confused with it; there is no well-marked color difference, though the adult Cooper's has a more clearly defined black cap. The chief difference is that the sharpshin has a square tail and Cooper's a rounded tail. Both have short, rounded wings and long tails, much longer than the Buteos; and both have the characteristic *Accipiter* flight, a few rapid wing strokes, alternating with short periods of swift gliding. As seen from below, both may be distinguished from the goshawk by the more conspicuously barred tail and primaries. A small male Cooper's appears to be not much larger than a large female sharpshin, but Dr. George M. Sutton (1928) has shown that the *average* male Cooper's *weighs* about twice as much as the average female sharpshin.

Fall.—Cooper's and sharp-shinned hawks make up a large part of the great fall flights of hawks that pass over the United States in

September, but they are often accompanied by ospreys and the Buteos. They fly high and seem to prefer to fly when the wind is from the northwest. Otto Widmann (1907) says that in Missouri "wholesale migration has been noticed from about the twentieth to the twenty-sixth of September, when singly or in pairs they have followed each other at intervals of a few minutes, from ten to twenty being visible to the spectator, but, as they are known to advance in a broad front, the whole movement must mean the depopulation of a large district."

DISTRIBUTION

Range.—North and Central America.

Breeding range.—During the breeding season, Cooper's hawk is found **north** to British Columbia (Fort St. James and probably Yellowhead Lake); probably rarely Alberta (Jasper Park and Edmonton); Manitoba (Oak Lake, Treesbank, and probably Kalevala); northern Michigan (Sault Ste. Marie); southern Ontario (Kenora, Sudbury, and Aylmer); Quebec (Montreal); and New Brunswick (Restigouche Valley). **East** to New Brunswick (Restigouche Valley and St. John); Maine (Calais, Bucksport, and South Warren); New Hampshire (Franklin Falls and Webster); Massachusetts (Andover, Taunton, and Fall River); Connecticut (Norwich and New London); eastern New York (Shelter Island); New Jersey (Red Bank, Vineland, Sea Isle City, and Cape May); North Carolina (Raleigh); South Carolina (Society Hill); Georgia (Savannah and probably Okefinokee Swamp); and Florida (Branford, Micanopy, Orlando, and Manatee). **South** to Florida (Manatee and St. Marks); Alabama (Greensboro); Louisiana (St. Francisville); Texas (Marshall and Kerrville); probably New Mexico (Mesilla Park and Silver City); Arizona (Huachuca Mountains and the Santa Rita Mountains); and northern Lower California (El Rosario.) **West** to northern Lower California (El Rosario and Guadalupe); California (San Diego, Escondido, Santa Ana Canyon, probably Santa Cruz Island, Fort Tejon, San Miguel, Paicines, and Mineral); Oregon (Klamath Falls, Elkton, Corvallis, and Dayton); Washington (Camas, Olympia, Tacoma, and Seattle); and British Columbia (Pemberton, Cariboo District, and Fort St. James).

The species has been detected casually in summer at still more northern latitudes. One was seen August 25, 1920, at Fort Albany in northern Ontario (Williams, 1921); C. W. Townsend (1913) recorded one seen at the mouth of the Natashquan River, Quebec, on July 30, 1912; while the same authority (1906) notes one seen on Cape Breton Island, Nova Scotia.

Winter range.—The winter range of Cooper's hawk extends **north** to Oregon (Salem); Colorado (Lay and Clear Creek); Kansas

(Hays, Manhattan, and Bendena); Iowa (Keokuk and Hillsboro); rarely southern Wisconsin (Sumpter, Shiocton, and Wauwatosa); southern Michigan (Kalamazoo); rarely southern Ontario (Oakville, Toronto, and Newmarket); New York (Geneva and Syracuse); and rarely Massachusetts (Taunton). **East** to rarely Massachusetts (Taunton); eastern New York (Orient); New Jersey (Princeton and Moorestown); Maryland (Cambridge); Virginia (Ashland); North Carolina (Raleigh and Fort Macon); South Carolina (Oakley Depot and Charleston); Georgia (Savannah and St. Marys); and Costa Rica (El Mojon). **South** to Costa Rica (El Mojon); Guatemala (Duenas); Oaxaca (Totontepec); Sinaloa (Mazatlan); and California (Salton Sea, Volcano Mountains, Pasadena, and Santa Barbara.) **West** to California (Santa Barbara, Paicines, Berkeley, Marin County, and Glen Ellen); and Oregon (Klamath Lake and Salem).

Migration.—Although the Cooper's hawk is found with fair regularity in winter throughout almost all of its breeding range, there is nevertheless a well-defined migration. In this respect its movements agree with those of its smaller relative the sharp-shinned hawk, with which it frequently travels. For the same reasons the usual method of showing the migration by early and late dates is not entirely satisfactory. Accordingly, the following brief comment, with a few dates in the West, will be sufficient to illustrate the spring and fall flights.

Spring migration.—The northward movement in spring takes place usually during the latter part of March and early part of April. Some observers have indicated that at this season the Cooper's hawks seen on Long Island, N. Y., may outnumber all other hawks put together. The earliest date of arrival of migrating birds at Ithaca, N. Y., is March 17, while the average date is March 25. Point Pelee, Ontario, a favorite area for the observation of hawk migrations, does not show a heavy spring flight, the small number seen usually passing through in early May. Apparently the heaviest known flight of these hawks at this season takes place at Whitefish Point, on the south shore of Lake Superior. The birds are most abundant during the first few days of May, only a few being seen after May 12.

Cooper's hawks have been observed to arrive in the western part of their range, as follows: Minnesota—Wilder, March 3; Hutchinson, March 12; and Fridley, March 18. Manitoba—Margaret, March 24; and Treesbank, April 4. Montana—Fortine, March 29; Billings, March 30; and Columbia Falls, April 10. British Columbia— Okanagan Landing, February 9; Comox, March 25; and Edgewood, April 21.

Fall migration.—The southward movement in autumn is frequently more conspicuous. Late dates of departure in the western part of the range are: British Columbia—Okanagan Landing, November 4.

Montana—Columbia Falls, October 21; and Fortine, October 22.
Manitoba—Margaret, September 20; and Treesbank, October 15.
Minnesota—Hutchinson, October 7; and Minneapolis, October 9.

At Mackinac Island, Mich., and Point Pelee, Ontario, the autumn
migration usually starts during the latter part of August. At the
latter point the movement reaches its height by the last of September
but continues through October, occasionally to the early part of
November. Migrating birds also have been observed at Ithaca, N. Y.,
as late as November 1.

At New Haven, Conn., the fall flight extends from about September 5 to October 15. On Fishers Island, east of Long Island,
N. Y., Cooper's hawks are not common in fall, but occasionally a
small flight is noted during the latter part of September, which also
is the time that the largest flights of this species are observed in the
vicinity of Stag Lake, N. J.

Egg dates.—New England and New York: 119 records, April 25
to June 26; 60 records, May 10 to 20.

New Jersey to Virginia: 48 records, April 7 to May 23; 24 records,
April 29 to May 11.

Ohio to Minnesota and Canada: 52 records, April 26 to June 21;
26 records, May 8 to 21.

Missouri to Colorado: 7 records, April 23 to May 30.

Washington to California: 58 records, March 31 to June 13; 29
records, April 19 to May 17.

Lower California to Florida: 21 records, February 22 to June 16;
10 records, April 15 to May 17.

ASTUR ATRICAPILLUS ATRICAPILLUS Wilson

EASTERN GOSHAWK

HABITS

From the heavily forested regions of Canada, the main summer
home of the goshawk, this bold brigand of the north woods, the
largest, the handsomest, and the most dreaded of the *Accipiter* tribe,
swoops down, in winter, upon our poultry yards and game covers
with deadly effect. He is cordially hated, and justly so, by the farmer
and sportsman; and for his many sins he often pays the extreme
penalty. But, as Herbert Ravenel Sass (1930) says—

We do not live by bread alone. Beauty and courage, swiftness and strength
mean something to us; and we shall find these qualities in high degree in the
hawks of the Accipiter clan. Especially is this true of the largest and strongest
of them, the goshawk, one of the deadliest, handsomest, bravest birds of prey
in the world.

None will dispute the goshawk's title to a place among the Kings of Winter.
A big hawk, longer but less bulky than the red-tail; broad-shouldered, compact,
yet clean of build; blue-gray above, with a coal-black crown, and white or

silver-gray below, with narrow black or slate-gray bars; proud and resolute of
mien, with brilliant orange eyes through which the fierce spirit of the fiery-
hearted warrior gleams at times like points of living flame—the goshawk ranks
second to none in martial beauty and in fearlessness.

Nesting.—The main breeding range of the goshawk is in the heavily
wooded portions of eastern Canada and a few of the extreme northern
States, although a few scattered pairs nest as far south as the mountainous regions of Pennsylvania and western Maryland. The only
nest I ever saw was found on June 10, 1912, on a steep, heavily
wooded mountainside, near the Fox Island River in Newfoundland.
While I was exploring the base of this mountain, I was attracted by
the screams of the hawks and started up the steep slope to look for
the nest, which I eventually found a long distance away from where
I first heard the hawks. The nest was about 20 feet up in a slender,
leaning paper birch, resting partly in a crotch of the birch and partly
on the trunk of a dead spruce that had fallen into the same crotch;
the latter was nearly horizontal, as it had fallen from a point higher
up on the steep slope (pl. 39). There was a fine view from the nest
over the valley of the river far below. It was a large nest, fully 3
feet in diameter and 18 inches thick; it was made of clean, fresh
sticks, nearly level on top, very slightly hollowed, and decorated
with fresh sprigs of balsam fir. It contained only one young bird,
less than half grown, on which the plumage was well started on the
wings and back. Both old birds were quite demonstrative while I
was hunting for the nest, flying about, alighting on the trees near
me, and screaming all the time; but while I was watching the nest,
on that and the following two days, I never saw anything of the
hawks; this was in marked contrast with the experiences of others.
Remains of rabbits and ptarmigan were found near the nest tree,
and in the nest one day was a rusty blackbird.

The goshawk nests regularly in the three northern New England
States and in the Adirondack region in New York, very sparingly
in Massachusetts, and rather often if not regularly in Pennsylvania.
Of its nesting habits in Maine, Ora W. Knight (1908) says: "They
nest in late April or early May, placing the large bulky structure of
sticks and twigs, lined with hemlock bark, green hemlock twigs,
willow and poplar twigs, with the expanding catkins attached, in
some convenient tree. Often the nest is in a small birch or maple
not over thirty feet up, at other times it is in an evergreen, either
pine, spruce or fir, but hard wood growth seems to be preferred by
them."

The first nest to be recorded in Massachusetts was found by Dr.
R. T. Fisher in the Harvard Forest, in Petersham, where a pair
of these hawks has been known to breed for at least three years,

perhaps longer.   J. A. Farley (1923) visited this nest on April 28, 1923, and thus describes it:

The nest was of enormous size but wholly new and hence free from woods dirt.   It was over 5 feet in length, 2 feet in breadth and 1 foot in depth.   It was very compactly made of sticks (mostly white pine and hemlock), many of them long and large; and it had a coniferous bark floor in its very slightly hollowed interior.   Quite a number of the longer and slimmer branches had green pine-needle bunches, but they were worked into the body of the nest, showing that they could not have been added recently.   The fresh fractures of many sticks showed that the Hawk had broken them from living trees. It was the most beautifully constructed large Hawk's nest that I have ever seen.   The nest was placed on horizontal limbs and against the trunk of a white pine of almost 2 feet diameter.   It was up 55 feet—two thirds the height of the tree.

More recent information comes to me from Albert A. Cross, who found a nest containing four eggs near Huntington, Mass., on April 19, 1931. He has sent me an excellent photograph of the nest (pl. 40) and some very full notes on it, from which I quote as follows:

The nest was in a sugar maple in a hemlock clump of considerable area and, roughly measured at the time, was placed at between 40 and 50 feet from the ground; however, afterward a more accurate measurement was made and the height was found to be in excess of 50 feet.   The nest itself was a huge affair placed against the body of the tree and measured 3 by 4 feet across the top and 35 inches in height.   Fresh hemlock spills had been added to the interior, but most of the needles had fallen from these, owing to the fact that fair weather had prevailed since April 7.   Besides the hemlock spills there was also a considerable quantity of hemlock bark with which the nest was partially lined.   The nest was well feathered.

Scattering hardwoods grew among the hemlocks, as beech, yellow birch, and sugar maple.   The feathers of grouse were littered promiscuously through the woods, indicating the activities of the hawks; excrement indicated their favorite perching trees.   While we were there the goshawks were flying about like bats, perching in trees or soaring about overhead and keeping up an incessant din with hardly a break in their calls.

In the days when passenger pigeons were abundant in Pennsylvania goshawks bred there regularly and commonly. But now this hawk is comparatively rare and breeds only in some of the mountainous counties, where it can find extensive forests of mixed conifers and hardwoods. R. B. Simpson (1909a) found a nest that year near Warren, Pa., in a region "heavily timbered with immense pines and hemlocks, and a good sprinkling of beech." The nest was 60 feet up in a white pine; it was lined with "leaves and a few fresh hemlock sprigs." He took three eggs from this nest on April 2, and on April 20 found the hawk incubating on a second set of eggs in an old red-tailed hawk's nest in the vicinity. These eggs were destroyed later, possibly by a crow; but on May 20 he found that the goshawk had laid a third set of two eggs in another old redtail's nest nearby, "high up in a big oak"; this set was allowed to hatch.

A. D. Henderson (1924) says that in the vicinity of Belvedere, Alberta, "the usual breeding place is in heavy poplar woods containing a scattered growth of spruce." The height of the nests from the ground varies from 25 to 75 feet. The nests are "made of dead sticks, with a lining of strips of dry poplar bark and a few green spruce twigs." He says further: "The Goshawk usually uses the old nest of another Hawk, building it up on top and relining it, but often builds a complete nest of its own. It seems to be attached to the locality in which it breeds and will sometimes occupy the same nest for a number of years. If the same nest is not occupied the bird will probably be found breeding in the same belt of timber not far away. I have never found a nest in an evergreen tree."

The goshawk evidently does not show the decided preference exhibited by its near relatives, the sharp-shinned and Cooper's hawks, for coniferous trees as nesting sites; in fact, most of the nests seem to be found in deciduous trees in mixed woods. Out of 62 records consulted, only 11 nests were in conifers, 7 in white pines, 2 in firs, and 1 each in a spruce and a hemlock. Of the 51 other nests, 18 were in beeches, 14 in birches, 11 in poplars, 6 in maples, and 1 each in an oak and a cottonwood. Lucien M. Turner (1886) says that in Alaska it sometimes nests on rocky cliffs. It does, however, require seclusion in some extensive tract of heavy timber, where it selects one of the largest trees. The height from the ground varies from 18 to 75 feet, but most of the nests are between 30 and 40 feet up. The nest is very large and is usually freshly built, but the hawk often uses an old nest of its own for successive seasons or appropriates and repairs an old nest of some other large hawk.

Eggs.—The goshawk lays ordinarily three or four eggs, sometimes only two and rarely five. These are ovate to elliptical-oval or oval in shape. The shell is rather rough, finely granulated or pitted. They are pale bluish white or dirty white and unmarked, except for occasional nest stains. I have never seen or heard of a spotted egg. The measurements of 50 eggs average 59.2 by 45.1 millimeters; the eggs showing the four extremes measure 65.5 by 47.3, 61.4 by 47.8, 52.7 by 43.9, and 56.1 by 42.9 millimeters.

Young.—The period of incubation is said to be about 28 days, but positive and accurate records seem to be lacking. Whether both sexes incubate is not known. The young apparently remain in the nest about four weeks and are fed at very infrequent intervals. I watched my Newfoundland nest for nearly all of one day and parts of two others, but no hawk came near it; Dr. George M. Sutton (1925), who spent a whole day watching a nest, says: "To the best of my knowledge the young bird in the nest was not fed all day, and the one on the ground certainly not before late afternoon." This may have been due to the presence of an observer, for I have noted that other hawks will not come near their eggs or young for hours if they know they are watched, and their eyes are exceedingly keen. Dr. Sutton found that "the young had evidently been fed almost altogether on chipmunks, although fur and some small bones of gray and black squirrels, weasels and white-footed mice were also found."

A. A. Cross, in his notes, mentions a variety of food found in the nest; on May 19 "the craws of the young were bulging with food, the weight of which seemed to cause them to pitch forward. In the nest was the foot of a grouse, and beneath the nest where it had lodged on a sapling was a handful of downy feathers, which we agreed had come from the breast of a barred owl"; on May 24 "the nest contained 11 chipmunks and a crow, all uneaten"; on June 12 it "contained the remains of two grouse, which had been cleaned to the bones and one partly eaten chipmunk"; and on June 13, there was one red squirrel in the nest. These three young hawks were certainly well fed. On June 18 two of them left the nest. He says that when the young were small their calls "could be likened to the peepings of day-old chicks", but later they resembled the notes of the old birds.

The young bird, less than half grown, that I watched in Newfoundland, was quite active, standing up in the nest most of the time and exercising occasionally by stretching one wing at a time or raising both together over his back until they almost touched; he gaped occasionally. His eyesight must have been very keen, for he turned his head to look at every bird that passed. He screamed several times, a note like the adult's but shriller and weaker.

Plumages.—The downy young goshawk when first hatched is well covered above and sparingly below with rather short, silky, white down; this is later replaced with longer, woollier down, with a grayish tinge on the upper parts. Before the young bird is half grown the juvenal plumage appears, first on the wings, then on the scapulars, tail, back, and sides of the breast. When between three and four weeks old it is fully fledged, except that the last of the down still persists on the belly and neck. It leaves the nest at about this stage.

In fresh juvenal plumage the upper parts are "clove brown", edged on the crown, upper back, lesser wing coverts, and upper tail coverts with "pinkish cinnamon" or "light pinkish cinnamon", darkest on the head; the scapulars, median and greater wing coverts, remiges, and rectrices are tipped with the same colors, and the greater coverts are broadly barred with the same; the under parts are white, strongly tinged or washed with "vinaceous-cinnamon" or "pinkish cinnamon" and broadly streaked on the breast, less broadly on the belly, with hastate spots of "bone brown." The cinnamon colors soon fade to pinkish white and eventually to dull white.

The juvenal plumage is worn for about a year, and it is not an uncommon occurrence to find one of a breeding pair in this plumage; rarely both birds of a breeding pair have been in this plumage. I have seen one young bird molting into the second year plumage in February, but usually this molt, which is complete, begins late in June or in July and is completed in October or early in November.

The second-year plumage is much like that of the adult, but the crown is streaked with white and the breast is heavily marked with broad shaft streaks and transverse bars or spots of brownish black. I have a young eastern bird in my collection showing the molt into this plumage. Harry S. Swarth (1926) evidently regards this type of markings as a racial rather than an age character, for he describes a young bird completing the molt from the juvenal plumage into an adult plumage, which is "pale colored and finely barred, as in *atricapillus.*" He says further: "The specimen just described (as well as another similar bird collected by Brooks) shows that differences of coarse or fine markings cannot be explained as different stages reached by the same individual."

Most other authors regard this as a second-year plumage. Major Brooks (1927) says:

> I entirely agree with Taverner that the heavily barred and striated adult plumage of the Goshawks is only one of age and is acquired the second year, the markings getting finer and more uniform with each successive year. This heavily marked stage may not be universal—it would be rash to say that anything was constant with such extraordinarily variable birds as the raptors. But that it does exist in a large proportion of cases is evident to anyone who has examined many Goshawks, not only in the dark colored race of the extreme northwest, but in the palest of eastern birds. Hence it cannot be regarded as a subspecific character. I have not seen the specimen taken September 5, so cannot say anything about its peculiarities. But the other "similiar bird collected by Brooks" distinctly supports Taverner's theory, as does another light-colored adult taken at Atlin which Swarth has forgotten.

The fully adult plumage, characterized by the clear black crown and pale, finely vermiculated under parts with narrow shaft streaks, is acquired at the next annual molt. Adults have one complete annual molt at variable times during late spring, summer, and fall. The sexes are alike in all plumages, but there is a decided difference in size. Dr. George M. Sutton (1928) has shown that females average about 9 ounces (nearly one-third) heavier than males.

Food.—Dr. A. K. Fisher (1893) summarizes the food of the goshawk as follows: "Of 28 stomachs examined, 9 contained poultry or game birds; 2, other birds; 10, mammals; 3, insects; 1, centiped; and 8 were empty."

During the invasion of goshawks, in the winter of 1926–1927, in Pennsylvania, Dr. George M. Sutton (1927) reported:

Of the 251 stomachs which were examined, 49 were empty; 41 held poultry (36 chicken; 1, domestic duck; 1, guinea-fowl; 3, domestic pigeon) ; 79 held game-birds (55 Ruffed Grouse; 1, Blue-winged Teal; 8, Ring-necked Pheasant; 15, Bob-white) ; 73 held game mammals (63, Cottontail Rabbit; 10, Gray Squirrel) ; 27 held small birds (19, sparrow-like birds, species not certain; 2, English Sparrow; 1, Robin; 1, Song Sparrow; 1, Meadowlark; 2, Blue Jay; 1, Hairy Woodpecker) ; 16 held small, non-game mammals (5, Red Squirrel; 1, Chipmunk; 3, Field Mouse; 7, White-footed Mouse) two held small snakes; and one held flesh of a dead sheep upon which the hawk was feeding when it was shot.

Lucien M. Turner (1886) thus describes its feeding habits in Alaska:

The tracts preferred by this Goshawk are the narrow valleys, borders of streams, and the open tundra, which it constantly scans for Ptarmigan and small mammals; the Lemming forming a considerable portion of its food. It will sit for hours in some secluded spot, awaiting a Ptarmigan to raise its wings. No sooner does its prey rise a few feet from the earth than with a few rapid strokes of the wing, and a short sail, the Goshawk is brought within seizing distance; it pounces upon the bird, grasping it with both feet under the wings; and after giving it a few blows on the head they both fall to the ground; often tumbling several feet before they stop; the Hawk not relinquishing its hold during the time. During the mating season of the Ptarmigans many males suffer death while striving to gain the affection of the female, for as he launches high in air, rattling his hoarse note of defiance to any other male of its kind in the vicinity, the Goshawk darts from a patch of alders or willows, or from the edge of the neighboring bluff, and with a dash they come to the ground, often within a few yards of the terror-stricken female, which now seeks safety in flight as distant as her wings will carry her. I have seen this hawk sail without a quiver of its pinions, until within seizing distance of its quarry, and suddenly throw its wings back, when with a clash they came together, and the vicinity was filled with white feathers, floating peacefully through the air. I secured both birds, and found the entire side of the Ptarmigan ripped open.

Audubon (1840) says of its hunting:

He sweeps along the margins of the fields, through the woods, and by the edges of ponds and rivers, with such speed as to enable him to seize his prey by merely deviating a few yards from his course, assisting himself on such occasions by his long tail, which, like a rudder, he throws to the right or left, upwards or downwards, to check his progress, or enable him suddenly to alter his course. At times he passes like a meteor through the underwood, where he secures squirrels and hares with ease. Should a flock of Wild Pigeons pass him when on these predatory excursions, he immediately gives chase, soon overtakes them, and forcing his way into the very centre of the flock, scatters them in confusion, when you may see him emerging with a bird in his talons, and diving towards the depth of the forest to feed upon his victim. * * *

Along the Atlantic coast, this species follows the numerous flocks of ducks that are found there during autumn and winter, and greatly aids in the destruction of Mallards, Teals, Black Ducks, and other species, in company with the Peregrine Falcon. * * *

I saw one abandoning its course, to give chase to a large flock of Crow Blackbirds (*Quiscalus versicolor*), then crossing the river. The Hawk approached them with the swiftness of an arrow, when the Blackbirds rushed together so closely that the flock looked like a dusky ball passing through the air. On reaching the mass, he, with the greatest ease, seized first one, then another, and another, giving each a squeeze with his talons, and suffering it to drop upon the water. In this manner, he had procured four or five before the poor birds reached the woods, into which they instantly plunged, when he gave up the chase, swept over the water in graceful curves, and picked up the fruits of his industry, carrying each bird singly to the shore.

W. E. Cram (1899) describes an interesting method of hunting; he followed tracks of a hawk through the woods on the snow; it walked much like a crow, but hopped for a few feet occasionally.

At times it followed in the tracks of rabbits for some distance. I have often known them to do this, and am inclined to think that they occasionally hunt rabbits in this manner where the under-brush is too dense to allow them to fly through it easily. I have sometimes followed their tracks through the brush until I came upon the remains of freshly killed rabbits which they had been eating.

E. S. Cameron (1907) says that a goshawk caught several of his fowls, and—

as the captured fowls weighed upwards of five pounds each it could not carry them off but ate the back or breast of its victims where they lay. A few days after the white hen episode the Goshawk killed a very fine cockerel and was observed by me almost in the act. To escape its enemy the terrified fowl had run under some young cedars which would have saved it from a Prairie Falcon or Peregrine, but were no protection against the relentless Goshawk which followed and seized its prey within the cover. So great was the strength of this cockerel that it ran an uphill distance of fifteen paces towards the fowl house, burdened with the clinging hawk, ere it fell dead. The Goshawk kills its prey by constriction of the feet, and it is quite certain that the squeeze combined with the shock is rapidly fatal to fowls.

Prof. R. T. Fisher, director of the Harvard Forest, told me that he rather liked to have the goshawks about "because they eat great quantities of red squirrels", which are injurious to the pine trees in the forest. I believe that these and other rodents, all more or less injurious to wild and cultivated trees and to the eggs and young of small birds, form a very large portion of the food of these hawks.

Robie W. Tufts, who has had extensive experience with nesting goshawks in Nova Scotia, writes to me:

I spent some time about a nest of this species last spring and made the following notes concerning their food: Chipmunks, 3; half-grown rabbit, 1; female hairy woodpecker, 1; ruffed grouse, 1 chick and 2 adult birds. The chipmunks invariably had their heads off, and I have on numerous other occasions seen chipmunks in the nests of this species and always decapitated. The nest that I had under observation intermittently last spring was near two large farms. One of these boasts of a poultry yard of no mean proportions, and the other supports a flock of hens of normal size. It seems note-worthy that neither of these farmers complained of losing a single bird all summer in spite of the fact that this nest was not over half a mile air line from said farms and furthermore the farmers did not even see any "hen hawks" about their premises. The poultry yard referred to produced about 300 chicks.

I wish to make it plain that the foregoing must not in any way be con-strued as an attempt on my part to prove that goshawks don't kill poultry. They do and often, far too often, but I do wish to make it plain that they don't make a regular practice of it. During the past 30 years I have examined many nests of this species, and I have never yet seen any evidence of do-mestic fowls in or about a nest. Prior to the spring of 1931 (the nest above mentioned) I recall seeing one grouse only in a nest, many robins and flickers, one blue jay, and a number of small rodents, mostly chipmunks.

All the following mammals, birds, and insects have been recorded in the food of the goshawk: Woodchuck, rabbits, hares, muskrat, squirrels, chipmunks, kitten, weasels, lemmings, shrews, mice, Brun-nich's murre, teals and other wild ducks, snipe, domestic poultry, quails, grouse, ptarmigans, pheasants, small hawks and owls, pigeons, doves, woodpeckers, crows, kingfisher, blackbirds, grackles, robin, a few sparrows, locusts, grasshoppers, and larvae of moths and beetles.

Behavior.—The flight of the goshawk is much like that of the Cooper's and sharp-shinned hawks. When traveling it flies high, with steady wing strokes and occasional sailings; but it very seldom indulges in soaring like the Buteos. When hunting it flies low, dashing along swiftly in the open, around the edges of woods or thickets, or through the tangles of branches in the forest; its broad rounded wings and long tail give it perfect control of its movements and enable it to dodge all obstacles and to drop suddenly on its prey in a surprise attack. It is a bold and intrepid hunter. Forbush (1927) writes:

Its attack is swift, furious and deadly. In the death grapple it clings ferociously to its victim, careless of its own safety until the unfortunate crea-ture succumbs to its steely grip. Its stroke is terrible. It is delivered with such force as sometimes to tear out most of one side of its victim, and its wing-power is so great that it can carry off rabbits and full-grown fowls. * * *

Dr. William Wood of East Windsor Hill, Connecticut, told of a Goshawk that followed a hen into a kitchen and seized her on the kitchen floor in the very presence of an old man and his daughter. The father beat off the hawk with a cane, while the daughter closed the door and finally killed the bold bird. Mr. J. A. Farley relates a similar tale from Lambert Lake, Maine. A Goshawk caught a half-grown hen. The hen, escaping, ran under a woman's skirts. The hawk followed right up to the skirt but was killed. They had to kill the hen, too, for its crop was torn open as a result of the hawk's fierce grip.

Many other tales have been told showing its audacity. Dr. Fisher (1893) tells of a farmer who chopped off the head of a fowl and threw it down beside him. "In an instant a goshawk seized the struggling fowl" and flew off with it.

Edwin Beaupre writes to me that on two occasions a goshawk attacked and attempted to carry off wooden duck decoys. Illustrating the fierceness and savagery of this hawk, Nuttall (1832) says that the female of a mated pair in captivity killed her mate. The Rev. C. W. G. Eifrig (1907a) writes of a battle between a goshawk and a barred owl:

One morning last February, Mr. Hugo Paeseler, a farmer, on going out into the woods adjoining his farm, noticed a space of about ten to fifteen feet square, where the snow had recently been much disturbed, deeply plowed up from some great commotion. That a fierce fight had been going on but a short while before was evident from the liberal quantities of blood sprinkled on the snow and the masses of feathers, single and in whole bunches, lying about and adhering to bushes and trees. On looking around for the principals of the fight, he found about ten feet away in one direction a Goshawk, lying on the snow with wings extended and frozen stiff. About ten feet away from the scene of hostilities, in the opposite direction, he found an owl, more damaged than the hawk, but still warm. It had alighted after the fight on a small spruce and fallen off, as the snow showed, and with its last strength crawled into a small log, lying with its hollow part conveniently near.

The fierceness of breeding goshawks, in defence of their eggs or young, has been noted many times. Clarence F. Stone writes to me as follows:

In 1921 I visited the goshawk nest on May 2. A new nest was made in a big yellow birch tree along the bog run to Nicks Lake.

The female was on nest, but when I was about 200 feet distant she swooped from nest and attacked me in a most savage manner. I could have killed her but did not wish to. However, the attack became so ferocious, with lightning rapidity of swoop after swoop, that I was obliged to protect myself with a club.

The nest was up about 45 feet. And there were no limbs for 25 feet. From the time I strapped on climbers it was a question in my mind whether I could reach the nest with a whole scalp.

Luckily the male goshawk did not appear, so I had but one angry and bold fighter to contend with. Before I reached the first limbs I was obliged to stall on my climbers and hug the tree with one arm while I flailed at the bird with a club. A score of times I missed but when she grabbed my cap and flew a hundred feet with it, I realized I must back down or else wing her.

Back she came with speed of an arrow, wings half closed, eyes blazing, and uttering angry "cacs", all of which meant that to save my scalp I must wing her. But she was so alert and quick that it was several minutes before I clipped her fore wing so that she fell to the ground, still full of fight and "cac-ing" loudly.

Albert A. Cross has given me his notes on some still more thrilling experiences that he, Harry E. Woods, and Lawrence J. Sykes had with a nesting goshawk near Huntington, Mass. Mr. Woods visited the nest before the eggs had hatched and "found the female goshawk in a very bad humor and hostile; she making four attacks on him while he was at the nest, coming at full speed and not uttering a sound. Woods was able to protect himself in a degree by pulling his coat over his head and dodging, but eventually the hawk lacerated him quite badly on the upper part of one hand and wrist. The male bird was not seen." Later on the female was shot by Mr. Sykes after she had attacked him on three different days while he was fishing in a nearby brook. The following account of it, sent to me by Mr. Cross, was published in the Springfield Union of June 6, 1931.

The bird caught the fisherman unawares the first time and tried to sink its talons, nearly an inch long, into his face and neck. He finally beat it off after it had circled and swooped at him a number of times.

Not many days afterward, the fisherman went back to the brook again and the bird gave him another battle. The third set-to was the day before yesterday. The bird this time was more persistent than ever and in one of its vicious dives struck the fish pole and broke it in two.

Yesterday the fisherman went to the stream with a gun. The hawk evidently saw him coming and met him some distance from the brook. The Springfield man who is an expert hunter as well as angler brought the bird down with two shots.

Dr. George M. Sutton (1925) who spent a whole day watching a goshawk's nest "was almost constantly attacked and screamed at by the female bird. For eight hours she remained at her post." He continues:

Before my companions left me I crawled into a rudely constructed blind where I crouched motionless, hoping that I would not be detected by the hawks. The female bird drove the departing group of men to the edge of the woods and then returned, calmer for an instant or two, apparently, and then, spying me without the slightest difficulty, redoubled her fury and bore down upon me with savage intent. Intrepid and insistent she swooped at me from all directions and only the branches of the blind kept me from the direct blows of her feet although the protecting boughs cracked and snapped at each onslaught. My being alone doubtless increased her daring and she perched at a distance of only twelve feet and screamed in my face, her bright eyes glaring, and her powerful beak expectantly parted. * * * With the Sept camera in hand I photographed the attacking bird, and while I tried to steel my nerve to accept the blows of her feet without flinching, I found I could not. Every time, when I saw her glowing eyes, partly opened bill, and loosely poised feet descending upon me I ducked and raised my arms in spite of myself. Had I not worn a strong cap and a cloth about my neck no doubt her talons would have brought blood more than once; and it was evident that the claw of the hind toe was most powerful and effective, since that nail dug in and dragged as the bird passed on. * * *

The most memorable thing about the day's experience was the method of attack of the female bird, which has partly explained to me the ease with which some of these birds capture their prey. When the Goshawk left her perch to strike at me her set wings and slim body were for several seconds almost invisible and the only actual movement perceptible was the increase in the size of her body as she swiftly approached. Three times at least I was looking directly at the approaching bird *and did not see her at all* because the lines of her wings and body so completely harmonized with the surroundings, and the front view was comparatively so small.

Mr. Tufts mentions in his notes a few cases where the female left the nest and disappeared silently and where the male seemed more concerned, but he says farther on:

I fear the impression you will gain from the foregoing notes is that the female goshawk is not fierce and bold and that the male is usually more bold in defence of the nest than is his mate. This is not so, and I realize in writing the foregoing notes I was stressing the unusual. The fact is that the female is usually much more in evidence about the nest than her mate and is often much bolder than the male and is much oftener vicious than timid at such times. Extremes are, however, noteworthy. I have had a female attack me while in the nest tree and tear my clothes and even my skin. Other times I have had her strike me a vicious blow with what I believed was a clenched talon. I have never had a male bird strike me.

Voice.—The voice of the goshawk shows its relationship to the Cooper's hawk, as it is similar, but louder and harsher, when attacking an intruder. This alarm note is usually written as *cac, cac, cac,* or *cuk, cuk, cuk,* or *kek, kek, kek.* Mr. Cross interprets it as "*keep, keep, keep,* with the *ee's* shrilly intensified." Dr. Sutton (1925) writes:

From 8:00 A. M. to 1:30 P. M. I heard her give only two call-notes. The most commonly given was the well known "*Ca, ca, ca, ca*" rapidly repeated and with a heavy goose-like quality that was noticeable. The first two syllables of such a series of screams were often hoarse and throaty with a sinister, double-toned character. Sometimes when the bird was passing swiftly through the trees to a perch she called "*Kuk, kuk, kuk*" in deep, somewhat milder tones. This note was decidedly like the warning note of the Blue Goose as heard in the James Bay region.

From 12:45 to 1:30 P. M. I remained almost absolutely quiet and feigned sleep as best I could in spite of the innumerable black flies which became increasingly annoying as the day advanced. But after this period of quiet the voice of the female bird suddenly changed, and her screams were so high, clear and plaintive that I was honestly startled. This new note sounded like "*Kee-a-ah*", and reminded me of the call of the red-shoulder, but was more powerful and at the same time more musical, and had a plaintive character that rather affected my sympathy. I sensed immediately that this call indicated a change of some kind in the bird's attitude. Suddenly the dark, swiftly flying male appeared, bearing in his claws a black squirrel. His scream as he approached the nest was long, high and thin, and not nearly so powerful as that of his mate.

The notes that I heard while the hawks were protesting at my approach to the nest were recorded as *krae, krae, krae,* or *kray, kray, kray,* loud, rapidly uttered, and repeated many times.

Mr. Tufts says in his notes:

The usual cry is a strident staccato *cac, cac, cac,* which has a piercing, menacing tone and is uttered rather deliberately. This is the common alarm note as an intruder approaches a nest, and I have never heard it except in the nesting season. In fact, I have never heard a goshawk make *any* note except at that time. It was not until the past spring that I learned of two new notes. One, a high-pitched, shrill *caw, caw, caw, caw,* uttered slowly and with a plaintive tone to it. I only heard this call twice during the whole time I was in the vicinity of the nest, and both times it was given under the same or similar conditions; viz, the female would come in with food and alight some 75 or 100 yards distant and then begin this call at more or less regular intervals. After perhaps 8 or 10 *caws*, there would be a pause of a few moments; then it would be repeated. It seemed as though she were suspicious that all was not well about her nest and this was a call to the mate (he was on hand and in close proximity on one occasion) or to the young that she was nearby with food. The other call that I heard for the first time was also last spring and during my observations about this nest. It was a peculiar *cluck, cluck, cluck*, uttered with deliberation.

Field marks.—The adult goshawk may be recognized as a large gray hawk, with broad, rounded wings and long, nearly square tail. The breast appears, at a distance, to be plain, pale gray. If near enough, the black cap and white stripe over the eye may be seen. The young bird cannot be distinguished from a young Cooper's hawk, except by size. Its flight is like that of a Cooper's hawk.

Winter.—Winter is the time when we look for the goshawks to swoop down upon us. The great winter invasions are supposed to be due to periodic failures in their food supply in the north woods, but probably there is some southward migration every year. Well-marked, heavy flights have been recorded in 1863, 1889, 1895, 1896, 1896–7, 1898–9, 1905, 1906–7, 1915, 1916, 1918, 1926–7, and 1927–8; of these the greatest invasions were in the fall and winter of 1896–7, 1906–7, 1926–7, 1927–8. Usually the goshawks begin to arrive in October, but in 1926 goshawks were taken in Pennsylvania on September 6, 9, 10, and 18; Dr. Sutton (1927) gives a full account of this flight. The great flight of 1906–7 is described in considerable detail by Ruthven Deane (1907). Some of these flights consisted almost entirely of adult birds. A great invasion of goshawks is quite likely to be followed by a marked scarcity of ruffed grouse; as large numbers of snowy owls are likely to be driven south by the same cause, our smaller wild creatures have a hard struggle for existence. Dr. E. W. Nelson (1887) says that many goshawks remain all winter as far north as northern Alaska, frequenting the patches of alders and preying on ptarmigan.

In 1934 the invasion began in Pennsylvania on October 10, and Maurice Broun (1935) says: "A pronounced movement ensued, with 24 observed on the 11th, 23 on the 12th, 19 on the 13th, 6 on the 14th, and 4 on the 15th. In the same period 330 Sharp-shins and 99 Cooper's Hawks were observed. This migration was attended by rain and snow and rapidly falling temperatures in northern New England."

Attempts to reduce the numbers of goshawks in Pennsylvania by the payment of bounties have not produced very satisfactory results. A law was passed in 1929 offering a bounty of $5 for each goshawk killed between November 1 and May 1. During the season of 1929–30, 9 sharp-shinned, 120 Cooper's hawks, and 76 goshawks were sent in, together with 296 harmless or beneficial hawks. And the following season the score was 180 harmful hawks, against 255 useful species. This seems a big price to pay for the few goshawks destroyed. Most men cannot, or will not, distinguish the good hawks from the bad.

DISTRIBUTION

Range.—North America, casual in the British Isles.

Breeding range.—The two races of the American goshawk (*A. a. atricapillus* and *A. a. striatulus*) breed **north** to Alaska (Bethel, Yukon Delta, Nulato, Bettles, Big Creek, and Fort Yukon); western Mackenzie (rarely Fort Good Hope, Fort Anderson, and the Grandin River); Saskatchewan (Beaver River, Wingard, and Pelican Narrows); Manitoba (Riding Mountain and Portage la Prairie); northern Ontario (Moose Factory); Quebec (Great Whale River and Fort Chimo); and Newfoundland Labrador (Hopedale). **East** to Newfoundland Labrador (Hopedale); Newfoundland (Nicholsville and Fox Island River); Nova Scotia (Pictou, Wolfville, and Halifax); Maine (Calais and Norway); New Hampshire (Alstead); Massachusetts (Petersham); Connecticut (Winchester Center); eastern Pennsylvania (Lopez); and rarely northern Maryland (once, at Jennings in 1901). **South** to rarely Maryland (Jennings); Michigan (Mackinac Island and Isle Royale); probably Colorado (Breckenridge and Lone Cone); possibly New Mexico (mountains west of Taos); probably northern Arizona (San Francisco Mountain); and central California (Lake Tenayo). **West** to California (Lake Tenayo, probably Pyramid Peak, Donner, Eagle Lake, and Mount Shasta); Oregon (Glendale); Washington (Bumping Lake and probably Lake Cushman); British Columbia (Chilliwack, Cariboo District, Dock-da-on Creek, and Hot Springs); Yukon (Lake Marsh); and Alaska (Chitina Glacier, Chulitna River, and Bethel).

The western form (*striatulus*) appears to breed only in the Pacific coast district from about Cook Inlet, Alaska, south to the Sierra Nevada of California, wintering east to Colorado. The eastern form (*atricapillus*) is more wide ranging, particularly in winter, when specimens have been collected west to California.

Winter range.—The normal winter range includes most of the breeding range, but in some seasons (depending largely upon the status of the food supply in the north) goshawks will work south to Florida (Lake Iamonia and St. Petersburg); Texas (Center Point); central Arizona (Fort Verde); and southern California (Lower Otay Reservoir, near the Lower California line).

Migration.—A study of the data available indicates that the migration of the goshawk can not be satisfactorily illustrated by dates of arrival and departure. Only casually does it travel south of the southern limits of the breeding range. Nevertheless, some years are characterized by heavy flights or invasions from more northern latitudes. During some autumn and winter seasons the species becomes unusually common, chiefly in the wooded section from Minnesota east to and including New England, and south to Pennsylvania. Notable flights in this part of the country occurred in 1895, 1896, 1897, 1906, 1907, and 1916. The birds arrived generally in October or November and remained common until the middle of March or the first of April.

The invasion of 1916 was notable in that large numbers of true *atricapillus* invaded California during November (Grinnell, 1917). Goshawks also were common that year in Kansas (Bunker, 1917), when several specimens were received by the University Museum at Lawrence.

Casual records.—A goshawk was recorded by Hantzsch (1929) as taken during the spring of 1901 at Ramah, northern Newfoundland Labrador. Reid (1884) records two specimens from Bermuda, one of which was shot on Somerset Island about 1862.

European records for this species appear to be limited to a few from Scotland and Ireland, as follows: A specimen was obtained at Brechin, Forfarshire, Scotland, in May 1869; one was taken at Strabane, County Tyrone, Ireland, on February 24, 1919; another was killed at Schiehallion, Perthshire, Ireland, during the spring of 1869; another was obtained in the Galty Mountains, Tipperary, Ireland, in February, 1870; and one was taken at Parsonstown, Kings County, Ireland, during the spring of 1869.

Egg dates.—Alaska, Canada, and Labrador: 100 records, April 5 to June 14; 50 records, April 23 to May 18.

United States: 20 records, April 1 to June 3; 10 records, April 17 to 30.

ASTUR ATRICAPILLUS STRIATULUS Ridgway

WESTERN GOSHAWK

HABITS

The western subspecies of the goshawk is none too well marked, and its breeding range is none too clearly defined. It is a darker-colored bird in all plumages. In the adult the upper parts are dark plumbeous inclining to sooty blackish; the under parts are more bluish ash, finely vermiculated with white, and with conspicuous black shaft streaks. In the young bird the upper parts are darker brown, with less of the buffy edgings; under parts are white, less buffy, with broader and blacker stripes.

Its breeding range is supposed to be in the boreal zones of the Pacific coast region from Alaska southward, possibly into Mexico.

But birds that are typical of the eastern race have been taken frequently in this region. It may be only a dark color phase. Its habits are evidently similar to those of the eastern bird. Major Bendire (1892) writes:

According to my observations, the general habits of the Western Goshawk are very similar to those of its eastern relative; it is equally destructive to small game of all kinds, particularly the Sooty, Ruffed, and Sharp-tailed Grouse, as well as to the fowls of the poultry yard. While nowhere abundant, it seems to be pretty generally distributed throughout the Blue Mountain region of Oregon and Washington, and breeds in suitable localities where food is plenty. During spring and summer it is seldom seen in the more open districts, though it is abundant enough later on, when the heavy snows drive the game into the foothills and lower valleys. I have shot quite a number of these birds at various times, and all, as far as I am aware, are referable to this subspecies, with one exception, which is intermediate between it and the preceding.

Nesting.—He records the finding of several nests, as follows:

My first acquaintance with the Goshawk dates back to 1870, and on April 21, 1871, while hunting in Lawyer's Cañon, 30 miles south of Fort Lapwai, Idaho, I found a nest of this subspecies containing a single egg. It was placed in the forks of a large cottonwood tree about 50 feet from the ground, and was a bulky affair, fully 2 feet in diameter and quite as deep. The nest was composed of sticks, some of them quite large and loosely put together. It was rather shallow on top and lined with weed stalks, a species of wild nettle, and a few pine needles. * * *

A nest found on April 18, 1876, was placed in the top of a tall and bushy juniper tree, only about 20 feet from the ground. It was not as large as the two former, and looked as if it had been newly built. It was situated in the fork of the main trunk and was well hidden. The female was on the nest and commenced screaming before we came within 20 feet of the tree, which caused the discovery. She defended her eggs valiantly, and did not cease her attacks on the climber till he finally succeeded in hitting her with a club, which caused her to leave. The male was not seen. The nest contained three slightly incubated eggs, and was sparingly lined with the dry inner bark of the juniper trees growing in the vicinity. On April 9, 1877, I found another nest not far from where the first was taken in 1875. This was built in a tall pine, at least 50 feet from the ground, and in addition to the usual juniper bark lining it contained a few green fir tops. This also contained three eggs, and incubation had already commenced. I shot the female, a handsome bird in the adult plumage, while it was circling about the climber and trying to strike him. The largest set obtained was one of five eggs. The nest was placed in a bushy pine in a cañon of the Blue Mountains, close to the road from the Umatilla Indian Agency to Grande Ronde Valley, Oregon. This nest, evidently used for years, was well out on one of the larger limbs and placed in a fork of it. It was quite large, and slightly lined with grass, tree moss "usnea," and a few scales of pine bark; distance from the ground about 50 feet. Both parents were present, and the female was shot, as she was too aggressive for the comfort of the climber. The male was also rather demonstrative, but not to the extent of his mate. The eggs were nearly hatched when found, April 17, 1881. All the cavities of the nests were very shallow, none being over 1½ inches deep. While none of the nesting sites were in the denser portions of the forests, they were all found in the heavy timber, and generally on the slopes of cañons not far from water.

Milton S. Ray (1926) took the first California set of eggs in the high Sierras of Eldorado County. The first nest he found contained three small downy young on June 19, 1922. It was "in a thickly timbered and very swampy section of the wood, 25 feet up in a rather small and slender tamarack." This nest was occupied again in 1923. The nest from which he took his three eggs on April 30, 1924, "was placed against the trunk of a lodgepole pine 35 feet up. The tree was about 65 feet in height and most of its lower branches were dead." He continues:

This nest, like the one found previously, was located in the heart of a swampy forest of pines [pl. 43]. Surrounding the nest tree were blackish, inhospitable pools of snow water, deep beds of pine needles, and thickets of fallen and standing dead timber. Ever there, below, was the dark, cool shade of the lofty pines and ever, above, the ceaseless roar of the wind in their swaying tops.

The nest, a gray, weatherworn, ragged-looking structure, was oval in shape and measured, in inches, 20 by 33; the long side was placed against the trunk of the tree. By measuring, the odd projecting twigs gave it a size of 34 by 60. The cavity proper was 9 by 2, and a very rough, uneven affair it was, with its lining of green tamarack sprays, strips of tamarack bark, and a few scattered goshawk feathers. The nest distinctly tapered towards its almost flat top and was 29 inches in height (or 39 inches, counting certain projecting twigs). The composition was entirely of small, smooth, dead tamarack twigs and branches. Most of these were one-quarter inch in diameter, some were three-quarters and some one-half, while a few were only one-eighth of an inch in diameter. I found, however, that the coarse-looking nest was very compactly built, and being supported by four branches of the tree and sheltered against the trunk, was well fitted to withstand the snow and gales that sweep through these altitudes in April and May.

Eggs.—The eggs are indistinguishable from those of the eastern goshawk. The measurements of 27 eggs average 59.3 by 45.6 millimeters; the eggs showing the four extremes measure 65.5 by 49.5, 62.9 by 50.1, 55 by 44, and 58.6 by 43.2 millimeters.

Behavior.—Major Bendire (1892) relates the following incident:

At the outskirts of the garrison, near the cavalry stable, was an old brush corral, much frequented by the fowls kept in the neighborhood. While walking past this fence I suddenly heard a great outcry and saw quite a commotion among a number of chickens in the place, which were squeaking and scattering in all directions at a lively rate. At the same instant a large Goshawk, an adult female, dashed through the inclosure, failing to get a chicken this time, however. I fired at her at short range, and, as it subsequently proved, peppered her well with dust shot as she went by, which possibly disconcerted her aim a little. Never dreaming for an instant that the bird would return after such a reception, I nevertheless inserted a heavier cartridge in my gun, and had scarcely done so when she came back to make a second and last attempt at a too venturesome chicken. This time I brought her down with a broken wing, and her flight was so suddenly arrested that she rolled over several times after striking on the ground. I never saw more vindictive fury expressed in a bird's eye than was shown by hers. She tried to attack me, and would have done so had she not been so badly wounded. The will and courage to do so were there, but her strength failed her. On skinning her I found a number of dust shot imbedded under the skin, showing that she had been hit the first time I fired. This, though, was not sufficient to cause her to leave without her intended victim, notwithstanding the fact that she saw me plainly enough the second, if not the first time. When its appetite for blood is once excited, the Goshawk is certainly devoid of all fear and discretion as well, while under ordinary circumstances there is no shyer bird to circumvent and bring to bag.

PARABUTEO UNICINCTUS HARRISI (Audubon)

HARRIS'S HAWK

HABITS

Only along our southwestern borders may we expect to find this conspicuously marked hawk. In southern Texas it is a common bird, in Arizona less so, frequenting the prairie regions, the chaparral, and the mesquite lands. Its range extends southward through Mexico. It is replaced in South America by a closely allied race. Audubon's type specimen was taken in Louisiana, but it is a rare bird there. Mrs. Florence M. Bailey (1903) writes of its haunts:

Fifteen miles west of Corpus Christi, Petranilla Creek throws a belt of rich vegetation across the prairie. Its walls are crowned with elms and live oaks whose serried branches are hung with waving gray moss, while encircling a floor massed with pink primroses grow a mixture of Mexican and United States trees and bushes—hackberry, ash, palmetto, all-thorns and cactus. Birds and mammals naturally flock here and also show southern admixtures, the clay banks of the creek being tracked up by coon, coyote, wild cat, and armadillo, while in April and May the trees are alive with such birds as the cuckoo, chat, wren, wood pewee, kingbird, cardinal, and a variety of warblers including the blackburnian, together with the golden-fronted woodpecker and nonpariel.

Nesting.—Mrs. Bailey (1928) sums up the nesting habits very well as follows: "A compactly made platform of sticks, twigs, weeds, and roots lined with green mesquite, elm shoots, and leaves, grass, bark, Spanish moss, and roots, placed in cactus, Spanish bayonet, chaparral, mesquite, hackberry, and other trees."

The only nest I ever saw was shown to us by a Mexican, near Brownsville, Tex., on May 24, 1923. It was only about 10 feet from the ground in a large, branching pricklypear cactus, in an extensive tract of dense chaparral. It had contained young but was then empty. The nest had been partially pulled down and the young killed, perhaps by a coyote or wild cat, both of which were common there.

George Finlay Simmons (1925) says that in Texas the nests range in height from 10 to 30 feet above ground, "in top limbs of chaparral bush or low tree (mesquite, elm, hackberry or blackjack) on prairie."

George B. Sennett (1879) mentions, among several nests, "one 25 feet high in an ebony tree, the other 20 feet high in a mesquite." Dr. James C. Merrill (1879) found a nest, near Brownsville, placed on the top of a Spanish bayonet some 8 or 9 feet above the ground.

Major Bendire (1892) writes of some Arizona nests:

Personally I met with the nest of this bird on but three occasions during the spring of 1872, while stationed near Tucson, Arizona. One of these nests, containing two fresh eggs, was found on May 17. It was a bulky structure, placed in a low bushy cottonwood tree, in a fork about 20 feet up, about 10 miles below Tucson, near the Laguna, the sink of the Santa Cruz River. It was composed of sticks, and sparingly lined with pieces of the dry inner bark of the cottonwood, and grasses. The bird made no hostile demonstrations, but sailed slowly around above the nest out of gunshot range. The inner cavity of the nest was slight.

The two other nests, each containing but two eggs, were found in low mesquite trees, about 15 feet above the ground, on June 4 and June 6, respectively. The first nest was a very slight affair, composed of mesquite sticks, as well as the dry seed pods of this tree, and a little grass. While standing directly under the nest I could see the eggs through the bottom of it. The third one was similarly situated, and both were found on the barren plains west of the camp.

Eggs.—Harris's hawk lays three to five eggs, usually three or four. The eggs are oval or ovate in shape, and the shell is smooth but without gloss. They are dull white or very pale bluish white and are usually unmarked. On close examination some of the eggs (Major Bendire says about one-half of them) are very sparingly and faintly spotted with small spots or dots of pale brown, buff, or lavender. The measurements of 52 eggs average 53.7 by 42.1 millimeters; the eggs showing the four extremes measure 57.5 by 44.5, 57 by 45.5, and 49 by 38.5 millimeters.

Young.—Incubation is said to last for about four weeks and to be shared by both sexes. Bendire (1892) says that "the eggs are deposited at intervals of several days, but incubation commences as soon as the first egg is laid." Both sexes assist in the care of the young.

Plumages.—The downy young is thickly covered with soft down, which at first is "pinkish buff" to "pinkish cinnamon" above and buffy white to grayish white below. In some the color of the upper parts is darker, approaching the bright "ferruginous" of the adult shoulders; but it fades out to buffy white before the plumage is assumed.

In full juvenal plumage the upper parts are "warm sepia" or "bister", with broad edgings of "warm buff" over the eyes and on the hind neck and with concealed bars of "ochraceous-tawny" on the scapulars; the wing coverts are conspicuously barred with "ochraceous-tawny"; the remiges are dark sepia above, with narrow darker bars, and whitish below, with narrow dusky bars; the rectrices are sepia above, with numerous darker bars, but lighter and tinged with rufous on the inner webs; the rectrices are grayish white below, with narrow dusky bars, and are broadly tipped with buffy white; the upper and under tail coverts are white, tinged with buff; the under parts are white, strongly suffused with "warm buff" on the breast, with large, tear-shaped spots of "warm sepia"; the thighs are usually pale buff, with a few very narrow dusky bars; in some birds the thighs are "Verona brown", spotted or barred with white and rufous.

This plumage is worn during the first summer and fall; I have seen it in birds taken from May to December. But usually during the fall, winter, and spring a change toward adult plumage takes place by means of a prolonged, gradual molt. The buffy edgings wear away, and much of the adult body plumage is acquired before summer, leaving the under parts dark and only narrowly streaked with white. I have seen young birds in full molt of body, wings, and tail in May and June. Apparently after this first complete molt, which has required so many months, the young bird has acquired a plumage that is practically adult, but I am not sure that it does not require another year to attain full perfection. I have seen adults molting wings and tail in December.

Food.—Mr. Sennett (1879) says: "I found in the crops of those I obtained mice, lizards, birds, and often the Mexican striped gopher (*Spermophilus mexicanus*), proving them active hunters, instead of the sluggish birds they appeared the year before at Brownsville."

Some observers say that it consorts with vultures and caracaras and lives largely on carrion; other good observers say that they have never seen it do so. It must be a good hunter, and it probably prefers fresh game, for various mammals, birds, and reptiles have been detected in its food, such as cottontail rabbits, wood rats, Florida gallinule, sora rail, night herons, green-winged teal, gilded flicker, and small snakes. Vernon Bailey (1902) found a nest containing young "to be fairly covered with bones of wood rats. There were a dozen skulls, and bones, legs, skin, and fur were strewn over the nest."

Dr. Loye Miller (1930), who found a nest nearly 40 feet up in a cottonwood in southeastern California, writes:

On the edge of the nest were two fresh bird bodies, one a Florida gallinule (*Gallinula galeata*) and the other a Sora (*Porzana carolina*). The Gallinule had been quite well plucked, most of the contour feathers and all the strong flight feathers having been removed. The hawk's talons had pierced the pelvis back of the acetabulum and the rib basket in the region of the posterior dorsals. The throat had been torn out, but otherwise nothing had been consumed. The Sora was beheaded, but otherwise was quite intact.

On a previous occasion some brief mention was made of the stomach contents of this species, the evidence being that it is an aggressively raptorial bird. Green-winged Teal and Gilded Flicker were identified in the stomachs. Add to these species the Florida Gallinule and the Sora, both secretive birds of rather dense cover, and the impression grows that the Harris Hawk is no mean hunter. Furthermore, he plucks his kill almost as completely as does a falcon, even wrenching out the strongly attached primaries.

Behavior.—Harris's hawk has been referred to as a sluggish, heavy bird, slow of flight and not graceful, but there is much evidence to the contrary. No very slow or sluggish hawk could catch the lively creatures recorded in its food. I have seen it chasing a western redtail and it has been seen to attack and drive away the big ferruginous roughleg. Vernon Bailey (1902) writes:

> In southern Texas the rich rufous marks and swift, clear-cut flight of the Harris hawk soon become pleasantly familiar, for he is one of the hawks that are both common and tame on the coast prairies. He is so tame that as you drive by a telegraph pole on which he is perching he will sometimes stand calmly on one foot looking down upon you with a statue-like indifference. In the mesquite thickets you may meet one at close quarters as he dashes under the thorny bushes in quest of wood rats, ground squirrels, and the small game that abounds in these dwarf forests; and sometimes, as happened one day when we drove along the Nueces River, you will see him sitting on a low branch feasting on a wood rat captured at the door of its stick house close by. If you chance near the hawks' nest a long harsh Buteo-like scream may make you look up to find one or both anxious birds circling overhead.

The following is from some notes sent by Maj. Allan Brooks to Dr. John B. May:

> Harris's hawk is a dual personality, a sort of Jekyll and Hyde character. A casual acquaintance with this species will probably show one, or more probably a pair, of these hawks sitting in the top of a tree that rises above the general scrub, sitting quietly like Buteos apparently taking little interest in their surroundings as they soak up the morning sun. Presently they will take flight, mounting into the air in easy spirals, higher and higher into the blue, and that will probably be the last you will see of them. But to see this hawk in action one has to be afield early while the mists still hang over the *resacas.* Then Mr. Hyde appears, a flutter of wings as a flock of teal rise in confusion with a dark shape striking right and left among them with all the dash of a goshawk. If unsuccessful, the next attack may be on a group of small herons, one of which may be singled out and followed until killed. Very often a pair of these hawks combine to secure their quarry, and I have seen a snowy heron shared amicably after it had fallen a victim to one of these raptores. In action and flight it combines many of the characteristics of the Buteos, marsh hawk, and goshawk.

Field marks.—Harris's hawk may be easily recognized, in adult plumage, by the dark-brown uniform color of body and head, appearing almost black, and by the bright rufous wing coverts and thighs; the white rump and long white-tipped tail are characteristic in any plumage. Its shape and flight are not unlike those of the marsh hawk.

Fall.—These hawks apparently gather into large flocks in fall and wander about. W. Lee Chambers (1924) reports two large flights, as observed by Frank Richmond, near Calexico, Calif. On October 23 between 400 and 500 Harris's hawks "were scattered over an area of about 80 acres in a field along a highway." As early as August 28, 1923, he saw about 250 of them; "some were perched in small trees

and some in the cottonwood trees directly over his head, and others
were hopping around on the bare ground."

Range.—Southwestern United States, and Central and South
America; not regularly migratory.

The range of Harris's hawk extends **north** to southwestern Cali-
fornia (San Diego, Brawley, and Palo Verde); Arizona (Yuma,
Pecks Lake, Tombstone, and the San Bernardino Ranch); southern
New Mexico (Carlsbad); and southern Texas (Eagle Pass, San An-
tonio, and Giddings). **East** to Texas (Giddings, Lavaca County,
Refugio County, Corpus Christi, and Brownsville); Tamaulipas
(Matamoros); Puebla (Tehuacan); Venezuela (Zulia, Margarita
Island, Cariaco Peninsula, and Angostura); Brazil (Villa Bella,
Chapada, Bahia, Cantagallo, and Rio de Janeiro); and eastern Ar-
gentina (Corrientes, Buenos Aires, Canuelas, La Plata, Cape San
Antonio, Santa Elena, and Santa Cruz). **South** to southern Argen-
tina (Santa Cruz and Chubut); and southern Chile (Chiloe). **West**
to Chile (Chiloe, Valdivia, Santiago, and Valparaiso); northwestern
Argentina (Concepcion, Tucuman, Oran, and Salta); Bolivia (Caiza,
Crevaux, and Santa Cruz); Peru (Ica, Lima, Trujillo, and Eten);
Ecuador (Puna Island, Guayaquil, Babahoyo, Chone, and Esmeral-
das); Colombia (Bogota); Panama (Santa Fe); Costa Rica (San
Jose); Oaxaca (Tehuantepec and Oaxaca); Colima; Jalisco; Na-
yarit; Sinaloa (Escuinapa, Rosario, and Mazatlan); Lower Cali-
fornia (Cape San Lucas, Miraflores, San Gregorio, San Juan, San
Quentin, and La Ensenada); and southwestern California (San
Diego).

The range above outlined is for the species, which has, however,
been separated into two races. The form occurring in the southwestern United States (*harrisi*) extends south through Central America to
western Ecuador, while the typical race (*unicinctus*) occupies the
South American range east of the Andes and probably also central
and northern Chile.

Migration.—Harris's hawk is not migratory in the usual sense,
but it is probable that, depending more or less upon climatic conditions, the individuals breeding in the United States withdraw into
Mexico during the winter season. On at least two occasions large
flights of these birds have been observed in southwestern California.
One of these flights, estimated to contain between 400 and 500 birds,
was observed on October 22, 1920, in the region between Heber and
Calexico, Calif. The second flight, estimated at 250 individuals,
was seen in the same region on August 28, 1923 (Chambers, 1921 and
1924).

Casual records.—The type specimen of *harrisi* was taken by Audubon between Bayou Sara, La., and Natchez, Miss. The species has been subsequently reported as occurring along the coast of Louisiana and on some of the larger islands, but additional specimens do not seem to have been taken.

A specimen was caught in a steel trap near Hillsboro, Van Buren County, Iowa, about 1895, and on December 24, 1917, one of a pair was shot by a farmer 4 miles southwest of Harrisburg, Ohio.

Egg dates.—California, Mexico, and Texas: 102 records, February 5 to June 21; 51 records, March 22 to May 7.

BUTEO BOREALIS BOREALIS (Gmelin)

EASTERN RED-TAILED HAWK

HABITS

The red-tailed hawk, with its various races, is the most widely distributed, most universally common, and best known of all our hawks, though in certain sections some other species may be much commoner. For example, in my home territory the red-shouldered hawk outnumbers it nearly ten to one; and on the prairies and plains of the Middle West Swainson's and ferruginous roughlegs are, or were, commoner than redtails. But this fine hawk, the largest and most powerful of our eastern Buteos, is no longer common over much of its former range. The widespread prejudice against all hawks is exterminating this useful species much faster than some of the most destructive hawks that are better able to take care of themselves, craftier, and swifter awing. It will be a sad day indeed when we shall no longer see the great redtail sailing over the tree-tops on its broad expanse of wing and ruddy tail, or soaring upward in majestic circles until lost to sight in the ethereal blue, or a mere speck against the clouds.

The distribution of this and the red-shouldered hawk in southeastern Massachusetts has always interested me. During my 50 years of experience with them, I have learned to regard them as competitive species, each intolerant of the other, antagonistic and occupying entirely separate ranges. In the western half of Bristol County, where the prevailing forest growth consists of hardwood trees, chestnut (formerly), oaks, and maples, with only scattering growths of white pine (*Pinus strobus*), the red-tailed hawk was until recently practically unknown; this region has always been the center of abundance of the red-shouldered hawk. On the other hand, in the Cape Cod region, comprising the southeastern part of Plymouth County and all of Barnstable County, where the prevailing forest growth is pitch pine (*Pinus rigida*) and stunted oaks, the redtail is the common species and the redshoulder so rare that

I have seen only one there in all my experience. In the intervening territory, where the prevailing forest growth is white pine, both species occur, but only in widely separated localities. In this latter region, during recent years, persecution under the bounty system has nearly exterminated all hawks. Meantime redtails began to invade the hardwood region in western Bristol County, supplanting the redshoulders in some of their long-established haunts. In 1929, 1930, and 1931, red-tailed hawks moved into three different tracts of hardwood timber that had been occupied by red-shouldered hawks for upward of 40 years, driving out the former tenants and in two cases appropriating their old nests. The larger and stronger bird seems to be the dominating species.

William Brewster (1925) noted the reverse of this replacement, for he writes: "That the Red-shouldered Hawk should have remained almost unknown in the Umbagog Region until after the Red-tailed Hawk had practically ceased to reappear, and that not long thereafter it should have apparently established itself as a summer resident in at least two localities, are matters of considerable interest, in view of the fact that throughout much, if not most, of Massachusetts there has been essentially similar and contemporaneous replacement of the greater by the lesser bird."

Spring.—Throughout the northernmost part of its range the red-tailed hawk is mainly migratory, a large majority of the birds wintering somewhat farther south. But a few individuals remain during winter, especially during mild seasons, not far from the northern limits of their summer range. I have seen them in Massachusetts during every winter month. Those that remain during winter or those that return early in the season begin their nest building late in February or early in March; I have seen a wholly new nest half completed and decorated with green pine twigs and down as early as February 18, over a month before the eggs are laid.

Courtship.—I believe that this and other large hawks remain mated for life, but, if one of the pair is killed, the survivor soon secures a new mate. The birds are apparently in pairs when they arrive on their breeding grounds, but they indulge in nuptial demonstrations more or less all through the nesting season. I have seen a pair of these hawks, in May when there were young in the nest, indulging in their joint flight maneuvers high above the woods where the nest was located; they soared in great circles, crossing and recrossing each other's paths, sometimes almost touching, and mounting higher and higher until almost out of sight; finally one partially closed its wings and made a thrilling dive from a dizzy height, checking its speed just before it reached the woods. E. L. Sumner, Jr., refers in his notes to such a flight: "About ten times, while they were circling near to-gether, the male would lower his legs and adjust his circles so that he came above his mate, and about four times he actually touched her back, or so it seemed." M. P. Skinner says in his notes: "These hawks at times performed wonderful evolutions high in the air, either one bird alone or several at a time. Such hawks would mount up to a high altitude, then half close the wings and drop down on an invisible incline at great speed only to open the wings again and shoot up at an equal angle. This was repeated again and again while the hawk described a series of deep V's and gradually passed out of sight in the distance."

Mr. Sumner (MS.) saw a male western redtail approach a female that was perched in a tree, hang for a moment just over her, then alight on her back and stay there about 40 seconds, with quite a bit of wing motion to balance himself; he then got off and perched beside her on the branch, but he soon flapped off and began to circle.

Clarence F. Stone writes to me about the mating antics of a pair of red-tailed hawks on a lofty horizontal limb of an elm tree near their nest:

Stopping to survey the woods before I entered, I beheld a pair of Red-tailed Hawks cavorting step by step, towards each other. Since they had not discovered my presence the performance continued to a finish. Stepping sideways until they were wing to wing and facing each other almost breast to breast, both birds suddenly dropped down backwards until there was physical contact below the limb, and thus the act of copulation took place. Immediately after, both hawks took to the air around and around each other in wide circles.

Another recorded note concerning the Red-tailed Hawk tells of a pair proceeding to reline their many years old nest, but before time for eggs one of the birds was killed by a farmer. All the remainder of that season, the bereaved hawk hunted and lived in the nest woods. On the following Spring this Red-tail returned alone and even did quite a bit of relining of nest—so much that I climbed up to see if there were eggs. As this nest was near home I visited it frequently during the season up to June, but always the Red-tail remained unmated. I think this instance shows "faithfulness" more than lack of opportunity to mate again.

Nesting.—My personal experience with the nesting habits of the red-tailed hawk in southeastern Massachusetts has been limited to the study of 19 nests over a period of 40 years, from which it appears that it is not a common bird here. Twice we found two nests in one season and one year we found three. The local distribution has been referred to above. Contrary to the experience of others elsewhere, we have found the redtail much less constant in its attachment to its nesting haunts than the redshoulder. In three cases we found them in the same patch of woods, but in different nests, for two years in succession, and once for three years. A popular nest at Blue Ridge, 35 feet up in a red oak in mixed woods, on a ridge between an open bog and a maple swamp, was occupied by a red-shouldered hawk in 1920; in 1928 it was occupied by a pair of broad-winged hawks; the following year a pair of redtails took possession of it and raised a brood of young; in 1930 it remained unused; in 1931 the redtails were back in it again and raised another brood; but in 1932 it was deserted again; raising a brood successfully did not encourage the hawks to return.

Our longest record covers a period of 13 years, during which time the nest was actually found in only four years. The territory covers a very extensive area in Mansfield and Norton in which there are a number of large patches of heavy timber of various kinds, white pines, oaks, and maples, interspersed with open bogs, swampy woods, cleared lands, and pasture. The redtail's nest was first discovered by my field companions, F. H. Carpenter and C. S. Day, in 1920; it was in an ideal situation, 54 feet from the ground on horizontal branches, against the trunk of a giant white pine that stood on the edge of a grove of heavy pines, overlooking an open meadow. We did not find the nest again until 1924, when we discovered it fully a quarter of a mile away; it was 52 feet up in one of a small group of scattered white pines in an open situation. Two years later the hawks were back in the old original nest in the big pine. This nest remained vacant until 1932, when it was again occupied. I have no doubt that the hawks nested somewhere in that big tract during all the intervening years, for we often saw them, but were unable to locate the nest in a region so difficult to hunt thoroughly. Mr. Day, who has all the eggs collected from this locality, is convinced that three different females presided over this territory, as shown by the three distinct types of eggs laid.

As mentioned above, red-tailed hawks invaded, in three successive years, three separate localities that had been occupied previously by red-shouldered hawks. I suspect that these three invasions were all made by the same pair of redtails, as the second and third localities are less than a mile and a half from the first. The "reservoir woods" in Rehoboth was once a fine, large tract of heavy chestnut, oak, and maple timber, partially swampy and drained by a small stream. A pair of red-shouldered hawks had nested continuously in these woods from 1882 to 1923, when the last nest we found there was built in a large scarlet oak 48 feet from the ground. In 1924 this nest was occupied by a pair of barred owls and in 1928 by a pair of red-tailed hawks (pl. 44) ; I did not visit the locality during the intervening years. The following year, 1929, we found the redtails nesting in the Blue Ridge nest referred to above. In 1930, they, or another pair, invaded another big tract of hardwood timber, Goff's woods, less than a mile away, where red-shouldered hawks had nested for nearly 50 years, and built a new nest 45 feet up in a red oak. And the next year they were back again in the Blue Ridge nest. Since then we have been unable to find any hawk's nests in any of the three localities, though much of the old woods is still standing.

All the nests found in the hardwood region were in oaks, varying in elevation from 35 to 48 feet. Those in the white-pine region were all in white pines and 35 to 70 feet above the ground. On Marthas Vineyard we found the lowest nests in the oak groves on the western part of the island; one huge nest was only 15 feet from the ground and another 30 feet. In the Cape Cod region the redtails nest in the largest pitch pines they can find, from 18 to 35 feet up, and occasionally in white pines where these trees can be found.

The nests of the red-tailed hawks will average somewhat larger than those of the red-shouldered; typical nests are from 28 to 30 inches in outside diameter, the inner cavity being 14 or 15 inches wide and 4 or 5 inches deep. The largest nest I ever measured was 42 inches in longest by 19 inches in shortest diameter. The nests are usually quite flat and shallow; but one that had been added to for an unknown number of years measured 3 feet in height. Dr. H. C. Oberholser (1896) gives the measurements of 7 Ohio nests that are somewhat larger than my averages; his largest nest measured 36 inches in height and 48 by 30 inches in outside diameter; the inner cavity was 7 inches deep.

The nests are well made of sticks and twigs, half an inch or less in thickness, and neatly lined with strips of inner bark, of cedar, grape-vine or chestnut, usnea, and usually at least a few green sprigs of pine, cedar, or hemlock. Some nests are profusely and beautifully lined with fresh green sprigs of white pine, which are frequently renewed during incubation and during the earlier stages in the growth of the young.

I have spent considerable time, with rather meagre results, attempting to watch the nest-building activities of these hawks. They "stake out their claim" late in February or early in March, a month before the eggs are deposited, by marking the nest they propose to use with a sprig of green pine. Nest building is a very deliberate process; the birds visit the nest at very infrequent intervals and are very cautious about it. If they suspect that the nest is watched they will not come near it. In order to watch them successfully it is necessary to have a blind that offers perfect concealment; a brush blind is utterly useless, as the hawks can see the slightest movement in it, and will not come near the nest again until the intruder departs. I believe that both sexes assist in nest building, though I have not proved it. Old nests are sometimes repaired in the autumn.

The nesting habits of the red-tailed hawk in other parts of its range differ somewhat from the above. Major Bendire (1892)

quotes Dr. William L. Ralph as to its nesting in Oneida and Her-
kimer Counties, New York, as follows:

In this vicinity the Red-tailed Hawk prefers birch trees above all others to
build in, and about 80 per cent. of their nests will be found in such situations.
The remaining 20 per cent. is about equally divided among beech, maple,
hemlock, elm, and basswood trees. Why these birds should prefer birch trees
I do not know, for they are usually not very hard to climb, while the most
difficult of their nests to reach were built in elm, hemlock, and basswood trees.
They generally select the largest and tallest trees they can find to build in, and
their nests are situated near the tops, in crotches formed by two or more large
limbs, or at the junction of large limbs with the trunks. They are usually
placed from 60 to 70 feet from the ground.

William A. and George M. Smith, of Lyndonville, N. Y., have sent
me data on 46 New York sets, showing very different preferences; 23
of their nests were in beeches, 9 in maples, 5 in oaks, 4 in elms, 3 in
basswoods, and 1 each in ash and hemlock. The heights from the
ground varied from 34½ to 78 feet, measured; and 24 were 60 feet
or over. There were 16 sets of three, but no larger sets. S. F.
Rathbun tells me that he has taken a set of four in central New
York, and about half of Dr. Ralph's sets were fours.

The largest nest I have heard of was found by Verdi Burtch
(1911) near Branchport, N. Y.; it was placed in a big pine tree and
measured 3 by 4 feet in diameter. He says: "My first set from these
woods was taken March 31, 1890 (20 years ago) and there has been
a nest in there or the adjacent woods nearly every year since that
time." A. D. DuBois mentions, in his notes, a nest found near
Ithaca, N. Y., that was 80 or 90 feet from the ground in a big pine
tree. He also sent me notes on three nests found in Sangamon
County, Ill. One was 50 feet from the ground "in the uppermost
main crotch of an elm tree"; another was at the same height in a
white oak; and the third was in the top of a big sycamore.

Throughout the greater part of its range the red-tailed hawk
seems to be more constant in its attachment to its nesting site than
we have found it in New England; it often returns year after year
to the same patch of woods. As it usually selects the tallest tree
it can find the nest is often at a great height, even over 90 feet from
the ground. It does not seem to be at all particular as to the choice
of a tree, except as to size; various pines, oaks, maples, hickories,
elms, sycamores, and poplars have been used. Small patches of
heavy tall timber are preferred, and the nest is usually on or near the
edge so that the bird can have a good outlook, and nests are often
built in more or less isolated trees in open situations. I believe that
the birds prefer to build a new nest each year, but they sometimes use
the same nest for consecutive years, though oftener they return to it
after an interval of a year or two. Lewis O. Shelley writes to me

that he has known a pair to use the same nest each season for four or five years. Often they appropriate a nest previously used by another hawk, owl, or crow or build on an old squirrel's nest. A. W. Brockway tells me that one of his nests was built on top of a gray squirrel's nest in which he could hear the young squirrels chatter as he pressed against the nest. For three seasons in succession J. A. Singley (1886) found a nest occupied by great horned owls early in the season and later by red-tailed hawks; this was in Texas where the owls nest early in the winter. If their first set of eggs is taken, the hawks will lay a second set, three or four weeks later, but usually in another nest; very rarely a third set may be laid; and Bendire (1892) says "on very rare occasions even a fourth."

Eggs.—In the eastern and southern portion of its range the red-tailed hawk lays almost invariably two eggs; I have never found three and twice have found incubated sets of one. In central and western sections sets of three are commoner, sets of four are not rare, and as many as five eggs have been found in a nest. The eggs are ovate, elliptical-ovate, or oval in shape, and the shell is finely granulated or smooth, without gloss. The ground color is usually dull or dirty white, sometimes faintly bluish white, or more rarely pale greenish white. The eggs average much less heavily marked than red-shouldered hawks' eggs. They are often nearly or quite immaculate, but they are usually more or less sparingly spotted; some are handsomely marked in even or irregular patterns, but very rarely heavily blotched. The markings are in various shades of dull reddish or yellowish browns, "snuff brown" to "ochraceous-tawny", more rarely "warm sepia", "auburn", or "russet"; some show underlying spots of "pale Quaker drab", or "pallid purple drab." A series of eggs from one female usually runs true to type, as to shape, color, and markings; and when a new female replaces her, a different type of eggs often results. The measurements of 59 eggs average 59 by 47 millimeters; the eggs showing the four extremes measure **66** by **50**, 64.5 by **51, 55** by 45.5, and 59.5 by **44** millimeters.

Young.—Incubation lasts for about 28 days; the male assists the female somewhat in this, brings food to her while she is incubating, and helps to feed the young. I have seen the male bring food to the nest, and his mate feed it to the young. The incubating bird is watchful and very shy; it is almost impossible to approach within 100 yards of the nest without flushing her, if she is watching. One of my nests was fully that distance from a rocky ledge, from behind which I often attempted to watch the nest; almost invariably, as soon as I showed my head above the crest of the ledge, if I could see her head on the nest, she would immediately stand up in the nest and fly away; and she would not return until after I left the woods.

On other occasions, when she was invisible on the nest, I could walk to within 10 yards of the tree before she would fly; I believe that at such times she was asleep on the nest. Even after the young have hatched these hawks are very cautious about returning to the nest; repeatedly I have waited in vain for their return, even when well concealed, after they had once seen me; and their eyes are exceedingly keen. They seem to be much more concerned about their own safety than about the welfare of their eggs or young.

The young hatch at intervals of one or two days and remain in the nest for four weeks or more. Often one of the eggs proves to be infertile, and oftener one of the young dies and is thrown out of the nest, or is forced out of the nest and is killed by the fall. Norman Criddle (1917) writes:

The number of eggs laid by each female varies somewhat and seems to depend, at least to some extent, upon the food supply. In 1917, the six nests under observation close to the writer's home, contained but two eggs each and in only one of the six did the parents succeed in rearing more than one young though both were hatched in every instance. The first nest was discovered on May 6, containing two eggs. Other nests with eggs were located as late as June 14. It is difficult to account for the mortality among the young, though it is noteworthy that the deaths occurred while they were still quite small, and that the latest hatched, and consequently smallest, was invariably the one to die. Dead examples presented no indication of violence but seemed to show that, in all probability, death was due to starvation, the lack of food being due in its turn to a scarcity of ground squirrels (gophers) and to the unusual number of hawks nesting in the district.

The curious habit of the old birds in gathering a green leafy bough and placing it in the nest, characteristic of Swainson's hawk also, is very marked in the Red-tail, a fresh bough being gathered at least once daily during the time when the young are small. There has been some doubt hitherto as to the cause of this habit, but by observing the nestlings I am led to believe that the bough acts as a sun shade, as the young have been seen to repeatedly pull the bough over themselves and crouch beneath it. Doubtless it also acts as a shield and hides the young from their enemies. The leaves are also occasionally eaten.

As the young develop they acquire a good deal of boldness and defend themselves with both beak and claws. They have a habit of closely watching the intruder backing up meanwhile at the approach of a hand; then suddenly they leap forward with wings outstretched, and it requires a rapid movement to escape their onslaught. The old birds make no efforts to defend their young, but fly high overhead uttering loud cries which are, at times, answered in a shriller key by the young beneath.

The young, when half grown, become very lively, walking about in the nest, stretching or flapping their wings, backing up to the edge of the nest to void their excrement in a long stream far over the edge; the ground under a nest of young hawks is well decorated with a circle of white. Their eyes are very keen, and they frequently raise their heads to watch passing birds or to look for the return of their parents. Their weak, peeping notes are heard occasionally, but when one of their parents is sighted they become quite excited and indulge in louder screams in feeble imitation of the adult's notes.

I have never happened to see the young leave the nest, but Mr. Sumner's notes, applying to the western race, describe such an event. Mr. Shelley writes:

The adults are quiet during the incubation period and until the young are on the wing. As soon as this stage is reached, they are brought east of the hill where the nest is situated to the broad, open fields and mowings of the nearby farms, where they spend the forenoons hunting their legitimate prey and nothing else. Afternoons as a rule they skirt the country to the west of the nesting hill. But on the east side their calls can be heard all forenoon for a month or more, during the period the young are being taught to fare for themselves. Many a time I have seen them catching mice. An adult plunges down 50 to 100 feet or so at a scuttling mouse, checks its rush a few feet above the ground, and, turning onto its back, gives a wheezy whistle of two syllables, whereupon one of the circling young dives, holds itself suspended clumsily over the spot marked by the parent, and, quite often, obtains the rodent when it moves again. The parents do, rarely, drop disabled mice from a good height as though discarding them, but in reality it is done so that the young may catch them in midair, which they attempt to do with fair luck; I have seen it done on several occasions.

Mrs. A. B. Morgan (1915) gives an account of a young red-tailed hawk which she raised in captivity that developed into a very interesting and most intelligent pet.

Plumages.—The small downy young red-tailed hawk is well covered above with long, soft, silky down, buffy white or grayish white in color; the white, hairlike filaments on the head are erected in life and fully half an inch long; the down on the under parts is shorter and scantier. This first down is replaced later by a whiter and woollier down. When about 17 days old the wing quills appear, closely followed by those of the tail. Before the young bird is half grown the feathers appear on the scapulars and the mid-dorsal tracts; the feathers come in next on the pectoral tracts. By the time the bird is four weeks old it is nearly fully grown and almost fully fledged, the last of the down persisting on the head, central belly, and legs. It is now ready to leave the nest and is able to fly.

In fresh juvenal plumage, in June and July, the upper parts are "warm sepia" to "bone brown", with narrow edgings of "tawny" or "ochraceous-tawny"; the tail is "bister", barred with brownish black, tinged and tipped with buffy white, and silvery white on the under side, with the bars showing through; in western birds the tail is often tinged with "tawny" or "orange-cinnamon", sometimes extensively so, but in eastern birds this color is seldom, if ever, seen; the under parts are largely white, more or less tinged with "ochraceous-buff", which fades out to white later in the season; the throat and sides of the neck are narrowly streaked with sepia, and the belly and flanks are heavily streaked or spotted with a dark sepia, suggesting the adult pattern. This plumage is worn throughout the first winter with little change except by wear and fading, the buffs being replaced by dull white.

A complete molt from the juvenal into the adult plumage begins very early in the spring, is very gradual, and is prolonged through the summer or into the fall, with much individual variation. I have seen a young bird with new red feathers in its tail in February, and birds with missing flight feathers are often seen during the nesting season. At the completion of this molt in fall young birds are practically indistinguishable from adults. Young birds raised in captivity have molted from the juvenal into the red-tailed adult plumage when a little over a year old. I have examined a large series of eastern birds and have not been able to recognize a second-year plumage, such as seems to occur in *harlani*; immature specimens of *calurus* often have reddish tails with numerous narrow black bars; these are probably first-year birds with erythristic tendencies. Neither erythrism nor melanism seems to occur in eastern birds, but cases of nearly, or quite, perfect albinism have been reported. Adults have one complete annual molt, which may begin in spring or early in summer and may be completed in September or October.

Food.—It is generally conceded that the red-tailed hawk is a highly beneficial species, as its food consists mainly of injurious rodents and as it does very little damage to domestic poultry or wild birds. Dr. A. K. Fisher (1893) writes:

Of 562 stomachs examined by the author, 54 contained poultry or game birds; 51, other birds; 278, mice; 131, other mammals; 37, batrachians and reptiles; 47, insects; 8, crawfish; 13, offal; and 89 were empty. It has been demonstrated by careful stomach examination that poultry and game birds do not constitute more than 10 per cent of the food of this Hawk, and that all the other beneficial animals preyed upon, including snakes, will not increase this proportion to 15 per cent. Thus the balance in favor of the Hawk is at least 85 per cent, made up largely of various species of injurious rodents—a fact that every thoughtful farmer should remember. * * *

The increase of any animal is always followed by a relative increase of its natural enemies. This is clearly shown on the river front in the vicinity of Washington, D. C., where the recent improvements have redeemed several hundred acres of ground from the tidal flats; and already in many places rank vegetation has grown up, affording shelter and sustenance for hordes of mice. At present in winter and early spring it is not uncommon to see ten or fifteen Red-tailed Hawks in different parts of this flat attracted hither by the abundance of their natural food. Prior to the reclamation of the flats not more than a pair or two were to be seen in the same neighborhood during the winter.

Of 173 stomachs of this hawk examined by Dr. B. H. Warren (1890) in Pennsylvania, 131 contained the remains of mice, 6 of rabbits, 3 red squirrels, 2 skunks, 18 small birds, 14 poultry, 3 insects, 3 snakes, and 4 offal or carrion. He says: "I have repeatedly found three and four mice in the viscera of one bird, oftentimes five, and in a few instances as many as seven of these destructive little rodents were obtained from the crop and stomach of one hawk."

Dr. George M. Sutton (1928) reports on the stomach contents of 32 redtails, taken in Pennsylvania in October, as follows:

Twelve stomachs were empty; in the twenty stomachs which held food were eleven Field Mice, four Short-tailed Shrews, three Red-backed Mice, three Chipmunks, three small Garter Snakes, two Red Squirrels, one Winter Wren, one Song Sparrow, one Hermit Thrush, one Gray Squirrel, one Brown Rat, one half-grown White Leghorn Chicken, one large grasshopper, two crickets, and one large beetle of the family Elateridae. Such an array of food items in only twenty-two stomachs is noteworthy. Only seven of these stomachs held but one item; the others had a variety in each. If the above stomach contents are at all normal the red-tail captures about five harmful or unimportant organisms to one economically valuable one.

The following mammals have been detected in the food of this hawk: House mice and various species of field and wood mice, rats, various squirrels, both arboreal and ground species, raccoons, gophers, prairie dog, spermophiles, woodchuck, rabbits, moles, bats, shrews, chipmunks, muskrat, porcupine, weasles, and skunks; as many as nine red squirrels have been found in a nest at one time. The following interesting account of a redtail attacking a cat is published by E. D. Nauman (1929):

A large Red-tailed Hawk (*Buteo borealis borealis*) came out of the timber and leisurely flew around over the meadow, hovering over one point a moment for special inspection. Then he flew back to the woods again. A few minutes later he flew out and hovered over the same place, then returned to the woods as before. After having performed this round trip movement several times, the Hawk finally flew to this point and plunged down into the meadow. Instantly there was a mighty commotion. Hissing, flopping, spitting, caterwauling; and one could see feet, claws, wings and tails whirling about just over the grass. The air was full of fur and feathers for a few moments, then the Hawk made his getaway, and with feathers much ruffled flew for the timber as fast as his wings could carry him. And an old gray tom cat went with great bounds in equal haste for the farm buildings! Both Tommy and hawk were licked but still able to go.

The bird list includes domestic poultry, young turkey, pintail, teals, and other wild ducks, gallinules, rails, pheasants, ruffed grouse, Hungarian partridge, various quails, doves, screech owl, kingfisher, woodpeckers, crow, starling, grackles, meadowlark, horned larks, orioles, various sparrows, juncos, thrushes, robin, and bluebird. Verdi Burtch (1927) found a freshly killed red-shouldered hawk and later saw a red-tailed hawk feeding on it. Lucy V. Baxter (1906) surprised an adult red-tailed hawk feeding on a freshly killed immature hawk of its own species. Probably most of the small birds are killed during the nesting season as food for the small young, though the young hawks are fed largely on mice and squirrels. Ralph J. Donahue (1923) writes: "Before the eggs of the red-tails hatched, the parents fed on rodents—mostly the striped ground squirrels (Spermophile). After the young got out of the shells, the whole bill of fare was young chicken. At different times we found chickens to the number of seven. There were times when we could not go to the nest for a week or two, and it may be there was other food fed to the young during that time."

Miscellaneous items of food include rattlesnakes, bull snakes and smaller snakes, lizards, turtles, frogs, toads, salamanders, crawfish, grasshoppers, crickets, beetles, grubs, caterpillars, centipedes, spiders, earthworms, and maggots.

Two common hunting methods of the red-tailed hawk are the lofty soaring flight, from which its keen eyes detect its prey far below, and its slow flapping or sailing flight low over the fields and meadows, much after the manner of the marsh hawk or roughleg; a third, and perhaps the commonest, method is watchful waiting on some commanding perch on tree or post from which it can quickly pounce on any moving object that it sees. Much of its hunting must be in the forests, for many woodland mice and squirrels are included in its food. To capture such active animals as red or gray squirrels, it is often necessary for these hawks to hunt in pairs; these lively animals can easily avoid the swoops of a single hawk by dodging around a tree; but, if there is a hawk on each side, the squirrel is doomed unless it can scamper into a hole. Col. N. S. Goss (1891) says that these hawks while "sailing often fill their craws with grasshoppers, that during the after part of the day also enjoy a sail in the air." Mr. Shelley says in his notes that "it is also a great experience to see these large Buteos alight in a newly hayed field to catch grasshoppers and crickets; as they hop along the wings are always maneuvered to give the bird a rising impetus and timed so that the feet no more than touch the ground when the insect is plucked and the bird is clear of the ground on the next bound for the insect ahead. More than anything else, this maneuver resembles the floppings of a hen with its head cut off, only more mathematical, to give a crude description."

Behavior.—The ordinary flight of the red-tailed hawk is rather slow and heavy, as it travels along in a straight line, with rather slow wing strokes. But its soaring flight high in the air is inspiring, as it mounts gracefully, gathering altitude rapidly, with no apparent effort, with its broad wings and tail widely spread and motionless except for occasional adjustments to changing air currents. Once, as I stood on the brink of a precipice looking down over a broad valley, I saw below me a red-tailed hawk floating over the valley and looking downward for game; it was facing a strong wind and was perhaps buoyed up by rising air currents, as it was poised as motionless as if suspended on a wire; it remained in one spot for three or four minutes and then sailed over to another spot a few rods away, where it hung for a similar period. Its spectacular "nose dives", referred to above, are thrilling and well illustrate its mastery of the air. Dr. B. H. Warren (1890) describes some interesting maneuvers as follows:

Red-tailed Hawks in their fall migrations are gregarious. One clear, cold autumn afternoon in 1876, I saw, near West Chester, a flock of these hawks. The sky was destitute of clouds, except a cumulus stratum directly beneath, and apparently about half way between the hawks and the earth. In the center of this vapor was an opening of sufficient size to enable me to watch the gyrations of the birds; two of them suddenly separated from the main body, approached each other screaming, and apparently in great rage. They descended screaming, and, to all appearances, clinched, to within about one hundred yards of the earth, when they parted. Evidently neither bird had received much injury, as they both, after taking short flights across the meadow, ascended in company with two or three of their companions that had accompanied them part way down, to the main body. Another individual closed his wings until the body presented a triangular outline, descended with almost lightning-like rapidity to the top of a sycamore, where it alighted, and remained for some seconds pluming itself. This party of hawks, after performing for nearly twenty minutes, these, and numerous other aerial antics, continued their southern flight.

Illustrating its marvelous powers of vision, he says: "A clear morning early in March, I saw a Red-tail circling over the meadows; every circle took him higher and higher in the air, until at an altitude where he appeared no larger than a blackbird, he stopped, and with nearly closed wings, descended like an arrow to a tree near by me; from this perch, almost the same instant he had alighted, he flew to the ground and snatched from its grassy covert a mouse. The momentum with which this bird passed through the atmosphere produced a sound not very unlike that of the rush of distant water."

This hawk is generally regarded as a sluggish, inactive bird, for it spends much of its time standing erect on some lofty perch, slowly scanning its surroundings. It is one of the shiest of our hawks; a man on foot can seldom approach one to within 100 yards, and often it will fly at twice that distance. But it seems to be less afraid of a man on a horse or in a vehicle; in regions where hawks are not much persecuted one can sometimes ride up within gunshot range.

A wounded redtail is a formidable object, as it throws itself on its back and presents its sharp and powerful talons; it will grab a gun barrel or stick and allow itself to be lifted up; or it will fasten its claws in the hand or arm of one who tries to handle it and can only with great difficulty be made to let go. Once, while I was hunting with John B. Semple in Florida, a wounded redtail dropped a long way off among some patches of saw palmetto; after a long search in vain we sent his springer spaniel to hunt for it; the plan worked successfully, but the dog was surprised and much frightened, as the infuriated hawk rushed out and attacked him.

These hawks are not at all courageous in the defense of their nest; they generally keep at a safe distance or disappear entirely; only on rare occasions has one been known even to attempt to attack a climber; I have seen it only once. Only twice have I seen one return to its nest when I was in plain sight near the nest tree; once when I was almost under the tree the hawk settled on the nest and would not leave until I rapped the tree.

Its behavior toward other birds is generally an attitude of stolid indifference. I have seen it drive away other hawks from the vicinity of its nest and, as stated above, have known it to preempt old-time nesting haunts of red-shouldered hawks. I have repeatedly seen it attacked by a party of crows; it often pays no attention to them but sometimes turns on its back and displays its talons, at which the crows beat a hasty retreat; occasionally the crows pay the extreme penalty for their temerity; crows have often figured in the food of this hawk. Kingbirds and blackbirds often attack the redtail and drive it away from their nesting sites, but I doubt if the hawk ever retaliates. Mr. Skinner says in his notes: "Once I found one near Southern Pines being tormented by four robins. It protected itself fairly well while in the top of a tall pine, but when it flew 26 more robins, which had been concealed in the foliage, gave chase and joined their efforts to the pecks of the first four tormentors."

Mr. Sumner (MS.) once saw a redtail attack and drive away a horned owl that had ventured too near its nest. Great horned owls habitually occupy old nests of the eastern redtail, probably preempting them before the hawks are ready to use them. I have always regarded these two as supplementary species, one hunting by day and one by night in similar regions and preying on similar victims. I once surprised one of these owls feeding on the remains of a freshly killed red-tailed hawk.

Voice.—The red-tailed hawk occasionally utters a note similar to that of the red-shouldered hawk, but usually it is quite distinct. The characteristic cry is described in my notes as a long drawn out, harsh, rasping squeal, *kree-e-e-e-e*, suggesting the squeal of a pig. It has also been written *cree-e-e*, *cree-e-ep*, or *pee-eh-h*. Bendire (1892) gives it as *kee-aah*, the redshoulder note, so often imitated by the blue jay; he also gives another note, *chirr* or *pii-chiir*, "when perched on some dead limb near their nest." The note has been said to resemble the sound made by escaping steam, but I could never quite see the resemblance.

Field marks.—Its outline, broad, somewhat rounded wings, and broad, rather short tail mark it as a *Buteo*. In adult plumage it should be easily recognized. As it flies straight away in the woods, or as it wheels in soaring flight, it shows a glimpse of its red tail, with no barring on the under side of it, in marked contrast with the conspicuously black and white barred tail of the redshoulder. The under side of the wing is whitish, without bars, but with a dark border formed by the dusky tips of the primaries and secondaries and there is usually a dark wrist mark near the bend of the wing. The sides of the head are very dark and the breast is largely whitish, with dark streaks only on the belly and flanks. The young bird looks very much like the young redshoulder; it has a faintly barred tail, and the streaking on the under parts is more like that of the adult redtail, very scanty on the breast, than like the young redshoulder, which is more uniformly streaked below.

Fall.—Early in September, red-tailed hawks begin to drift southward from New England and other northern parts of their range. These fall flights are very spectacular and usually contain a variety of species; they are seen to best advantage on clear cool days with a northwest wind. These large mixed flights often contain hundreds of individuals, spread out over a wide area and continuing to pass for several hours. Dr. Fisher (1893) has seen a flock containing 65 red-tailed hawks "flying in a comparatively compact body, probably not more than a few feet from each other." H. S. and H. B. Forbes (1927) thus describe a flight as witnessed in New Hampshire on September 14, 1926:

Far out to the northwest two Hawks, perhaps a mile away, were seen wheeling over the valley at a slightly lower level than our point of observation. Then, as if from nowhere, other Hawks rapidly appeared, swooping, turning and soaring upwards in irregular steep spirals. More and more individuals appeared until the specks resembled a swarm of large insects, black against the pearl gray clouds. The total number was estimated to be between thirty and forty. Now they soared slowly, now flew with rapid wing beat at great speed. Each individual chose his own course without evidence of leadership. In from five to ten minutes (the exact time unfortunately was not noted) the flight had gained great altitude and to our astonishment the highest birds began to disappear in the clouds, some of them reappearing and again diving into the mist. Finally the whole flight had spiraled upward into the cloud mass and was lost to view. Once, half a minute later, a few specks wheeled out toward us and for a moment could be dimly seen through the edge of the cloud. That was the last glimpse.

Dr. Thomas S. Roberts (1932) writes from Minnesota:

While driving from Ten Mile Lake, Otter Tail County, to Breckenridge on the Red River, on October 7, 1927, the writer, accompanied by Mr. Kilgore and Mr. Breckenridge, passed through what was evidently a large migration of Redtails. There were a few scattered all over the country, but on the open prairie between Nashua and Campbell, in Wilkin County, many of the fence posts, telephone and telegraph poles, and straw-stacks and hay-stacks, were occupied by birds, while others circled in the air, and a few were walking about on the ground. Forty-eight were counted, most of them in a limited area.

Maurice Broun (1935) says of the fall migration at Kittatinny Ridge, Pa., in 1934:

It may come as something of a surprise to learn that these splendid birds made up fully 50% of the entire Hawk migration. The first Red-tails recorded were two on September 30. No conspicuous movement took place until October 12, when 205 birds were counted. Thereafter during the month there were nine days of relatively heavy flights, the greatest number of 427 birds occurring on October 23. The first part of November, however, brought the major flights, with an average of 244.5 birds per day for 12 days. On November 1, I recorded 592 Red-tails—as many as 213 in a single hour; on November 2, 853 Red-tails. Kramer reported diminishing numbers of Red-tails during the latter part of November, except for 67 on the 24th. He saw 9 on December 2, and 4 on the next day.

DISTRIBUTION

Range.—North and Central America and the islands of the Caribbean Sea.

Breeding range.—The red-tailed hawk breeds **north** to Alaska (Innoko River, Salcha Slough, and Joseph); Yukon (Forty Mile); Mackenzie (Fort Good Hope, Fort Norman, and the Grandin River); northeastern Manitoba (Fort Churchill, and probably York Factory); probably northern Ontario (Missinaibi River, Mattagami River, and Moose Factory); and Quebec (probably English Bay, Mingan Island, Piashti Bay, and probably Natashquan). **East** to Quebec (probably Natashquan and Gaspé County); Prince Edward Island (North River); Nova Scotia (Kentville and probably Digby); Maine (Bucksport, probably Lewiston, and Portland); eastern Massachusetts (Danvers, Boston, and Cape Cod); New Jersey (Princeton, Vineland, and Sea Isle City); eastern Virginia (Spottsville and Dismal Swamp); North Carolina (Raleigh and Pinehurst); South Carolina (Columbia); Georgia (Savannah, Blackbeard Island, and St. Marys); Florida (San Mateo, Fruitland Park, and Fort Pierce); the Bahama Islands (Little Abaco); probably northern Haiti (Terrier Rouge); Puerto Rico (Mayaguez, Manati, near Cayey, and probably Hacienda Catalina); and the Virgin Islands (Vieques Island, probably Culebra Island, and formerly St. Croix Island). **South** to the Virgin Islands (formerly St. Croix Island); probably southern Dominican Republic (Beata Island); Cuba (Trinidad); Jamaica; probably Panama (Chiriqui); Costa Rica (Santa Maria de Dota and Cartago); and Colima (Socorro Island). **West** to Colima (Socorro Island); Nayarit (Tres Marias Islands); Lower California (Guadalupe Island and San Pedro Martir Mountains); California (San Diego, San Clemente Island, Santa Catalina Island, Los Angeles, Santa Barbara, Port Harford, Monterey, Santa Cruz, Alameda, Petaluma, Cazadero, Mount Sanhedrin, and probably Crescent City); Oregon (Glendale, Bandon, Elkton, Newport, and Olney); Washington (probably Tacoma, Seattle, Everett, and probably Bellingham) ; British Columbia (Beaver Creek, probably Cumshewa Inlet, and Porcher Island) ; and Alaska (St. Lazaria Island, probably Yakutat, Chitina River, Nushagek, Iditarod River, and Innoko River).

The range as above outlined is for the entire species (*B. jamaicensis* of some authors), which has, however, been separated into several geographical races. True *borealis* occupies the greater part of this vast area from Yukon, Mackenzie, Manitoba, and Quebec south to Texas, Oklahoma, Arkansas, Alabama, and northern Florida. The western redtail (*B. b. calurus*) is found from Alaska and central western Mackenzie south to Lower California and. east to the edge of the Great Plains. Krider's hawk (*B. b. krideri*) breeds from south-central Canada, North Dakota, and Minnesota south in winter to Louisiana and Mississippi; accidental in Georgia and Florida. Harlan's hawk (*B. b. harlani*) breeds in northwestern British Columbia, southeastern Alaska, and southwestern Yukon, wintering south to the Gulf coast; casual in California. The Florida redtail (*B. b. umbrinus*) is found in the Florida Peninsula, Cuba, the Isle of Pines, and probably the Bahama Islands. *Buteo b. jamaicensis* occupies Hispaniola, Jamaica, Puerto Rico, the Virgin Islands, and probably the Leeward Islands; *B. b. fumosus* is found on the Tres Marias Islands off the west coast of Mexico; *B. b. socorroensis* is confined to Socorro Island, also off the west coast of Mexico, and *B. b. costaricensis* is found from the highlands of southern Mexico south to Costa Rica and probably western Panama.

Winter range.—The species winters throughout the southern part of its breeding range and north to southern British Columbia (Chilliwack and Okanagan) ; Utah (Provo) ; Colorado (Boulder) ; southeastern South Dakota (Vermillion) ; Iowa (Sioux City and Keokuk) ; central Illinois (Rantoul) ; southern Michigan (Detroit); New York (Rochester, Geneva, Auburn, and Rhinebeck) ; Connecticut (Hartford) ; and eastern Long Island (Gardiners Island). It is occasionally noted at this season and may sometimes winter north to Alaska (Admiralty Island and Eagle) ; North Dakota (Jamestown) ; Minnesota (Lanesboro) ; Wisconsin (Viroqua, Madison, and Princeton) ; southern Ontario (Coldstream, London, Mill Brook, and Ottawa) ; Vermont (Montpelier) ; New Hampshire (Meriden and Monadnock) ; and Maine (Cumberland County).

Spring migration.—Early dates of arrival in the spring are: Maine—Lewiston, March 15; Portland, March 17; Auburn, March 20; and Avon, March 21. Quebec—Montreal, April 22. New Brunswick—Scotch Lake, March 25; and St. John, March 27. Nova Scotia—Wolfville, March 22. Prince Edward Island—North River, April 20. Ontario—Toronto, February 14; Port Dover, February 26; and London, March 12. North Dakota—Larimore, March 20;

Grafton, March 21; Wahpeton, March 23; and Argusville, March 29. Minnesota—Minneapolis, February 22; and Becker County, March 11. Manitoba—Treesbank, March 15; Margaret, March 23; and Winnipeg, April 10. Saskatchewan—Dinsmore, March 17; Wiseton, March 21; Muscow, March 30; and Skull Creek, April 11. Wyoming—Laramie, March 25; and Yellowstone Park, March 29. Idaho—Rathdrum, March 20; Coeur d'Alene, March 25; and Meridian, March 27. Montana—Missoula, March 12; Fortine, March 17; and Fort Custer, March 21. Alberta—Camrose, March 17; Stony Plain, March 30; Alliance, April 1; and Flagstaff, April 5. Yukon—Forty Mile, April 11. Alaska—Kupreanof Island, March 5; Craig, March 12; Wrangell, April 15; Beaver Mountains, April 20; and Tanana Crossing, May 7.

A redtail (no. 309391) banded on March 27, 1926, at McGregor, Iowa, was killed on May 16, 1926, at Howard Lake, Minn.

Fall migration.—Late dates of fall departure are: Alaska—Savage River, September 18; and Bettles, October 5. Yukon—Selkirk Settlement, October 12. Alberta—Belvedere, October 14; Glenevis, October 14; and Athabaska Landing, October 31. Montana—Missoula, October 1; Anaconda, October 9; Gallatin County, October 12; and Three Forks, October 17. Idaho—Priest River, September 29. Wyoming—Yellowstone Park, November 17. Saskatchewan—Muscow, October 3; Indian Head, October 20; and Eastend, October 21. Manitoba—Winnipeg, October 19; Treesbank, October 29; and Aweme, October 30. Minnesota—Minneapolis, November 23; and Isanti County, November 20. North Dakota—Charlson, October 19; Argusville, October 21; and Red River Valley, October 29. Ontario—Ottawa, October 22; Kingston, October 30; Point Pelee, November 1; and Toronto, November 7. Prince Edward Island—North River, October 24. Nova Scotia—Pictou, October 12; and Wolfville, November 20. New Brunswick—St. John, October 2; and Scotch Lake, October 25. Quebec—Montreal, October 24; Quebec, October 24; and Hatley, October 29. Maine—New Vineyard, October 23; North Livermore, November 3; Matinicus Island, November 8; and East Hebron, November 30.

Records of banded birds throw some light on the general migration routes of this species. Five young redtails banded during June and July at Muscow, Saskatchewan, were recaptured later in the same year as follows: No. 661180 was taken on September 22 near Grafton, N. Dak.; no. 309016 was killed on October 12 at Butterfield, Minn.; no. 200645 was shot near Chetopa, Kans., on November 6; no. 309019 was recovered on November 3 at Brumley, Mo.; and no. 200641 was shot at Chelsea, Okla., on November 24. One banded at Milwaukee, Wis. (no. 233936) in November was retaken the fol-

lowing month at Blue Island, Ill.; while another banded in May at
Denzer, Wis. (no. 235860) was shot in November at Rusk, Tex. One
banded in May at Chester, Mass. (no. 312005) was killed the following November at Brandywine Summit, Pa.; and one banded at Mid-
dlefield, Mass. (no. 386652), in June was recovered in November at
Springfield, Va.

Casual records.—The red-tailed hawk was listed by Macoun (1909).
as breeding in Newfoundland, but no supporting evidence was cited;
Noble (1919) states that it is found in Labrador but not in Newfoundland, while Austin (1932) does not list it among the birds of
Newfoundland Labrador.

A specimen of *B. b. borealis* is said to have been shot in Notting-
hamshire, England, in the autumn of 1860.

Egg dates.—Alaska and Canada: 53 records, April 3 to June 12;
26 records, May 4 to 18.

New England and New York: 148 records, March 25 to June 21;
74 records, April 4 to May 17.

Maryland to West Virginia: 15 records, March 26 to April 26; 7
records, March 27 to April 8.

Ohio to North Dakota: 85 records, March 6 to June 30; 42 records,
April 7 to May 3.

Iowa to Colorado: 44 records, February 28 to June 28; 22 records,
April 3 to 28.

Washington to California: 292 records, February 14 to May 29;
146 records, March 19 to April 1.

Arizona and Texas to Florida: 97 records, February 18 to June
17; 48 records, March 7 to April 3.

BUTEO BOREALIS KRIDERI Hoopes

KRIDER'S HAWK

HABITS

Krider's hawk is a well-marked pale race of the red-tailed hawk,
occupying the plains and prairie regions of the Middle West. It
was described and named by Bernard A. Hoopes (1873) from a pair
of immature birds taken by John Krider in Winnebago County, Iowa,
in September 1872. An excellent colored plate, published with the
description, illustrates the extreme white phase in immature plumage.
The adult is much like the eastern red-tailed hawk, but lighter
colored; there is much white on the upper parts, the tail is pale
rufous, and the under parts are nearly pure white, with very few
markings and with only a pale buffy tinge in the thighs. Krider's
hawk is easily recognizable in all plumages by extreme lightness,
although the immature plumages of the light phase of Harlan's hawk
are nearly as light colored and closely resemble it.

Nesting.—The nesting habits of Krider's hawk are similar to those of other red-tailed hawks, due allowance being made for its environment. It evidently prefers to nest well up in big trees, but in the prairie regions, where heavy timber is scarce, it is often obliged to nest at a low elevation. It apparently nests somewhat later than the eastern redtail, as the records given below show. The Rev. P. B. Peabody (1895) has published data on eight nests found in southern Minnesota, on dates ranging from April 22 to May 11, six of the dates being in May. This hawk shows no partiality for any particular kind of tree; the eight nests were divided thus: 2 in elms, 2 in white oaks, and 1 each in basswood, rock maple, black oak, and black walnut. The heights from the ground were 30, 40, 50, 60, and 75 feet. Two interesting nests he describes as follows:

Locality, a heavily wooded island. Nest in a great elm, nearly inaccessible, far out on horizontally spreading branches of a large main bough, at the very top; an old, broad and flat nest, roughly made of large sticks, with hollow, twelve inches in diameter. Lining, fibrous bark, twigs, feathers of small birds. * * * Locality, the very steep, deep, and heavily wooded bank of river, fringing a cultivated plateau. One mile from nest III. A flat, old nest, far out, nearly over the water, on leaning branch of rock maple, sixty feet up. Large sticks. Lining, soft fibrous bark and grass.

Dr. R. M. Anderson (1897) describes five nests found by him in Iowa. One of these was an old Swainson's hawk's nest, and another had been previously occupied by a red-tailed hawk and the following year by a great horned owl. Three of his nests were in burr oaks, 46, 50, and 57½ feet up, and one was in a black oak 35 feet from the ground. He noted that the nests all contained green, leafy twigs, mainly cottonwood and poplar, which appeared to be renewed daily.

M. A. Carriker, Jr. (1902), found an interesting nest of this hawk, near Warbonnet, Nebr., on a ledge about 20 feet from the base of a cliff in a canyon. He says of it: "The site had evidently been used by the birds for several successive years, for the pile of sticks composing the nest was at least one and one-half feet in thickness and three feet in diameter, occupying a pocket on the ledge. Fragments of skulls, vertebrae, and feet of various rodents lay scattered about, together with the vertebrae of a large snake and some fragments of a recently killed prairie-dog."

The only nest I ever examined was found on June 1, 1901, near Stump Lake, N. Dak. It was about 30 feet from the ground in the topmost branches of a small elm. One of the parents, a very light colored bird, was sailing about overhead and screaming anxiously *kreeah, kree-a-a-ah*, a prolonged squealing whistle. It was a large nest of sticks, lined with dry grass, and contained three very young hawks, a pipped egg, parts of a cottontail rabbit, two ground squirrels, and two field mice. Within a few feet of the tree was an old elm stub, in which a goldeneye had a set of 10 eggs and a house wren a set of 7 eggs (pl. 48).

Eggs.—The eggs of Krider's hawk are practically indistinguishable from those of other redtails, though Mr. Peabody (1895) says that some of them are more like red-shouldered hawk's eggs in their markings. The measurements of 52 eggs average 59.8 by 49.9 millimeters; the eggs showing the four extremes measure **63.4** by 48.3, 62.7 by **50.1**, **55.7** by 44.6, and 58.4 by **44.2** millimeters.

The sequence of molts and plumages is the same as in other redtails, but the racial characters are always evident. Its food habits, general behavior, and voice are all similar. It migrates away from the northern portion of its range for the winter. A. G. Lawrence tells me that it is very rare in southern Manitoba, departing in September and returning in April.

BUTEO BOREALIS CALURUS Cassin

WESTERN RED-TAILED HAWK

HABITS

From the western edge of the Great Plains westward this very variable form of the red-tailed hawk breeds from southeastern Alaska and central western Mackenzie southward to Mexico. Its chief characters are its interesting color phases, varying from a light phase, which is practically indistinguishable from the eastern bird, to a melanistic phase, which is wholly dark sooty brown, except for the red tail. Between these two extremes are numerous intermediate plumages, to be referred to later. The immature bird is consistently darker, more heavily streaked below and with a more or less distinctly streaked throat, and more spotted tibiae.

Because of its similarity in the light phase to the eastern bird, some very good ornithologists think that the western bird should not be given a separate name. The color phases are no less variable than, or strikingly different from, those that occur in Swainson's hawk, the ferruginous roughleg, and some other hawks.

Spring.—Throughout the northern portion of its range the western redtail is migratory. M. P. Skinner's records for 10 years in Yellowstone National Park show that it was first seen there very regularly between March 29 and April 3, departing again about the middle of October. He thinks its arrival in spring is dependent on the appearance of the ground squirrels, which come out of hibernation about the first of April.

That these hawks know where to find a good food supply while migrating is shown by their spring visits to the Farallon Islands, of which Walter E. Bryant (1888) writes: "Every spring the island is visited by numbers of these hawks. In 1882 they came in April, about the time of the arrival of the murres, leaving again in May.

During their short stay they fed almost exclusively upon the murres, killing, in the estimation of Mr. Emerson, several dozen a day. In 1887 the lighthouse man killed about seventeen of these hawks, and during the month of May, 1885, twenty-eight, mostly of this species, were destroyed."

Nesting.—After making due allowance for the difference in environments and in available nesting sites, the nesting habits of the western redtail and all its habits are similar to those of its eastern relative. It prefers to place its nest at some lofty situation in the tallest tree it can find, but, as its distribution is governed more by the food supply than by suitable nesting sites, it is often obliged to nest at low elevations or on cliffs.

I have recorded in my notes only 12 nests, 8 seen in Arizona and 4 in California. Of the Arizona nests three were on rocky cliffs, two in abandoned nests of the golden eagle, and one in an old raven's nest. Only one of the eagle's nests was closely examined; this was on the face of a bulging rocky cliff on the steep side of a mountain, 75 feet from the bottom of the cliff and 25 feet from the top, giving a fine outlook over the valley far below. It could be reached only by going over the cliff on a rope. The old nest had been repaired somewhat and lined with strips of yucca and other soft fibers. I collected two eggs from it.

Two nests were found in open country as we drove along the roads, one 40 feet up in a cottonwood and one 25 feet up in a mesquite. In the deep canyons of the Catalina Mountains, where the giant cottonwoods and sycamores grow, we saw the loftiest nests, approximately 90 and 100 feet above the rocky beds of the streams.

The lowest nest, and one of the most interesting, was only 10 feet from the ground in a double-headed soapweed yucca, which stood out alone on an open plain; it was a bulky old nest that had been in use for years and was securely held between the branches of the yucca. It held two pretty eggs (pl. 49).

Of the California nests only one was on a cliff; this may have been an old raven's nest, but it had been extensively rebuilt. The other three were in large sycamores, 40, 60, and 70 feet from the ground. All the above 12 nests were in commanding situations where the birds could have a good view of their surroundings.

In Arizona the western redtail often nests in the giant cactus, or saguara, placing its nest where one or two main branches project from the trunk and bend upward. J. H. Clark (1900) describes four such nests at heights varying from 6 to 30 feet from the ground. He also mentions two nests in palo verdes 10 and 12 feet up.

An unusually lofty nest is described and fully illustrated by William L. Finley (1905) and Herman T. Bohlman. The nest was 120

feet from the ground in a giant cottonwood on the bank of the
Columbia River in Oregon. It required a vast amount of energy and
daring to take the fine series of photographs of eggs and young at
different ages that they secured.

J. A. Munro (1919), referring to the Okanagan Valley in British
Columbia, writes: "This is the characteristic hawk of the lower mountains. They are equally at home in the dense coniferous forests at
the edge of cultivated land, in the open park country of the yellow
pine (*Pinus ponderosa*) or in the midst of deep canyons and rock
cliffs. * * * The same nests are used for several years, usually
built in tall coniferous trees, forty to sixty feet above the ground. A
site commanding a view of open range or valley is preferred."

W. Leon Dawson (1923) shows a photograph of a redtail's nest
only a few feet from the ground in an ocotillo bush; this was in the
Imperial Valley desert, where trees are scarce.

E. L. Sumner, Jr., writes to me of a nest in which the three eggs
were surrounded and partially covered by a piece of white wrapping
paper; how the paper came there is unknown, but it resulted in the
desertion of the nest. James B. Dixon (1902) reports a somewhat
similar case in which "the hawks had secured a large piece of barley
sack and with this made a lining for the nest, the eggs being covered
by it."

Referring to a thickly populated area in San Diego County, Joseph
Dixon (1906) writes:

We found the western red-tail and sparrow hawks and the Pacific horned
and barn owls especially abundant. In one valley in a distance of six miles
we found twenty-two hawks' nests. Seven of these nests were occupied by redtails, three by horned owls and one by a red-bellied hawk. Each pair of redtails usually had two and sometimes three nests, for they seem to occupy
different nests from year to year. Two nests were often found built close
together and in one instance there were three nests in one clump of trees.

These twenty-two nests were all located in sycamores which often stood at
a bend in the creek or near the edge of the grove. By actual measurement we
found that the average height from the ground of twenty-two nests was fifty-
five feet. The extremes were seventy-five and forty-three feet. We estimated
that there was a pair of hawks to every one-half square mile of territory.
What becomes of the offspring in this densely populated district is a problem
that I have been unable to solve. But some of them evidently stay near their
birthplace, as we found that out of seven pairs, two pairs had moved in since
last year.

Eggs.—The western redtail usually lays two or three eggs, perhaps
oftener two; four eggs are occasionally laid, and five or even six have
been recorded. The eggs are practically indistinguishable from those
of the eastern redtail, although they may average a little more
heavily marked. The measurements of 48 eggs average 59.2 by 46.4

millimeters; the eggs showing the four extremes measure **64.5** by 48.3, 61.5 by **49.6,** and 53 by **43.5** millimeters.

Young.—Incubation lasts for 28 days and is shared by both sexes. Bendire (1892) says that "the eggs are deposited at intervals of a couple of days." Both parents assist in the care of the young, which remain in the nest for about six weeks. In the nest that Mr. Finley (1905) studied the young were hatched on April 20 and they left the nest on the first of June.

Mrs. Irene G. Wheelock (1904) tells the following remarkable story of a young hawk that was thrown out from a nest and fluttered helplessly to the ground:

It fluttered about on the grass, and after resting a time managed to scramble into a low bush, where it felt more secure, though it really was much more exposed. In the meantime the adults had circled wildly about with discordant screams, and the mother still remained near. Curious to see how she would manage to get that unlucky youngster back into his nest, we moved off fifty yards and watched through the glasses. Both parents swooped down and looked at him, from on the wing, again and again, screaming when away, but silent whenever near him or the nest. At length a more sudden swoop and a momentary flutter, as a butterfly flutters over a flower. Then she rose carefully and slowly, with the young in her claws, and carried him to the nest. It was impossible to see whether she was holding him between them or grasping him by them.

Mr. Sumner, in his notes, thus describes the departure of a young bird from its nest:

One bird, when frightened, walked to the edge of the nest and facing the breeze spread his wings and then, half balancing, walked out on a small dead branch until 4 feet from the nest, where he remained, now with folded wings, quite at ease, although swaying in the wind. All at once the branch broke under his weight, whereupon he sprang into the air with vigorously flapping wings and flew partly against, partly beaten back and buoyed up by, the wind for 100 feet. He lit on the slough bank, stayed there a moment, and then, of his own accord, jumped off and flapped low over the water across the slough and lit again on the other side. Presently he began to run, head low, wings partly unfolded, in typical hawk fashion, always putting distance between himself and me. Presently he squatted down, in the same prone position as in the nest—a move that may well be one of self protection, and here he stayed, even though I hid beneath a dead tree limb 100 feet away, with his eyes fixed on me, until after 15 minutes I got tired and left.

Leslie L. Haskin says in his notes that these young hawks "are the noisiest of all young land birds. This is especially true just after they have left the nest. They follow the old ones around at feeding time—which seems to be all the time that it is light—screaming at the tops of their voices. Two or three young redtails hungrily following the old ones sound like the squalling of a litter of pigs. In feeding the young birds the old hawk often mounts high in the air

while the young ones circle, squalling, below her. Then she drops
the food, and the young birds catch it before it reaches the ground.
Snakes are a favorite article of diet, and when dropped in this way
are easily seen."

Plumages.—The downy stage and the sequence of plumages and
molts are the same as in the eastern red-tailed hawk. Therefore it is
necessary to consider here only the interesting color phases that seem
to occur only in these western birds. In the extreme melanistic phase
the entire adult plumage, except the tail, is very dark brown, varying
from "bone brown" above to "warm sepia" or "bister" below; the tail
varies from "Sanford's brown" to "hazel", a deep rich red, with a
broad subterminal band and numerous more or less broken bars of
black. In a common variation from the above the breast and belly
are more or less tinged with tawny or rufous shades and the tibiae
are spotted or barred with these colors. The erythristic, or red,
phases are quite variable. An extremely red adult has "tawny"
edgings on the head and neck; the scapulars are notched with buffy
shades; the entire under parts from chin to tail are rich reddish
brown, varying from bright "hazel" or "cinnamon-rufous" on the
breast to "amber brown" on the tibiae; the upper breast is lightly
streaked and the belly heavily spotted with black; the tibiae are
faintly barred with a darker brown; the tail is much as in the black
phase but slightly paler. Intermediates between these two phases
have duller, sootier browns on the under parts, the breast and belly
showing two quite distinct colors. There are also intermediates
between both of these phases and the common light phase. Immature birds of the dark phase are much darker above and much more
heavily marked below than in the light phase. They look very much
like young *harlani*, but can be recognized by the scarcity or almost
complete absence of the conspicuous white spots on the upper parts
so prominent in *harlani*.

I am not sure that we can recognize the young of the red phase,
though we might expect them to show richer buff or tawnier shades.

Food.—The feeding habits of the western redtail are similar to
those of the eastern bird, but the western form is even more beneficial to the agriculturalist, for it lives in a region where injurious
rodents are very abundant and troublesome. Joseph Dixon (1906)
says: "Each pair of hawks had its own squirrel pasture and the
birds resented the trespassing of other hawks on their domain. The
remains of gophers, ground squirrels, meadow mice, young cotton-
tails and two species of snakes, the striped racer and gopher snake,
were found in red-tails' nests, but ground squirrels seemed to be
their principal diet. I found as many species of small mammals

in hawks' and owls' nests in two days as I did by trapping for a week."

Mr. Sumner found in a nest parts of seven ground squirrels, one pocket gopher, and two cottontail rabbits. J. Paul Miller (1931) made a study of five red-tailed hawks' nests in the Big Bend country in eastern Washington and summarizes the food results as follows: "The food brought to the nests consisted entirely of Columbian Ground Squirrels (*Citellus c. columbianus* (Ord)) with the exception of one meadow mouse (*Microtus* sp.), and although birds of various species were numerous in the vicinity of all the nests, they did not seem to be disturbed by the hawks. About six squirrels per day from the time the young hatched until they were nearly feathered seem the average number provided. This is strong evidence as to the benefits which were locally derived from the activities of these birds."

Mr. Finley (1905) says of the food found in the nest he studied: "On the first visits we found the remains of quail and pheasants in the aerie. One morning we saw the mangled body of a screech owl; almost a case of hawk eat hawk. Later in the season when the banks of the Columbia overflowed, and covered most of the surrounding country, the old hawk did not abandon his own preserve. He turned his attention entirely to fishing. Where the carp and catfish fed about the edges of the ponds he had no trouble in catching plenty to eat. Twice we found carp over a foot in length in the aerie. On our last visit we picked up the head bones of seven catfish in the nest."

The following quotation from one of A. W. Anthony's (1893) Lower California papers is doubly interesting:

At La Grulla a pair of redtails were nesting near our camp. The male was a very light bird, while the female was so dark as to be several times mistaken for the dark phase of *swainsoni*. On May 16 the female was shot as she rose from the nest, and on skinning her I found in her stomach the remains of a *Cyanocephalus* and a nearly complete rattlesnake that must have measured over two feet in length. On the following day the male was seen flying about the nest with another female fully as dark as his former mate, and I was surprised to see her feeding young ten days or two weeks old. I had supposed the nest still contained eggs. As it was such a clear case of adoption I concluded to leave them undisturbed, but the unfortunate male was doomed a few days later to lose his second mate which was shot by a member of our party; upon dissection this bird was also found to have a large rattlesnake coiled up in her stomach. We frequently saw redtails sailing about over the meadows with large snakes hanging from their talons.

In connection with the above dangerous feeding habit, it is interesting to note that J. S. Hunter (1898) reports that a red-tailed hawk was seen attacking a rattlesnake, which bit the hawk twice and killed it. As this was in Nebraska it probably was one of the eastern redtails and perhaps not accustomed to rattlesnakes.

Behavior.—My experience with red-tailed hawks has taught me that they are very shy birds; they usually keep well out of gunshot range even when they have young in the nest. But I once saw an unusual exhibition of boldness and aggressiveness shown by a western redtail. While out with A. M. Ingersoll and J. B. Dixon, near Escondido, Calif., his climber, Gus Hanson, attempted to collect a set of three eggs from a nest about 70 feet up in a tall sycamore. One of the hawks attacked him, darting down at him time after time and looping the loop above him several times. We all agreed that it was the greatest exhibition of the kind we had ever seen.

Mr. Skinner says in his notes:

Here in the Yellowstone National Park, where they are protected, these big hawks become so tame they can be readily studied. Often, I have passed them on stubs and telephone poles without disturbing them in the least, although I might be less than 50 feet distant.

They prey almost exclusively on rodents, and I have never seen one attack a bird. The larger birds like the ducks and geese are indifferent to a red-tail's presence, but the attitude of the smaller birds is even more astonishing. I have seen a red-tailed hawk on a river bank with an unconcerned robin on a nearby bush. I have seen a redtail fly over a flock of conspicuous rosy finches on the ground without alarming them by either its shadow or its presence.

Other small birds, such as bluebirds and juncos, have shown similar indifference. On the other hand he has seen the hawks attacked by crows, nutcrackers, sharp-shinned hawks, sparrow hawks, kingbirds, Brewer's blackbirds, and once by an Audubon's warbler. In some cases the small birds were probably driving the hawk away from the vicinity of their nests. He once saw a flock of seven robins drive a red-tailed hawk to cover in a fir tree. He has seen the hawks fighting each other quite often, usually on the wing. Once he "watched a redtail flying high that would at intervals make a long, swift shoot down toward another redtail flying below that would turn and present its talons to meet the attack, real or pretended."

Mr. Sumner's notes record an attack by a redtail on a horned owl; the hawk dived at the owl from a height of 75 feet; the owl made no effort to turn over but "received the blow of the hawk's talons in the middle of its back." Another time he saw a pair of hawks executing their flight maneuvers near their nest. "About ten times, while they were circling near together, the male would lower his legs and adjust his circles so that he came above his mate, and about four times he actually touched her back, or so it seemed." As illustrating the confidence that small birds have in these hawks, he noted an occupied kingbird's nest and an occupied Bullock's oriole's nest in the same tree with a red-tailed hawk's nest containing young.

BUTEO BOREALIS HARLANI (Audubon)

HARLAN'S HAWK

HABITS

Harlan's hawk has always been regarded as a subspecies in the *Buteo borealis* group and is so recognized in our latest A. O. U. check-list (1931). James L. Peters (1931), in his new check-list, treats it as a full species, and several good ornithologists agree with him. I am inclined to accept this view, because the chief character of *harlani*, the mottled tail of the adult, seems to be a qualitative rather than a quantitative character. I shall not attempt to discuss here, or come to any conclusion regarding, the status of the many puzzling hybrids or intermediates that have been shown to exist, but shall merely try to present some of the facts and some of the theories.

P. A. Taverner (1927) has made some extensive studies of the *borealis* group and has published an excellent paper on the subject, beautifully illustrated with colored plates. He has suggested that *harlani* is a color phase of *calurus*, and that *krideri* is a color phase of *borealis*, stating that "*krideri* occurs along the western boundary of the *borealis* range and *harlani* at the northern extent of *calurus* territory, neither having any centre of distribution where they occur in purity." This seems like an untenable theory and to be based on incorrect premises, for it is a well-known fact that pure *krideri* occupies quite an extensive breeding range on the prairies and plains of the Midwestern States; and Harry S. Swarth (1926) says of *harlani*, in the Atlin region of northern British Columbia: "The birds were abundant and nesting over a wide expanse of territory, and within that region they were the only form of *Buteo borealis* that was seen."

Recently, Mr. Taverner (1936) has published another paper on the complicated relationships in the *Buteo borealis* group, to which the reader is referred for his latest views on the subject. Dr. Louis B. Bishop has also made quite an extensive study of this group, based on his large collection of the various races and on his study of other collections, and does not wholly agree with Taverner's conclusions. I have discussed the matter extensively with both of these gentlemen, and have examined an immense amount of material in various museums and private collections; but I must confess that there are many perplexing problems yet to be worked out before the relationships can be fully understood.

Mr. Swarth (1926) says of the haunts of Harlan's hawk in the Atlin region:

These dark-colored Buteos were seen by us almost daily through the summer and in all parts of the region that we visited. On May 21 several were observed soaring low over the snow-covered slopes on the east side of White Pass. During the next week, at Carcross, they were seen daily; apparently several pairs were settled on their nesting grounds near the town.

About Atlin these hawks were distributed throughout the lowlands; there were nesting pairs at intervals of a few miles in whatever direction one traveled. Although the species was thus relatively numerous, specimens were hard to obtain; the birds were remarkably wary.

The Harlan hawk is in the Atlin region mostly a bird of the timber. The sort of perch most often chosen is the top of one of the taller spruce trees, often in fairly dense woods but always with such a commanding view as to make approach unseen out of the question. With the exception of the dark colored hawks seen in White Pass early in the season and supposed to be of this species, none was observed in the open country above timber line. The abundance of ground squirrels might have been supposed to be an attraction to that region, too. They were extremely wary always, so much so that although both birds of a pair might circle about, screaming, as long as an intruder remained in their territory, it was generally impossible to approach within gun shot.

Spring.—H. V. Williams, of Grafton, N. Dak., who has collected some 50 Harlan's hawks for Norman A. Wood and a few more for the author, says that he never sees them in that State except on the migrations; his spring dates range from April 3 to May 6. A. G. Lawrence's records for southern Manitoba average around April 8, the earliest April 1. He says in his notes:

On April 9, 1916, I witnessed at St. Vital, Manitoba, a large flight of red-tailed hawks and Harlan's hawks. It was a fine day, and the snow had partly melted away in open places, though still deep in the woods. In the afternoon a redtail was seen high overhead, then several, then more and more, until when I came to a clearing I counted between 60 and 70 redtails and Harlan's circling around in two groups, the Harlan's numbering about 15 to 18. There was little wind, and the birds had to flap their wings fairly frequently, circling round and round, making no sound, now close together, now scattering and spreading over a wide area, then reforming into a flock to continue circling over the clearing. This performance continued for over an hour, the birds still circling when I left.

Nesting.—Audubon (1840) was, of course, mistaken in thinking that Harlan's hawk bred in Louisiana. Its breeding range was formerly supposed to include certain south-central States in which it is now known to occur only in winter. There are several sets of eggs in collections, said to be *harlani*, which are undoubtedly something else. These may be eggs of the ferruginous roughleg, Swainson's, or western red-tailed hawk, all of which have melanistic color phases. Positively identified eggs of Harlan's hawk have apparently never been taken. All we really know about the nesting habits of this hawk is contained in the following brief account by Mr. Swarth (1926):

One nest was found. It was in the valley a few miles from Atlin, in rather open spruce woods, just above a stretch of marsh land. The nest was near the top of an isolated spruce, on a branching limb, about sixty feet from the ground. It was a huge mass of sticks, a platform that had been flattened to such an extent that the young birds were in plain sight from the ground nearby. On July 6 it held two young, with feather rows showing through the down on the breast. Returning on July 20 we found the young birds gone, but discovered them in nearby trees. They had evidently just left the nest; wing and tail feathers were not yet full grown, and they could make but short flights. On August 11 a second brood, again of two birds, was found, obviously just out of the nest. These birds could fly but feebly; when found they were on the ground in dense spruce woods. One young bird and one parent were shot.

Plumages.—The downy young and early nesting plumages of Harlan's hawk seem to be unknown. Mr. Taverner (1927) has illustrated two nearly fully fledged nestlings, nearly fully grown, taken with their two parents in the Mount Logan area, Chitina River Glacier, Alaska, on July 19, 1925. These two specimens are quite unlike (see his pl. 3, figs. 4 and 5), one being the darkest and the other the lightest colored of the brood of three. The male parent is a dark bird, which Mr. Taverner calls "an almost typical black-phased *calurus*" but which Dr. Bishop calls the second-year plumage of *harlani* in the dark phase (see his pl. 3, fig. 2). The female parent (his pl. 3, fig. 3) we can all agree to call a typical adult *harlani* in the light phase. The reader may form his own conclusion, but it seems to me that Dr. Bishop is probably more nearly correct.

There seems to be no doubt that *harlani* has two very distinct color phases, an extremely dark phase and an extremely light phase. These two phases evidently interbreed, causing considerable confusion. The spotted tail is characteristic in the adults of both phases, but the immature birds are not so easily recognized. Norman A. Wood (1932), who has made a careful study of a large series of Harlan's hawks, tells me that in the dark phase of the first-year plumage *harlani* can be distinguished from the similar stage of the dark phase of *calurus* by numerous and conspicuous white spots, some of them rather large, on the webs of the feathers of the back, scapulars, and wing coverts, and by the generally blacker tone in the entire plumage. In *calurus*, the general tone is browner and the dark-brown feathers are white only basally.

The immature plumages of the light phase are much like similar stages of *krideri* and are nearly as light colored. Specimens of immature birds in this plumage have been collected in Alaska and British Columbia, which have suggested the extension of the range of *krideri* into that region, which seems to be too far removed from its normal range. I believe that the two young birds collected by Mr. Dixon, at Flood Glacier on the Stikine River, on July 31, 1919, and so well described by Mr. Swarth (1922), were *harlani* in this plumage; it is interesting to note that Mr. Swarth was "disinclined to regard" them "as examples of *krideri*, thereby extending the range of that form far to the westward."

Dr. Bishop seems to be convinced that Harlan's hawk has a well-marked second-year plumage, and I am inclined to agree with him, though Mr. Taverner denies it. This plumage, in the dark phase as we recognize it, is characterized by a dull red tail, crossed by dark bars and often showing some spotting; the white-spotted plumage of the upper parts has been replaced by a uniform, sooty-brown plumage, somewhat darker than in *calurus*. I have not seen any birds in the light phase that I could recognize as second-year birds. If, as seems likely, *harlani* has a distinct second-year plumage, which I have been unable to trace in other red-tailed hawks, it tends to strengthen the theory that *harlani* is a distinct species.

Food.—Swarth (1926) says on the food of this hawk: "Of the six specimens I collected four had crop or stomach or both well filled. Two contained rabbit (*Lepus americanus macfarlani*), one held ground squirrel (*Citellus plesius plesius*) and chipmunk (*Eutamias borealis caniceps*), and one held rabbit and chipmunk."

Norman A. Wood writes to me that his collectors in Arkansas tell him that these hawks feed mostly on rabbits and quail, but also on squirrels, field rats, and mice, and more or less on small birds.

Behavior.—Audubon (1840) writes:

This species, although considerably smaller than the Red-tailed Hawk, to which it is allied, is superior to it in flight and daring. Its flight is rapid, greatly protracted, and so powerful as to enable it to seize its prey with apparent ease, or effect its escape from its stronger antagonist, the Red-tail, which pursues it on all occasions.

The Black Warrior has been seen to pounce on a fowl, kill it almost instantly, and afterwards drag it along the ground for several hundred yards, when it would conceal it, and return to feed upon it in security. It was not observed to fall on Hares or Squirrels, but at all times evinced a marked preference for common Poultry, Partridges, and the smaller species of Wild Duck.

Fall.—Mr. Swarth (1926) says: "During September, Harlan hawks were migrating in numbers. They were seen near Atlin daily, and between Atlin and Teslin (September 7 to 15) a number were observed drifting southward. On September 21, I saw two, the last observed."

Mr. Williams gives fall dates for North Dakota from September 28 to November 20, and says: "The flight usually starts with a few stragglers and gradually increases to its height (about 200 birds for a day or so), then gradually decreases. I see usually from 700 to 1,000 every spring and fall. They fly with the red-tails and circle the same way, usually very high, but after the first few days a number of them stop to feed, but are very shy and wild."

Winter.—Harlan's hawk seems to be a bird of rather restricted range at all seasons. Mr. Swarth (1926) words it as follows: "Breeds in extreme northern British Columbia, east of the coast ranges, north into the valley of the Yukon, and eastward for an undetermined distance. Migrates southward east of the Rocky Mountains, through the Mississippi Valley to a winter home in the Gulf States."

According to N. A. Wood (1932) "its winter range includes a territory embracing all of Arkansas, southern Missouri, Oklahoma, and northern Texas." One of his collectors in Arkansas, Clyde Day, says: "These hawks seem to make their winter quarters in a strip about 100 miles north and south and 300 miles east and west. I see on an average ten a day, half of them Harlan's."

BUTEO BOREALIS UMBRINUS Bangs

FLORIDA RED-TAILED HAWK

HABITS

More than 30 years ago Outram Bangs (1901) gave the above name to the red-tailed hawk of southern Florida. His description was based on only a single specimen collected in April 1888 at Myakka in Manatee County. The characters given were: "Size and proportions as in *Buteo borealis borealis;* color, above, darker; throat and middle of belly marked with broad, conspicuous striping and banding of deep chocolate-brown; tail-feathers with dark brown markings (the remains of bands) near the shafts. From *B. borealis calurus* the new form differs in being less suffused with reddish below, and in different general tone of coloration."

At the request of Mr. Bangs and with the expert help of John B. Semple, we were able to collect, during the winter and spring of 1930, a fair series of these hawks. With the exception of one bird, probably a migrant from farther north, this series shows that this race is uniformly well marked and is permanently resident in the southern half of Florida. Birds collected by William G. Fargo in Brevard County are clearly referable to this race, but the characters are not so pronounced. Intergradation with northern *borealis* may occur in northern Florida. The range of this race doubtless includes Cuba and the Isle of Pines, and possibly Jamaica.

Red-tailed hawks are fairly common, as hawks go, in Florida. They are widely distributed throughout the flat pine woods, locally known as "piney woods" or "flatwoods", with which a very large portion of Florida is covered. One may drive for many miles along roads and see nothing but an apparently endless expanse of flat country covered with an open growth of tall, slim Caribbean or long-leafed pines, rough-barked and scraggly, but in their perfection sturdy, grand, and impressive. The stand is so open that dense undergrowths of low-growing saw palmettos, with their curious, half-buried root-stems trailing along the ground, form large and almost impenetrable thickets. Many shallow grass-lined ponds and open grassy savannas are scattered through them. The soil is mostly sandy, but in many places it is rich enough to support a luxuriant growth of shrubbery and many beautiful wild flowers. To the casual observer a drive through such an endless, bewildering maze of pines, stretching away into the dim distance, becomes monotonous and tiresome. But I can share the sentiments so well expressed by Dr. Charles T. Simpson (1923):

> There is a nameless charm in the flatwoods, there is enchantment for the real lover of nature in their very sameness. One feels a sense of their infinity as the forest stretches away into space beyond the limits of vision; they convey to the mind a feeling of boundless freedom. The soft, brilliant sunshine filters down through the needle-like leaves and falls in patches on the flower covered floor; there is a low, humming sound, sometimes mimicking the patter of rain-drops, as the warm southeast wind drifts through the trees; even the loneliness has an attraction. To me it all brings a spirit of peace, a feeling of content-ment; within the forest nature rules supreme.

Nesting.—I have examined only two nests of this hawk, in two quite different situations (pl. 52). While driving over the Kissimmee Prairie, near Bassinger, Fla., on March 21, 1925, we saw a large nest about 40 feet up in a lone cypress that stood near a small cypress hammock. The tree was heavily draped with Spanish moss. We could see the tail of the sitting hawk projecting over the edge of the nest, and its mate was perched on a nearby tree. As I climbed to the nest the old birds attacked me, screaming vigorously; one nearly struck me in its swoops. The nest was made of large sticks, lined with fine twigs, grasses, straws, and gray moss; it measured 40 by 27 inches in diameter and was 18 inches high. It contained a pipped egg, a recently hatched young hawk, and the remains of a young cottontail rabbit.

The other nest, found on February 15, 1930, in Glades County, was fully 60 feet from the ground in the topmost crotch of a big long-leafed pine (pl. 52); the tree towered above the tops of all the surrounding trees in open flat pine woods, and the nest could be plainly seen from the road at a long distance. The nest was made of pine sticks and twigs, deeply hollowed and profusely lined with dead and green pine needles; it measured about 24 inches in diameter, the inner cavity being 10 inches wide and 4 or 5 inches deep. The hawk flushed off the nest but did not offer to attack me while I was collecting the two eggs. Both parent birds were secured.

Mr. Fargo has sent me some notes on some large nests that he found in Pasco and Brevard Counties, some "nearly as large as bald eagles', but wider than high. All the nests seen were in some tree where a good outlook was to be had at the loss of hiding the nest.

These birds purposely select a pine tree or cypress that stands a bit away from the forest or if in a swamp, then at its edge, so that they can see out on the open country."

Joseph C. Howell has sent me data on 10 nests that he has examined. Two of these were in oaks and only 20 feet from the ground. The others were all in pines ranging in elevation from 17 to 60 feet, most of them from 40 to 50 feet up. One nest that he describes as a large one measured 2 feet in diameter and 2 feet in height; the inner cavity was 9 inches across and 3 inches deep; it was an old nest that had been repaired. "The foundation of the nest was largely of good-sized pine twigs, with one oak limb and a piece of palm boot here and there. The lining was of pine needles, mostly green, pure white down from the parent, live Spanish moss, bits of dead palm frond, and a few strips of green palm leaf."

Eggs.—The eggs of the Florida redtail apparently do not differ materially from other redtails' eggs. The set I collected happened to be more heavily marked than usual, but I have seen others nearly, or quite, immaculate. The measurements of 32 eggs average 60 by 47.3 millimeters; the eggs showing the four extremes measure 65.1 by 45.3, 61.3 by 49.3, 54.6 by 47.5, and 61.7 by 43.8 millimeters.

Behavior.—We noticed nothing in the habits of this hawk that differed from those of its northern relatives. It is very shy, but that is true of the species elsewhere. Its food seems to consist mainly of mammals, such as rabbits and cotton rats; no evidence of bird killing was seen.

BUTEO LINEATUS LINEATUS (Gmelin)

NORTHERN RED-SHOULDERED HAWK

HABITS

The well-known red-shouldered hawk, with its various races, is widely distributed throughout the eastern and southern portions of the United States and southern Canada, wherever it can find suitable environment in well-watered woodlands scattered through open country, even in quite thickly settled farming country, or in wooded river bottoms. In such localities it is generally the best known and commonest hawk. It is less common than the broadwing in extensive heavily forested regions and less common than the redtail in the open country of the Middle West. It is very rare, or practically unknown, on the western prairies and arid plains, or in the northern coniferous forests. The northern form ranges, as a breeding bird, west to the eastern edge of the Great Plains and south to North Carolina.

Spring.—Although the red-shouldered hawk has been called the "winter hawk", it hardly deserves that name in New England, where it is much less hardy than the red-tailed hawk and is seldom seen in winter. In the northern portions of its breeding range it is mainly migratory, though it is usually listed as a permanent resident. C. J. Pennock tells me that a specimen taken as late as February 14 in Charlotte County, Fla., was referred to this form. Dr. Winsor M. Tyler's notes give the average date of arrival in Lexington, Mass., over a period of 20 years as March 16, the earliest being March 3 and the latest March 30. My dates are somewhat earlier, as I have found them working on their nests, in Bristol County, as early as March 5. Dr. Tyler has sent me the following notes on a migratory movement he observed on March 22, 1911, at Lexington:

This morning we saw three or four red-shouldered hawks circling at no great height. We soon lost sight of them off toward the north. Later, about noon, a single bird flew in great circles overhead, making good progress, although intermittently, in a northerly direction. A few minutes later another bird flew over in a similar manner, screaming as he passed along. Then we saw three others—two a little behind the first, and one very high above him—all four circling on outspread wings—the tail wide-open like a fan. The birds cut great circles in the air, or rather, as they passed northward, they described immense loops, and each circuit carried them farther away toward the north. Sometimes one would lead, or would appear from the ground to do so, then as he swung southward he would appear larger again until he turned first eastward and then northward. Then as he moved away, we lost him in the distance.

As we watched the four birds circling together a fifth joined them. His wings were bent back at the shoulder, and his tail was shut and he sailed in this position, slanting downward, I think, moving rapidly toward the others. Upon overtaking them he spread his wings and tail to their utmost extent and continued with the other birds in their great slow circles. The fifth bird shot out from just under the sun and holding a straight course, passed over our heads. On this day and at this hour (noon) the sun is directly south; hence the route of the six birds was astronomically due north.

Courtship.—One of the delights of early spring, on one of the first balmy days of March, when the genial warmth of the advancing sun is thawing out the hibernating butterflies, the early wild flowers are showing signs of new life, and the first hylas are peeping in the marshy pools, is to walk through the now leafless woods, breathe the fresh fragrance of awakening spring, watch for the early migrating birds, and listen for the courtship cries of our favorite hawks, old friends of many years' standing. The blue jays can closely imitate their cries, but there is a difference we can recognize. And soon we see them soaring in the air in great circles high above the same old woods where they have nested for many years. We believe that they are mated for life, and we like to think that this is the same old pair that we have known so long. But probably it is only the continuation of a pair, for if one of a pair is killed, the survivor promptly brings a new mate into the territory, or feudal domain of the pair. These hawks are very noisy and conspicuous at this season, in marked contrast with their behavior at other seasons.

Their loud, wild cries of *kee-ah, kee-ah* are frequently repeated as they circle overhead, their wings and tails broadly extended and stiffly held with only slight adjustments. Frequently they swing near each other and then far apart, or, mounting high in the air, one may make a thrilling dive downward toward the other. These evolutions are indulged in every year, even by mated pairs, and constitute, I believe, their principal courtship display. Dr. Tyler tells me that he has seen four birds thus engaged, perhaps in a spirit of rivalry, or perhaps merely in play, as an outlet for surplus energy. Lewis O. Shelley has sent me the following account of a courtship performance:

On March 24, 1930, while crossing a back-lying mowing between two plots of woodland, a red-shouldered hawk was seen to come from the west and begin to mount higher by spiraling, until it had gained an altitude of about 1,000 feet, screaming the common *kee-you, kee-you* note every little while, usually on the outer swoop for the next vault in the rise. At the zenith of its flight the calls were loudest, two-syllabled screams.

Just at this time another hawk, the female, came from the west, crossed 30 feet overhead, and alighted in a bare oak 200 yards away at the edge of the woodland. The male had evidently been watching the female's approach, as, several moments before I knew of her presence, he began shooting downward with swift lunges for several hundred feet at once, checking the rush and sweeping a wide spiral before again dropping down. No sooner had the female alighted than the male, from a height of at least 200 feet, made a last rapid drop that landed him on the female's back. Just the second before this contact she had spread her wings and crouched down close to the branch and crosswise of it. Copulation was immediate, occupying about 30 seconds. Then the male hopped along the branch and they sat facing opposite directions, immovable, a foot or so apart, for 10 minutes. At the end of this period, the male launched forth and flew back toward the west, where he proceeded to climb beyond the range of the naked eye. Soon after he left the oak, the female followed, but did not go near his location.

Nesting.—Our experience with the nesting habits of the red-shouldered hawk in southeastern Massachusetts has been quite extensive, covering a period of 50 years. I find in my notes the records of 173 nests that I have examined personally; if I include the nests examined by my field companions, F. H. Carpenter and C. S. Day, the list would run up to nearly 250. This is not, however, a remarkable record, for my correspondence with others in the Northeastern States shows that many of them have found these hawks equally common. Nearly all collections have big series of the eggs of this hawk, which speaks well for its popularity. I shall never forget the thrill I experienced when I found my first hawk's nest and how I prized those handsome eggs! I have never forgotten what Henry D. Minot (1877) wrote in one of my earliest bird books: "Size has always a fascination for the world. The young collector prizes a hawk's egg more than that of the rarest warbler. The egg is big,

the bird that laid it is big, the nest in which it was laid is big, the
tree in which the nest was built is big, and the wood in which the
tree grows is big." And I have never quite lost that thrill. As I
walk down the old familiar cart path into the well-known woods,
the home of many generations of Buteos, I am filled with keen expectancy; the warning cry of the hawks is heard, and I am soon
gazing at a well-feathered nest in a lofty crotch. Perhaps the bird
has already flown, or perhaps a blow on the tree trunk will send her
sailing off through the woods. Sometimes she may return to circle
overhead and scream defiance, but oftener not.

The outstanding feature of our experience with the red-shouldered
hawk has been the constancy with which each pair, or its continuation, has clung to its chosen territory, in spite of the annual robbing
of its nest and the cutting down of one portion after another of its
woodland home. As long as there are any trees suitable for nesting
purposes the hawks will remain in the vicinity. We have had under
observation at various times over 30 pairs of these hawks within a
radius of 15 miles of my home. We have not had time to visit all
these pairs every year but have located as many as 22 pairs in a
season as recently as 1922. There has been a marked decrease during
the past 10 years. To illustrate the continuity and permanency of
localized pairs, I submit brief histories of four of our oldest pairs.

The "chestnut hill" pair was first located in 1882 in an extensive
tract of magnificent chestnut timber, where trees 4 feet in diameter
at the base and 60 feet to the first limb were not uncommon. The
hawks nested in this section for 8 years until extensive cutting of the
big timber forced them to move into an adjacent tract of swampy
woods. Meantime one of the hawks was shot by my companion, but
the survivor secured a new mate and occupied the same old nest the
following year. After that the hawks were forced to move every
few years, until the last of the woods were cut off. The last nest
of this pair was found in 1922, a lapse of 41 years, during which
we actually found the nest 20 times.

The most continuous record is that of the "reservoir woods" pair.
From 1882 to 1907, inclusive, we found the nest every year, an unbroken record of 26 years. From that time on our records were more
irregular, as more or less cutting was done in various parts of the
woods, until the last nest was found in 1923, after a lapse of 42 years,
during which we found the nest 31 times. In 1924 this last nest was
occupied by a pair of barred owls, and in 1928 we found red-tailed
hawks had appropriated the same old nest (see pl. 44). The woods
have been nearly ruined since then, and no hawks have nested there.

The history of the "Goff's woods" pair is similar, but there are
more breaks in the record. It also began in 1882, and the last nest

was found in 1926, a lapse of 45 years, during which time the nest was found 29 times. The hawks were seen or heard in the woods for two years longer, but in 1930 a pair of red-tailed hawks moved in and that ended the record of this pair.

The "Dighton big woods" pair has shown the longest record, a lapse of 47 years, but its early history is much broken and we have actually found the nest only 15 times. We found the first nest in 1886 and for 4 years in succession. The main tract of heavy chestnut and oak timber was cut soon after that, so we did not hunt the locality again until 1909. From that time on we found them, whenever we hunted for them, in an adjoining tract of oak woods that was growing up to respectable size. Here we found them every year for the 8 years, 1927 to 1934, and we hope they will continue to nest there. I believe that all these pairs, or their successors, have nested continually in all these localities. The breaks in our records are mainly due to our failure to visit all the localities every year, though in many cases there has been much shifting about, due to extensive cutting in different parts of the woods. My field companion, Mr. Day, has the records of two pairs, near Boston, for 37 years each, and the "Weld farm" pair, which he found from 1909 to 1923, is said to have nested in that tract as far back as 1872. Fred H. Kennard (1894a) gives an interesting 10-year history of four pairs of hawks in Brookline, Mass., and vicinity, including the Weld farm pair, which is well worth reading.

I have already explained the distributions of the red-tailed and red-shouldered hawks in southeastern Massachusetts under the former species, the latter being much commoner in what I call the hardwood region. Here the red-shouldered hawk shows no very decided preference for any particular species of tree, but usually selects one of large size. Of 177 nests in the hardwood region 49 were in chestnuts, 46 in red oaks, 26 in white pines (usually scattered among the hardwoods), 19 in white oaks, 15 in swamp white oaks, 13 in scarlet oaks, 8 in maples, and 1 in an ash. Certain pairs seem to prefer to nest in pines, even where suitable hardwoods are available. Of 41 nests in the white-pine region, 31 were in pines, 4 each in beeches and red oaks, and 1 each in a maple and a chestnut. The heights from the ground varied from 20 to 60 feet; 27 were 30 feet or less, 36 were 50 feet or more, and a majority were between 35 and 45 feet. Very few of the nests were actually in swampy woods, although many were in the dry parts of woods near swamps or streams; but some were in high, dry woods, far from any water.

In a hardwood tree the nest is usually placed in the main fork of three or more branches, seldom on horizontal branches against the trunk and very rarely in the fork of a branch. In a pine tree it is almost invariably on three or more branches against the trunk. Mr.

Day once found a nest a foot out from the trunk of a pine in a cluster of twigs and small branches, a very unusual and insecure location; but the same nest was used again 10 years later! This same pair once built in a red cedar.

The nest of the red-shouldered hawk is usually well and firmly built and securely placed on a solid foundation, so that it lasts for a number of years. I believe the hawks prefer to build a new nest each year, as they usually do so; occasionally we have found them using the same nest for two, or even three, years in succession; oftener they return to the old nest after a lapse of two or more years. Whether the nest is robbed makes very little difference; I have known a pair to raise a brood successfully for three years in succession in the same nest and then abandon it the next year for no apparent cause. On the other hand, I have known a pair to use the same nest three years in succession, including a second set one season, although the eggs were taken each year. Occasionally red-shouldered hawks use nests built by other species; I have never found them using an old crow's nest, but they sometimes alternate with barred owls; we have three times found them using old nests of Cooper's hawks, and they often build over old squirrels' nests; in four cases we have seen a gray squirrel run out from the lower portion of an occupied hawk's nest. Probably the hawks seldom molest these large squirrels; once we found the hawk's eggs broken, for which the squirrels may have been to blame.

About the first week in March, in a normal season, our red-shouldered hawks begin repairing their old nests, or building new ones, about a month or more before the first eggs are laid. If an old nest is to be used it is plainly marked thus early in the season with a fresh sprig of pine, hemlock, or cedar, as a sign that the owner has claimed possession. Nest building is a deliberate process, lasting four or five weeks, and sometimes nests are built that are never used. One pair, or possibly an unmated bird, has for several years built a nest that was not used; one year they built two nests, using material from the first, which was a fine large nest, to build the second, a quite inferior nest; although I flushed the bird off this second nest, no eggs were ever laid in it.

The nest of the red-shouldered hawk is usually recognizable as a substantial, well-built structure, filling the crotch to a considerable depth and rather flat on top. It is usually smaller than a redtail's nest, which it resembles otherwise, but it differs from the nests of the Accipiters in containing much more soft material, and it is larger than the carelessly built nest of the broadwing. It is usually well decorated with bits of white down, which increases as incubation progresses; more or less down is also scattered about in the woods near the nest. The nest is made of whatever small sticks or twigs are easily available, mixed with strips of inner bark, dry leaves, usnea, mosses, lichens, twigs of evergreens with needles, and other light materials; the inner cavity is neatly lined with finer shreds of inner bark, softer mosses or lichens, and fresh sprigs of pine, cedar, or hemlock, producing a pretty effect, with a sprinkling of white down from the hawk's breast. The nests vary considerably in size; an average nest would measure about 24 by 18 inches in outside diameter, the inner cavity about 8 inches in diameter and 2 or 3 inches deep, and the height, or thickness, 8 to 12 inches. The largest nest that I have measured was 36 inches in longest diameter, 18 inches high, and hollowed to a depth of 6 inches; the smallest measured only 16 by 14 inches in diameter, 7 inches high, and hollowed only 2 inches. Dr. Harry C. Oberholser (1896) gives the measurements of 10 Ohio nests; they exceed my average in size but not my largest except in height; two of his nests were 27 and 28 inches high.

Our experience is that the red-tailed and red-shouldered hawks never nest near together and that the same is true of the Cooper's and sharp-shinned hawks, these being two pairs of competitive and antagonistic species; we have, however, occasionally found one of the Buteos nesting in the same woods with one of the Accipiters. But Dr. William L. Ralph (Bendire, 1892) says: "I have known the Red-tailed and Red-shouldered Hawks and the Great Horned Owl to nest near one another in a small wood, and on one occasion I found a pair of each of the Sharp-shinned, Cooper's, and Red-shouldered Hawks, and of the Long-eared Owl breeding so near together that I could stand beside the nest of a Ruffed Grouse, which was close by also, and throw a stone to any of the others."

I have always considered the red-shouldered hawk and the barred owl as tolerant, complementary species, frequenting similar haunts and living on similar food, one hunting the territory by day and the other by night. We often find them in the same woods and using the same nests alternately, occasionally both laying eggs in the same nest the same season, resulting in a mixed set of eggs on which one or both species may incubate. Walter A. Goelitz (1916) has described and photographed a tree containing nests of both hawk and owl within a few feet of each other.

Some slight differences are noticeable in nesting habits in different parts of the country. The choice of a tree depends on the type of heavy timber prevailing, as the hawks choose whatever large trees are available. In New York State, Dr. Ralph (Bendire, 1892) says that they nest in birch, ash, maple, and beech trees, with a preference for the first two. William A. and George M. Smith have sent me data on 45 New York nests; of these, 26 were in beeches, 12 in maples, 3 in ashes, 2 in basswoods, and 1 each in an oak and a hemlock.

I have never found fresh green leaves of deciduous trees or plants in a nest until after the eggs had hatched, but it seems to be a commoner practice elsewhere. J. Claire Wood (1906) says that a majority of the nests found in Michigan are so decorated after the eggs are advanced in incubation, and that "a nest found in Genesee county on April 12, 1903, containing three fresh eggs, presented a beautiful green interior the hollow being lined with 'boxberry' leaves and the surrounding platform concealed beneath a profusion of spruce twigs with their covering of green needles. Not only are the leaves of various trees used but entire plants of such as night shade and violet. Have found the latter so fresh that the adhering flowers had not commenced to droop."

S. F. Rathbun writes to me that he has found nests lined with "the dry blades from cornstalks and the dried webs of the tent-caterpillar's retreat." He thinks that the green sprigs of hemlock, so often found in nests, are usually picked up from the ground after having been blown off the trees. He has noticed that this material was used freely when windy weather prevailed during the nest building period and was sometimes lacking during very calm seasons. The use of green lining is, in my opinion, for sanitary rather than ornamental purposes.

Eggs.—The ordinary set of the northern red-shouldered hawk consists of three or four eggs; sets of three are much commoner than four, only about one set in three or four consists of four eggs. Sets of five are very rare; I have taken only one and heard of only two others in my home territory. Bendire (1892) records two sets of six. The eggs are the usual hawk shape, ovate to rounded-ovate or oval. The shell is smooth but without gloss until much worn by incubation. The ground color is dull white or pale bluish white, sometimes clear white when fresh, but often much nest stained. They are perhaps the handsomest of the eggs of the Buteos and show an almost endless variety of types and colors of markings. Some are boldly and irregularly marked with great blotches of "warm sepia", "bay", "chestnut", "auburn", "amber brown", "tawny", "russet", or "ochraceous-tawny"; sometimes two or three shades of light and dark browns appear on the same egg; some are spotted or blotched with various shades of "purple-drab" or "ecru-drab", with or without overlying spots of the different browns; sometimes any of these colors are splashed longitudinally at one end. Some eggs are more evenly covered with small spots or fine dots of any of the above colors. Some are very sparingly marked, but wholly immaculate eggs are very rare. The measurements of 50 eggs in the United States National Museum average 54.7 by 43.9 millimeters; the eggs showing the four extremes measure 59 by 47, 51 by 41, and 53 by 40 millimeters.

Young.—The period of incubation is about 28 days, and it is shared by both sexes. I have seen the pair change places on the nest; the male comes flying low through the woods, screaming, and alights on the edge of the nest with an upward glide; the female rises up in the nest (sexes assumed) and flies away, and the male settles on the eggs at once. The male also feeds the female while she is incubating. Once, after I had been watching a bird on a nest for over an hour, her mate flew up, bringing a mouse in his bill; he stood beside her on the nest, gave her the mouse, and then quickly flew away; she rose up off the eggs, stood on the side of the nest, and ate the mouse, pulling it to pieces and swallowing it in small bits; this took about five minutes, after which she settled down on the eggs as before. I have also seen a hawk come to a nest, where his mate was incubating, leave food on the nest, and fly away, after which she stood up and ate the food. I have spent many hours in a blind watching hawks at their nests. The blind must be well hidden and offer perfect concealment; a brush blind is useless, as their eyesight is very keen; they will detect the slightest movement, leave the nest at once, and will not return to it if they know that they are watched. Much of the time while incubating the hawk is alert, with head raised and constantly looking about; at such times she will fly if anyone approaches the tree; at other times I have seen the same bird apparently asleep on the nest, with her head invisible, when I could walk up and pound the tree before she would leave. Once a hawk that I had been watching for some time rose up and stood on the edge of the nest for 10 minutes, looking around, then settled down on the eggs and went to sleep; I had to pound the tree to make her fly.

On May 20, 1932, I saw the whole process of feeding a brood of three young; they were about one-third grown but still all downy; they were very lively, constantly moving about, voiding their excrement over the edge of the nest, and looking up into the sky, apparently hungry. At 9.55 a. m. a hawk flew over screaming but did not come to the nest, and at 10.15 the same thing happened again, the young watching eagerly. At 10.30 a hawk came to the nest, apparently left some food and flew right away; about 10 minutes later this happened again. The young made no attempt to eat the food. At 11 a. m. the female, presumably, alighted on the edge of the nest and proceeded to feed the young; she tore the food (I could not see what it was) into small bits and fed the three young as nearly in succession as she could; they took the pieces from her bill, and she swallowed a piece occasionally herself. Competition was keen, and in the struggle the largest one got more than its share, taking several pieces in succession; when it had enough it turned its back on its mother and moved over to the other side of the nest, facing out, with its crop much distended. She continued feeding the others until the second and finally the third turned away satisfied. When all their crops were stuffed full, she stood on the nest watching them for a few minutes and then flew to a nearby tree. The feeding process required about 20 minutes. Evidently they are fully fed, with all they can hold, at very infrequent intervals, for often I have watched a nest for an hour or two without any results.

For the first week or so the small downy young are quite inactive, lying quietly in the nest, but when about 10 days old they begin to sit up and move around. As the eggs are laid at intervals of two or three days, and as incubation usually begins before the set is complete, there is generally quite a difference in the sizes of the young. When about two weeks old the plumage begins to appear and the young from that time on become more and more active. My young friend Robert W. Harding watched a nest of young hawks one day from 10.10 a. m. until 12.30 p. m. and recorded happenings for me; they were never still for 10 minutes and generally not for more than one or two; they were constantly standing up, lying down, flapping wings, preening their budding plumage, frequently screaming, and occasionally squirting their excrement far over the edge of the nest; the ground and bushes under the tree looked as if sprinkled with whitewash; twice, at intervals of about an hour, one of the old birds brought in food, once a frog and once a small snake. When about five or six weeks old the young are nearly fully fledged and ready to leave the nest. They begin by climbing out on the branches and, perhaps, returning to the nest at night. But they soon learn to flutter down to the ground or flap awkwardly through the woods. They are guarded and fed by their parents for some time after they leave the nest, until they learn to shift for themselves. That the nestlings are well fed is shown by the following full meal found in the crop and stomach of a young hawk taken by F. Seymour Hersey (1923): "A garter snake fifteen inches long; the head and about four inches of another snake of similar size; both hind legs of two frogs of good size; several small pieces of flesh probably of these frogs; a small turtle about the size of a silver dollar; three legs and the bill of Ruffed Grouse chicks; a large quantity of mouse hair mixed with green leaves."

Only one brood is raised in a season, but, if the first set of eggs is taken, a second set will be laid, usually in another nest, about three or four weeks later. Occasionally, if necessary, a third attempt is made.

Lewis O. Shelley sends me the following note on two young red-shouldered hawks that were taken from the nest when about three

weeks old and mounted: "When the young hawks were being skinned
both birds were found to have the ears affected with maggots; 20
were collected from the four ears and preserved; the maggots were
similar to *Protocaliphora* larvae commonly infesting bluebird and
tree-swallow boxes and rarely some nests such as the phoebe. These
maggots had eaten about the ears not only to disfigure the outer
ear cavity, leaving the marks of their attacks so that they are preserved in the mounted specimens, but had completely destroyed
both ear drums of both birds. It is to be wondered how long these
young hawks would have survived had they lived to reach maturity."

Plumages.—The small downy young is thickly covered with long,
soft, silky down, longest on the head, yellowish white above, tinged
with "vinaceous-buff" on the back and wings, and whiter below. This
is succeeded by a dense covering of short, thick, woolly down, thickest
on the belly, where it is pure white, and grayish white above. The
wing quills are the first to sprout, when the young bird is about two
weeks old, followed by the scapulars, wing coverts, and then the contour plumage. The back is fully feathered first, then the sides of
the breast. The head, center of the breast, and thighs show the last
of the down. This has all been worked out in detail, from birds
raised in captivity, by Mr. Kennard (1894b), with much information
on the growth, food, and behavior of the young hawks.

In fresh juvenal plumage the crown, back, and wing coverts are
dark "bister", with "tawny" edgings; the tips of the primaries are
brownish black without barring, but inwardly they are extensively
patterned with dusky and "pinkish cinnamon"; the secondaries are
notched with the latter color, and the tertials and upper tail coverts
are broadly tipped with it; the tail is tipped with white and has a
broad subterminal band of brownish black and seven to nine narrower bands above it, the spaces between being variegated with "hair
brown" and light gray on the outer portion and with much "orange-
cinnamon" on the basal half; the breast varies from pale buff ante-
riorly to buffy white posteriorly, with elongated ovate spots of sepia,
largest on the breast and smallest and palest on the belly and tibiae.

This plumage is worn for about 18 months without much change,
except by wear and fading; the tawny edgings wear away and the
under parts fade out to whitish, but the cinnamon tints in the
remiges and rectrices generally remain. The complete postjuvenal
molt begins in summer and is prolonged through the fall. Some
birds, perhaps all, acquire during this molt numerous triangular
spots of sepia on the breast, which persist while the rufous plumage
of the breast is being assumed. The wings and tail are apparently
molted late in fall or in winter, and I doubt if the fully adult plumage is acquired until after the next annual molt.

Adults have a complete annual molt, beginning sometimes in April and in some cases not completed until October. Molting birds are very scarce in collections, probably because these hawks seek the seclusion of the woods during summer. We have often remarked that, although we could find from 20 to 30 breeding pairs during any spring, I could almost count on the fingers of one hand all I have ever seen in summer.

Food.—The red-shouldered hawk is one of our most beneficial and least harmful hawks. It certainly does not deserve its common name, "hen hawk"; Dr. A. K. Fisher (1893) found that its diet consisted of 65 percent small rodents and only 2 percent poultry. Its diet is most varied, as it includes mammals, birds, snakes, frogs, fish, insects, centipedes, spiders, crawfish, earthworms, and snails, 11 classes of animal life. Of 220 stomachs examined by the Biological Survey, 3 contained poultry; 12, other birds; 102, mice; 40, other mammals; 20, reptiles; 39, batrachians; 92, insects; 16, spiders; 7, crawfish; 3, fish; 2, offal; and 1, earthworms. Dr. B. H. Warren (1890) says: "In my examinations of fifty-seven of these Hawks which have been captured in Pennsylvania, forty-three showed field mice, some few other small quadrupeds, grasshoppers, and insects, mostly beetles; nine revealed frogs and insects; two, small birds, remains of small mammals, and a few beetles; two, snakes and portions of frogs."

Dr. Fisher (1893) quotes J. Alden Loring as saying: "The pair reared their young for two years in a small swampy piece of woods about 50 rods from a poultry farm containing 800 young chickens and 400 ducks, and the keeper told me he had never seen the hawks attempt to catch one."

The mammal food recorded includes mice of various kinds, shrews, moles, squirrels, chipmunks, rabbits, muskrats, opossums, and skunks. Birds are not so often taken, but the list includes sora rail, pheasant, bobwhite, chickens, mourning dove, woodcock, screech owl, sparrow hawk, flicker, crow, blackbirds, meadowlark, robin, and various sparrows. Other items are lizards, toads, various frogs, salamanders, turtles, grasshoppers, crickets, mole crickets, beetles, wasps, katydid, cicada, spiders, centipedes, earthworms, snails, and various lepidopterous larvae. Two of these hawks, said to have been destroying birds, were brought to Prof. J. E. Guthrie, of Ames, Iowa; he says (1931): "I found in the stomach of one, a striped ground squirrel, a young rabbit, and twenty-four full grown grasshoppers. The other one's stomach was completely filled with our largest common species of grasshoppers, and one that perhaps has been doing the most damage of any in the central states this year. I identified the remains of forty-nine specimens. It is of interest to know that we have these helpers with us this summer when they are so much needed."

Edward H. Forbush (1927) writes:

Mr. Wilbur Smith informed me that a Red-shouldered Hawk appeared in Birdcraft Sanctuary, Fairfield, Connecticut, on January 1, 1920, and was seen by the superintendent daily for two months thereafter. There were nine pheasants in the sanctuary and also a bantam hen with several chicks in a coop, open at the top, but neither chickens nor pheasants were molested by the hawk. About February first it entered a hen-yard, where it found a skinned deer's neck and a dead opossum, and it fed on the carrion every day for about two weeks, while the hens merely withdrew to the coop. Mr. Frank Novak, the superintendent, saw the rooster standing one day within four feet of the hawk. During the worst kind of weather this bird did not molest a bird or a chicken, but it was repeatedly seen to catch rats and mice.

On the other side of the question Fred H. Kennard (1894a) says:

In each of the Hawks of this species that I have examined, I have invariably found feathers and birds' bones, and lots of them. The frogs alone, of which they eat great numbers, would seem to more than balance the injurious rodents of which they are also fond; and as for insects, I do not believe that the Brookline Red-shouldered Hawks eat as many in a year as an ordinary frog could in a day. They must differ in their habits, and accommodate themselves to their surroundings. Perhaps they are, as a species, beneficial, particularly where they hunt in open country; but in such country as we have around Brookline, I am sure they do more harm than good.

These hawks often use their old nests as feeding stations; these often show more or less down, which tempts us to climb to them, only to find that they have not been repaired and that they contain only remnants of animals and birds that the hawks have eaten.

Behavior.—The soaring flight of the red-shouldered hawk, oftenest seen during the early part of the nesting season, is much like that of the other large Buteos. It is powerful and graceful, often protracted to a great height and occasionally ending in a thrilling nose dive. Often, while hunting, it glides swiftly along on rigid wings just over the tops of the forest trees or even through the woods; again it glides low over the marshes or meadows in search of frogs or mice. Its coloration is concealing in the lights and shadows of the forest, where it can slip up unawares on the squirrels in the trees, or pounce down upon its humbler prey on the ground. When flying from its nest it swoops downward and flaps away in rather heavy flight, quite unlike the bulletlike dash of the Accipiters. On returning to its nest it flies low and glides up to it in an easy curve. When circling above the intruder at its nest and screaming, it does a great deal of flapping, interspersed with short sailings, and then it may glide off out of sight. As with all hawks, the feet are extended behind, a little below the tail.

There is considerable variation in the behavior of different individuals about their nests, and we have noticed that these individual characteristics are apparent year after year. Mr. Kennard (1894a) has noticed the same individual traits in his 10-year records of four pairs near Brookline, Mass. Some individuals always leave the nest quietly before we are near enough to see them and do not show themselves afterward. Others fly as soon as we come within sight, and still others wait until we rap on the tree or start to climb it. A common behavior is for one or both birds to circle about overhead, screaming, for a while and then gradually to drift away. On rare occasions I have had one remain close by, flying from tree to tree and swooping at me as she passed, but I have never had one strike me. Once one perched in a tree above me, stretched out her head, ruffed up the feathers of her neck in a menacing pose, and screamed angrily. E. B. Williamson (1913), of Bluffton, Ind., describes the actions of a very bold and savage pair that he encountered for two seasons; he writes:

One of the parents remained in the top of the tree calling fiercely but not moving. No attention was paid to it or to the other parent which was not noticed at the time. Just as I stood up on tiptoes to look in the nest this other parent gave me a hard blow on the side of my head, fortunately striking the heavy felt hat I wore in which three sharp cuts about half an inch long were made. My scalp was slightly cut by the unexpected attack, which resulted in a decided headache. Being thus put on my guard, I watched this parent, which soon returned to the attack, flying from the top of a tall tree about one hundred and fifty feet from me, straight at my head. I struck at it, but missed and the bird swerved, missing my face by about a foot. A third similar attack was made, but in this case the bird missed me by about three feet. All this time the other parent remained possibly fifteen feet directly over me, calling shrilly.

The following year his experience with this pair was equally interesting, for he (1915) says:

Both birds met me at the edge of the woods and flew about with noisy screaming at some elevation as I walked westward. At the west side of the woods I turned and walked in a northeasterly direction directly towards the beech tree in which the first set of eggs were taken in 1913. The female was in a tree top near this beech and when I was possibly 200 feet away she launched herself directly at me. I could hardly conceive she would attack me as I stood on the ground, but she came straight on and I had to drop to my knees to avoid her blow. She alighted west of me and I walked on toward the nest, watching her over my shoulder. I had hardly stepped forward when she again dashed to the attack with more fierceness possibly than before and I again was compelled to drop to my knees. She came to rest about 30 feet from me in a small maple where she rested in a threatening attitude for some time while I stood admiring her. Her plumage was perfect, her breast being almost red, and her attitude of fearless defiance as she stood leaning toward me made a picture impossible to forget. She made no further attacks till I began climbing the tree when she struck at me viciously four times. It is needless to say I kept her in sight all the time, keeping the tree between us as much as possible, and jerking my head out of the way to avoid her outreached claws.

Mr. Rathbun writes to me: "We think it a habit of the male red-shouldered to perch during the night near the nest on which his mate is sitting. We know this to be a fact in the case of a pair with which we had an acquaintanceship over a period of years. For on a moonlight night we climbed the tree when the nest was in use by the birds, and one of the hawks flushed from its perch on a limb close by the nest, the other from the nest itself. Both birds slipped away as noiselessly as shadows."

Red-shouldered hawks make good pets, but they must not be taken too young; it is difficult to get proper food for them, and if fed on a pure meat diet when young they are likely to develop rickets and die; they should not be taken until they are nearly grown and well feathered. Dr. Louis B. Bishop (1901) has published an interesting note on this subject, and Mr. Kennard (1894b) has related his experience. I once had a beautiful adult, that had been wing-broken, in my aviary; it became very tame and gentle and seemed to know me; when I entered its cage it would fly up to me and take food out of my hand. Its favorite, and perfectly natural, food consisted of bird bodies, English sparrows, and mice, with an occasional red squirrel or frog. It came to a tragic end, as a great horned owl that I had in the next cage broke in, killed it, and partially ate it. Mr. Shelley tells me of an immature bird that was captured and confined in a large packing box:

Thereafter its fierceness grew apace; it became more than willing to fight in distrust some of its visitors, particularly those in the household where it was held a captive, yet to others was calm and made no overtures of hatred or unfriendliness. To those of whom it was a captive, never did it show but the keenest of distrust, even though they fed it. My eldest brother, having a broken wrist at this time and being somewhat of a sportsman with leisure time to spend, shot squirrels for the hawk as a dietetic change from poultry. The bird came to know his voice, and expectancy showed in its eyes when my brother came near, prompted, no doubt, by its desire for its natural food.

Voice.—During the breeding season the red-shouldered are the noisiest of our hawks. Their characteristic note is a loud, shrill scream, similar to one of the notes of the blue jay, but different in quality. It sounds like *kee-aah* or *kee-oow*, the first syllable on a higher key and strongly emphasized, the second dropping off in pitch and prolonged. This call is usually repeated rapidly from two to four times; but I once saw a bird sit in a tree, watching me, and give four series of calls, repeating this note from 18 to 21 times in each series. On two or three occasions I have heard and seen the red-shouldered hawk give a long, plaintive call, all on one key, much like the well-known note of the broad-winged hawk, *ke-wee-e-e-e*, but on a lower key and not so prolonged, with the accent on the second syllable. Ora W. Knight (1908) says he has "heard the birds utter a scolding *cac*, *cac*, *cac*", which I have always attributed to the Cooper's hawk.

Enemies.—The larger hawks have few enemies except man, but Verdi Burtch's (1927) experience in finding a redtail feeding on a freshly killed redshoulder and the murder of my pet hawk by a great horned owl show that this species has at least two avian enemies. But sportsmen, farmers, and poultry and game breeders are all sworn enemies of all hawks and will not be convinced that there is any good hawk but a dead hawk. The bounty system is far too prevalent and leads to the killing of far too many old and young hawks in or near their nests, which the farmers hunt up and watch until the young hatch; the old birds are then more easily shot and the heads of the young secured. I believe we have saved the lives of many a family of hawks by taking the eggs in April; they lay a second set in May and stand a better chance of raising a brood when the leaves are out; then the nests are harder to find and the farmers have ceased to look for them.

Field marks.—The adult red-shouldered hawk is easily recognized by the more or less pale, ruddy underparts and by the conspicuous black-and-white barring on the wings and tail; the broadwing has a barred tail, but the bands are fewer, broader, and less conspicuously black and white. The young redshoulder is much like the young broadwing, but larger; it is smaller than the young redtail, and the markings on the underparts are more evenly distributed, whereas the young redtail has a largely white breast and dark markings on the belly and flanks. Mr. Forbush (1927) says: "The *only necessary* field mark when bird is soaring (even at a height or distance which may require a glass) is the apparent translucent spot in the wing near its tip formed by the short black and white wing-barring. This is diagnostic and no other New England hawk has anything like it."

Fall.—About the middle of September and during October these hawks become more in evidence and begin slowly drifting southward. Dr. Charles W. Townsend (1920) mentions a large flight that occurred at Amesbury, Mass., on September 18, 1886, when a "flock of about 300" passed. Referring to northeastern Illinois, Dr. E. W. Nelson (1877a) writes:

Mr. R. Kennicott speaks of an immense flight of this species, consisting of thousands, which passed over Chicago, "in October, 1854." The main fall migration of hawks in this vicinity takes place the last of September or first of October, and a statement of the numbers which pass in a single day, to one who has not observed them, would be received with incredulity. Choosing a day when there is a strong south or south-west wind, the hawks commence moving south early in the morning and continue flying the entire day, and so numerously that, taking a stand at a good point, one would have from one to fifty hawks in view, with but very few intermissions, throughout the day. Among these occur all the migrants, but by far the greater number consist of the smaller species.

Winter.—While most of these hawks migrate to a milder climate, a few remain all winter as far north as southern New England. William Brewster (1906) says that they "are often seen" in the marshes and along the rivers near Cambridge, Mass., in winter. "At this season, when they are nearly or quite silent, they are given to haunting level, open country sprinkled with large, isolated trees. In some of these the Hawks have favorite perches to which they resort day after day and year after year, to bask in the winter sunshine and to watch for meadow mice."

Referring to Pennsylvania, Dr. B. H. Warren (1890) writes:

During the winter these hawks frequent principally the large water courses, meadow-lands, and the vicinity of ponds, and not unfrequently an individual of this species can be observed on its perch overlooking a spring-head. When the streams and meadows are frozen I have noticed that they especially resort to such localities as last named. When disturbed from its perch it utters, in a plaintive and impatient voice, the note, *keeo, keeo.* Its flight, generally short, is graceful and very owl-like. This hawk, like its relative, the Red-tail, may be observed sitting by the hour on some favorite tree or stake adjacent to swampy or boggy ground, watching for small quadrupeds and batrachians, which constitute its principal fare.

DISTRIBUTION

Range.—North America, exclusive of the Rocky Mountain region, from southern Canada south to northern Mexico. The range of the red-shouldered hawk is discontinuous, the species occurring over the eastern part of the country west to Nebraska, Kansas, and central Texas and reappearing (as the red-bellied hawk) in the Pacific coast regions from Oregon south to northwestern Mexico.

Breeding range.—The breeding range extends **north** to eastern Oregon (Camp Harney); Nebraska (Neligh, Linwood, and Omaha); Iowa (Boone and La Porte City); Wisconsin (La Crosse, Honey Creek, New London, and Sturgeon Bay); northern Michigan (probably Isle Royal, McMillan, and Sault Ste. Marie); southern Ontario (Parry Sound, Reaboro, and Ottawa); southern Quebec (Montreal and Quebec); and Prince Edward Island (North River). **East** to Prince Edward Island (North River); New Brunswick (Grand Manan); Maine (Bucksport, South Warren, and Portland); New Hampshire (Webster); Massachusetts (Andover, Salem, Boston, and Dighton); Rhode Island (Newport); Connecticut (New London and New Haven); New York (Ossining and New York City); New Jersey (probably Princeton and Cape May); Maryland (Cambridge); Virginia (Ashland and probably Dismal Swamp); North Carolina (Walker and Lake Ellis); South Carolina (Charleston); Georgia (Savannah, Riceboro, and St. Marys); and Florida (Palatka, San Mateo, Tomoka, Titusville, St. Lucie, Fort Pierce, Miami, Cape Sable, and Key West). **South** to Florida (Key West, Logger-

head Key, Tallahassee, Whitfield, and Pensacola); Louisiana (New
Orleans and Baton Rouge); Texas (Beaumont, Columbus, and probably Corpus Christi); Tamaulipas; and Lower California (San
Rafael). West to Lower California (San Rafael); California (San
Diego, Escondido, San Onofre, Los Angeles, Santa Paula, probably
near Buena Vista Lake, probably Visalia, Palo Alto, Sonoma, and
probably Tehama); and Oregon (Camp Harney).

According to L. B. Potter (MS.) two pairs of these hawks nested
at Eastend, Saskatchewan, in 1909 and 1910. This is, however, well
outside of the normal breeding range.

The range above outlined is for the entire species, which has been
separated into five subspecies. The northern red-shouldered hawk
(*Buteo l. lineatus*) breeds from southern Ontario, Quebec, and
Prince Edward Island south to Kansas, Missouri, and North Carolina, wintering south to the Gulf and South Atlantic States; the
Florida red-shouldered hawk (*B. l. alleni*) is resident in the Southern States from Oklahoma and eastern Texas east to South Carolina
and south to central Florida; the insular red-shouldered hawk (*B. l.
extimus*) occupies the southern portion of the Florida Peninsula and
the Florida Keys; the Texas red-shouldered hawk (*B. l. texanus*)
is resident in the Coastal Plain region of southeastern Texas and
adjacent parts of the State of Tamaulipas, Mexico; and the red-
bellied hawk (*B. l. elegans*) is the Pacific coast race resident chiefly
in the Sacramento and San Joaquin Valleys of California, ranging
south to Lower California and Colima.

The nonmigratory status of the southern forms is indicated by an
adult (A625381) banded on January 31 at Lafayette, La., which
was killed in the same area on October 2 of the same year. Two
others (210898 and 210899) banded as juveniles in Leon County,
Fla., on May 7, 1924, were recaptured in the same region on October
20, 1924, and April 18, 1930. On the other hand, three birds from
the same nest (309826, 309827, and 309828) banded at Windsor,
Conn., on June 15, 1924, were all recovered in this State after the
lapse of several years, the recovery dates being May 11, 1927, October
8, 1928, and November 14, 1928.

Winter range.—In winter the red-shouldered hawk is found **north**
to California (Tehama); western Texas (San Angelo); Oklahoma
(Oklahoma City); Missouri (La Grange); southern Illinois (Odin);
Indiana (Logansport and Richmond); Ohio (New Paris, Oberlin,
and Youngstown); northwestern Pennsylvania (Meadville); New
York (Geneva, Ithaca, and Rhinebeck); and eastern Massachusetts
(Boston). **East** to Massachusetts (Boston); Rhode Island (Provi-
dence); Connecticut (New London); Long Island (Orient); New
Jersey (Princeton, Camden, and Newfield); Maryland (Cambridge);

Virginia (Wallops Island); North Carolina (Raleigh); South Caroline (Marion, Georgetown, Mount Pleasant, and Charleston); Georgia (Savannah); and Florida (Palatka, Daytona, Orlando, St. Lucie, Royal Palm Park, and Key Largo). **South** to Florida (Key Largo, Sanibel Island, and St. Marks); Mississippi (Biloxi); Louisiana (Lake Catherine, New Orleans, and Vermilion Bay); Texas (Citrusgrove and Brownsville); Tamaulipas; and rarely Colima (Plains of Colima). **West** to rarely Colima (Plains of Colima); Sinaloa (Mazatlan); Lower California (Colorado River Delta); and California (Snelling, Marin County, Marysville, and Tehama).

Red-shouldered hawks are sometimes noted in winter north to Kansas (Independence); Iowa (Hillsboro); central Illinois (Rantoul); southern Michigan (Grand Rapids, Ann Arbor, and Detroit); Ontario (Hamilton and Toronto); and Vermont (Bennington and Montpelier).

Spring migration.—Early dates of spring arrival in the regions north of the normal winter range, are: New Hampshire—Concord, March 13; South Manchester, March 19; Tilton, March 20; and Sanbornton, March 29. Maine—Lewiston, March 4; Waterville, March 7; Auburn, March 11; and Portland, March 15. Quebec—Montreal, March 5; Isle Jesus, March 13; and East Sherbrooke, March 19. Michigan—Three Rivers, February 11; Detroit, February 14; Grand Rapids, February 16; and Kalamazoo, February 19. Ontario—London, February 7; Strathroy, February 17; Point Pelee, February 23; and Ottawa, March 13. Iowa—Brooklyn, February 5; Springville, February 13; and Sigourney, February 15. Wisconsin—Durand, March 3; New London, March 5; Milwaukee, March 10; and Superior, March 13. Minnesota—Minneapolis, February 21. Nebraska—Badger, March 25. South Dakota—Huron, March 16.

Fall migration.—Late dates of fall departure are: South Dakota—Yankton, November 14. Nebraska—Neligh, October 6. Wisconsin—Ladysmith, October 15; New London, October 28; and Greenbush, October 30. Iowa—Keokuk, October 13; and Marshalltown, October 25. Ontario—Point Pelee, November 1; Ottawa, November 6; Toronto, November 12; and London, November 20. Michigan—Schoolcraft, November 3; Grand Rapids, November 8; Detroit, November 13; and Ganges, November 18. Quebec—Hatley, October 25; and Montreal, November 6. Maine—Portland, October 16; Winthrop, October 28; North Livermore, November 3; and Buckfield, November 11.

A red-shouldered hawk (311766) banded on June 17 at Belchertown, Mass., was trapped and released at Hanover, Pa., on December 25 of the same year; another (312011) banded on June 1 at Worthington, Mass., was caught in a trap at Seagrove, N. C., on

November 28; while a third (201381), banded on October 11 at Harper, Kans., was killed at Caddo, Tex., on September 7 of the following year.

Casual records.—Brooks (1917) reports that he has twice seen the red-bellied hawk at Chilliwack, British Columbia; Fannin states that he took it at Burrard Inlet (Kermode, 1904), and Macoun (1909) reported that W. B. Anderson had found it at Port Simpson. Nevertheless, no specimen of the species is known from this Province. A specimen was taken on November 17, 1853, at a camp on the Little Colorado River, N. Mex., and Dr. Henry claimed that he saw it at Fort Thorn during the winter of 1856–57 (Bailey, 1928). It also has been reported from Colorado, Wyoming, North Dakota, Montana, and Alberta, but in none of these cases has a specimen been collected. A specimen was taken on February 26, 1863, at Kingussie, Scotland.

Egg dates.—Southern Canada: 41 records, April 16 to May 25; 21 records April 24 to May 7.

New England and New York: 383 records, March 5 to May 31; 192 records, April 18 to 29.

New Jersey to Virginia: 99 records, March 19 to June 28; 49 records, April 10 to 25.

Ohio to North Dakota and Colorado: 75 records, March 13 to June 21; 38 records, April 13 to May 1.

Washington to California: 185 records, February 12 to June 19; 93 records, March 23 to April 13.

South Carolina and Florida to Texas: 196 records, January 20 to June 3; 98 records, March 2 to April 4.

BUTEO LINEATUS ALLENI Ridgway

FLORIDA RED-SHOULDERED HAWK

HABITS

Ridgway (1884), in naming this southern race of the red-shouldered hawk, characterized it as "smaller than *B. lineatus*, the adult much paler in color, with no rufous on upper parts, except on lesser wing-coverts; the young decidedly darker than in true *lineatus*." He says further: "The very decided ashy coloration of the upper parts, relieved only by fine shaft-lines of black on the head and neck, dusky clouding on the back, and white streaking on the occiput, combined with the pale coloration of the lower parts, serves readily to distinguish this race from the true *B. lineatus*."

This small pale race might more properly be called the southern red-shouldered hawk, for it is widely distributed throughout the Southern States, from South Carolina to Arkansas and Oklahoma, and a more recently described form occupies the southern part of

Florida. In Florida it is decidedly the commonest hawk and quite evenly distributed in all kinds of timbered regions; it seems to be equally at home in the extensive flat pine woods and in the dense live-oak hammocks. It is much more abundant than hawks are elsewhere, is quite tame and conspicuous, and, during the breeding season, very noisy. It seems to be less of a forest bird and is oftener seen in open country than is its northern relative. It is most abundant in regions like the Kissimmee Prairie, where wide open prairies or savannas are dotted with small hammocks of live oaks and palmettos. In the flat pine woods it is more widely scattered and seems to prefer the smaller tracts or the vicinity of small cypress swamps.

Courtship.—Donald J. Nicholson (1930) writes:

Early in December the birds begin their wild courtship "songs", which consist of loud, piercing, shrill calls, or screams, given while circling in the air. With loud .cries they either soar or flap their wings rapidly, going in a circle higher and higher. From one to four individuals may be seen in the air at a time over the chosen nesting site. Spirited swoops and long dives through the air are often seen, they calling sharply the while. These cries are given also flying from one place to another. They are most noisy at this period, and keep it up throughout the entire day at intervals.

Nesting.—The Florida red-shouldered hawk nests in a variety of situations and is not particular as to the choice of a tree. My first nest was found on April 24, 1902, at Oak Lodge, across the Indian River from Grant. It was about 25 feet from the ground, in a nearly horizontal crotch of a wide-spreading live oak, in the middle of a dense hammock of live oaks and palmettos. The nest was a handsome but bulky affair, measuring 24 inches in diameter and 18 inches high, the inner cavity being 10 inches across and 3 inches deep. It was made of sticks, profusely draped with Spanish moss hanging down in a long festoon on one side and decorated with white down and two sprigs of evergreen; it was lined with green leaves of the live oak, Spanish moss, a snake's skin, and strips of inner bark. It contained only one egg, nearly ready to hatch.

A different type of nesting, more typical of the northern or the Texas varieties, was seen in the heavy, river-bottom forest along the Hillsboro River. This magnificent forest contains some of the finest timber I have ever seen in Florida—live oaks, pin oaks, hickories, locusts, palmettos, pines, and cedars, with an undergrowth of hawthorn, ironwood, and dogwood. High up in one of the largest pin oaks, fully 50 feet, was the hawk's nest, much as we should expect to find one in our northern woods. Although the hawks were flying about and screaming on February 22, 1925, the nest was empty at that time; but my companion, Oscar E. Baynard, collected a set of eggs from it later.

On March 8, 1925, while we were walking along the edge of a cypress swamp in Polk County, we heard a hawk scream and saw it fly out from the swamp. We waded in, where the water was less than knee deep, and found the nest about 60 feet up in a tall cypress. It was a new nest but still empty. Bendire (1892) mentions two nests found by Dr. Ralph in pine trees in cypress swamps; one was 57 feet and one 40 feet from the ground.

These hawks nest more abundantly in the small mixed hammocks along the Kissimmee River than I have ever found them elsewhere (pl. 58). Walter B. Savary found no less than 65 nests in this region, in an area about 10 miles long by 5 miles wide, during a single season. Of these, 35 were in cabbage palmettos, 15 in live oaks, 10 in gums, 3 in bays, and 1 each in a maple and a myrtle. The highest nest was 60 feet from the ground in a tall, slender gum, and the lowest was only 9 feet up in a myrtle. Practically all the nests were in small, mixed hammocks of an acre, or less, in area; some were in trees in small clumps of bushes, not tall enough to screen the nests from view. As to the placing of the nests, he says in his notes that in palmettos the nests are invariably placed "on the dead leaf stubs just beneath the living fronds; a caracara always builds among the live stems, but I never have found a hawk's nest so situated. If an oak is chosen, the bird seeks either a very leaning trunk, on which she can set the foundation, or a slanting limb."

As Mr. Savary made an extended stay in this region, he was able to learn some interesting facts and has sent me some elaborate notes. One discovery showed the length of time that the nest is occupied before the eggs are laid; the hawk "stakes out its claim", as it were, long in advance and guards its chosen territory against all intruders. "Once a location is settled upon the birds cling to it year after year. So attached are they to their home site that to hold it against intrusion they mark the nest with green leaves several weeks before laying, thus letting others know that the premises are preempted." In one striking instance he flushed a hawk off its nest on January 18 and on climbing to it found only a fresh spray of airplant in the nest. Expecting to find eggs soon, he climbed to it again five days later but found only the airplant in place. After a further lapse of two weeks the airplant marker was still in the nest, but dry and crisp; thinking the nest was deserted, he threw out the marker and "left the nest for at least a month." About March 5 he found that the hawk "had renewed her signature by placing a spray of myrtle" in the nest. Finally, on March 14, nearly two months after noting the first sign, he found two eggs in the nest.

Eggs.—The number of eggs laid by the Florida red-shouldered hawk is usually two, occasionally three, and very rarely four. The eggs differ from those of the species elsewhere only in size; they show similar wide variations in types and colors of markings. The measurements of 105 eggs average 52.6 by 42.7 millimeters; the eggs showing the four extremes measure 58.4 by 44.5, 56 by 45.8, 49.1 by 42.3, and 50 by 40.5 millimeters.

Food.—This hawk lives on much the same kind of food as other red-shouldered hawks, a small percentage of poultry and other birds and a large percentage of insects and small vertebrates. Out of 20 stomachs reported on by Dr. A. K. Fisher (1893), only one contained poultry and only two other birds, a sora rail and sparrows. Other items mentioned are mice, a catfish, frogs, lizards, snakes, a turtle, dragonflies, crickets, grasshoppers, cicadas, beetles, cockroaches, spiders, crawfish, earthworms, and various larvae. Bendire (1892) adds, on the authority of Dr. William L. Ralph, "their food consists principally of mice, Florida rats, young rabbits, the small gray squirrel found in this State, and probably an occasional frog or small snake." Arthur T. Wayne (1910) says: "During the breeding season this hawk frequently catches chickens and even grown fowls, but its principal food is mice, frogs, and snakes. It is very fond of water-snakes and will sit on a dead tree by a pond of water for hours waiting to prey upon them."

Charles J. Pennock writes to me that "they may often be seen to feed high up at the borders of a dense pall of smoke from a swamp grass fire, where large, winged grasshoppers abound." One that he shot was feeding on a fox squirrel.

Behavior.—The habits of the Florida redshoulder are not essentially different from those of its northern relative, except that it is much less shy, often stupidly tame and unsuspicious. Often it will sit on a pole or tree by the roadside and allow one to drive by within a few yards. Only once have I had one offer to attack me near its nest, and Mr. Savary had only one such experience. But Mr. Nicholson (1930) says:

The birds are very bold and fearless in the defense of their nests, either while the eggs are fresh, or with young. One day I visited five nests, and the first bird carried away my cap in her talons and struck me such a severe blow that it gave me a bad headache, and left a scratch on my forehead. At two more nests I was attacked and struck upon the head. Many other times this has happened. This bird coming swiftly as an arrow directly for your head, screaming wildly, gives a timid soul the shivers, and unless you wildly wave your arms and shout, most likely she will give you a stiff blow that will put fear into you, and respect for their bravery.

Mr. Savary says in his notes: "For a hawk it is a very gentle bird among others of its kind and, not intolerant of neighbors, often nests in the same grove with a crow or a caracara. In three instances I have found its nest and a caracara's within 50 feet of each other. There is one exception, however—it does not like the Florida barred owl, and I have seen it in hot pursuit of one that had come near its domicile. Sometimes in this pursuit it is joined by a crow, and I have seen the two hustling an owl's departure with considerable spirit."

BUTEO LINEATUS ELEGANS Cassin

RED-BELLIED HAWK

HABITS

In a large grove of big cottonwoods near San Jacinto, Calif., on March 8, 1929, I made the acquaintance of this beautiful hawk. As we sat on a log, eating our lunch, we heard and saw three different birds; of those seen clearly, one was immature and one a handsome adult. I could well imagine that I was away back home in the good old New England woods in April, listening to the screams of our familiar Buteos and watching their graceful soarings over the leafless treetops. As it lives in similar haunts, its habits, appearance, and voice seemed identical with those of our eastern red-shouldered hawk. The birds seemed to be interested in the locality, and we found what proved to be their new nest, 40 or 50 feet up on a branch and against the leaning trunk of a large cottonwood. My companion, Wright M. Pierce, visited this nest on April 9, with the result shown on plate 60.

The red-bellied hawk is about the same size as its eastern relative but much more brilliantly colored, the deep rufous of the breast being nearly, or quite, unbroken in adults; young birds are much darker, the deep brownish markings prevailing.

I was told that this hawk had become very scarce and that I could hardly hope to see one in southern California. It has undoubtedly been greatly reduced in numbers in the more thickly settled regions, but we seldom failed to find it in suitable localities, wooded river bottoms and lowland forests, remote from civilization. It is not an open-country bird like the redtail and so is less in evidence. It prefers the sheltered groves along the streams in the lower interior valleys, extending its hunting range into the adjoining fields and marshes.

James B. Dixon (1928) writes:

The typical range of a pair of these birds usually contains a central grove of oak, willow, or cottonwood trees in a river bottom, in which to build the nest. The birds are particularly partial to such a location when the surrounding canyon sides are heavily wooded and the stream bed is surrounded by open meadows of wet pasture land and alfalfa fields. They have a habit of sitting low on some dead snag or telephone post from which they can dart suddenly down and capture their prey. Their sense of hearing is extremely keen and I think they hunt as much by it as by sight. They do not descend from a great height in a grand swoop to strike their unsuspecting prey as does the Western Red-tail or the Golden Eagle, their hunting tactics being much more like those of the Marsh Hawk and the American Long-eared Owl.

Courtship.—Mr. Dixon (1928) says that this hawk is so noisy during the mating season that it becomes very conspicuous.

The usual program is for the bird leisurely to ascend in wide spirals to an elevation of 1500 to 2000 feet above the nest grove, where it will give a few preliminary flaps of its wings, the signal for the noise to begin, and squalling and diving it will descend to the same place from which it started or to the nest grove nearby, in a series of nose dives and side slips. I have seen eagles doing this same stunt without the noise, but have also noted that always in the offing there is an interloper in the form of another eagle, to whom it is perhaps given as a warning. The Red-bellied Hawk seems to do this stunt for the sheer joy of the thing.

Nesting.—The same writer, who has "a record of twenty-three nesting locations within a radius of thirty miles of Escondido", says:

The Red-bellied Hawk, like the Cooper Hawk, selects as a nest site, not some commanding view of its hunting grounds, but a location in a densely wooded grove. Preferably, the nest is placed about one-half way up the main stem of the tree, upon horizontal limbs and braced against the main trunk. This is a distinctive trait where nesting groves have not been disturbed by clearing of land or been washed away by floods. Rather than leave a chosen hunting ground, however, the hawks will accommodate themselves to almost any kind of a location. Considering their size, the birds build the smallest structure of any of the raptores hereabouts. I have often found nests which from the ground looked as though they could not possibly contain eggs, let alone conceal a sitting bird, but upon climbing the tree, the bird would leave and the nest would be found to contain four eggs. After incubation is well begun it is almost impossible to flush the sitting bird by any other method than climbing to the nest, and in several instances I have known the bird to remain until the climber reached it.

The nest is composed outwardly of dead twigs of the trees common to the river bottom, such as sycamore, willow and cottonwood, the inner part of frayed-out bark of the cottonwood and willow. This bark makes a soft mat upon which the finishing touches of green leaves and downy feathers are placed. The green leaves are constantly replenished during the incubation period and long after the young are hatched. After incubation has progressed somewhat a large number of downy feathers will make their appearance on and around the nest. This becomes so noticeable in some cases as to be a sure sign of occupancy and one which I have never noticed to such a marked degree in any of our other raptores.

A pair of hawks often has two or more nests, usually in the same tree or in adjoining trees, and if undisturbed they will remain year after year in the same grove. If an old nest is used, very little is done to it with the exception of relining with bark and green leaves; so the structure does not take on such a large size as with other hawks. The determining factor in a location seems to be the food supply, and if that is to be had the hawks will use whatever trees are available. I have found nests in willow as low as twenty-five feet from the ground and in large sycamores as high as eighty-five feet. I have never found these hawks using any nest but one constructed by themselves, though I have found other birds using theirs.

On April 9, 1929, I spent a most interesting day in the field with Mr. Dixon in his territory, visiting seven nests of red-bellied hawks.

A brief description of these nests will illustrate the variety of situations chosen (see pl. 59). The first was in a small, densely wooded, swampy swale, such as our eastern bird sometimes chooses; and the nest was 40 feet from the ground in a leaning sycamore. The second was found after a long search through an extensive cottonwood flat, open in some places and thickly wooded in others; we finally flushed the hawk off her nest, about 70 feet up in a tall cottonwood; this nest held three downy young. As we walked down a cart path close to the bank of a river, among an open growth of tall sycamores in a narrow valley, the third nest was seen 68 feet from the ground in one of these tall trees; it was new but empty. In a patch of smaller sycamores and willows we found the fourth nest, from which we flushed the bird; this nest was 40 feet up in a slender leaning sycamore, so slender that we had to rope it to a larger tree before it was safe to climb it; this held three eggs. The fifth nest required a thorough search in a thick patch of large willows and other dense growth in a swampy hollow; we finally rapped the hawk off a very large old nest only 30 feet up in a spreading willow and collected four eggs from it. While we were driving along the road we saw a nest about 50 feet up in a tall sycamore, which towered above all the surrounding trees; we supposed it was a redtail's nest, being in such an exposed situation, but were surprised to see a red-bellied hawk fly from it; this yielded a set of three eggs. The seventh and last nest was fully 75 feet from the ground in the top of a tall slender eucalyptus in the center of a grove of these trees; it was a small nest, and the tail of the sitting bird projected over the edge of it; as the tree was swaying badly in a strong breeze we did not care to climb it.

C. S. Sharp (1906) says that these hawks "have a great fondness for Eucalyptus groves, making their nests at times on the masses of bark that have sloughed off and collected in some large crotch of the main branches." He continues:

Since 1898 I have had good opportunity for observing an isolated pair. These birds have occupied six different nests—all in Eucalyptus trees—either in groves or as shade trees on sides of the road, the extremes being about a mile apart. Every year but one they have been levied on for one set of eggs. On one year only was a second set taken from them. After the removal of the first clutch the birds have gone to the nearest nest—generally to a nest in the same grove and only a few rods away and have occupied it for a second, never going from one extreme limit of their range to the other.

One nest was for three years occupied first by a pair of Pacific horned owls. In 1899 I found the hawk on the nest which held two fresh eggs, and two young owls were in the branches of the next tree. As that was then the only nest in the grove it looked as if there had been a rather hasty eviction. In another nest of this pair in 1898 I found three eggs of the hawk and one of the long-eared owl.

Major Bendire (1892) found two nests near Camp Harney, Oreg.; one was "in a young pine on some limbs close to the top and the trunk of the tree, * * * on the outskirts of the heavy timber"; the second nest "was placed in a tall juniper tree, likewise near the trunk and about 20 feet from the ground." He also mentions nests found by A. W. Anthony in giant cactus and candlewoods.

Eggs.—Three or four eggs constitute the usual set laid by the red-bellied hawk, three being commoner than four; sets of two are uncommon and sets of five very rare. Bendire (1892) says that Dr. B. W. Evermann found as many as five eggs in a nest. The eggs are similar to those of the eastern race but are more often richly, heavily, and handsomely marked though showing all the usual variations. The measurements of 46 eggs average 53.4 by 42.1 millimeters; the eggs showing the four extremes measure **56.2** by 44.8, 54 by **45.7**, 48.3 by 40.6, and 50.2 by **39.6** millimeters.

Young.—Mr. Dixon (1928) writes:

After the eggs are laid and incubation begins, the two birds seem to share equally in this duty. Incubation period varies from twenty-three to twenty-five days, varying according to the care with which incubation was conducted in the first few days, during the laying of the eggs and directly afterwards. As incubation starts usually with the laying of the first egg, the young emerge from the shell over a period of several days. Quite a difference in size is often noted when they are first hatched, but this disappears as they reach the age of four or five weeks. In several instances where I have observed that the heavily marked eggs of a set were laid first, they were the first to hatch, and in all cases where infertile eggs were noted, these were the lightly marked or plain eggs of the set. Infertile eggs are not at all uncommon and it is rarely that all of a set of four eggs are fertile. The young birds do not develop very fast the first week, but thereafter they increase rapidly in weight up to five weeks from hatching. Then the feathers begin to make their appearance and from this time on the feathers develop rapidly.

Food.—The food habits of the red-bellied hawk proclaim it a very useful bird, living largely on injurious rodents, amphibians, reptiles, and insects. It very seldom attacks poultry or other birds. Mr. Sharp (1906) gives us some good evidence of this:

One of my friends in San Pasqual Valley, where these hawks are common, told me the red-bellied and red-tailed hawks had nested on his ranch as long as he could remember (he is a very old resident) and it was very seldom they would touch a chicken tho the latter were running free all the time. * * *

All the time I was at the nest some 200 chickens of all ages and sizes were working around the barn yard, in the corral and out on the stubble beyond, many of them fully 200 yards from shelter but they never even gave a warning cry when the old hawk flew from the nest across the yard.

Mrs. Irene G. Wheelock (1904) says:

In most parts of California where they breed, the records show them to have eschewed everything with feathers, and to have dined upon small snakes, lizards, frogs, insects, and crawfish. Fur and feathers are caught only as a last resort, when there are hungry young in the nest.

Dr. Harold C. Bryant (1921) found in the stomach of one bird a number of insect remains, including larvae of a hawk moth (*Pachysphynx modesta*):

This caterpillar when full-grown is from two to two and one-half inches in length, of a light green color, with yellow lines on the head and along the sides of the body, and feeds on various species of willow. Eight of these caterpillars were found in the hawk's stomach together with the remains of two mole crickets (*Stenopelmatus*, sp.), one beetle (*Coniontis*, sp.), one ground beetle (unidentified), and some grass and pieces of wood that doubtless were picked up with the food.

W. Leon Dawson (1923) says that if this hawk "rises on occasion to a ground squirrel or a brush rabbit, he oftener descends to fence lizards and frogs, or even insects."

Behavior.—The red-bellied hawk does not differ materially in habits or voice from its eastern relative; the resemblance is striking. Mrs. Wheelock (1904) says:

The Red-bellied Hawk is exceptionally fond of bathing, and in California it usually builds within a hundred yards of water. Both adults indulge in a daily bath, returning to the same place at about the same hour for it.

Laurence M. Huey (1913) saw a pair of these hawks make an attack on a nest of Pacific horned owls containing young:

The three young were rather large and partly feathered. As the old bird left the nest a pair of Red-bellied Hawks set out in pursuit. One continued to chase the old owl, while the other hawk returned and robbed the nest of one of the young owls This was torn to pieces and eaten in a nearby tree.

In view of the prevailing impression that these hawks are disappearing rapidly, the following remarks by Mr. Dixon (1928) are encouraging:

In 1907, I personally visited and either collected a set of eggs from, or located, the nests of seven pairs of Red-bellied Hawks in the northern end of San Diego County, and in 1927, twenty years later, I made it a point to renew my acquaintance with these seven locations. In every instance I found a pair of hawks still resident in the same general locality. * * *

Twenty years ago it was a common practice for everyone traveling through the country to carry along a shotgun, and any bird of prey was considered a good target. This condition does not exist at present, as the cost of ammunition has increased, the game laws are being enforced, and, last and most important, the people are becoming educated to the fact that our hawks and owls have their economic place in the well-being of the farmer, and they are seldom shot. Other changes are taking place which make the outlook in this section more cheerful for a continuing number of these beautiful birds to live here. A few years ago not far from where I live there was a long strip of river bottom in which resided a single pair of Red-bellied Hawks. Today, this same stream has been dammed and where the river bottom used to be there is a lake, and along the shores of this lake, in the same area which used to support a single pair of birds, three pairs now live, and all of them seem to thrive and find plenty of food. This has proven conclusively to me that if food supply conditions are right, the existing birds will breed up to fill in this favored area or less favorably situated birds will move in to fill the gap.

BUTEO LINEATUS EXTIMUS Bangs

INSULAR RED-SHOULDERED HAWK

HABITS

The above common name of this small race of the red-shouldered hawk is based on the erroneous impression that it is confined to the Florida Keys, and that is the only range given for it in the latest A. O. U. check-list (1931). During three seasons I have traveled extensively over many of the keys. Although I have seen these hawks on some of the larger keys, I have never seen a hawk's nest on any of the keys. The type specimen was collected on Key Biscayne, opposite Miami in Dade County, which is practically a part of the mainland and a long way from the Florida Keys proper. As a matter of fact, it is a widely distributed and very common hawk all through the southern third of Florida and for an undetermined distance farther north. Birds that I have collected in the southern counties, as far north as Lake Okeechobee, are all typical of this form. How much farther north it ranges, or where it intergrades with *alleni*, is yet to be learned; a gradual diminution in size makes it difficult to draw the line.

Outram Bangs (1920), in describing this form, gave as its characters: "Similar to *Buteo lineatus alleni*, and not much different in color though perhaps averaging in general a little darker and richer, but much smaller." The striking color characters of both *extimus* and *alleni* are the extreme grayness on the head and upper parts generally and the paleness of the under parts; these are quite noticeable in the field. The "darker and richer" colors referred to by Mr. Bangs are not noticeable in my specimens.

The center of abundance of *extimus* seems to be in Monroe County and around the southern edge of the Everglades, where it is exceedingly abundant for a hawk. Everglades red-shouldered hawk would have been an appropriate name, for it is in no sense insular. As one drives along the Tamiami Trail these little hawks are much in evidence and very tame, perched on the telegraph poles and allowing a close approach; they seem to realize that no shooting is allowed within a mile of this road. They are oftenest seen in and about the small cypress swamps, where they probably find abundant food. They are less often to be found in the flat pine woods and about the hardwood hammocks on dry ground.

Nesting.—My first nest of this hawk was shown to me on April 27, 1903, near Flamingo, at the southern tip of Florida. It was about 30 feet from the ground in a black mangrove in a grove of these trees near the shore; it was the usual nest of sticks lined with mangrove leaves. A single young bird, fully grown, was sitting up in the nest, but it flew away as I started to climb.

During the winter and spring of 1930, I climbed to and examined six nests of this little hawk and saw a number of others. On January 23 we saw the hawks building their nest in an unusual situation in a small clump of buttonwoods and other small trees and bushes in the Everglades. The site chosen was only 10 feet above the ground in the leafy top of a small buttonwood where the top of another fallen tree rested against it. The nest was so well concealed that I was not sure that it was a nest until I looked into it. It was made of sticks, weeds, and grasses and lined with green leaves from the surrounding trees. On January 31 this nest contained one very pretty egg, but when I visited it again, on February 10, it was empty and deserted.

The southern part of the Everglades is dotted with small mottes, or islands, an acre or two in area, of small or medium-sized cypresses, growing in water a foot or so in depth. These were favorite nesting sites for these hawks, and most of the many nests that we saw were in such situations; early in the season, before the cypresses were in full leaf, the nests were conspicuous at a long distance. A low nest of this type, found on February 27, was only 15 feet above the water on some horizontal branches of a small cypress on the very edge of the motte. It was made of sticks and twigs of cypress and lined with weed stems, strips of cypress bark, green twigs, and green leaves; it was profusely decorated with white down and contained three eggs; it measured 24 inches in outside and 7 inches in inside diameter, the inner cavity being 2 inches deep. Other nests were well within the mottes and higher up, 20 to 30 feet, in larger cypresses, but generally in plain sight. Once, while I was watching a nest on which I could see the head of the incubating bird, I heard a hawk scream and saw it come sailing along through the trees and alight on the edge of the nest; the sitting bird, apparently the male, immediately arose and flew away; the newcomer settled on the nest and began incubating. I climbed to the nest and found only one egg; this was the second nest on which we had found a hawk incubating on one egg, perhaps for protection against crows.

On January 31 I visited a nest that I had previously located in some flat pine woods on a large island in the Everglades; the hawk had flown from the nest when I rapped the tree and returned to it within five minutes, while I sat in plain sight only 50 yards away; and this time she swooped at me when I climbed the tree; two eggs nearly ready to hatch might have made her unusually anxious; these eggs must have been laid very early in January. The nest was at least 45 feet from the ground in a slender Caribbean pine; it was made of pine twigs and grasses and lined with green and dry pine needles; it measured 15 inches in diameter and 10 inches high and was 4 inches deep inside.

Still another type of nesting was seen on February 15 in Glades County. The nest, containing the usual two eggs, was only 15 feet from the ground in a small live oak at the very edge of a mixed hammock of cabbage palmettos and oaks; it was lined with inner bark and oak leaves. Birds shot in this vicinity are clearly referable to this race.

Eggs.—Two eggs form the usual set for this hawk, but occasionally three are laid. They are indistinguishable from those of *alleni*, with surprisingly little average difference in size. The measurements of 39 eggs average 51.8 by 41.5 millimeters; the eggs showing the four extremes measure 57 by 43, 55 by 44.8, 44.5 by 41.2, and 50.3 by 37.5 millimeters.

What has been said about the food habits and behavior of the Florida red-shouldered hawk applies equally well to this smaller race. If anything, the southern Florida birds are tamer, less shy, and more in evidence along the roadsides. Both races are resident in Florida and begin nesting in midwinter.

BUTEO LINEATUS TEXANUS Bisho

TEXAS RED-SHOULDERED HAWK

HABITS

Based on a series of 16 adults and 6 immature birds collected near Corpus Christi and Brownsville, Tex., Dr. Louis B. Bishop (1912) gave the above name to the red-shouldered hawk of southern Texas, describing it as—

Similar to *Buteo lineatus elegans*, but breast usually more spotted with buffy, the dark shaft lines of chest more conspicuous and the head and back more rufous. * * * These Texas birds are much more richly colored below than fall specimens of *B. l. lineatus* from Connecticut, having the chest and breast uniform bright cinnamon rufous and the abdomen, tibiae and lower tail-coverts bright buff heavily barred with cinnamon rufous. They are larger than *B. l. alleni* from Florida and have the head and neck not grayish but even more rufous than *lineatus*.

Six young birds collected at the same time differ from the description of young *B. l. elegans* by having the pale spaces on the outer webs of the primaries as large as in *B. l. lineatus*. From the latter they differ by having the lower parts, especially the tibiae, more buffy and the dark markings larger—sagittate or cuneiform instead of oval—and numerous even on the tibiae, which are slightly if at all spotted in *B. l. lineatus*. Young *B. l. alleni* is smaller and has less buff in the plumage, and the dark markings below are even heavier than in the Texas race.

The 1931 A. O. U. check-list gives the range of this race as "central southern Texas south into Tamaulipas, Mexico." Some confusion exists in previous literature; the red-shouldered hawks in different parts of Texas have been referred to as *alleni* or as *elegans* by earlier writers. The range of *alleni* extends into eastern and northern Texas, but just how far does not seem to be known. The evidence seems to show that *texanus* ranges at least as far north and west as Corpus Christi and Austin. That some of the earlier writers referred to this bird as *elegans* before *texanus* was named is not surprising, as the two birds are much alike.

Unlike the Florida bird, the Texas redshoulder is essentially a bird of the heavily timbered river bottoms. It is decidedly the commonest large hawk in Texas and in certain favorable localities is really abundant. William Hahn, Jr. (1927), says that along the Nueces River one "can often count anywhere from 10 to 25 in the air at one time." This doubtless refers to the courtship season, which begins in February and which Walter B. Savary tells me lasts for about a month. At this season the birds may be seen circling in pairs over the treetops, calling almost constantly. These river-bottom forests are often extensive and very dense, with many trees of enormous size—elms, pecans, hickories, cottonwoods, live oaks, pin oaks, and hackberries.

Nesting.—The nesting habits are very similar to those of the northern red-shouldered hawk, except that they are often much more concentrated in favorable localities. Mr. Hahn (1927) writes:

I noticed that these birds will sometimes build their nests real close to each other. In one instance I collected four sets in one clump of trees that was not a half-mile square. It also had the fifth set in it when I had to leave. * * *
Most of the nests found were in elms, live oaks, and hackberry trees, all were in forks of the limb. The nests measured anywhere from 14 inches to over 2 feet across and from 6 inches to 15 inches deep. The larger ones were those used year after year, as very few were new ones. These hawks often dart at you when you go to their nests and on two occasions I had them strike me in the back and fly away uttering a very harsh scream. The nests are made of twigs, Spanish moss, lined with green leaves or some green substance all the time, and some moss, also feathers from the bird's body.

George F. Simmons (1915) describes a nest, found in Harris County, Tex., that was only "thirty feet up in a small pine tree"; it was "neatly lined with quite a quantity of fresh, green and fragrant pine needles." The other nests which he "located were all in pines, from 40 to 80 feet from the ground, generally in open pine woods with little under brush." Elsewhere he says (1925) that the nest is placed from "30 to 55, rarely 75, feet up in tallest bottomland trees, in topmost crotch where several limbs branch out from main trunk to form a heavy fork, generally in cedar elm, but often in tall pecan, cottonwood or live oak trees."

Eggs.—The Texas red-shouldered hawk lays two to four eggs, generally three. These are indistinguishable from eggs of the northern race. The measurements of 50 eggs average 53 by 42.9 millimeters; the eggs showing the four extremes measure 57.2 by 44.8, 53 by 46, and 49.9 by 39.1 millimeters.

Food.—The food of this hawk is much like that of other red-shouldered hawks. It seldom attacks poultry but lives mainly on small mammals, snakes, and frogs. It has been recorded as killing some birds, such as quail, cardinals, and various sparrows. Its feeding habits are mainly beneficial.

BUTEO ALBONOTATUS Kaup

ZONE-TAILED HAWK

HABITS

The zone-tailed hawk is a Central American species that reaches the northern limit of its range in our Southwestern States. My acquaintance with it is limited to a brief visit to two of the picturesque canyons of the Catalina Mountains in Arizona. After a long drive over the rolling plains east of these mountains, we dipped down a sharp decline into Apache Canyon, where we pitched camp for a few days. This is one of the most picturesque canyons in Arizona. It is a broad, deep, rocky canyon, well watered by a stream of good clear water flowing over a wide stony bed. The sides of the canyon are rough and rocky, in some places very steep or even precipitous, and more or less overgrown with small giant cactus, hackberries, thorns, mesquite, and mountain misery. In the steep rocky walls are numerous small caves, crevices, and ledges where we found nests of the turkey vulture, golden eagle, raven, and canyon wren. The broad bed of the stream is heavily wooded with large picturesque sycamores and giant cottonwoods, with lofty spreading branches that reminded me of New England elms, towering over the tops of the other trees, including a variety of oaks, maples, and walnuts. In one of these big cottonwoods near our camp was an apparently new hawk's nest, fully 100 feet from the ground, about which a pair of zone-tailed hawks showed considerable concern (pl. 61). Our companion, Frank C. Willard, told us that formerly there were two pairs of these hawks in this canyon, but we could not locate the second pair. We did, however, locate two pairs of Cooper's hawks and two pairs of western redtails with nests in the lofty treetops. Perhaps the redtails had driven away the other pair of zonetails.

Nesting.—The nest near our camp, referred to above, was in the topmost branches of the giant cottonwood. It was a difficult and hazardous climb; and as the hawks had apparently not yet laid, we did not care to attempt it. The next day, April 19, 1922, we explored Edgar Canyon, a few miles farther north in the same mountains. This is a similar canyon but narrower; it is heavily wooded with large sycamores, cottonwoods, and other trees growing along the rocky bed of the stream and with a dense growth of oaks, maples, walnuts, hackberries, thorns, and mesquites on the drier banks. While we were walking down the bed of the stream we were delighted to see a zone-tailed hawk fly from the leafy top of a tall cottonwood (pl. 61). Its nest was barely visible in the thick foliage near the end of a slender branch in the very top of the tree, at least 60 feet from the ground. The hawk began screaming and was soon joined by its mate; both birds circled about in the vicinity as long as we were there. There was no doubt about its identity, but, to make doubly sure, I shot the female; I could easily have shot both. The nest looked inaccessible, but we made a scoop out of a tripod leg, a handkerchief, and a piece of barbed wire; and Mr. Willard made a spectacular and daring climb, tying the upper branches together with ropes, and getting near enough to the nest to scoop out the single fresh egg. When I skinned the bird the next day I found an egg in her oviduct fully formed and ready to be laid. The nest could not be closely examined on account of its position, but it was at least partially lined with green leaves.

Major Bendire (1892) has shown that some of the earlier accounts of the supposed nesting of this bird are open to question and may refer to the Mexican black hawk, which might be mistaken for the zone-tailed. The nest that he found on April 22, 1872, on Rillito Creek, about 40 feet up in a crotch of a big cottonwood, was undoubtedly authentic; his hasty descent from the tree, with an egg in his mouth, when he discovered some Apache Indians watching him, is historic. He also mentions two nests found by Dr. Edgar A. Mearns in Arizona, the parent birds being shot in each case. The nests were both in forks of large cottonwoods, one 25 feet and one 50 feet from the ground; the nests were "coarsely built of rather large sticks, with considerable concavity, * * * lined only with green leaves of cottonwood attached to the twigs."

A. W. Anthony wrote to Major Bendire (1892) that he found these hawks "not uncommon" on the San Pedro Martir Mountains in Lower California at an elevation of about 7,000 feet. Two pairs were seen on April 24, both nesting in tall pines. "The birds were greatly worried at our presence, flying about overhead and constantly uttering a loud querulous cry, not unlike that of *Buteo borealis*. One of the nests, examined from the ground, was rather a bulky affair of sticks, and placed in the very top of a pine about 70 feet up. Several shots from our rifles failed to drive the birds away. Shortly afterward a second pair were seen, and one of these was secured."

Frank Stephens, while collecting for William Brewster (1883), found a nest near Tucson in a mesquite "well hidden by bunches of mistletoe."

Eggs.—Sets of two eggs are the rule for the zone-tailed hawk, occasionally only one or as many as three. The eggs are ovate, short-ovate, or nearly oval in shape, and the shell is smooth or finely granulated. The color is dead white or faintly bluish white. They are usually entirely unmarked, but apparently some are faintly and sparingly marked with small spots of pale lavender or pale yellowish brown, sometimes concentrated about one end or the other. Some collectors claim that this hawk never lays spotted eggs, but I see no reason why it should not do so occasionally, as the marsh hawk and Cooper's hawk are known to do. J. H. Bowles writes me that he has a slightly spotted set, the parents of which were shot. The eggs that we collected in Arizona were unmarked, and I have another unmarked set in my collection. Any eggs supposedly of this species that are heavily marked are probably referable to the Mexican black hawk. Major Bendire (1892) discusses this question quite fully. The measurements of 37 eggs average 55.6 by 43.5 millimeters; the eggs showing the four extremes measure 63 by 45, 56 by 49.6, and 52.4 by 38.9 millimeters.

Young.—That both sexes incubate has been definitely proved. Incubation probably lasts for about four weeks. Nothing seems to be known about the young.

Plumages.—So far as I know, no downy young, nestling, or young juvenile zone-tailed hawks have ever found their way into collections, and I have never seen any. I have, however, seen a number of immature birds taken in January, March, April, May, July, and December. These are evidently juvenal or first-year birds, and this plumage is apparently worn without much change all through this year. The contour plumage is much like that of the adult, but the concealed portions of the feathers are more extensively white, and more or less white spotting shows on the breast and back. The under sides of the primaries are whiter than in adults, with narrower dusky bars; in adults these are grayer, with more numerous and broader bars. The tail is also quite different; on the upper side it is "fuscous" or "hair brown", broadly barred with black and narrowly tipped with white; on the under side it is "pale neutral gray" and white, with a broad subterminal bar and about half a dozen narrow bars of dark gray or dusky. On the under side of the adult tail the three pure white zones, on the inner webs of all but the central pair of feathers, are very conspicuous against the otherwise black tail. Material is too scanty to outline the molts.

Food.—The zone-tailed hawk evidently feeds mainly on lizards, frogs, and small fishes, which it finds along the beds of the streams where it lives. It also eats a few small mammals and an occasional bird. Harry S. Swarth (1920) shot at one which dropped a "desert quail" it was carrying, and another had in its stomach the remains of a Gila chipmunk. He writes:

Although, as he circles about on lazy wings, or drifts slowly across a canyon, the zonetail appears too sluggish for any rapid action, the capture of the chipmunk by this individual speaks volumes for the speed that can be attained when necessary. If there is any small mammal that is harder to see in the brush or that can get out of sight with greater speed than the Gila chipmunk it should be safe from any danger. Merely to catch sight of one of these animals, though they be heard chipping in the bushes all around, is no small feat; while for a hawk to lay talons on one, in his chosen haunt of underbrush, logs, and rocks, bespeaks a swoop of lightning speed.

Behavior.—After recording in my field notes the resemblance of the zone-tailed hawk to the turkey vulture in its flight, I was interested to read that several other observers had noted the similarity. Its flight is apparently lazy and sluggish; it usually holds its wings at an angle above its body when soaring, often carries its tail partially closed, and tilts its body from side to side after the well-known manner of the vulture; this is not a universal rule, however, for it often sails on flat wings with spread tail. The dark body and the lighter pattern of the primaries and secondaries, as seen from below, add to the resemblance. The white zones in the tail do not show at all angles and are conspicuous only from below.

These hawks are evidently not shy about their nesting sites. I could easily have shot both of the pair we found nesting, but I was satisfied with one. Dr. Mearns (1886) shot both parents at one of his nests and one of the pair at the other. In both cases one of the birds came screaming at him before he began to look for the nest. One, which he saw standing on her nest, "gave a loud whistle and came skimming towards him." The mate of the other flew from the nest, circled over the canyon a few times and disappeared. Again he writes: "One day, when examining the work of beavers beside the Verde, a Zone-tailed Hawk emerged from the dark shade of a neighboring belt of cottonwoods, moving straight towards me on motionless wings and passing within a few feet, scanning the water beneath with intent interest and paying no attention to me, but moving its head with a restless side movement."

Voice.—I recorded its cry as an incessant and somewhat peevish whistle, halfway between the notes of the red-tailed and the broad-winged hawks. Dr. Mearns (1886) called it a "loud" or "shrill whistle."

Field marks.—Its resemblance to the turkey vulture is referred to above. It has often been confused with the Mexican black hawk, which it closely resembles. The latter has a broad white band on both the upper and the under surfaces of the tail. The zonetail has no white on the upper surface of the tail but has three pure white bands on the under surface; the outermost is the broadest and most conspicuous and the innermost hardly shows at all in flight (see account under Mexican black hawk).

Fall.—The zone-tailed hawk disappears from the northern portions of its range in winter. W. E. D. Scott (1886) says: "On two occasions I have seen from the railroad, while passing through the country between Casa Grande and Bowie stations, flocks of at least fifty birds of this species, evidently migrating and closely associated together. This was in the early part of September, 1882, and as the train was going very slowly, and I was close to the birds, and had become very familiar with them in life about Riverside in the months just preceding, I could be very certain of my identification."

DISTRIBUTION

Range.—Southwestern United States, Central America, and northern South America.

The range of the zone-tailed hawk is very poorly understood. It breeds **north** to Arizona (Big Sandy Creek, Prescott, Salt River, and Paradise); southwestern New Mexico (Gila River and Tyrone); and southern Texas (Guadaloupe Mountains, Davis Mountains, Devils River, Fredericksburg, and Comal County). Most of the eggs that have been collected have, in fact, been taken in the Southwestern United States.

From this northern part of the range the species occurs **south** through Lower California, the mainland of Mexico, and other Central American countries to Dutch Guiana (Surinam); British Guiana (upper Pomeroon River); Venezuela (La Guaira and Macuto); Colombia (Bonda and Mamatoco); Panama (Pearl Islands); and the coast of Peru (Griscom, 1932). The southern part of this range is occupied chiefly by a doubtfully distinct race (*B. a. abbreviatus*), but true *albonotatus* apparently breeds south as far as eastern Panama (Perme), where a specimen was taken in the tropical lowlands on August 7, 1929.

Over much of its range it would appear that the zone-tailed hawk is more or less resident, but during the winter season most individuals withdraw from the United States, and in some Central American localities (as Costa Rica) the species is known chiefly as a rare migrant.

Egg dates.—Arizona and Texas to Central America: 14 records, February 2 to June 7; 7 records, April 13 to May 21.

BUTEO ALBICAUDATUS HYPOSPODIUS Gurney

SENNETT'S WHITE-TAILED HAWK

HABITS

The northern representative of this South American species extends its range into the United States only in the open and prairie regions of southern Texas. George B. Sennett (1878) and Dr. James C. Merrill (1879) were the first to record this handsome hawk **as a** breeding bird in Texas. Both discoveries were made in 1878 near

Brownsville, Tex. The latter says of its haunts: "This fine Hawk is a rather common resident on the extensive prairies near the coast, especially about the sand ridges that are covered with yucca and cactus."

D. B. Burrows (1917) says of its haunts: "The White-tailed Hawk is known in southern Texas as the prairie hawk and the White-breasted Hawk. It is plentifully distributed over the lowlands wherever it is open or sparsely covered with bushes and stunted trees, but does not frequent the rough, hilly portions nor the timbered river bottoms."

Nesting.—Dr. Merrill's (1879) two nests, found May 2, 1878, were placed in the tops of yuccas on the prairie; they "were not more than eight feet from the ground, and were good sized platforms of twigs, with scarcely any lining"; each contained a single egg. Mr. Sennett (1878) also found two nests on May 16, 1878; one he describes as "an immense nest on the top of a large Spanish bayonet, and some twelve feet from the ground." The other was similarly situated but only 8 feet up. "The locality was a sandy ridge, dividing a lake from the salt marshes."

Capt. B. F. Goss found this hawk nesting near Corpus Christi, Tex., and wrote to Major Bendire (1892) as follows:

We found the favorite breeding places of the White-tailed Hawks to be a strip of open bushy land lying between the thick line of timber and chaparral along the coast and the open prairie. Any bush rising a little above the surrounding level seemed a suitable nesting site, and no attempt was made to conceal the nest. In most cases it was very prominent, and could be seen for a long distance. I examined fifteen; they were all placed in low bushes, generally not higher than 6 feet. In a few cases I had to stand upon the wagon to reach them. They were composed of sticks, dry weeds, and grasses. A coarse dry grass entered largely into the composition of most of them. They were poorly constructed, but moderately hollowed, and usually lined with a few green twigs and leaves. Taken as a whole, the nests looked ragged in outline and slovenly finished. About one nest in four contained three eggs, the rest but two.

Herbert W. Brandt has sent me his notes on two nests found by him in Nueces County in 1919. The first, found on March 19, was—

located in a white thorny Armagosa bush standing as a lonely clump in a great flat prairie, with an unobstructed view of the horizon on all sides. The nest was exposed to view and plainly noted three-eighths of a mile away from the machine on the road. It was typical redtail type, being a sharp triangular compact V in outline. The male bird was circling over the road 500 feet or so up. We were watching him show the white on the top of his tail when the nest was seen. On our approach, the female left at a quarter of a mile— a very large bird—and sailed silently away joining its mate. They remained up nearly out of sight until we left the nest, whereon they circled 100 feet above the nest and, on noting the eggs gone, departed. During the entire time at the nest perhaps an hour they did not utter a sound.

The nest was 7½ feet up and made of thorny sticks intermixed with bunches of grass. The inside was neatly and compactly made, the lining being of bunch grass neatly placed and a few white breast feathers of the incubating bird were in evidence. No food remains were seen anywhere about the nesting site.

The above nest measured 33 by 26 inches in outside and 11 inches in inside diameter; it was 18 inches in height and was hollowed 4½ inches. The second nest, found 10 days later, was larger, measuring 36 inches outside; it was also "located in an Armagosa bush, just waist high to the top and the bottom of the nest 18 inches above ground. The site commanded a view for miles of country around, being on a ridge." He says that the nests are used year after year, as the successive layers indicate.

The record low seems to be a nest found by G. B. Benners (1887) on Padre Island, "built in the most peculiar situation, being on a Scrub Oak, not more than a foot and a half from the ground. The bush was the only one in sight." And the record high seems to be a nest 15 feet from the ground in a crotch of a large mesquite, from which a set in my collection was taken by E. F. Pope. D. B. Burrows (1917) says:

Most of the nests are built in the tops of thick clumps of thorny bushes commonly known as the black bush. Two of these nests were placed in the tops of thick clumps of the catclaw, and it is almost as much of a feat to secure the eggs from such nests as it would be to vanquish the angry feline with no weapon of defense in hand. All of the higher nests were placed in small trees, and the mesquite, huisache and hackberry being used.   *   *   *

The nests of the White-tailed Hawk are composed of coarse sticks at the base with finer sticks in its upper structure, fairly well depressed and lined with small tufts of dry bunchgrass pulled from the ground. The latter nests I have found to be similarly constructed, but the lining in them will consist in part or wholly of green leaves from the mesquite and from the huisache trees. Like the other buteos this hawk will use the same nest year after year, adding to it each year until it becomes quite bulky.

Eggs.—The white-tailed hawk lays usually two eggs, sometimes three and occasionally only one. Of 30 sets collected by Mr. Burrows (1917), three contained one egg each, 26 contained two, and only one set contained three. The proportion of sets of three in collections is greater than this. The eggs are ovate, oval, or elliptical-oval in shape, and the shell is smooth or finely granulated. The ground color is dull white or very pale bluish white. About one-third of the eggs are unmarked. Others are faintly and sparingly marked with small spots of pale, dull browns, or buffs, "clay color", "cinnamon-buff", or "pinkish buff", rarely with "chestnut-brown" and occasionally with underlying lavender spots. The measurements of 50 eggs average 58.9 by 46.5 millimeters; the eggs showing the four extremes measure 65 by 50 and 52.7 by 42.2 millimeters.

Plumages.—The downy young of the white-tailed hawk is an odd-looking chick quite different from other young hawks. It is entirely covered with short, cottony down, yellowish white, with a black space around the eye; the crown and occiput are thickly adorned with long, silky, hairlike down, from half to three-quarters of an inch in length, which probably is erected in life; it is basally whitish, but varies from "warm sepia" to "bone brown" toward the tips; similar, but shorter, silky down tipped with "sepia" adorns the back and wings. Mr. Burrows (1917) says: "The young of this species when in the down are mouse color, differing in this respect from most other hawks. This I presume, is a protective coloration, for the nests are seldom protected in the least by branches or foliage above."

The juvenal plumage appears first on the scapulars, then on wings, tail, and body. The down disappears last from the throat, central breast, flanks, and tibiae. In fresh juvenal plumage, in May, the entire upper parts are brownish black, broadly tipped with "pinkish cinnamon", on the scapulars and wing coverts; the sides of the breast are the same with buffy tips; the upper breast is "cinnamon-buff", with black streaks; the belly is dark sepia; broadly tipped with "cream-buff"; the tibiae are broadly barred with "cream-buff" and dusky; the tail is "hair brown" to "drab", inner webs largely white, with numerous faint dusky bars on the upper side and with a buffy white tip. One very dark October bird, in which the light edgings have mostly worn away, is almost wholly a deep, rich, brownish black, or very dark "warm sepia", the feathers of the belly and tibiae being tipped with light buff; this may be a melano.

During the first winter the upper parts become faded to "bister", and considerable new white plumage appears on the under parts from throat to belly; by spring the breast and belly are largely white.

Apparently a complete molt occurs in summer and fall, though I have not been able to trace it, at which a second-year plumage is assumed. In this the upper parts are similar to the first plumage; the anterior, lesser wing coverts are dull "tawny", and some of the scapulars are edged, barred, or notched with this color, forecasting the adult color pattern; the rump is white, barred with dull "russet" and dusky; the new tail is grayish white, with about 10 narrow dusky bars on the central pair of feathers; the other rectrices are narrowly barred with dusky and more or less clouded with gray; all rectrices have a broad, subterminal black band; the throat and sides of the neck are "bister" and the upper breast clear white; the belly, flanks, and tibiae are white, barred with dusky, heavily on the flanks and lightly on the tibiae.

I believe that at least another year is required to assume the fully adult plumage and perhaps more than that to reach its highest perfection. Apparently, as the bird grows older, the head and mantle become grayer, a soft plumbeous-gray; the lesser wing coverts become more extensively rufous, "cinnamon" to "orange-cinnamon"; the tail and rump become whiter; the under parts become whiter, finally including the throat and chin; and the dusky barring disappears, or becomes very faint, except on the flanks.

I have been unable to trace the molts, as I have seen no summer or fall specimens. Probably a complete annual molt occurs during that time.

Food.—Mr. Burrows (1917) writes: "This species feeds upon rabbits and wood rats, in fact I have found that its diet is largely confined to rabbits which are found in immense numbers in that section. If the nest contains young birds it is sure to show that they are amply provided for by the amount of fur and the number of rabbits' feet found in and about the nest. * * * I have never found any evidence that they feed at any time upon other birds."

Other observers have noted cotton rats, quails, snakes, lizards, frogs, grasshoppers, and beetles among the food of this hawk.

Behavior.—Mr. Burrows (1917) says of the behavior of the white-tailed hawk near its nest:

The bird will always leave the nest while the intruder is quite a distance away, often at a distance of a quarter of a mile, for as I said above the nest commands a broad expanse and the bird is always on the watch. When the female leaves the nest, the male bird usually joins her at once and the two often disappear and do not come in sight while the nest is being examined; at other times they mount high in the air, far above the reach of a shotgun and directly above the nest, where they will often stand poised in one spot for several minutes at a time coolly watching development below. At such a time the birds always face the strong sea breeze which blows so steadily and strong that it is possible for them to appear perfectly motionless and stand suspended in mid-air.

Voice.—Burrows says on this point: "The note is peculiar, somewhat like that of Cooper's Hawk when disturbed at the nest, but in a much higher key and with a tinkling, musical sound. It consists of the syllables ke-ke-ke-ke-ke-ke-ke repeated many times."

Dr. Merrill (1879) thought the cry sounded "much like the bleating of a goat." And Mr. Brandt says in his notes: "The call of the female was not a Buteo-like scream, but an uttered *cut-a, cut-a, cut-a*, resembling very much the call of the laughing gull."

Field marks.—Mrs. Florence M. Bailey (1916) notes these very prettily as follows:

When opportunity afforded I noted eagerly its immaculate breast, white rump, and white tail with black subterminal band; but the impression of the bird is what is recalled to-day when a level prairie comes to mind. At a distance one sees a large statue of a Hawk on the prairie floor; on nearer approach, a King of Hawks looking up with calm enquiring gaze, both gaze and pose bespeaking the silent power of the race. The white of the Hawk, by Mr.

Thayer's view of protective coloration, has been worked out to the undoing of its prey, the small mammals that look up at it against the light of the sky into which its whiteness enables it to fade; while on the other hand the small mammals have become colored like the prairie to protect them from furred and feathered hunters that look earthward.

DISTRIBUTION

Range.—Southern Texas south through Central America and through South America from the Colombian Andes, east to Trinidad and Surinam, and south through Brazil, Uruguay, and Argentina to the Rio Negro. Accidental in Arizona.

The range of the white-tailed hawk extends **north** to southern Texas (Marfa, Boquillas, Bee County, and Calhoun County). **East** to Texas (Calhoun County, Rockport, and Brownsville); Vera Cruz (Jalapa); probably Yucatan (Merida); Trinidad (Port of Spain); Dutch Guiana (Surinam); eastern Brazil (Matto Grosso, Bahia, Minas Geraes, Sao Paulo, and Parana); Uruguay; and Argentina (Buenos Aires, Cape San Antonio, and the Rio Negro). **South** to Argentina (Rio Negro). **West** to Argentina (Rio Negro, Mendoza, Tucuman, and Formosa); eastern Bolivia; Colombia (Rio Negro and Bonda); Panama; Costa Rica (San Jose, Cerro de Santa Maria, and Laguna de Ochomogo); Guatemala (San Geronimo and Duenas); western Mexico (Oaxaca, Puebla, Colima, and Zacatecas); and southwestern Texas (Marfa).

The white-tailed hawk has been separated into three geographical races, all of which are included in the outline presented above. Sennett's white-tailed hawk (*Buteo a. hypospodius*) is found from the Lower Rio Grande Valley in the United States south through Mexico and other Central American countries to the Andes of Colombia and Venezuela; *B. a. colonus* occurs from Colombia to Surinam in Dutch Guiana, including also the islands off the north coast of South America, as Aruba, Bonaire, Curaçao, and Trinidad; *B. a. albicaudatus* is found in southern Brazil, Uruguay, and in Argentina south to the Rio Negro.

Casual records.—In the spring of 1897, between Florence and Redrock, Ariz., G. F. Breninger collected an egg that he reported to be of this species. The adult was flushed from the nest but was not obtained. Two years later (January 15, 1899) he collected a male at Phoenix, Ariz. If no mistake was made in the identification of these specimens they remain unique as being the only records for the United States outside of Texas.

Egg dates.—Texas and Mexico: 69 records, February 1 to July 4: 34 records, March 30 to April 24.

BUTEO SWAINSONI Bonaparte

SWAINSON'S HAWK

HABITS

This highly beneficial and almost entirely harmless hawk enjoys a wide distribution over the western half of North America and a large part of South America, where it is one of our commonest Buteos. It is essentially a bird of the wide open spaces, prairies, plains, and even deserts. Major Bendire (1892) says: "On the arid wastes and table lands of southern Arizona, as well as in the sage and bunch grass districts of Nevada, Oregon, Washington, and Idaho, Swainson's Hawk is especially abundant, outnumbering, perhaps, all the other Raptores of these regions combined. It is eminently a prairie bird, shunning the densely timbered mountain regions, and being more at home in the sparingly wooded localities found along the water courses of the lowlands."

Spring.—Swainson's hawk is a highly migratory species, wintering mainly south of the United States and returning to its northern breeding grounds in spring in spectacular flocks or waves. Referring to Escondido, Calif., C. S. Sharp (1902) writes: "The Swainson hawks arrive here from the south about the 10th to 20th of March, sometimes in large flocks or in bands of a dozen or two. The earliest and largest flocks all go north, the summer residents not coming until a couple or three weeks later, and going at once to their quarters which they refit preparatory to permanent occupancy later on."

In El Paso County, Colo., Charles E. H. Aiken (1914) noted these hawks migrating in numbers as early as March 11, during a light snow fall, and as late as April 20, "a bright sunny day succeeding a period of stormy weather." These were all melanistic birds, though "nearly all Colorado breeding birds are of normal coloration." M. P. Skinner says in his notes that in Yellowstone National Park it comes later and departs earlier than the redtail. Its arrival in spring varied from April 18, 1920, to May 3, 1921, sometime after the appearance of the ground squirrel.

E. S. Cameron (1907) gives the following account of a remarkable flight observed in Montana:

My first introduction to these hawks was in April, 1890, when an extraordinary invasion of them—probably nearly two thousand birds—alighted around the ranch where I was staying on the west bank of the Powder River. They came in the afternoon from a southerly direction and, for a time at least, followed the downward course of the river, as a neighbor living above reported the enormous hawk army which flew over. The wide river bottom where the ranch is situated is thickly overgrown with cottonwoods, and the fence of the saddle horse pasture all but joins the buildings. When the last birds had arrived, the trees inside this pasture were simply black with them; but as there

appeared to be numbers beyond, I saddled my horse in order to reconnoitre
further. * * * Having ridden round the fence I found that not only were
the trees filled with clusters of buzzards, but that the ground below was covered with them sitting in rows among the cattle, the sight surpassing anything
I had hitherto seen in bird life. All were obviously worn out and appeared
asleep; but those on the ground, if closely approached, were not too tired to
fly up and join their comrades in the trees. * * * I gave the estimated
number of buzzards at about a thousand; but it became obvious afterwards
that two thousand would have been nearer the true count, as twenty trees
each containing fifty birds give a total of a thousand without including all
those on the ground and in more distant cottonwoods.

Nesting.—My experience with the nesting habits of Swainson's
hawk has been mainly in North Dakota and Saskatchewan, where I
recorded in my notes some 25 nests. Of the four nests found in
North Dakota in 1901, two were in timber belts or groves of large
trees near Stump Lake; one of these was 40 feet from the ground
in the main crotch of an elm, near the top; the other was only 15
feet up in a leaning swamp oak on the edge of the woods next to the
lake. Another was only 8 feet from the ground in a small tree on the
open shore of the lake. And the fourth was 14 feet up in a slender
little cottonwood, hardly strong enough to bear my weight, in a little tree claim near a ranch.

In southwestern Saskatchewan in 1905 we found 13 nests, eight of
which were found in a one day's drive up Maple Creek. In the same
region the following year we found eight nests, two of which were
the same nests used the previous year. Most of these nests were in
the timber belts along the small streams, where the trees were small
or of moderate height. They were placed in cottonwoods, other
poplars, willows, or boxelders, mostly at about 20 feet above the
ground; one was 35 feet up, and some were as low as 10 or 12 feet.
One was only 7 feet up in a clump of small willows near Crane Lake.
Another was on a shelf of a cutbank in open country. Perhaps the
most interesting nest of all was in a little patch of large bushes on a
steep hillside; it was a very large nest, resting on the ground, held
in place by the surrounding bushes and built up to a height of 4 feet
on the outer side, but level on top; our driver told us that it had been
in use for several years. One nest was found in a solitary poplar
on an open plain (pl. 63).

Many of these nests were in commanding situations, and practically all of them were so located that the incubating bird could have
a good outlook. They were typical *Buteo* nests, made of large sticks,
finished off with twigs, weeds, or grasses, lined with inner bark and
fresh green leaves, often attached to the twigs, from the surrounding
trees; many were decorated with the flower clusters of willows, or
other trees, with lichens and with down or feathers shed by the hawk.
The measurements of several nests varied from 21 to 28 inches in

outside diameter; the inner cavity was from 8 to 9 inches in diameter and was hollowed to a depth of from 2 to 5 inches. All the nests found around June 1 had incomplete sets, and fresh sets were found up to June 14, showing that most of the eggs were laid during the first two weeks in June. The first young, just hatched, were found on June 25. These hawks desert their nests on slight provocation; several new freshly lined nests and nests with incomplete sets that we examined early in the season were later found to have been abandoned. Hunting for nests of Swainson's and ferruginous rough-legged hawks is a simple and easy matter in this open country; all one has to do is to drive along near the timber belts and watch for the conspicuous nests; and climbing irons are seldom needed.

Mr. Cameron (1913) says of Montana nests that he has seen:

The nests of *B. swainsoni* are made entirely of sticks, or of sticks combined with other materials, such as sage-brush, wild-rose brambles, and cottonwood or cedar twigs. There may be an elaborate lining of green weeds, or quantities of wool—perhaps only a scanty layer of grass. Some birds line their nests with fresh leaves, which are renewed at intervals, but, in my experience, this does not occur until after the full clutch of eggs has been laid. The parent birds roll back the eggs and replace them on the leaves, which is not a difficult feat, as many nests are almost flat. As the hawk apparently mates for life, the nest, which is very strongly put together, increases in size with the yearly repairs. In my own experience I have known disused nests to be practically intact after a period of seven or eight years. Since 1889, I have seen a great many occupied nests, but only kept notes of fourteen. Of these six were in ash trees, six in cottonwoods, one in a low cedar, and one in a wind-swept pine-top. This last, on a dominant scaur of the pine hills, was the most picturesque of all, but could not, of course, endure long without renewal, and is the only nest I have seen thus exposed.

S. F. Rathbun tells me of a nest he found in eastern Washington that was placed on "a rather large pinnacle rock. It was about 8 feet in height, and its top was almost flat. The color of the rock and that of the nest were so much alike that at a little distance the two blended, but anyone who knew what he was looking for would see the nest. It was about 2 feet high and nicely made. Outwardly it was made of dry branches of sage, with a lining of soft strips of the sage." The nest contained only one egg. When he visited the nest the next day, to see if more eggs had been laid, he "found there had been placed across the top of the inside of the nest a tuft of bunchgrass", concealing the egg. This may have been done to protect the egg against marauders.

Mr. Skinner mentions in his notes from Yellowstone Park "a nest on a tall fir at the edge of a cliff." A. Dawes DuBois describes in his notes a nest "35 or 40 feet from the ground in the top of a cottonwood" in Montana. "It was composed of sticks and coarse twigs (the largest about three-eighths of an inch thick), most of them freshly broken. There was a doubled piece of baling wire in the structure, perhaps 3 feet long in its doubled form. The lining beneath the eggs consisted of fresh green twigs from the nest tree, with the leaves and catkins attached. Below this was some half-dry grass or hay; below that, dry weed stalks; and below these, a lower layer of green leaves (now wilted) and a quantity of coarse bark. A few downy feathers adhered to the outer sticks of the nest."

Bendire (1892) mentions a nest that he found in Oregon, "fully 50 feet up" in a large cottonwood; also some nests found by Capt. B. F. Goss in North Dakota that were "in the high timber along the streams from 40 to 60 feet up." He says further: "In southern Arizona, especially in the vicinity of Fort Huachuca, where this Hawk is a resident and exceedingly common, Lieut. H. C. Benson, Fourth Cavalry, U. S. Army, found forty-one of their nests between May 12 and June 18, 1887. All of these were placed in low mesquite trees, from 3 to 15 feet from the ground. A few found by me near Tucson, in the spring of 1872, were located in similar trees from 10 to 18 feet from the ground."

In California the nesting season is earlier than it is farther north. Mr. Sharp (1902), referring to San Diego County, says:

Nesting begins in April. My earliest record is April 15. Fresh eggs may be taken until the middle of May, but the later ones are second sets. Third sets are very unusual and show an amount of perseverance in the birds that should be respected.

The nest of the Swainson hawk is the usual bulky, unsightly mass of sticks of the raptores, and is placed near the top or on a small outlying branch of a cottonwood or sycamore at an elevation of about 50 feet. (My records run from 35 to 75 feet.) Occasionally a live oak will be taken but as I know of only one such instance, it can hardly be considered regular in this section, at least.

* * * Although the birds—even if their eggs are taken—will return to the same locality year after year and generally to their first nest I have never known them to attempt a second set in a nest just disturbed.

They sometimes will occupy an old nest nearby, but in almost every instance in my experience have built a new nest quite near to the old one but a little higher up and a little further out towards the end of the branch, as though they had learned wisdom by experience. On May 5, 1901, a set of two eggs was taken from a sycamore about fifty feet from the ground, nowhere near the top of the tree. The birds moved to another sycamore 200 yards away and by May 12 a week later, had built another nest at the *top* of the tree and seventy-two feet from the ground. This also contained two eggs which were taken. The birds then moved on a few hundred yards to a much taller sycamore and built a nest in the top of that, and well out of reach and raised their young in peace.

The highest nest of which I can find any record is mentioned by W. L. Dawson (1923), "100 feet up in a giant yellow pine", in Modoc County, Calif. He also shows a photograph of a nest in a giant cactus, or saguaro, in Arizona. Wright M. Pierce sent me photographs of a nest in a Joshua tree in the Mojave Desert (pl. 64).

Eggs.—The Swainson's hawk lays two to four eggs, usually two. Bendire (1892) says: "About one nest in four contains three eggs, and a set of four is rarely met with. I found but one such in over thirty nests." The eggs vary in shape from short-ovate to oval. The shell is smooth or finely granulated. The ground color is pale bluish or greenish white when fresh, fading to dull white. About one-fifth of them are immaculate or nearly so. Others are irregularly and more or less sparingly spotted with various shades of brown, buff or drab, "chestnut-brown", "cinnamon-brown", "cinnamon", "clay color", or more rarely with shades of "Quaker drab." Very rarely are they heavily marked. The measurements of 166 eggs in the United States National Museum average 56.5 by 44 millimeters; the eggs showing the four extremes measure 62 by 46.5, 60.5 by 47.5, 50 by 41, and 53 by 39.5 millimeters.

Young.—Incubation is shared by both sexes and lasts for about 28 days. The young hatch at intervals of a day or two and are bountifully supplied with food by both parents. Mrs. Irene G. Wheelock (1904) says that after the "young are fledged, you may see them jumping with raised wings through the grass in brisk pursuit of crickets and grasshoppers. This they learn to do by imitating the parent, and it is probably their first lesson in pursuing prey. In the nest, they are fed upon small mammals and, even before their down has changed to feathers, they will tear their food with all the ferocity of a young puppy."

Mr. Cameron (1913) writes:

The young birds as soon as they were able, sat about in the branches, but returned to the nest at night, and also on hot days, during which the parents shaded them. * * * The nestlings have enormous appetites, and consume more in proportion to their size than any other raptorial bird which I have studied or kept in confinement. When hungry they set up a piercing kitten-like cry until they are supplied with food. * * * As observed in this instance, the female buzzard acquired the power of flight in twenty-eight days, and the male only after thirty-five days.

Plumages.—The downy young, when first hatched, is thickly covered with white down with a slight yellowish tinge. The down becomes whiter with advancing age. The juvenal plumage is acquired in about the same sequence as in other young Buteos; in a half-grown nestling the throat, neck, center of the breast, the flanks, tibiae, and rump are still downy, while the rest of the plumage is well developed. In this fresh juvenal plumage the upper parts are "warm sepia", broadly edged with "cinnamon" or "pinkish buff"; the tail is "hair brown" or "fuscous", banded with black, with a broad, subterminal black band and with whitish tips; the breast and tibiae vary from "ochraceous-tawny" to "cinnamon-buff", with hastate or sagittate spots of "warm sepia." This plumage is worn throughout the first year with no change except by wear and fading; the buffy edgings above wear away and the buffy tints below fade out to pure white. The sexes are alike in this plumage.

Subsequent molts and plumages are quite puzzling and very much complicated by the three color phases and their intermediates, as well as much individual variation and some slight sexual differences. Coues (1874) recognized only a first-year and an adult plumage in light-phase birds; he noted a melanistic phase but not the erythristic phase. E. S. Cameron (1908a and 1913), who made a careful study of this subject and raised birds in captivity, describes four successive plumages of each sex and states that this bird does not become fully adult until the fifth year. This would be a decided departure from the procedure in other Buteos. A study of his descriptions suggests that he confused individual variations and seasonal changes with progressive age developments.

I can recognize only three plumages in the normal or light phase, a first-year, a second-year, and a third-year, which is practically adult but perhaps subject to some modification with advancing age. Toward the end of its first year the young hawk begins to molt from the first-year into the second-year plumage. This molt is probably complete, but very irregular and quite prolonged. It begins with the molt of the primaries and tail in April or May, continues with the body molt during summer, and is not completed until September or later. I have seen a bird in worn first-year plumage as late as September 9, which was probably over 15 months old.

In the second-year plumage the sexes begin to differentiate, and the color phases, which I have not been able to recognize in first-year birds, become evident. In light-phase birds the chin and throat are white, more or less streaked with dusky, and sharply contrasted with the breast band, which is acquired with this plumage; this breast band is much like that of the adult "tawny" to "russet" in males and "drab" to "hair brown" in females, while the belly and tibiae are more or less barred, spotted, or clouded with "tawny", "russet", "hazel", or "warm sepia" in a variety of shades and patterns in different individuals; females are more heavily marked than males; birds that will eventually develop the melanistic phase are much darker and more heavily marked at this age; and a tendency toward the erythristic phase may be indicated by a preponderance of reddish shades. The feathers of the mantle have buffy edgings fading later to white.

Coues (1874) included the above in the variable adult plumages, and Mr. Cameron (1913) regarded it as a third-year plumage; but I cannot agree with either, though their remarks are well worthy of study.

At the next summer molt, when two years old or more, the bird assumes a plumage that is practically adult and much lighter everywhere. The mantle is more uniform, dull, light brown, without the buffy or white edgings; the throat is pure white, without the dusky streaks; the breast band is lighter in color and is sometimes spotted with white; the lower under parts are whiter with fewer dark markings; in old males the belly, flanks, and tibiae often become nearly pure white, but females always have more dark markings on these parts than males.

In all these plumages there is a seasonal change due to wear. In fresh fall plumage the feathers of the mantle and breast band and the flight feathers are suffused with a grapelike bloom, giving a bluish, ashy tinge to the plumage; as this wears away during winter the plumage appears darker or more brownish, producing a very different effect.

Extremes of the other two color phases are not common, but intermediates between them and the normal, or light, phase are often in evidence, especially in immature birds. In the extreme melanistic phase of the adult, the entire body plumage is "warm sepia", or dark sooty brown, except for some white bars and tips on the upper tail coverts and white, tawny, and dusky barred under tail coverts; the wings and tail are merely somewhat darker than in the normal phase. In adults of the extreme erythristic phase the upper parts, wings, and tail are as in the normal phase; the breast band is similar but mixed with "tawny"; the remaining under parts are deep, rich "tawny" or "hazel", streaked on the belly with dusky or black, and the upper and under tail coverts are as in the dark phase. Immature birds show the phase characters in a variable degree.

Food.—The food habits of Swainson's hawk are highly beneficial; it is one of the farmer's best friends, for it feeds almost entirely on injurious rodents and insects, with a minimum of birds and poultry. Dr. A. K. Fisher (1893) says that of 18 stomachs examined, 7 contained small mammals, rabbits, gophers, spermophiles and mice, 8 contained insects, 3 lizards, and 3 frogs. One stomach contained 68 locusts and another 50 grasshoppers. None contained traces of birds or poultry. He quotes Dr. C. Hart Merriam as follows:

Driving along the crest of the plateau just south of the Umatilla River, at about sundown, we were astonished to see a very large number of large hawks hopping about on the ground, catching grasshoppers. We counted about 150 of these hawks, and there must have been at least 200 in the immediate neighborhood. At first we took them to be roughlegs, but later ascertained that nearly if not all were Swainson's hawks (*Buteo swainsoni*). The period between sundown and dark in that region is so short that the birds were still catching grasshoppers when overtaken by darkness.

About 6 o'clock the next morning I visited the same place and was gratified to find the hawks engaged in making their breakfast of grasshoppers. They were scattered over a larger area than when we saw them the previous evening. Before 8 o'clock most of them had left the hills and settled down for

the day in the poplar trees along the river bottom. Here I found the trees
literally full of hawks, and counted as many as thirteen in one tree. Two
of the three whose stomachs were examined contained grasshoppers and no
other food. The third contained, in addition to grasshoppers, the head of a
meadow mouse of the genus *Arvicola* (subgenus *Chilotus*). One contained
88 grasshoppers, another 96, and the third 106. * * *
Assuming that each hawk captured 200 grasshoppers a day and that there
were 200 hawks, the daily catch would be 40,000 grasshoppers. At this rate
these hawks would destroy 280,000 grasshoppers in a week and 1,200,000 in a
month. * * * When in southern California about a month later I was
told by Mr. Edward Merriam that on three occasions he had noticed similar
gatherings of hawks in San Diego County. Once he saw a flock of several
hundred large hawks catching crickets in cracked adobe soil in the San Mar-
cos Valley. At night the hawks came into the live oaks at the head of the
valley to rest. He shot one and found its stomach packed full of large black
crickets.

He also quotes H. W. Henshaw on another useful habit as follows:
"Camping here [San Fernando Valley, Calif.] one evening our attention was directed to the great number of gophers (*Spermophilus
beecheyi*) which in large colonies inhabited some barren hills near the
station. Toward dusk the place was visited by at least a dozen of
these birds, which took up their positions on the hillocks thrown up
by the animals in front of their burrows, and awaited with patience
the moment when a favorable opportunity should occur to snatch a
supper. Elsewhere I have frequently seen them thus employed, and
their persistence in destroying these pests should entitle them to due
consideration at the hands of the farmer."

John V. Crone observed a large number of Swainson's hawks
moving about in the air and has sent me the following interesting
note: "Soon someone observed that one and then that others were
constantly extending their feet forward and apparently picking at
them with their beaks. This seemed a curious habit so we watched
closely, when the interesting and significant fact became apparent
that these splendid birds were busily engaged in catching and eating
insects while flying through the air. When we discovered this we
spent more time in finding out how it was done. The insects were
seized by the bird's feet. Nearly all the volplaning, somersaulting,
circling, and other gyrations of the hawk ended with a quick thrust
of the foot, which seemed rarely to miss the prey."

I can find no evidence that this hawk ever attacks poultry or game
birds, and most observers agree that it seldom, if ever, kills birds of
any kind. Coues (1874) says: "I scarcely think they are smart
enough to catch birds very often. I saw one make the attempt on
a Lark Bunting. The Hawk poised in the air, at a height of about
twenty yards, for fully a minute, fell heavily, with an awkward
thrust of the talons—and missed. The little bird slipped off, badly
scared no doubt, but unhurt, while the enemy flapped away sulkily,

very likely to prowl around a gopher-hole for his dinner, or take pot-luck at grasshoppers." McAtee (1935) says: "It is evident that this species is by no means a 'bird hawk' as only 1 grouse and 9 small birds were found in 111 stomachs."

But Mr. DuBois tells me that he found in a nest, beside a sitting bird, "the bodies of two young birds, seemingly longspurs." These young birds crouching in the grass may have been mistaken for mice. Rats, mice, and snakes are often taken. The evidence given below shows that small birds have little to fear from these hawks. J. A. Munro (1919) says that Maj. Allan Brooks "found seven downy Ruffed Grouse in the crop of a breeding female." Mr. Cameron (1913) says that this hawk often attempts to catch lark buntings but seldom succeeds; it usually swoops at them on the ground; once he saw one chase a bunting in the air, but the small bird escaped. He found frogs greatly preferred by a young hawk he had in captivity; he writes: "So voracious was the bird's appetite that he would account for six large frogs at a meal, and was often compelled to disgorge those which he had swallowed whole to avoid being choked. I have known him to devour an entire rattlesnake at one time."

Dr. John B. May (1935) saw several Swainson's hawks catching dobson flies (adult hellgrammites) on the wing; these flying insects were caught in the hawk's talons and eaten in the air after the manner of kites. Swainson's hawk seeks its prey by soaring and circling over the open prairie, often high in the air, or watches for it while perched on some dead branch, telegraph pole or fence post, or even on some little eminence on the ground. Mr. Skinner says that on the ground it "can walk quite easily and even run expertly. It even hunts grasshoppers and crickets by running them down. At such a time a lot of Swainson's hawks look much like a flock of small turkeys." S. F. Rathbun tells me that he has seen Swainson's hawks following a man on a tractor, and close behind the harrows with which he was summer-fallowing. The man stated that these hawks often do this for the purpose of catching the meadow mice and gophers disturbed by the harrows.

Behavior.—Swainson's hawk is a gentle, unobtrusive bird, living in harmony with its feathered neighbors both large and small. Mr. Skinner's notes contain several references to the confidence in these hawks shown by sparrows, robins, and bluebirds that hopped about and even sang in close proximity without showing the least sign of fear. Bendire (1892) writes: "It is no unusual sight to find other birds, such as the Arkansas Kingbird, *Tyrannus verticalis*, and Bullock's Oriole, *Icterus bullocki*, nesting in the same tree; and the first-mentioned species goes even further than this, sometimes constructing its home immediately under the nest of these Hawks or in the sides of it. Two such instances came under my personal observation."

English sparrows and house finches and also mourning doves have been known to build their nests in the lower parts of a Swainson's hawk's nest. Frank Stephens (Bendire, 1892) says: "On one occasion I took a set of eggs of this species, and a set of *Icterus cucullatus nelsoni* from a nest pendant from some of the twigs composing the Hawk's nest; another time I found nests of *Tyrannus verticalis* and *Carpodacus mexicanus frontalis* built in the mass of the Hawk's nest, all occupied at the same time."

Grinnell and Storer (1924) speak of a tree, regularly used as a perch by a pair of these hawks, that "was tenanted by a pair of Plain Titmouses, with their brood of young, and a pair of Western Bluebirds with a completed set of eggs; while a likely looking hole higher up was being prospected by a pair of Violet-green Swallows."

But the small neighbors are not always friendly to the good-natured hawks. Mr. Cameron (1913) writes:

In one instance during 1899, a pair of Kingbirds had built their nest in some choke cherries immediately below that of the hawk, which was in an ash tree growing amidst them. Yet another Swainson's Hawk, nesting close by, was so unfortunate as to have a pair of Sparrow Hawks (*Falco sparverius phalaena*) domiciled alongside. Neither of the Swainson's Hawks could flap out of the nesting tree without being immediately attacked by one or other of these aggressive birds—sometimes by all of them together. * * * When his mate was sitting, I have seen a male Kingbird (*Tyrannus tyrannus*) alight on the hawk's back and be carried round for several seconds, while he vented his rage by pecking at her. No matter how high the hawk might soar, the small aggressor would keep above her, renewing his attacks at intervals until both were lost to view. The hawk responded to each assault by merely giving four sluggish, downward flaps after which she would sail on motionless wings as before.

Mr. Munro (1919) says: "In trying to escape from their tormentors, they sometimes turn completely over, sideways, in a 'loop the loop' movement. I once saw two Swainson's Buzzards fly towards each other, fasten their claws together and drop several yards, rolling over and over."

Swainson's hawk is one of the tamest hawks; while perched on a tree, a pole, or a fence post it will often allow a close approach; it will sometimes return to its nest while the observer is standing under the tree. But it is not aggressive and has rarely been known even to threaten to attack an intruder. It appears sluggish and lazy, as it spends much time sitting erect on some convenient perch. It shows a decided preference for some favorite perch or observation point, which it regularly occupies; this may be a dead tree, dead branch, post, stump, or the crest of some cliff or cut bank; such places are well marked by the profusion of droppings, bits of down, and feathers. I have seen a pair of these hawks show as much resentment at the invasion of such a domain as they do when a nest is approached. As it launches into flight, or rises from the ground, this hawk appears awkward and heavy, but when well under way, its flight is strong and graceful, as it sails upward in great spirals until almost out of sight. It has all the full flight powers of the best of the Buteos. Facing a strong wind it can hang almost motionless or glide swiftly down a long incline. A pair often indulge in graceful evolutions, apparently in play or for exercise. Migrating flocks often move along in a series of long ellipses.

Voice.—I recorded the cry of Swainson's hawk as a prolonged, shrill, somewhat plaintive whistle, *kree-e-e-e*, suggestive of the cry of the broad-winged hawk, but not so prolonged and not so plaintive. The call is given while the hawk is in flight or while perched on a tree. Taylor and Shaw (1927) describe the notes as follows: "One call of the Swainson hawk was set down as a loud *squeeuk!* Another call note closely resembles the *peear!* of the western red-tail. The loud scream-call is often followed by a series of strongly whistled syllables resembling *tsip! tsip! tsip! tsip!*"

Mr. Cameron (1913) says of the voice of the young bird:

The young buzzard's cry differed according to his age. When he was quite young it resembled a kitten as stated, but by the end of August, when he was seven weeks old, it became loud and shrill like the scream of a sea-gull, though more piercing. At two months old he developed a musical cry, the appealing tone of which never failed to create a deep impression upon all who heard it. It consisted of four notes insistently repeated like Ē Ŭ, Ē Ŭ, the second Ē being a half tone lower than the first, and may be described as long sustained wails followed by short staccato notes. While these four notes are difficult to express in words they could be easily reproduced upon the violin, and are not unlike the plaintive but shriller tones of the British Lapwing (*Vanellus vanellus*) when hovering over its breeding grounds.

Field marks.—In the normal adult, or common, light phase of plumage, Swainson's hawk is easily recognized by the color pattern of the under parts, pure white throat and belly, broad chest band of dull brown, grayish tail, with many narrow bands, and the unmarked buffy under wing coverts. Immature birds, melanistic birds, and birds in the many intermediate stages and phases are very difficult to recognize, except by a process of elimination. Young birds are generally darker below than other young Buteos and show a suggestion of the adult color pattern. Erythristic and partially melanistic birds often show a similar suggestion. The wings are narrower and slightly more pointed than the redtail's; the wing beats are somewhat quicker and more frequent; and there are light patches on the sides of the rump.

Fall.—Late in August Swainson's hawks gather into large flocks, wheeling and circling high in the air as they gradually drift southward. Mr. Skinner's earliest date for the last one seen in Yellowstone Park is August 23, and his latest certain date is September 20.

Many observers have noted the spectacular fall flights of these hawks, constantly passing in small bands, or in flocks of hundreds. H. W. Henshaw (1875) writes:

At Camp Grant, Ariz., in the latter part of September, this hawk was present in very large numbers. About a mile below the post, out on the plain, the stream was bordered by some large cottonwoods; and these were habitually used as roosting-places by the Turkey Buzzards and Hawks conjointly, as the whitened appearance of the branches and the ground below testified, as well as the fetid odor in their vicinity. Hawks and buzzards appeared to be on terms of the most intimate companionship with each other, and one tree often held seven or eight of either birds. The buzzards seemed if anything rather the shyer of the two, and were generally the first to start, when immediately the whole band would leave their perches, and begin circling in the air, gradually ascending higher and higher till out of danger. Thus they would continue wheeling about till the coast was clear, when all would again resume their perches.

Winter.—Most of these hawks spend the winter south of the United States, but a few remain in the southern portions of their range in this country. Major Bendire (1892) says: "On the eastern slopes of the Rocky Mountains it winters from about latitude 39° southward, a few remaining in favorable localities still farther north. On the Pacific coast I have observed a few wintering in southeastern Oregon in about latitude 42°, the majority passing southward, and the birds remaining are probably such as breed much farther north, replacing the regular summer residents, which in turn move south on the approach of cold weather."

DISTRIBUTION

Range.—North and South America.

Breeding range.—Swainson's hawk breeds **north** to Alaska (probably Mount Sischu, Fort Yukon, and latitude 66° 43′ N. on the Alaska-Yukon boundary); Mackenzie (Onion River and Fort Anderson); Saskatchewan (Quill Lake, Touchwood Hills, and Qu'Appelle); and Manitoba (Oak Lake, Treesbank, Aweme, Carberry, Oak Point, and Winnipeg). **East** to Manitoba (Winnipeg); western Minnesota (Hallock and Browns Valley); Iowa (Forest, La Porte City, Grinnell, and Sigourney); rarely Illinois (Philo); rarely Missouri (Pierce City); Oklahoma (Norman); Texas (Henrietta, San Angelo, Rocksprings, and Cotulla); and central Mexico (San Diego). **South** to central Mexico (San Diego); Durango (Rio Sestin); probably Sonora (Opodepe); and possibly Lower California (near Ensenada de Todos Santos). **West** to possibly Lower California (near Ensenada de Todos Santos); California (San Diego, Capistrano, Santa Monica, Santa Paula, Santa Barbara, Mount Pinos, Buttonwillow,

Alila, Paicines, Petaluma, Shasta Valley, and probably Hornbrook);
Oregon (Swan Lake, Fort Klamath, Prineville, and The Dalles);
Washington (probably Cleveland, Yakima, probably Tacoma, and
probably Bellingham); British Columbia (probably Shuttleworth
Creek, Okanagan Landing, and probably Kispiox Valley); and
Alaska (Onslow Island, probably Juneau, and probably Mount
Sischu).

Winter range.—The winter range of the Swainson's hawk appears
to be located entirely in the Republic of Argentina. Visual winter
records for this species in the United States are almost invariably
errors in identification for melanistic roughlegs or for one of the
races of *Buteo borealis*. The few specimens taken in North America
during this season probably represent disabled birds prevented by
their physical condition from performing the migratory flight to
normal winter quarters.

The winter range in Argentina is not fully known, but it extends
north at least to Tucuman and south to San Pedro, Cordoba,
Platanos, Buenos Aires, and Barracas al Sud.

Migration.—Because of much misidentification in the field, there
is a dearth of authentic data illustrative of the migrations of this
hawk. Griscom (1932) refers to the great flights that pass through
Central America as "one of the sights of the bird world." He says:
"Apparently the great majority of the individuals in existence pass
over Central America in a comparatively few days in a few enormous flocks which take hours to pass a given point. The birds
alight only casually, and only three specimens have been collected in
Central America." The birds pass south through this region in
October and return north in April.

Spring migration.—Early dates of arrival in the United States
and Canada are: Iowa—Sigourney, March 8; Grinnell, March 21;
La Porte City, March 23; and Iowa City, March 29. Minnesota—
Heron Lake, March 27; Jackson, April 5; Wilder, April 8; and Elk
River, April 12. Texas—Corpus Christi, March 16; Somerset, March
20; Refugio County, March 28; Kerrville, April 3; San Angelo,
April 3; and Houston, April 4. Oklahoma—Norman, March 14;
and Okmulgee County, April 10. Kansas—McPherson, March 29;
and Fort Riley, April 3. Nebraska—Badger, March 16; Lincoln,
March 19; and Alda, April 3. South Dakota— Vermillion, April 4;
Forestburg, April 4; and Petrodie, April 7. North Dakota—
Marstonmoor, March 8; Chase Lake, March 10; Grafton, March 21;
and Grand Forks, March 31. Manitoba—Margaret, April 2; Aweme,
April 5; and Pilot Mound, April 15. Saskatchewan—Eastend,
March 31; and Indian Head, April 1. New Mexico—Glenrio, April
9. Arizona—Huachuca Mountains, April 3; and Tucson, April 19.

Colorado—Denver, March 9; El Paso County, March 11; and Boulder, March 31. Wyoming—Yellowstone Park, April 18; Laramie, April 23; and Cheyenne, April 27. Montana—Terry, March 23; Fortine, March 28; and Billings, April 3. Alberta—Flagstaff, March 15; Alliance, March 28; and Camrose, March 31. California— Escondido, March 10; Fresno, March 14; Pala, March 17; and Pomona, April 4. Oregon—Lake Malheur, April 17. Washington— Seattle, March 7; Grand Dalles, April 16; and Tacoma, April 27. British Columbia—Okanagan Landing, April 11.

A late date for departure from winter quarters is April 17 (1921), when one was seen over the summit of the Sierra San Xavier above Tafi Viejo, Tucuman, Argentina.

Fall migration.—Late dates of departure in the autumn are: Alaska—Juneau, September 7. British Columbia—Okanagan Landing, October 15. Washington—Walla Walla, October 16. Oregon— Cold Spring Bird Reserve, October 14. California—Saticoy, October 2; and Fresno, October 15. Alberta—Flagstaff, September 17; Andrew-Beaver River, September 23; and Calgary, September 29. Montana—Fortine, October 29; and Bridger Mountains, November 18. Idaho—Spring Creek, October 31. Wyoming—Powder River, September 24; Fort Laramie, September 26; and Yellowstone National Park, October 3. Colorado—Baxter Pass, September 23; and Beulah, October 20. Arizona—Camp Grant, September 28. New Mexico—Bear Springs Mountains, September 29; Laguna Magdalena, September 29; and Datil, October 15. Saskatchewan—Indian Head, September 28; and Eastend, October 12. Manitoba—Shoal Lake, October 2; Aweme, October 6; Reaburn, October 17; and Margaret, October 18. North Dakota—Charlson, October 19; Antler, October 23; Harrisburg, November 1; Kindred, November 9; and Grafton, November 14. South Dakota—Arlington, October 19; Harrison, October 26; and White River, November 2. Nebraska— Hillsdale, October 10; Nebraska City, October 12; Gresham, October 15; and Lincoln, October 27. Kansas—Lawrence, October 10; and Cimarron, November 1. Oklahoma—Tulsa, October 15; Norman, October 17; and Fort Sill, November 1. Texas—Fredericksburg, October 11; Bee County, October 23; Somerset, October 26; and Corpus Christi, November 15.

The southward flight of these hawks is well illustrated by the record of one banded at Red Lodge, Mont., on September 19, 1916, and killed near Bogota, Colombia, on October 29, 1916.

Casual records.—While there are several records for Swainson's hawk in the eastern part of the continent, the species can be considered only as a casual east of the Mississippi River. Among these occurrences are: Ontario, a specimen was collected at Moose Factory in 1881, and four or five specimens have been taken in both spring and fall in the vicinity of Toronto; Quebec, several records for the region about Montreal; Michigan, a specimen was taken at White-fish Point, May 19, 1923, one was collected in Cheboygan County in October 1883, and another was obtained at Hessel on October 13, 1908; Maine, one was taken at Glenburn, May 19, 1888, another at Gouldsboro on September 15, 1886, and a third at Calais about October 8, 1892; Vermont, an adult male was taken near Hartland on May 23, 1915; Massachusetts, one taken at Hamilton on April 20, 1872, a young male taken at Wayland about September 12, 1876, one shot at Salem on October 28, 1889, and another was taken at Essex on May 29, 1892; New York, one in October 1877 in Onondaga County, one at Brockport on October 1, 1889, one at Cornwall on October 14, 1892, and two taken at Lake George on September 9, 1920; Florida, a specimen collected at Key West on November 28, 1895, and another at Miami Beach on December 7, 1922.

The only known record for Ecuador is that of a specimen taken on January 15, 1921, at Zambiza, about 10 miles northeast of Quito, and now in the Zoological Museum at Stockholm, Sweden.

Egg dates.—British Columbia to Saskatchewan: 13 records, May 10 to June 18; 7 records, May 30 to June 14.

Washington to Illinois: 29 records, April 19 to August 17; 14 records, May 23 to June 12.

Iowa to Kansas and Colorado: 12 records, May 10 to July 14; 6 records, May 17 to June 5.

California and Oregon: 68 records, March 17 to June 10; 34 records, April 24 to May 11.

Arizona to Oklahoma and Texas: 38 records, March 6 to July 9; 19 records, April 13 to May 25.

BUTEO PLATYPTERUS PLATYPTERUS (Vieillot)

BROAD-WINGED HAWK

HABITS

In May, when the tender, freshly opened oak leaves are as big as a crow's foot, when the farmer goes out to sow his corn, and when the hosts of warblers are migrating through the treetops, then may we look for the home secrets of the broadwings. They are gentle, retiring, quiet birds of the deep forests. They are seldom seen in the open country except when migrating or soaring in great circles over their woodland homes. In my home territory, southeastern Massachusetts, we find them oftenest in the extensive hardwood forests of chestnut (formerly), oaks, beech, and maples; occasionally a pair makes its home in a maple swamp or in a narrow strip of mixed woods along a stream; and often we have found them in forests of white pine, mixed with hardwoods, and more rarely in the stunted forests of pitch pines and oaks on Cape Cod. Generally they are quite remote from human habitations in quiet woods.

J. W. Preston (1888) writes: "In hidden retreats, where the tangled wilderness of lakes and forests guards in lonely silence the streams which feed the Red River of the North in Minnesota, I found the Broad-winged Hawk (*Buteo pennsylvanicus*), breeding abundantly. At home with the Barred Owl, and unmolested by stealthy tread of wild cat or lynx, he is in this region indeed a 'bird of the wilderness'."

Frank L. Burns (1911) in his excellent monograph, which far overshadows this brief life history, says:

In Pennsylvania it haunts the wild rocky wooded ravines above the small streams and close to small ponds and swamps. While it is not unknown to the large grove, it loves the continuous woods over which it can pass undisturbed and unseen from one feeding ground to another; shunning the cultivated area altogether or traversing it only to visit some nearby swamp or pond.

The ideal station, and I refer to a definite existing locality in S. E. Pa., would seem to be a tract consisting of upland, hillside and swamp, well covered with mixed hardwood timber, with here and there an unoccupied clearing, an unfrequented public highway, cart road or path, with room enough for the silent deadly swoop after the unfortunate mouse or red squirrel; also a small stream or pond, in the shallow reaches of which it can snatch the crayfish, and surprise the frog or trout-eating water snake on the borders of the pools. The verdant foliage supplies it with numberless insects and fat larvae. As it sits upon one of its favorite perches, well hidden by a leafy screen, should a chance human intrude, it cranes its neck and utters its plaintive whistle, which the uninitiated attributes to the cry of one of the numerous small woodland birds or the creaking of two opposing branches in the wind.

William Brewster (1925) says of its haunts in Maine:

Within the forest the Broad-winged Hawk leads, for the most part, an untroubled and sedentary life contenting itself with such animal food as may be had with the least possible exertion and confining its hunting operations to areas of no very great extent. Although not averse to frequenting flat lands where spruces, balsams, and other evergreens flourish to the general exclusion of deciduous trees, it is most likely to be met with on the crests or flanks of ridges heavily timbered with intermingling hemlocks, beeches, yellow birches, and rock maples of the largest size, or in deep glens watered by sluggish brooks flowing beneath dense canopies of overarching foliage.

Spring.—Mr. Burns (1911) gives the following general outline of the spring migration:

My data tends to the conclusion that the vast bulk of migrating Broad-wings ascend the Mississippi valley, distributing its quota near the mouth of every river valley. Natives of Minnesota and Manitoba region have a comparatively simple journey after entering the United States, but the vast horde pouring into the Ohio valley enroute for Ontario, Quebec and possibly northeastern United States and southeastern Canada, have a more intricate and fatiguing journey. As few if any migrants appear from the West Indies, the Gulf States must receive their supply from the Mississippi valley also; and the Atlantic

States from North Carolina to Pennsylvania and New Jersey, overland from Tennessee, Cumberland and possibly Ohio valleys. More complete data may show a South Atlantic coast migration of which I have no hint.

Many observers have noted and published accounts of the great spring flights of these and other hawks. They have regular flyways that they follow year after year, along river valleys or the shores of large lakes or even along the coast line; they seem to dislike to cross large bodies of water and deviate from their main northward or northeastward course to avoid such crossings. Heavy spring flights occur along the coast of New Jersey, northeastward across New England and eastward and northeastward along the shores of Lakes Erie and Ontario. The birds usually fly high, sometimes almost out of sight, in loose straggling flocks or small parties. On favorable days, with westerly winds, they may be seen passing in a steady, widely scattered stream all day long. Some days the sky seems to be full of them, as far up as one can see. Broadwings are often accompanied by redtails, redshoulders, roughlegs, Accipiters, small falcons, and ospreys. When they fly low, as they often do, many are killed in regularly organized hawk shoots, for sport or with the mistaken idea that they are vermin.

Dr. Thomas S. Roberts (1932) thus describes a big flight that occurred in Minnesota:

Toward evening on April 21, 1925, a vast flight of Hawks arrived from the south and settled in the numerous trees of both towns and all the groves and tree-claims, from several miles west of Wheaton to several miles east of Herman—a front of at least twenty-five miles. They remained until the following evening when all the survivors left, going northward. The appearance of the Hawks on the evening of the twenty-first brought out every man who owned a gun in both towns and most of the farmers in the surrounding country. It was stated that residents of Wheaton stood on their lawns and shot dozens from the trees and as they circled about. At the same time a similar fusillade was in progress at Herman and on the farms between and adjoining. It was estimated that at least three thousand Hawks were killed at Wheaton and one thousand at Herman. A "Crow-shoot" happened to be in progress at Wheaton and the participants brought in one thousand five hundred of these Hawks. There is no way of estimating how many were killed by the farmers, but the number was probably many hundreds, if not thousands. No one can venture a guess as to how many Hawks were included in this vast flight! It would seem as though all the Broad-wing population of the north country must have been traveling in company.

Courtship.—As this and other Buteos are probably mated for life, the love making is largely expressed in nuptial flights in which both birds flap or soar in small circles, frequently passing close together and occasionally darting down at one another in a playful mood. Mr. Preston (1888) says: "During the mating season (which begins about the first week in May), the clear, shrill scream constantly echoes in the dim woods, as one answers back to another from some chosen perch."

Lewis O. Shelley has sent me the following note on the mating performance of the broad-winged hawk, as observed in the vicinity of three old nests, at least one of which had been reconditioned:

On May 1, 1934, as I was passing, a male broadwing was seen perched on a limb beside the extreme northern nest; and a female sat on a limb beside the center nest, both birds facing east. The male commenced his moderate, whiny, screamed call—the mating call—and after six quick utterances flew over to the female, who, seeing his act, turned to face the westward—part of the mating maneuver. He alighted on her and mating, or copulation, was immediate, lasting one full minute, the male continuing the mating call throughout and balancing himself by leisurely half-flaps of the wings. He then flew to a perch nearby and sat there in a noncommittal attitude. The mating call might be called a wheezy whistle, with an intake of breath and then its expulsion, this giving a 2-toned call that has a rather musical sound, as *whee-ooou*. This call was given each time he lowered the tail.

Nesting.—Our experience with the nesting habits of the broad-winged hawk in southeastern Massachusetts has been rather peculiar. Although our records for the other hawks date back to 1882, it was not until 1899 that we found our first nest of this species in Bristol County. During those 17 years we were frequently hunting in suitable localities and finding numerous nests of red-shouldered, Cooper's, and sharp-shinned hawks, but we never even saw a broad-winged hawk to recognize it. We were young then, and eyes and ears were keen; it seems hardly likely that we overlooked it; and I am inclined to think that it moved into our territory about the close of the last century. Betwen 1899 and 1928 we found 23 nests, generally one or two nests in each season that we looked for them; in 1928 we found four nests; but since then we have been unable to find another nest and rarely see a bird, though we have hunted the same territory more carefully than ever. I now think they have left us or become very rare.

My notes fail to record the kind of tree in which two of the nests were located, but of the 21 others 7 were in oaks, 6 in white pines, 4 in chestnuts, 2 in pitch pines, and 1 each in a maple and a gray birch. From this it appears that they show no great preference for any particular species of tree. The nests were mostly at moderate heights; the highest was 40 feet from the ground in a large white pine; the lowest was only 24 feet up in a gray birch, so slender that it would hardly bear my weight; and all the others ranged in height from 25 to 36 feet. In a deciduous tree the nest is usually placed in the main crotch, supported by three or more branches, or against the trunk on horizontal branches; in a pine tree the latter is the usual situation. I have never seen a nest out on a branch away from the trunk.

The nest is usually small and rather poorly built. I believe that the hawks usually build a new nest each year, though they occasionally make over an old squirrel's nest, or an old nest of the crow or another species of hawk. One pair used, for one season only, a large nest originally built by a pair of redshoulders and subsequently occupied for two seasons by a pair of redtails. A typical nest in a 3-branched main crotch of a chestnut was made of dead hardwood sticks and dead leaves, lined with a few strips of inner bark, lichens, and a few chips of outer bark; it was built up to a height of 12 inches on what was probably an old squirrel's nest, and measured 17 inches in greatest diameter. Another nest in a white pine was made of pine sticks and twigs, lined with chips of outer bark. All the nests of which I have the records were more or less profusely lined with chips of outer bark of oak or pine, which sometimes filled the whole center of the structure; a few were also partially lined with fresh sprigs of pine and many with green oak leaves; in one nest a sprig of fresh oak leaves had been laid over the eggs. The largest nest measured 21 by 17 inches in outside diameter and the smallest 14 by 12 inches; the inner cavity varied from 6 to 7 inches in diameter and from 1 to 3 inches in depth; the outside height varied from 5 to 12 inches according to the location, the flattest nests being in pines.

Only once have we known a pair to use the same nest for two seasons. In only a few cases have we known them to nest in the same tract of woods for more than two successive seasons and never for more than three or four. This is in marked contrast to the stability of red-shouldered hawks in their established homes.

In other parts of its range a great variety of forest trees has been chosen by the broad-winged hawk as nesting sites; the most abundant and characteristic large tree of the region seems to be the one oftenest chosen. Mr. Burns (1911) lists pines, hemlocks, spruces, larch, chestnut, oaks, birches, maples, beech, linden, poplars, cottonwood, balm-of-Gilead, hickories, walnuts, magnolias, ashes, wild cherry, and elm. He also says: "The height from the ground varies from 3 feet in the broad-forked bole in Minnesota as recorded by George Cantwell, to the 87 foot oak of Delaware Co., Pa., essayed by Harry G. Parker, and the 90 foot black ash of Kalamazoo, Mich., made famous by Dr. Gibbs."

He took the trouble to dissect a newly built nest that contained the following material:

20 white oak twigs, 6 to 10 inches long; 26 chestnut twigs, 4 to 16 inches; 50 chestnut oak twigs, 5 to 16 inches long and many-branched; 77 dead sticks probably principally chestnut; 2 chestnut blossoms, 46 chestnut bark scales, 1 x 2 to 2 x 6 inches; and a few leaf sprays. It was placed upon a foundation consisting of a Crow's nest, from which it was separated. * * *

An almost invariable custom of the Broad-wing is that of placing sprays of fresh green leaves and sometimes blossoms, of the chestnut, oak, poplar, maple, wild cherry, basswood, cottonwood, elm, pine, spruce, hemlock, balsam, and in one instance, evergreen vine and swamp grass, in the nest, under and around the eggs or young; seldom more than one kind of leaf used in the individual nest, though it is frequently renewed. The sprays are broken from the tops of trees and carried to the nest by means of the beak. In rare instances when the leaves have not appeared, green twigs with buds and blossoms attached, have been employed.

J. H. Riley (1902), who has watched broad-winged hawks at their nest building, says: "The nest is often finished and left for some little time before eggs are deposited. In building, the birds never seem to be in a hurry, and several days will elapse without apparently anything being done. A few sticks a day, at the most, seem to be the limit of their exertions, and at this slow rate, it takes them fully three to five weeks or more to complete their domicile."

Eggs.—Two eggs seems to be the commonest number for the broad-winged hawk to lay. I have found twice as many sets of two as I have of three. Four eggs is an unusual number, and incubated single eggs have been found. Bendire (1892) quotes O. C. Poling as saying that sets of four are not uncommon in Illinois and that he found one set of five. Mr. Burns (1911) has been unable to locate this set and doubts it. His data for 406 sets show 15 sets of one, 183 sets of two, 190 sets of three, and 18 sets of four eggs, with "a substantial increase in the number of eggs in a set from the south northward." The eggs vary in shape from elliptical-ovate or ovate to nearly oval. The shell is finely granulated. The ground color is dull white, pale bluish white, or creamy white. The color, shape, and size of the markings show endless variations, and many eggs are very beautiful. Some eggs are boldly and irregularly blotched with rich or bright browns, "burnt sienna", "amber-brown", or "chestnut"; some are irregularly spotted or finely sprinkled with these colors, often concentrated at one end or in a ring; others are evenly covered with small spots or minute dots of the same colors. All the above types of markings appear in other paler colors, such as "hazel", "tawny", "cinnamon-buff", "vinaceous-fawn", and "ecru-drab." Some of the prettiest eggs have great washes or splashes of the paler browns or drabs overlaid with blotches or spots of the darker browns. Underlying markings of purple or ecru drabs are so common as to be almost characteristic of the species and some eggs have only such markings. Occasionally an egg is nearly immaculate. The measurements of 51 eggs average 48.9 by 39.3 millimeters; the eggs showing the four extremes measure 54 by 40.5, 50.5 by 42.2, 44.5 by 39.1, and 47.2 by 36.3 millimeters. The largest eggs of this species seem to be larger than the smallest eggs of the red-shouldered hawk, which suggests the possibility of some mistakes in identification.

Young.—The period of incubation is between 21 and 25 days; unless the nest is visited daily during the laying period and again during hatching days and the eggs marked, it cannot be determined accurately. Both parents assist in incubation and in the care of the young. If the female is killed, the male will continue incubation and rear the young alone. Mr. Burns (1911) says: "Early in May, '93, a nest was found just completed. No eggs were ever deposited and but one bird seen in the vicinity. At every visit he showed as much solicitude as if it was occupied, and several times upon ascending, fresh green poplar leaves had been added to the lining. The nest was not deserted until the latter part of June; the conclusion that it was built by an unmated or bereaved male, seems well founded."

He noted that the young remained in the nest for 41 days in one case, and again that one was prematurely flushed from the nest 29 days after hatching. He writes:

Abundance of food is provided and the nest supplied daily with green leaf sprays, by the parents. The tender young are protected from the hot summer sun, inclement weather and cool nights. I have found the male covering 5 days old hawklets. Even when they have become fairly well fledged, one or the other of the birds seem always in attendance in a nearby tree top. The whistled protest of the parents as they shadow one through the woods, is all the hint one often has of their presence and unceasing vigilance. How long they are guarded after leaving the nest, I am unable to say, but for a week or two after the nest is vacated, a protesting whistle from a hidden form in the neighboring foliage informs one of the jealous care of the juveniles doubtless also hidden nearby. The immatures are unmercifully driven out of the adult's territory the following spring, should they attempt to invade it. Parental care does not survive the winter's frost. * * *

The chick utters a peeping cry as soon as out of the shell and appears hungry as soon as its down has dried. I have seen it turn its head and bite at my thumb, when less than a day old. For some days the adults dismember the food and the young soon learn to snatch it piecemeal from their beaks. When from five to eight days old it sits erect and its mouth flies open at every sound; it is able to disgorge a pellet the size of a hazelnut, scratch itself and behave as well to its mates as the best tempered of birds.

I have seen a pair of adults still attendant on a brood of fully grown young as late as July 29. Young broad-winged hawks will make good pets, if not taken from the nest when too young. One that I took when about a week old and fed on raw meat finally sickened and died. Probably it did not get the right food. It would be safer to wait until the bird is three or four weeks old and nearly ready to fly. See an article on this subject by Dr. Louis B. Bishop (1901).

Plumages.—When first hatched the chick is covered with buffy white down, basally grayish. Later the down becomes whiter. I have a brood of three young in my collection taken out of the nest at different ages. One, 9 days old, is still wholly downy. On an-

other, 16 days old, the down is whiter and the remiges and rectrices
are growing out and beginning to burst their sheaths; and there are
a few feathers appearing on the scapulars and upper back. The
third, 21 days old, is still downy on the head, neck, central breast,
and belly, but elsewhere it is well feathered; the feathers of the upper
parts are "warm sepia", with "tawny" edgings; those of the breast
are "warm buff", with broad streaks of sepia; and the tail is 2 inches
long and partly in sheaths.

A fully grown young bird in full juvenal plumage, July 29, shows
the complete development of the above plumage. The edgings on
the upper parts are narrower; the wings are much as in the adult
but whiter below; the under parts are whiter but with a buffy tinge
and with large hastate spots or streaks of dark sepia on the breast
and with rounder spots on the tibiae; the tail is dark brown, "fuscous", above with indistinct darker bars, the inner webs being whiter
with more distinct bars; the under surface of the tail is gray, with
bars and a broad subterminal band of darker gray.

This plumage is worn for nearly a year with no change except by
wear and fading. Beginning with the wings in April or May, and
continuing through the summer, a complete molt produces by September a plumage that is practically adult. Adults have one annual
complete molt from April or May to August or September. Mr.
Burns (1911) has given a very full account in detail of all the molts
and plumages, to which the reader is referred.

Mr. Riley (1908) has called attention to the fact that "birds from
the eastern United States exhibit apparently two phases of plumage;
a light grayish brown backed bird with little or no reddish edges
to the feathers, and with the bars below prout's brown; and a dark
bird with the feathers of the sides of neck and upper back strongly
edged with cinnamon-rufous, and the bars below of the latter color,
heavier, and sometimes confluent on the chest." There is also a
melanistic phase, which Robert Ridgway (1886a) describes in part as
follows: "Plumage of head, neck, and body, entirely continuous dark
sooty brown, without the faintest indication of markings, even on
the lower tail-coverts or lining of the wing; back darker, with a
chalky cast in certain lights."

Dr. B. H. Bailey (1917) described as a new subspecies, *Buteo
platypterus iowensis*, a bird evidently exactly like the one described
by Mr. Ridgway, and mentions several others taken in Manitoba,
Minnesota, and Iowa. These are all evidently melanistic individuals
of *Buteo platypterus platypterus*. As normal broad-winged hawks
are common in the same region, and as melanism or other color
phases are known to occur only in limited portions of the ranges of
other species, there is no reason for recognizing this as a subspecies.

Food.—Mr. Burns (1911) gives the following interesting account of the hunting and feeding habits of this hawk:

The rather sedentary Broad-wing most frequently waits for its prey while perched on a convenient stub or dead limb. A slight stir below and it bends forward with dilating pupils, cat-like, with twitching tail, swaying body, light foothold; it springs forward with marvelous quickness, snatching up the object with its talons; if its captive is not too heavy it carries it to one of its favorite perches, there to devour it unless disturbed, when it reluctantly retires after a whistled protest. Very small mammals are swallowed whole, and the larger skinned and even the leg bones clean-stripped and left attached to the hide. Birds are plucked of primaries, rectrices and a few breast feathers, flinging them aside with a quick flirt of the bill; after tearing off and devouring the head, the body is ripped open and the intestines eaten, piece by piece the limbs and body follow. Large snakes, toads and frogs are usually skinned, and smaller ones torn in sections after the head has been disposed of. Crawfish are eaten piecemeal, and insects, spiders, etc., usually disappear intact. I have seen it fly toward its nest with a mouse dangling from a single sharp talon in its throat, and a medium-sized snake grasped firmly with both feet; yet kite-like, it will sometimes securely hold an uninjured beetle, grasshopper or earth-worm, in one foot bent forward to breast, and resting on the other, delicately pick the tidbit to pieces. At times it is said to hunt on the wing, circling in the air, upon sighting its quarry it becomes stationary for an instant and then descends with considerable velocity, thrusting forward its feet with lightning-like rapidity, securely grappling its victim only when its body follows or its legs bend in the rise; unlike the Osprey which seems to have the power to clutch with extended leg, its hold is secure only upon this movement. Digestion is comparatively rapid and the indigestible parts, consisting of the nicely-cleaned bones enveloped in the hair, feathers, etc., are regurgitated in the form of pellets before fresh food is taken.

In his long list of species known to be eaten he includes the northern hare and cotton rabbit, various mice and shrews, red squirrel, chipmunk, and weasel among mammals; flicker, phoebe, sparrows, ovenbird, brown thrasher, and wood thrush make up the bird list; the reptiles and amphibians include lizards, small snakes, frogs, and toads; other items are various larvae of large moths, beetles, locusts, grasshoppers, dragonflies, thousandlegs, spiders, ants, fiddler crabs, crayfish, and earthworms. His summary of the contents of 115 stomachs states that "11 contained birds; 31, mice; 17, other mammals; 17, reptiles; 22, batrachians; 45, insects; 10, crawfish; 2, spiders; 1, thousand-legs; 2, earth worms; and 7 were empty."

Dr. A. K. Fisher (1893) says of its food:

Among mammals the smaller squirrels and wood mice are most frequently taken, though field mice and shrews also are found in the stomach contents.

During August and September a considerable portion of the food consists of the larvae of certain large moths which are common at this season, * * * and it is the exception not to find their remains in the stomachs examined. Grasshoppers, crickets, and beetles are also greedily devoured. * * *

The only act of the Broad-winged Hawk which seems injurious to agriculture is the killing of toads and small snakes; the former of which are exclusively

AFPI LIBRARY

insect-eaters, the latter very largely so. In one respect its enormous value ranks above all other birds, and that is the destruction of immense numbers of injurious larvae of large moths, which most birds are either unable or disinclined to to cope with.

Lewis O. Shelley sent me the following note:

A broad-winged hawk a few years ago was unusually bothersome to a pair of bluebirds that nested on a pole. Many times a day it swooped from some nearby tree as a bird entered the box. It did one day succeed in getting the female, under interesting conditions. I had seen the hawk alight in a tree and the female bluebird enter its box without molestation. Then the hawk flew at the pole and struck it with the wings in a forward-check attack that set the pole a-quiver. Usually this would be unavailing with a tight-sitting bird such as the bluebird, but the young were nearly ready to leave and this might explain it. Anyhow, the female stayed in the box for some moments, and in the meantime the hawk had taken a perch on the box, so that when the bird's head appeared in the entrance, with a half flop sidewise the hawk grasped the bluebird's head and pulled her from the entrance. I later shot the hawk.

William Brewster (1925) writes:

The visits of the Broad-winged Hawk to the shores of the Lake are made oftenest late in May or early June when toads (*Bufo americanus*) are spawning plentifully. It seems to prefer these unattractive batrachians to any other prey, perhaps because they are so easily secured; for at all times when not diverting itself by aerial flights the Broad-wing is one of the most sluggish and indolent of birds, rarely undertaking any vigorous exertion which can well be avoided. Of this its predilection for toad-hunting and manner of pursuing it afford evidence no less amusing than convincing. After alighting on a low branch or stub overlooking some shallow reach of calm water besprinkled with innumerable floating toads absorbed in the cares and pleasures of procreation, and rending the still air with the ceaseless din of their tremulous voices, the Hawk will often gaze down at them long and listlessly, as if undecided which particular one to select from among so many, or dreamily gloat over the wealth of opportunities for such selection. It may finally glide swiftly, yet without effort, along a slight downward incline to a toad forty or fifty yards away, or may drop more abruptly and awkwardly on one closer at hand, flapping its wings at the last moment to check the impetus of its descent. In either case it is almost certain to capture the unheeding quarry which may be borne off to a distant nest or quickly torn asunder and devoured on some near by perch. If this be much frequented for such a purpose, the ground beneath it is likely to become strewn with glutinous strings of toad spawn which the Hawk apparently never eats.

A man who owns a trout farm once brought me a broad-winged hawk that he said had been catching his trout, but its crop contained a frog, recently swallowed, and in its stomach was a partially digested field mouse. This hawk does occasionally catch small fishes; Mr. Forbush (1927) refers to one that had 17 minnows in its gullet. I can find very little evidence that it ever attacks poultry; most observers say that it never does. Ellison A. Smyth, Jr. (1912), says: "An incubating female was brought to me on May 18 by a farmer, who said that it had a nest in a large oak tree near his home, and that it was killing his chickens, and that he had shot it just after it had eaten a chicken. I skinned it and opened the stomach in his presence, and showed him, to his astonishment, that its crop contained the remains, easily distinguishable, of a young rat."

Behavior.—The broad-winged hawk is generally considered a sluggish bird, quiet, gentle, unobtrusive, and unsuspicious; it is the tamest of all the hawks; one has no difficulty in approaching it, as it sits on some low limb in the woods calmly watching the intruder with apparent indifference. If forced to fly it flaps along through the trees, much after the manner of an owl, and alights again at no great distance. But above the treetops it is far from sluggish in its soaring flight, fully equal to the best of the Buteos in sailing on its broad wings. Mr. Shelley has sent me the following interesting notes on one of its spectacular flight maneuvers:

The soaring of the broad-winged hawk, in 1926, was watched on several occasions. A family group of six birds had been noted about a densely wooded tract and a hill known as Smith's Hill, where I often observed adults earlier in the year as they crossed over its rocky summit to hunt over the lower valleys to the west. Little time was available to spend with them, but with the young fledged and on the wing, their hunting excursion as a family unit was always a spectacular sight. A still more pleasing exhibition was when, toward the period of the fall migration, they met in what I considered a spirit of play. In this performance they resembled more than anything a batch of dry leaves lifted and tossed and whirled on a zephyr of brisk autumn wind. A low call would be given, believed to be from an adult, whereupon the birds if separated would congregate at the spot where the first bird wheeled and sailed and called some 200 feet in the air. Then, with the family together, more calls could be heard, growing fainter as the birds rose in their display. Slowly at first, but gradually gaining momentum, the six birds on set pinions soared in and out among each other, round and round in a radius not greater than a quarter mile, lifting and ducking, volplaning and diving steeply toward earth at varying angles, constantly rising, nevertheless, into the clear blue sky. As height was gained and maintained, the dives and sails became swifter, in the forms of arcs and a series of dips and rises; a lower bird rising above them all, only to side-skip, arc, dive, and rise again, another repeating the maneuver, then another, and another. As leaves on the wind current, there seemed no advantageous goal to their actions, except to rise, slowly at first and then with the gain of altitude, swiftly, up, up, and finally, lost to sight. Then in from 5 to 20 minutes they reappeared as tiny dots, by the aid of binoculars, as they shot down plummetwise, banked, regained altitude, but slowly lowering, in spectacular sweeps through the air, growing clearer until the entire physique could be made out, and, finally, on set wings, a sail that would take them to the summit of Smith's Hill and the dark wilderness fastness of the Fuller Wood beyond.

An example of extreme tameness or stupidity is the incident related by Audubon (1840) when the hawk sat quietly on its nest while Bakewell covered it with his handkerchief and brought it down; afterwards it sat unafraid while Audubon measured it and drew its picture. A similar incident is described in the following notes received from J. Hooper Bowles:

The tail of the sitting bird could be plainly seen sticking over the edge of the nest, but no amount of pounding on the base of the tree would move her. Consequently my brother climbed up, and much to our surprise she still remained on the nest when he reached it. I then climbed up and joined him, but the hawk stayed perfectly still and did not show the least sign of fear or anger. In fact, she showed rather less emotion than a "broody" hen sitting. We stroked her and finally lifted her off the nest and tossed her into the air, when she flew to a tree not far away where she was soon joined by her mate. They then flew about among the trees uttering their creaking, wheezy notes, never showing a sign of the anger that is common with most of the other hawks. I have seldom seen a bird of any species that was so docile as this female hawk when we were handling her.

Mr. Shelley writes to me, as follows:

Two immature broad-winged hawks early on the morning of September 18, 1932, were noticed flying about the edge of a sugar-maple woods, some 200 yards in extent. They could be individually identified by one having much the lighter breast and belly; also their accent on the calls were of varied pitch and tone and emphasized their amateurish attempts. Their loud cries were stressed by the *thief, thief* and softer call notes of blue jays, of which six or eight congregated and gyrated through the woodland with the hawks, as in some spirit of play. From one end of the woods to the other they flew, calling in turn followed by the jays, now and then circling to some tall tree on the wood's edge to perch a few minutes and call and preen the plumage. Their flights were mostly semicircular or spiral, round about over the maples and among the upper branches, occasionally out over my position and two nearby mowings, usually flapping as they flew but sometimes on a rising current of air sailing on set wings, nearly every such flight followed by alighting for a period of some three minutes, when the activities would again be entered, random in purpose with no apparent reason for the display other than testing of their powers. At no time were they seen to molest the blue jays—forever noisy, associating in the whole performance. Each day following up to and including the September 25, early in the forenoon, the hawks went through their performance along the edge of this woods, with the jays always attending, spending half an hour or more at their activities. On the twenty-fifth they were shot at by an ignorant person and thus frightened away.

About its nest I have always found it solicitous but never aggressive. Often it will betray the presence of its nest by its peculiar plaintive notes and will continue its doleful complaint long after the nest has been robbed. But I have never known one to attempt or even threaten to attack me. Mr. Burns (1911), however, cites several cases where the intruder has been attacked. And Dr. B. H. Warren (1890) says that A. G. Boardman (probably George A., of Calais, Maine) "finds it to be courageous and spirited. A man whom he had employed to obtain a nest, was attacked with great fury, while ascending the tree; his cap was torn from his head, and he would have been seriously injured if the bird had not been shot. Another instance is mentioned by Dr. Wood, where the hawk had attacked a boy climbing to her nest, fastened her talons in his arm and could not be removed until beaten off and killed with a club."

The gentle broadwing would live in peace and harmony with its neighbors; it apparently never attacks anything but its prey. But its neighbors are not always kindly disposed toward it. I have seen it chased and attacked by crows and once by a red-tailed hawk. Others have seen it attacked by Cooper's hawk, sparrow hawk, kingbird, and martin. Probably any bird might attack it if it came too near a nest containing young, for it is well known to eat young birds. But it has often been known to nest in the same patch of woods with other Buteos, Accipiters, owls, crows, and various small birds on apparent good terms. Dr. Charles W. Townsend has sent me the following note: "I have several times seen at Ipswich, Mass., in the autumn migrations a broad-winged hawk flying toward the southwest momentarily enveloped in a dense flock of starlings. The starlings performed evolutions, first on one side then on the other of the hawk, and finally closed in on it from all sides. On one occasion, after these maneuvers, the hawk dropped to the ground and at once rose and entered the great flock of starlings flying above. Neither species appeared to attack the other. Was this play or an attempt on the part of the starlings to confuse the hawk and prevent it from doing damage to them?"

Voice.—The cry of the broad-winged hawk, when alarmed near its nest, is very peculiar. I have written it in my notes as a shrill whistle in a high key, long drawn out and plaintive, diminishing in force, like *kwee-e-e-e-e*, or *ker-wee-e-e-e-e*. It has been likened to one of the killdeer's notes or the song of the wood pewee; it somewhat resembles both of these but is more monotonous, less accented than either. To my ear it is quite distinctive and not to be mistaken for anything else. "Burroughs calls it the smoothest, most ear-piercing note he knows of in the woods" (Burns, 1911). It has been variously recorded by others as *kill-e-e-e*, *siggee*, *tig-g-e-e-e*, *che-wee-e-e*, *peeo-we-e-e-e*, *ku-e-e-e*, and various other syllables, all expressing it fairly well.

Field marks.—The broad-winged hawk may be recognized as a *Buteo* by its shape and its broad, rounded wings, which are broader in proportion to its size than in other Buteos. It is much smaller than the other common Buteos. The adult, when soaring overhead, has a distinctive color pattern, a white throat, brown-barred under parts, tail barred with three black and three white broad bands, and wings mainly white with dusky tips and a black spot near the bend. The upper parts are dark brown, and the upper surface of the tail shows gray and black bands in the same proportions as on the under side. In immature plumage it is much like the young red-shouldered hawk and cannot be recognized except by size and shape.

Fall.—From the northern portion of its range the autumnal migration starts during the latter part of August, but the main flight passes through the Northern States in September. Mr. Burns (1911) says: "The retrograde movement can be traced in the east through Rhode Island, Connecticut, southeastern New York, northern New Jersey, eastern Pennsylvania, and Maryland and Virginia, when it is lost."

Dr. C. C. Trowbridge (1895) writes:

Always during the last few days of August, and even later, before the brisk fall winds commenced to blow, a few stragglers of the Accipiters and Buteos would be seen soaring southward in Connecticut, some drifting with the wind far above in the clouds, while others were sailing low down over the fields.

But in the middle of September, when the stronger winds blew from the northwest and north, and the temperature lowered, the number of hawks which were passing greatly increased. Sometimes, however, when there was little or no wind, and the day was warm and dull, or if the prevailing winds had been southerly for several days, very few hawks were observed. But suddenly, when a fair breeze had sprung up from the northwest, the sky above the land near the sea-coast became almost clouded with hawks of various species, active and restless, circling and soaring about.

Flights in which there were many hundreds of birds I have seen many times, and I have on certain occasions counted several hundred hawks soaring together in one flock, looking like an immense swarm of gigantic insects. Often on a day after a flight, the wind having turned again to the south, many species of hawks were found in the woods and about ledges of cliffs, some perching on old trees, others lazily feeding, while a few were seen soaring about in a sluggish manner, showing the presence of an unusual number of hawks, although few of them appeared to be migrating.

Describing a day's flight, he says:

On the 16th day of September of the following year (1887), there occurred another great flight of hawks, and I was again fortunate enough to witness it. There was little wind at first, and the hawks did not appear until nine o'clock in the morning, when a few Sharp-shinned Hawks were observed. But later on in the day, the wind increased in force. Thousands of hawks of different species flew past New Haven, and Broad-winged Hawks (*Buteo latissimus*), both adults and young, appeared soaring in immense clusters. In one great flock alone there must have been three hundred hawks, the greater part of which were undoubtedly *Buteo latissimus* although with field glasses I distinguished several species in the flock. I also observed several Bald Eagles (*Haliaeetus leucocephalus*) in various plumages, circling high. The flight continued from nine o'clock in the morning until darkness set in in the evening. The day was cool and fine and the wind blew very briskly from the north. On the next day there was a flight for a short time early in the morning, but the direction of the wind changed and the flight ceased soon after.

From Sussex County, N. J., comes the following account by Mr. von Lengerke (Burns, 1911):

On Sept. 22, '07, the number exceeded any ever observed before. I was on the top of a mountain near Stag Lake, about 1200 ft. above sea level. I was armed with a Hensoldt binocular eight power glass. The day was clear, and at one time late in the forenoon, several thousand hawks, Broad-wings mostly, were in view. They came from a northeasterly direction. A constant stream, very high up, could be seen for a long while, and they were going in the direction of the Delaware Water Gap. Over the valley to the S. W. the birds seemed to collect into an immense flock, while hundreds, if not thousands, of birds were gyrating around and around; describing smaller and larger circles in the air, in height of from, I should judge, 600 to 2000 ft. above the earth. Most birds were Broad-wings. There were, however, other hawks, such as Red-tails and Red-shoulders among them, while the *Accipiter* genus was represented by some Cooper's and more Sharp-shinned, which, however, were mostly flying lower and took no part in the general evolution.

Similar flights doubtless occur through the Middle West, but the records are not so complete, and the birds are apparently more scattered. Heavy flights occur in Texas, for George F. Simmons (1925) noted on "November 10, 1918, 2000 in three connected flocks, moving southward in teeth of norther, a steady stream passing over western edge of Austin for about 20 minutes, sailing and circling in peculiar gyrating motion, flying low, ever circling and circling, but always leisurely along with the forward movement of the stream, the sky seemed full of hawks."

Dr. Roberts (1932) quotes the following from Frank Blair, superintendent of a State game farm in Minnesota, in regard to a big flight that occurred on September 14, 1924:

The flight began at 4.30 P. M., coming from the north and moving southward. They were flying high in the air. We estimated their numbers at six to seven thousand. It took about twenty-five minutes for them to pass. When over the Farm, some four or five hundred left the main flock and, descending in a rapid, almost vertical plunge, alighted in trees about the Farm, remaining while the others passed on. My assistants and myself shot 102 of these birds in about four hours, but in the meantime twelve or fifteen young Pheasants had been captured. The entire flock consisted of Broad-wings and never before had I seen anything like it.

Philip A. DuMont (1935) witnessed in Louisa County, Iowa, on September 23, 1934, a flight of some 400 individuals in a somewhat different formation, of which he says:

The behavior of these birds was of considerable interest to the writer. The flock looked like a long, slightly weaving streamer, with the birds gliding along on set wings, two or three abreast and in close formation, one behind the other.

DISTRIBUTION

Range.—Eastern North America, the Caribbean region, and northwestern South America.

Breeding range.—The broad-winged hawk breeds **north** to central Alberta (Belvedere, Edmonton, and Camrose); probably Saskatchewan (Hudson Bay Junction); Manitoba (probably Aweme, Portage la Prairie, and Winnipeg); Ontario (Kenora, probably Lake Seul, probably the mouth of the Pagwachuan River, and the Timagami Forest Reserve); and Quebec (probably Inlet, Quebec, and Mont Louis River). **East** to Quebec (Mont Louis River); New Brunswick (Scotch Lake and St. John); Maine (Holden, Ellsworth, and Portland); Massachusetts (Boston, Taunton, and Cape Cod); Rhode Island (near Newport); Long Island (Miller Place); New Jersey (Morristown, Salem, and Cape May); Maryland (Easton); Virginia (Prince Edward County); North Carolina (Raleigh); South Carolina (Greenwood); Georgia (Athens and Atlanta); Florida (Micanopy, Lake Harney, and Palm Key); probably formerly Puerto Rico and the Lesser Antilles (Antigua, Dominica, St. Lucia, the Grenadines, and Grenada). **South** to the Lesser Antilles (Grenada); Cuba (Isle of Pines); Florida (Manatee, St. Marks, and Pensacola); Alabama (Mobile); Louisiana (Hammond and St. Francisville); and eastern Texas (Houston and possibly rarely Austin). **West** to Texas (possibly rarely Austin); Oklahoma (Vinita and Copan); western Missouri (Kansas City); eastern Nebraska (Omaha); northwestern Iowa (Sioux City); North Dakota (Fargo and probably the Turtle Mountains); probably southeastern Saskatchewan (Moose Mountain); and Alberta (St. Anne, Glenevis, and Belvedere).

The range as above outlined is for the entire species, which has been separated into five geographic races, all being confined to the Caribbean region, except typical *platypterus*, which is the only form in continental North America. *Buteo p. cubanensis* is found in Cuba, the Isle of Pines, and probably occurred formerly in Puerto Rico; *B. p. rivieri* occupies the Lesser Antilles from Dominica to St. Lucia; *B. p. insulicola* is restricted to the island of Antigua; and *B. p. antillarum* occupies the Lesser Antilles on the islands of St. Vincent and Grenada and the larger Grenadines.

Winter range.—Despite many published statements to the contrary, a careful study of the available data fails to show that the broadwinged hawk is a regular winter resident anywhere in the United States. A few may occasionally winter in Florida (near St. Marks Captiva Island, Sanibel Island, Monroe and Volusia Counties, and the Florida Keys), and there are records of winter occurrence, supported by specimens, from more northern localities, but these can be considered only as casuals.

The winter range is chiefly in northwestern South America and in Central America. It appears to extend north, casually to southern Mexico (Santa Efigenia). From that region the species winters south through Guatemala (Coban, Salama, Duenas, San Geronimo, Secanquim, and Barillas); Nicaragua (Escondido River); Costa Rica (San Jose and Escaso); Panama (Gatun, C. Z., Boquete, and the Bananas River); Colombia (Santa Marta, Minea, Bonda, Valparaiso, San Antonio, Mamatoco, and other points); Venezuela (Merida,

Valle, Escorial, Culata, and Conejos); Ecuador (Gualaquiza, Pallatanga, Chimbo, Bucay, Naranjos, Baeza, Zambiza, Mindo, and Oyacachi); western Brazil (Matto Grosso, Upper Amazonia, and Rio Jabary); to northern Peru (Chamicuros).

The West Indian races are, of course, resident in their respective regions.

Spring migration.—In the words of Griscom (1923), "the Broad-winged Hawk can unquestionably claim the dubious distinction of being the most misidentified of our local birds, and it is no exaggeration to state that ninety per cent of the entries in the note books of students regarding this species up to a few years ago were either unreliable or unsatisfactory." The truth of this assertion has been kept in mind in the compilation of the following series of dates of arrival and departure. So far as possible, dates of the collection of specimens have been utilized. When these were lacking, dates of observation have been included only when obtained by observers who are believed to be thoroughly familiar with the species. Broad-winged hawks are remarkably regular in their migrations and are now believed to be the last of the hawks to arrive in spring and among the first to depart in autumn. Despite all precautions, however, it is recognized that errors of identification may be responsible for some of the following dates.—F. C. L.

Late dates of spring departure from winter quarters are: Peru—Huambos, March 9. Colombia—Bonda, March 26; Santa Marta, April 10; and Fusagasuga, April 13. Panama—Boquiti Chiriqui, April 25. Costa Rica—Carrillo, April 5; and San Jose, April 20. Guatemala—Patulul, April 3.

Early dates of spring arrival are: Georgia—Chatham County, March 11; Atlanta, March 29; and Macon, March 31. North Carolina—Raleigh, April 4; and Boone, April 14. Virginia—Ashland, March 15. District of Columbia—Washington, March 15. Maryland—Easton, March 21. Pennsylvania—Herrick, March 11; and Renovo, March 22. New Jersey—New Brunswick, March 13; Bloomfield, March 25; and Pennsville, April 11. New York—New York City, April 24. Connecticut—Portland, April 10; and Fairfield, April 17. Massachusetts—Dalton, April 2; Harvard, April 8; and Huntington, April 24. Vermont—Rutland, April 9; Woodstock, April 12; and Wells River, May 5. New Hampshire—East Jeffrey, April 22; Kensington, April 25; and Amherst, May 1. Maine—Winthrop, April 16; Auburn, April 25; and Waterville, April 27. Quebec—Montreal, April 26. Louisiana—New Orleans, March 20; and Hammond, April 4. Mississippi—Bay St. Louis, March 24; and Biloxi, April 1. Arkansas—Pike County, March 10; and Delight, March 21. Kentucky—Danville, April 14; and Russellville, April 30. Missouri—St. Louis, March 15; Mount Carmel, March 23; and

Monteer, April 5. Illinois—Rantoul, March 21; Glen Ellyn, April
10; and Chicago, April 11. Indiana—Bloomington, March 12; Bick-
nell, March 15; and Waterloo, March 25. Ohio—Columbus, March
20; Youngstown, March 27; and Oberlin, March 28. Michigan—
Ann Arbor, March 24; Vicksburg, March 27; Port Huron, April 1;
and Sault Ste. Marie, April 18. Ontario—London, April 6; and
Ottawa, April 11. Iowa—Sigourney, March 27; Tabor, March 28;
Emmetsburg, March 29; and Marshalltown, April 1. Wisconsin—
Unity, March 21; Nashotah Lake, April 24; and Madison, April 24.
Minnesota—Minneapolis, March 11; and Elk River, March 15.
Texas—Corpus Christi, March 15; Houston, March 21; and Hidalgo,
March 26. Kansas—Independence, April 1; and Blue Rapids, April
28. Nebraska—Omaha, April 8; and Neligh, April 26. South Da-
kota—Vermillion, April 18. North Dakota—Fargo, April 2; and
Talma, April 4. Manitoba—Margaret, April 4; and Aweme, April
8. Alberta—Carvel, April 18; Fort McMurray, April 18; and
Camrose, April 25.

Fall migration.—In common with other hawks of this and other
genera, the broadwing sometimes travels southward in large loose
flocks. Observations of these flights have been made from several
points, notably in the vicinity of New York City and in New Jersey.
They generally occur in September, when as many as 1,000 birds
have been estimated in one flock.

Late dates of fall departure are: Alberta—Glenevis, September
4; and Athabaska Landing, September 5. Manitoba—Shoal Lake,
October 2; and Margaret, October 4. North Dakota—Jamestown,
October 16; and Talma, October 20. South Dakota—Sioux Falls,
September 26; and Yankton, October 3. Nebraska—Greenville,
September 28; and Red Cloud, October 2. Texas—Corpus Christi,
November 10; and High Island, November 11. Minnesota—St. Paul,
November 11. Wisconsin—New Richmond, October 1; Unity, Octo-
ber 8; and New London, October 19. Iowa—Emmetsburg, October
26; and Keokuk, November 9. Ontario—Point Pelee, October 14;
Ottawa, October 16; and Toronto, October 29. Michigan—Ann
Arbor, October 10; and Sault Ste. Marie, October 11. Ohio—Aus-
tinburg, October 11; Weymouth, October 20; and Columbus, Novem-
ber 12. Indiana—Waterloo, October 7; and Richmond, November
10. Illinois—Mallard, October 8; Chicago, October 20; and La
Grange, October 25. Missouri—Concordia, September 24; and
Jasper City, October 19. Arkansas—Fayetteville, September 28;
and Delight, October 16. Quebec—Montreal, September 24. Maine—
Embden, September 4; Portland, September 17; and Machias,
September 30. New Hampshire—Jaffrey, September 10; Dublin,
September 20; and Jefferson, October 9. Vermont—Woodstock,

September 27. Massachusetts—Wellesley, September 8; and Harvard, September 28. New York—New York City, September 27. New Jersey—Bloomfield, October 31. Pennsylvania—Erie, September 12; and Renovo, October 6. District of Columbia—Washington, September 22. North Carolina—Raleigh, September 14.

Early dates of fall arrival in the winter quarters are: Nicaragua—Escondido River, September 30. Colombia—Santa Marta, October 12; and Rio Toche, October 23. Venezuela—Merida, October 29.

Casual records.—The broad-winged hawk has not been reported on many occasions outside of its normal range. The following records must, however, be considered as casuals: Nova Scotia, one was seen at Ingonish, Cape Breton Island. on August 29, 1905, and one was taken at Sherbrooke in August 1903; Ontario, a specimen was collected at Moose Factory in 1862 by J. McKenzie; South Carolina coastal plain, one was seen by Dr. A. K. Fisher near Charleston on April 26, 1886, and another was taken by Wayne in this same general region on January 15, 1889; Colorado, a specimen was killed at Manitou on or about May 15, 1926; Dominican Republic, Dr. Alexander Wetmore on May 31, 1927, at Santiago saw a recently mounted specimen said to have been killed nearby.

Egg dates.—New England and New York: 72 records, April 19 to June 28; 36 records, May 16 to 31.

New Jersey to Maryland: 57 records, April 13 to June 14; 29 records, May 11 to 25.

South Dakota to Alberta and Saskatchewan: 45 records, May 15 to June 27; 22 records, May 23 to June 5.

Ohio to Minnesota: 8 records, May 2 to 21.

Georgia and Florida to Missouri: 5 records, April 15 to May 26; 3 records, May 2 to 17.

BUTEO BRACHYURUS Vieillot

SHORT-TAILED HAWK

HABITS

This is another South American hawk, tropical or subtropical in distribution, that appears in the United States only in Florida. It always has been extremely rare and local even there, and now I believe it has almost, if not quite, disappeared from that State. I saw it many years ago in the extensive mangrove swamps of extreme southern Florida, where there may be a few still left. Arthur H. Howell (1932) has published a number of records for the species in various parts of Florida and says that it "occurs locally in small numbers from Cape Sable north to Palatka, Gainesville and St Marks." The records for central and northern Florida are all old, and I can find no recent ones for any points north of Lake Istokpoga; this lake is the only recently known breeding locality and I now believe that these hawks have been extirpated even there.

Robert Ridgway (1881) first recorded this hawk as a bird of the United States, based on a specimen taken by W. S. Crawford at Oyster (Estero) Bay, Fla., on January 28, 1881. At that time some doubt existed as to whether the little black hawk (*Buteo fuliginosus* Sclater) and the short-tailed hawk (*Buteo brachyurus* Vieillot) were color phases of the same species or were distinct species. But when W. E. D. Scott (1889) discovered a mated pair building a nest and secured both birds it was definitely proved that both of these strikingly different color phases belong to one and the same species. *B. fuliginosus* Sclater then became a synonym of *B. brachyurus* Vieillot, the earlier name. As the male of Scott's pair was black and the female white breasted, he was misled in assuming that the difference was sexual. It has since been proved that both phases occur in both sexes.

Nesting.—Scott's (1889) nest, found near Tarpon Springs on March 16, 1889, was the first nest recorded in Florida; he says of it: "The locality was on the edge of a 'hammock', and the nest, the foundation of which was finished, was in a gum tree some forty feet from the ground. Both birds were seen in the act of placing additional material on the structure."

During the next month, that same year, C. J. Pennock (1890) found a nest near St. Marks, Wakulla County, the farthest north breeding record; he writes:

April 3, I noticed a small black Hawk fly to a nest in a pine tree about three miles back from the coast. On climbing to the nest I found the tree had formerly been occupied by Herons, there being three old nests besides the one occupied by the Hawk, which also I took for an old Heron's nest. It had evidently been added to recently, and contained two or three fresh twigs of green cypress on the bottom. At this time there were no eggs. I again visited the nest April 8. The old bird was seen near, and this time she showed some concern, flying around us above the tree tops as we approached, and several times uttering a cry somewhat resembling the scream of the Red-shouldered Hawk, but finer and not so prolonged. The nest had received further additions of cypress twigs, but was still empty.

The latest and most complete account of the nesting habits of the short-tailed hawk is given by Herbert W. Brandt (1924) as follows:

Lake Istokpoga is the second largest lake in Florida, lying northwest of Lake Okeechobee in the central part of the state. It is roughly twelve to fifteen miles across and is entirely surrounded by a large cypress growth. To the south, reaching nearly to Lake Okeechobee, is a very dense impenetrable swamp, said to be one of the worst in Florida, and one through which very few white men have gone. It is in this swamp and in the big cypress bordering the lake that we found the Short-tailed Hawk.

During the latter half of March, 1923, we spent considerable time watching these birds, and on the 29th of that month, Mr. Howell found a nest in the dense swamp, three hundred yards from the lake. The male bird would sit by the hour in a big cypress tree near the mouth of Istokpoga Creek, evidently using this tree as a lookout perch. Mr. Howell patiently watched this bird on a number of occasions, and finally, about five o'clock, on the evening of the 29th the Hawk left his perch, circled upward a couple of times and dived into the swamp about one hundred yards from the lookout tree. A careful search of the swamp, in the direction taken by the bird, resulted in locating a nest in the top of a tall, slender magnolia, and on rapping the tree the female flushed from the nest [pl. 71].

The following day, March 30, I took a set of two nearly fresh eggs from this nest, which was in a swamp magnolia up fifty-eight feet from the ground. The tree was one foot in diameter at the base and very heavily overgrown with poison ivy, making the ascent rather difficult. It stood in a dense jungle of small trees and undergrowth, with water and mud knee deep. * * *

The nest was two feet in diameter and nearly a foot in height and was very large for the size of the bird. It was an entirely new nest. In a nearby tree was another nest, similar to this one, which may have been used by the pair in a previous season. The occupied nest was placed in the topmost part of the tree in a three-pronged fork among the heavy vines, and just eight feet below the highest leaf. The tree was only two inches in diameter at the nest, which was built entirely of cypress twigs, freshly broken off, with a small amount of moss and lichens remaining on them. It was lined with finer sticks of the cypress tree, dry magnolia leaves, and a few sprays of green cypress.

He also quotes the following data referring to a single fresh egg taken by Dr. W. L. Ralph at San Mateo, Fla., on April 4, 1893: "Nest in a very dense cypress swamp about half a mile from Dunns Creek and about the same distance from pine woods. It is in the top of an immense cypress tree, about twenty feet from the trunk at the end of the largest limb, 95 feet from the ground (measured). Male killed was dark phase—female light. Nest composed of small sticks and Spanish moss thinly lined with leaf-covered cypress twigs; it was flat without much of a hollow and placed in front of and connected with a former one."

Donald J. Nicholson gave him the following unpublished notes on two nests that he found:

April 12, 1910, Istokpoga Lake, Fla. Nest on extreme outermost branches of a large cypress limb overhanging Istokpoga Creek; 30 feet above water; two eggs heavily incubated. Nest of sticks and moss, lined with green oak and gum leaves and a plentiful supply of green cypress boughs. April 12, 1910, Istokpoga Lake (south of creek). Nest in a tall slender cypress among the uppermost branches of the top, about 50 feet up. Composed of cypress twigs and moss, lined with leaves and moss. Contained two young birds about a week old in white down. Only one parent seen, which was of an entire sooty black cast and had a white beak and light yellow claws. While in the tree it screamed the entire time, sometimes circling in the air or sitting in a nearby tree. Heard screaming at night in moonlight.

Eggs.—The short-tailed hawk lays one to three eggs, regularly two. In shape they are nearly oval or short-ovate; the shell is finely granulated and without gloss. The ground color is pale bluish white in the unmarked eggs, or dull, dirty white in the heavily marked eggs. Some eggs are immaculate or show only a few scattered, minute dots or scrawls of pale buff, or a few scattered brown spots. Some are irregularly spotted and some heavily blotched, about one end or the other, with dark browns, "Vandyke brown" or "warm sepia." One set is extensively washed with "tawny" and spotted with "chestnut-brown." The measurements of 27 eggs average 53.4 by 42.8 millimeters; the eggs showing the four extremes measure 57.5 by 40.9, 53 by 45.5, 48.6 by 41.8, and 49.2 by 40.3 millimeters.

Plumages.—The young found in the nest by Mr. Nicholson (Brandt, 1924) were "in white down." I have seen no nestlings. In juvenal plumage the upper parts are brownish black, mixed with "light ochraceous-buff" on the head and neck; the feathers of the scapulars, wing coverts, and upper tail coverts are edged with "tawny"; the tail is blackish brown above and silvery gray below, crossed by 9 or 10 bars of black above and dark gray below; the under parts are white, strongly tinged with "light ochraceous-buff." I have seen birds in this plumage in November and in April, taken in Venezuela. Other birds, taken there in June and October, are similar, but the upper parts are clear brownish black, without tawny edgings; these may be older birds, but they still have juvenal tails; in the adult tail there are about half as many bars and these are broken or incomplete, except the broad subterminal bar. Another immature bird, collected in Venezuela in August, is dark brown above, without edgings, and white below, streaked on the sides of the neck and breast and heavily spotted on the flanks and belly with blackish brown. Some birds in the black phase show more or less white in the under parts; these are probably immature. The two striking color phases of adults are too well known to need any description here, but one of Robert Ridgway's (1881b) early papers on this subject is instructive.

Food.—Very little has been published about the food of the short-tailed hawk. One that Harold H. Bailey (1925) had in captivity "ate readily of hamburger steak, small bits of meat, mice and rats." Mr. Howell (1932) says that a stomach "examined in the Biological Survey contained the feet and other remains of a Sharp-shinned Hawk." Probably snakes, frogs, lizards, small mammals, and small birds are also eaten.

Behavior.—Mr. Brandt (1924) has this to say about behavior and voice:

As we approached the tree the male, a bird in the black phase, flew up and circled above, uttering a few cackling notes, somewhat like the Red-shouldered Hawk. This was heard but once. When we struck the tree the female, a beautiful bird in the light phase, immediately flew off the nest. As I climbed the tree they both uttered a single high pitched squeal, not unlike the alarm note of the Broad-winged Hawk, but a little bit harsher. They continued to utter this note during all the time that I was on the tree, which was at least an hour. * * * They were very unsuspecting and gentle in appearance, being most dove-like in their attitude, and having none of the ferocious features that are characteristic of the Buteos. Scott, however, speaks of them as being extremely wary and difficult to approach. * * * The short-tailed Hawk is a very expert flier and sails by the hour, high in the air above the swamps, without a beat of its wings. It is often seen in company with the Swallow-tailed Kite or with either of the Vultures. During nest building the male accompanies the female, which carries sticks to the nest, while he hangs stationary, with motionless wings, fifty feet over the tree. This is the same trait that is shown by the White-tailed Hawk in Texas.

Field marks.—It should be easy to recognize the short-tailed hawk in either phase of plumage. In the dark phase it is a medium-sized black hawk with no light markings showing from above; as seen from below the body and under wing coverts appear black and the wings and tail very light with indistinct barring. In the light phase the pure-white under parts and dark-brown upper parts are quite distinctive.

DISTRIBUTION

Range.—The Florida Peninsula and south through Central America and South America to northern Argentina; nonmigratory.

The short-tailed hawk appears to be a rare species throughout its range. It occurs and breeds locally in small numbers on both sides of the Florida Peninsula from St. Marks and San Mateo south to probably Chatham Bay, Fort Myers, Miami, and Cape Sable. The species is not found elsewhere in North America. Its range extends **south** through eastern Mexico (Tamaulipas, Veracruz, Tehuantepec, and Cozumel Island); probably Guatemala; Nicaragua (Escondido River); Costa Rica (Irazu, San Antonio, La Palma, Cartago, and San Jose); Panama (Chiriqui, Veragua, and Gatun, C. Z.); Colombia (Santa Elena, Bonda, Quindiu Pass, and Mamatoco); Venezuela (Merida, Escorial, Valle, Monte Sierra, and Culata); French Guiana (Cayenne); Ecuador (Chimbo); Peru (Amable Maria, Tinta, and La Merced); Bolivia (Tilotilo); Brazil (Cantagallo, Taquara, Para, Matto Grosso, Piracicaba, Chapada, Bahia, and Rio de Janeiro); to northern Argentina (Alto Parana).

Egg dates.—Florida and Mexico: 14 records, February 12 to June 10; 7 records, March 15 to May 1.

URUBITINGA ANTHRACINA ANTHRACINA (Lichtenstein)

MEXICAN BLACK HAWK

HABITS

From an extensive range in South and Central America this well-named, coal-black hawk crosses our southern border to a limited extent from the Lower Rio Grande, in Texas, to southern Arizona. It is much less common here than the zone-tailed, with which it might be easily confused. And, like that bird, it is only a summer resident in the United States. In Arizona it is found in the heavily wooded canyons and arroyos, watered by mountain streams, or in the river-bottom forests, always near water. Gerald B. Thomas (1908) says that in British Honduras, where this "is by far the most abundant hawk of the region", its favorite haunt is "the long stretches of sand dunes and savannas studded here and there by clumps of palmetto and gnarled pines."

Nesting.—I have seen only one nest of this species. It was found by my companion, Frank C. Willard, on May 19, 1922, in the mesquite forest near Tucson, Ariz. This was once a magnificent forest extending for several miles along both banks of the Santa Cruz River; but the Papago Indian woodchoppers had been cutting down the larger trees all over it and making a network of cart roads all through it. There were only a few large trees left, which were more or less scattered, with many open spaces between them; a few very large hackberry trees still remained, and there were many thickets of small mesquite and thorns and some large patches of medium-sized hackberry and mesquite. White-winged doves fairly swarmed through the thickets, and their tiresome notes were the dominant sounds, mixed with the softer notes of mourning and ground doves. The forest was rich in bird life and the air was filled with their music, rich-voiced cardinals and hooded orioles, mockingbirds, desert wrens, Arizona vireos, Lucy's warblers, phainopeplas, and noisy Gila woodpeckers. Overhead, turkey vultures soared lazily and the beautiful Mexican goshawks glided gracefully.

Mr. Willard had found the black hawk's nest while I was busy skinning birds in camp. The birds were very tame when he first climbed to the nest; the female did not fly off until he was in the tree, and the male came and alighted in the next tree. But when I came, armed with gun and camera, it was very different; the female flew off before we came within range and the male was not seen at all. I waited a long time in vain for a chance to secure one of the birds, but they never came near enough for a shot. One circled within binocular range so that we could positively identify it by the distinctive tail pattern.

The nest was about 35 feet up in the topmost crotch of one of the larger mesquites, which stood out by itself in one of the large open spaces (pl. 71). It was built on the top of a large bunch of mistletoe and was made of rather large sticks and pieces of mistletoe, lined with fine green leaves, grass, and a piece of string. It was a shallow nest, measuring over 20 inches in outside and about 10 inches in inside diameter. It contained two typical eggs.

Dr. Edgar A. Mearns (1886) was the first to record the nesting of the Mexican black hawk within our borders. On June 19, 1885, he found a nest on Fossil Creek, 30 miles east of Fort Verde, Ariz. He says: "The nest was built in a cottonwood tree in the same grove in which we first found the birds. The nest had evidently been the birthplace of many generations of these Hawks, for it measured four feet in depth by two feet in width. It was lined with a layer of cottonwood leaves several inches deep, was very slightly concave, and composed of large sticks, much decayed below, showing that they had been in position for a number of years. The nest was about thirty feet from the ground. The female parent remained too shy to return to the nest until I began to climb the tree."

Major Bendire (1892) describes a Texas nest as follows: "Mr. D. B. Burrows writes me that he found a nest of this species in Starr County, Texas, on April 25, 1891, containing a single egg. The female was shot from the nest, and dissection showed that no more eggs would have been laid. The nest, a newly constructed one, was placed in a dense willow grove in the main forks of a tree of this species, about 30 feet above the ground, and growing about 80 yards from the banks of the Rio Grande. It was about 15 inches wide by 8 inches deep and rather shallow. It was composed of dry twigs and was well lined with green willow leaves."

F. H. Fowler (1903) says: "At the Natural Bridge near Fort Verde, I saw several nests of this bird in 1893, some of which were old, but several new and containing young. One or two were in cups in the rock of the bridge; the others in giant sycamores; that grew in the narrow canyon."

G. B. Thomas (1908), who has had considerable experience with this hawk in British Honduras and has examined some 27 nests, says: "The nest itself is a huge platform of sticks often measuring four feet across and two feet in depth, sometimes deeply and other times only slightly cupped, lined with pieces of green leaves and green pine needles. Their location I always found was in a pine tree, the distance from the ground varying from fifteen to fifty or sixty feet. More often, however, they were between twenty and thirty feet up, in small pines."

Eggs.—The Mexican black hawk lays from one to three eggs, oftener two or one. Of 13 sets of which I have the records, from northern Mexico and the United States, 3 consisted of three eggs, 6 of two, and 4 of one. Farther south the sets are apparently smaller, for Mr. Thomas (1908) says: "According to several good authorities the usual complement of eggs is two and three, but in only one instance out of the twenty-seven nests examined was there more than one egg, and this exceptional nest contained two. In some cases they are beautifully marked with lavender, umber and light brown, and in other cases they are totally unmarked; however the greater majority show distinct markings. * * * Like many other hawks, if the nest is robbed, they at once go to work on another nest, and I have taken three sets in one season from the same bird."

The eggs that I have seen are ovate, short-ovate, or nearly oval. The shell is finely granulated, and the ground color dull white. They are sparingly spotted with dull or light browns, "sepia" to "tawny-olive" or lighter; and some are nearly immaculate. The measurements of 60 eggs average 57.3 by 44.9 millimeters; the eggs showing the four extremes measure **66.5** by 44.5, 61.2 by **48.3**, 50 by 45, and 53.1 by **42.3** millimeters.

Plumages.—I have never seen the downy young, but Dr. Mearns (1886) describes one, on which the feathers are just emerging, as "Covered with dense woolly down, nearly white on head and breast, passing into grayish posteriorly upon the head, throat, sides of breast, tibiae, and back." On this bird the feathers appeared first on the scapulars, wings, and tail, then on the body, as in other young hawks.

The juvenal plumage of *Urubitinga* is quite distinctive, very different from any other North American hawk. The head, neck, and entire under parts are from "cream-buff" to "cinnamon-buff" in fresh plumage, heavily marked on the head and neck, nearly concealing the buff, less heavily on the belly and throat with elongated or tear-shaped spots or streaks of brownish black; the tibiae are paler buff, irregularly barred with dusky; the back, scapulars, and wing coverts are brownish black or dark sepia, narrowly tipped, deeply notched or barred with "tawny" or "russet"; the primaries are barred on the outer webs with dark gray and black, the secondaries with "fuscous" and black; the under wing coverts are pale buff, sparingly spotted, and the under side of the remiges is mainly whitish, irregularly barred and broadly tipped with dusky; the tail is broadly barred with black and white, four or five bars of each, tinged near the end and on the under side with "pinkish buff" and some grayish.

As I have seen specimens in this plumage in every month in the year, except January and May, it is evidently worn for at least the first year without much change. A body molt begins in December, at which more black plumage is acquired during winter; the brown edgings wear away and the buffs fade; the whole plumage becomes

,faded before summer. A complete molt continues through summer and fall, and by January the adult plumage is practically complete. Adults apparently have a complete annual molt, which is either much prolonged or very irregular; adults have been seen in complete molt in October, in body molt in February, and molting wings and tail in May.

Food.—Prof. F. E. Sumichrast (Lawrence, 1876) says: "The kind of food is varied; being naturally voracious, they despise no living prey, and I have taken out of their stomachs small quadrupeds, young birds, reptiles, crustacea, and insects. They are fond of fish, and on the borders of shallow brooks they easily catch the smaller kinds."

Mr. Thomas (1908) says that in the sand dunes and savannas of British Honduras "the ground is honey-combed by thousands of holes of various sizes, the abode of countless numbers of huge land crabs." He continues:

In the evening, as soon as the sun is down, they come out from their holes by thousands, hurrying here and there and always fighting, brandishing their big claws in the air like a pigmy wielding a huge scoop-shovel.

It is then that the hawks are seen busily engaged in their pursuit of food, as these crabs form almost their sole diet in this particular locality. They always catch and kill more than they can eat at the time in order that they may not be wanting on the morrow when all the crabs have gone deep in their holes to escape the heat of the day. Occasionally I noticed a hawk flying to the nest with a large lizard or snake, but more frequently they were satisfied with the crabs obtained the night before. In not one instance did I see them in pursuit of any birds, nor do their nests with young show any signs that birds are ever taken as prey.

Mr. Fowler (1903) saw one capture a green-tailed towhee, carry it to the top of a stump, and proceed to tear out its feathers. Dr. A. K. Fisher (1893) reports that of six stomachs examined three contained frogs, two contained fishes, and one a snake.

Behavior.—Mr. Thomas (1908) writes:

In flight they excel every one of the hawks, kites, or falcons except possibly the Swallow-tailed Kite. Their flight is really marvelous, excelling in some particulars even the far-famed Frigate or Man-o-war Bird. The greater part of the year they are rather dull and sluggish but when nesting time comes they are ever on the wing until the young are able to take care of themselves.

It is very interesting to see them obtain material for the nest. They circle high in the air sending out their queer whistling cry, when suddenly one of them folds its wings very close to its side and plunges towards the ground with the speed of an arrow. One almost holds his breath expecting to see the great bird strike the earth with such force that he will be transformed into a lifeless mass of bone and feather. But suddenly just before he reaches the dead tree, thru whose branches you expect to see him crashing, he throws open his wings to their full extent, his tail spreads and flattens against the downward rush and the great talons hang loosely down. Then gliding swiftly over the topmost branch, the swinging and apparently useless feet suddenly stiffen, a faint crack is heard and he slowly fans his way over to the nearby nest, firmly grasping in his talons a twig from the tree on which he seemingly so nearly escaped destruction. * * *

The old birds are very bold when the nest contains young and often perch on a branch five or six feet from the nest while one handles the young. Often, too, the male, circling high in the air with dangling legs, a marked peculiarity of this species, will suddenly make one of his awful plunges straight at the intruder, swerving just in time to avoid the shock which would undoubtedly kill the bird and knock the intruder out of the tree.

Dr. Mearns (1886) says: "Always extremely shy, they were usually found hidden in the foliage near the water in some low situation. whence, when surprised, they generally managed to escape through the foliage of the cottonwoods without affording a good opportunity for a shot. Their flight is swift and powerful."

Voice.—My field notes record the cry as a weak, hoarse, squealing note. Dr. Mearns (1886) says: "Their loudly whistled cry is different from that of any bird of prey with which I am acquainted, but is difficult to describe, although rendered with great power."

P. L. Jouy (1893) writes: "I was attracted from quite a distance by a curious harsh squawk which I at first took to be the note of a night-heron; following up the sound my surprise was great to see a large dark-colored hawk perched on the branch of a tree and uttering, at frequent intervals, a harsh and prolonged cry like *Kā-ā-ā-āh, Kā-ā-ā-āh!*"

Field marks.—The only bird with which the Mexican black hawk is likely to be confused is the zone-tailed hawk. As seen from above the former is wholly black except for a broad white band across the center of the tail and a narrower one near the base; the white-tipped upper tail coverts and the narrow tips of the tail feathers are not conspicuous. The latter appears wholly black, or nearly so, from above. As seen from below the former is wholly black, including the wings, except for two white bands across the tail, which is also narrowly white-tipped. The latter shows light-colored primaries and secondaries, and three white bands across the tail, one broad, one narrow, and the basal one inconspicuous. The Mexican black is also a heavier bird, with broader wings and shorter tail.

DISTRIBUTION

Range.—Southern Arizona, New Mexico, and Texas south to Ecuador.

The Mexican black hawk has been found **north** to southern Arizona (Agua Fria, Tonto Creek, Santa Catalina Mountains, Los Cabesos, Tombstone, and Fort Huachuca); southern New Mexico (San Luis Mountains); and southern Texas (Starr County and Brownsville). From this region it ranges **south** through Mexico (Sonora, Chihuahua, Tamaulipas, Nuevo Leon, Nayarit, Yucatan,

Quintana Roo, Puebla, Jalapa, Sinaloa, and Tehuantepec); British Honduras (Belize); Guatemala (San Guatemala, Vera Paz, Chiapam, San Geronimo, Duenas, San Jose, Finca Sepaciute, La Montamita, San Antonio, Finca El Cipres, and Ocos); Honduras (Tela, Choloma, and Omoa); Salvador (La Libertad); Nicaragua (Escondido River, San Carlos, Bluefields River, Greytown, Matagalpa, San Emilio, Lake of Nicaragua, and San Rafael del Norte); Costa Rica (Guayabo, Pigres, San Jose, Cerro de Santa Maria, Bolson, Guacimo, El Hogar, Pozo Azul de Pirris, Talamanca, Los Sabalos, San Juan del Sur, La Palma, Puntarenas, Angostura, Palmares, Miravelles, and Nicoya); Panama (mouth of the Rio Juan Diaz, Ancon, C. Z., Chiriqui, Chepo, Veragua, Chitna, Bugaba, and Rio Tuyra); the Windward Islands (St. Vincent, Grenada, St. Lucia, and probably Trinidad); Guiana (mouth of the Waini River, Citaro, Cayenne, and Surinam); Venezuela (La Pedrita, Rio Uracoa, the Orinoco region, and Carraquito); Colombia (Santa Marta, Bonda, Cinto, Rio Atrato, and Playa Concha); and Ecuador (Puna Island and Guayaquil).

The birds of the Windward Islands, probably including those of the north coast of Colombia and Venezuela, have been separated subspecifically as *U. a. cancrivorus*, while those inhabiting the region from eastern Panama to southern Ecuador are recognized by some authors under the name *subtilis*, sometimes as a full species and sometimes as a subspecies of *anthracina*.

The species has been detected in February in southern Arizona, indicating that at least some individuals are not migratory. On the other hand, it has been observed to arrive at Tucson, Ariz., on March 13; at Agua Fria, on March 26; and at Apache Tejo, N. Mex., on April 12.

Egg dates.—Texas to Canal Zone: 22 records, February 8 to May 30; 11 records, March 31 to May 5.

ASTURINA PLAGIATA PLAGIATA Schlegel

MEXICAN GOSHAWK

HABITS

One of the greatest delights of my days spent in the mesquite forest near Tucson, Ariz., was the frequent glimpses we had of this beautiful little hawk sailing gracefully over the treetops. Its mantle of pearly gray and its breast finely barred with gray and white were well contrasted with a tail boldly banded with black and white. The exquisite combination of soft grays, black, and white made it, to my mind, one of the prettiest hawks I had ever seen.

The mesquite forest, where these hawks were quite common, was on the banks of the Santa Cruz River and is more fully described under the preceding species. Major Bendire (1892) also found them common in "the timber in the Rillitto Creek bottom near Tucson" and says that Otho C. Poling found them "in a deep wooded cañon" in the Huachuca Mountains, where he was camped "among some thick spruce and sycamore woods." He says further: "It seems to be found only in the vicinity of water courses, and not, like many of the other Raptores, on the dry and comparatively barren desert-like plains."

Courtship.—Major Bendire (1892) observed a flight maneuver in April, which was probably a courtship activity, he writes:

From that time on not a day passed without my seeing two or three pairs of these handsome little Goshawks (which were readily recognized by their light color) engaged in sailing gracefully over the tree tops, now sportively chasing each other, or again circling around, the female closely followed by the male, uttering at the same time a very peculiar piping note, which reminded me of that given by the Long-billed Curlew in the early spring (while hovering in the air in the manner of a Sparrow Hawk), rather than the shrill cries or screams usually uttered by birds of prey. To my ear, there was something decidedly flute-like about these notes. After they were paired they became more silent.

Nesting.—In the mesquite forest, referred to above, we found, on May 19 and 20, 1922, three nests of the Mexican goshawk. The first nest was 40 feet from the ground in a large mesquite; we were attracted to the spot by the cries of one of the birds, and when we rapped the tree the incubating bird flew off; both birds circled about in the vicinity but did not come very near; it was a small nest made of sticks and branches of mistletoe, lined with a few green leaves of mesquite; it held three eggs. We found another nest later in the day that was fully 60 feet up in a giant hackberry tree, the largest of a group of big trees. The female was standing on the nest when we first saw it, and the male soon flew up and alighted in the next tree; I shot both birds. The nest was made of sticks and was lined with green leaves, apparently plucked within reach from the nest; it measured about 20 inches in outside and 10 inches in inside diameter and contained three fresh eggs. The third nest was found the next day, 30 feet up near the end of a branch in a large mesquite; it was made of sticks and lined with green twigs and leaves of elder; it also held three eggs. Near it was an Arizona cardinal's nest with young and a white-winged dove's nest with eggs (pls. 72 and 73).

Major Bendire (1892) writes:

About the last week in April several pairs had selected their nesting sites within a radius of 10 miles from my camp, and commenced building. All the nests found by me, four in number, were placed in cottonwood trees, usually the largest to be found in the vicinity, and as near their tops as they could be placed with security.

The first nest was obtained on May 17, and the male, who was sitting on a limb close by, was shot. This nest was located in the topmost branches of a large cottonwood tree near the laguna, the sink of the Santa Cruz River, not less than 70 feet from the ground, and contained three fresh eggs, the only set I found which contained this number.   The nest, not a very substantial affair, consisted of a shallow platform, composed principally of small cotton-wood twigs, a number of which were green and had been broken by the birds themselves.   I have seen them do this, selecting a suitable twig, then flying at it very swiftly, grasping it with their talons, and usually succeeding in breaking it off at the first trial.   *   *   *

The nests are rather frail structures, and were all apparently newly built. They were shallow and but slightly hollowed, not more than 1½ inches deep. The last two found were very difficult to get at, resting as they did on very slender limbs, and from the fact that they were composed principally of green twigs it was no easy matter to detect them.

Eggs.—The Mexican goshawk lays two or three eggs.   All the three sets we collected were of three eggs, but Bendire (1892) says that only "about one set in four contains three eggs"; and most of the sets in collections seem to consist of two.   Most of the eggs are oval in shape, but some are ovate or elongate-ovate.   The shell is quite smooth but not glossy.   The color is white or very pale bluish white; and they are almost always unmarked, though usually more or less nest stained.   Bendire (1892) mentions two eggs "marked with a few buffy brown spots" about the larger ends, and another in which the spots were "scarcely perceptible to the naked eye."   The measurements of 50 eggs average 50.8 by 41 millimeters; the eggs showing the four extremes measure **56** by 41.2, 51 by **43.5, 47.3** by 39.4, and 49 by **38.2** millimeters.

Young.—Otho C. Poling watched a female of this species bringing food to her young, and sent the following notes to Major Bendire (1892):

She made half a dozen or more trips daily to the nest, and whenever she arrived her presence was at once hailed by the hungry nestlings.   I watched her closely; she would make daily trips to the mesquite plains for cotton tails (*Lepus arizonae*), some 6 or 8 miles out in the valley.   After the first week a neighbor came to my camp and during my absence shot the female, and presented it to me on my return.

Up to this time I had not seen the male, or at least had seen only one individual at a time, but noticed on the following day that another bird, evidently the male, appeared and carried on the feeding of the family as regularly as if nothing had happened.   The young were now growing rapidly, and their cries were much louder while being fed.   One day, on glancing up at the nest, I saw one of them perched upon a limb beside it.   The parent bird was near by with some game, and seemed to be urging the young one to fly to it, if it would have its meal.   Although it demanded its regular allowance loudly, I observed it was left out of reach by the old bird until its first lesson of flying was learned.   The young were three in number, and all were out of the nest the following day, but returned to it at night.   They remained about for several days and finally disappeared.

Plumages.—I have not seen the downy young or any nestlings of the Mexican goshawk.   In the fully grown juvenal the entire upper parts are "fuscous", the feathers of the head and neck showing considerable concealed white; the feathers of the back, scapulars, and wing coverts are tipped, edged, or deeply notched with "pinkish cinnamon"; the tail is "fuscous" above, broadly barred with black; the under parts are white or buffy white, heavily marked with club-shaped spots, or streaks, of "fuscous"; the tibiae are buffy white, profusely barred with dusky. I have seen birds in this plumage in nearly every month from July to April, showing that it is worn throughout the first year without material change except by wear and fading. A complete molt between April and September apparently produces the adult plumage. I have seen a bird just beginning this molt on the back and breast on April 15 and another just starting to molt the scapulars and wing coverts as late as September 18. A bird taken June 19 was in full molt in body, wings, and tail. I have seen only two molting adults, one molting its tail in June and one completing its wing molt in February.

Food.—The stomach of one of the birds I shot contained the remains of a small snake and that of the other a lizard. Others have noted among its food small rabbits, squirrels, mice, quails, young doves, fishes, and beetles. Major Bendire (1892) says: "When in search of food their flight is powerful, active, and easily controlled. I have seen one of them dart to the ground with arrow-like swiftness to pick up some bird, lizard, or rodent, continuing its flight without any stop whatever. A good proportion of their food consists of beetles, large grasshoppers (a species of which about 3 inches long was especially abundant), and other insects; these are mostly caught on the wing, and I believe small birds also form no inconsiderable portion of their food, as I have seen them chasing such."

Col. A. J. Grayson, as quoted by George N. Lawrence (1874), says in his notes:

Although its flight is vigorous and quick, and the feet and claws sharp and strong, yet its prey consists of the more humble and weaker creatures of the woods and field, such as lizards, small snakes, field mice, etc. These are indeed his staff of life. But in the general breeding season of birds, when the young are not fully fledged, and are just beginning to try their tender wings from the parent nest, it is then it finds opportunities to change its usual diet of lizards and snakes, to the more tender young birds of any species that may come in his way. He often visits the poultry of the natives and carries off young chickens; will follow the Chachalaca (or wild tree chicken) and her young brood through the densest woods, in his skulking way, until an opportunity offers to pounce upon one of them, thus keeping the mother constantly on the alert; she will defend her brood with great courage when he makes the attack, but he is often too quick for her vigilance, and carries off one at a time from day to day, until her precious chicks are mostly all devoured by the sly manoeuvering of this hawk.

Behavior.—To me its flight seemed swift, active, and graceful, more like that of an *Accipiter* than a *Buteo*. It must be very swift on the wing for it catches the small lizards, which were very common in the mesquite forest and are the swiftest moving reptiles I have ever seen. Mr. Stephens regarded its flight as "Falcon-like and very swift." Colonel Grayson, however, says: "It is rather sluggish in its habits, sitting for hours upon the limb of some decayed tree in apparent listlessness, permitting you to approach very near without moving from his perch, then flying but a short distance before again alighting."

Voice.—I recorded its cry as a loud plaintive *cree-ee-ee*, suggestive of the cry of the broad-winged hawk, but louder and not so prolonged. Frank Stephens (Bendire, 1892) "compares their cry to a loud 'creer', repeated four or five times, and says that at a distance it sounds much like the scream of a peacock."

Field marks.—The Mexican goshawk is easy to recognize by its conspicuous field marks. No other hawk looks much like it. The gray breast, the black and white banded tail, and the white under side of the black-tipped wings are all distinctive when seen from below. The view from above shows a darker-gray back, a white rump, and a black tail with white bands.

DISTRIBUTION

Range.—Southwestern United States, Central America, and northern South America.

The Mexican goshawk has been recorded breeding **north** to Arizona (Gila Bend, the Santa Catalina Mountains, and probably Tombstone); southwestern New Mexico (Fort Bayard); and southwestern Texas (probably Lometa). From this region its range extends **south** through Mexico (the States and territories of Tamaulipas, San Luis Potosi, Nayarit, Vera Cruz, Oaxaca, and Quintana Roo); British Honduras (Cayo District); Guatemala (Naranjo, San Lucas, Finca Carolina, Finca El Espina, Hacienda California, Finca El Cipres, Pueblo Escuintla, Chimalapa, Savana Grande, Retalhulen, and San Geronimo); Honduras (Tela and Roatan); Salvador (La Libertad); Nicaragua (Sucuya, Lake Nicaragua, San Juan del Sur, and Chinandega); Costa Rica (La Barranca, La Palma, Nicoya, San Mateo, Acajutla, Santo Domingo, Bolson, and Bebedero); Panama (Panama Railroad); Colombia (Remolino, Magdalena River, Villavicencio, Barrigon, Bonda, Don Diego, Fundacion, Mamatoco, and Dibulla); Guiana; Ecuador (Babahoyo, Gualaquiza, and Zamora); Bolivia (Lower Beni and Rio Surutu); and Brazil (Sao Paulo and Matto Grosso). Also recorded from the Lesser Antilles (Trinidad).

The range above outlined is for the entire species (*nitida* Latham of some authors), which has been separated into several rather poorly defined races, all confined to tropical regions. *Asturina p.*

plagiata has the most extensive range occurring from southern Arizona and southern Texas south to Guatemala.

While generally nonmigratory, *A. p. plagiata* generally withdraws entirely from the United States during the winter season. It is then found north regularly to Sinaloa (Mazatlan and Escuinapa) ; Puebla (Chietla) ; and Yucatan (Chable, Merida, and Espita).

Early dates of spring arrival in Arizona are: Huachuca Mountains, March 31; and Tucson, April 4.

Casual records.—The species has been recorded from Illinois (Fox Prairie, August 17, 1871) and from Iowa (Van Buren County, May 25, 1895), but neither of these records is considered satisfactory. A winter specimen was collected at Brownsville, Tex., on December 5, 1885, and identified at the Biological Survey.

Egg dates.—Arizona and Mexico: 48 records, March 16 to July 2; 24 records, April 19 to May 31.

BUTEO LAGOPUS S. JOHANNIS (Gmelin)

AMERICAN ROUGH-LEGGED HAWK

HABITS

Contributed by Charles Wendell Townsend

To anyone who has been to the summer home of the rough-legged hawk in the North, or has seen it in its winter migrations, the mention of its name brings up visions of a splendid bird, one of the largest and finest of our hawks. Past master in the use of air currents, whether it is poised motionless in a breeze over a cliff, or scaling close to the ground and quartering it like a harrier, or swinging proudly in great circles up and up into the blue sky, this great hawk is always a thing of joy and beauty. Limited in its diet almost exclusively to rodent pests, and therefore of the greater value to the agriculturalist, this hawk is still pursued by man with his keen and cruel hunter instincts and his unreasoning prejudice against all hawks. Where a Japanese cabinetmaker would take his block and rapidly sketch the graceful poises of a hawk, the western barbarian takes his gun and kills and hardly glances at his beautiful and blood-stained victim, as he leaves it where it has fallen.

As a consequence, this magnificent and most beneficent of hawks has been growing scarcer in the past 50 years or more, not only in its breeding range but in its winter flights into the United States.

Spring.—As the smaller rodents constitute the chief food supply of the rough-legged hawk, the northerly migration of this bird in the spring follows the retreating snow, for not until the snow melts are the runways of meadow mice revealed. In eastern Massachusetts it goes north in the latter part of April and early in May; May 14 is my latest date. On several occasions in April and May I have seen two of these birds, probably a pair, circling together to a considerable height and then striking out in direct flight for the northeast.

As is the case with most hawks, migration of considerable numbers of this bird may sometimes be observed in the spring. E. A. Doolittle (1919) records a flight of 100 broad-winged hawks with about 20 roughlegs in Lake County, Ohio, on April 27, 1919. He says: "All the roughlegs were sailing with the wind and flying in a straight northeasterly direction, while the broad-wings kept in bunches and circled about to some extent while still progressing steadily in the same direction."

[Frank L. Farley writes to me that on his arrival at Churchill on June 7, 1936, large numbers of these hawks, perhaps a hundred or more a day, were passing in migration; this migration had evidently been going on for some time, for he found six nests with eggs during the following week. He says: "I should say that at least 1,000 of these hawks must have quietly passed over our camp during the first 10 days we were there; they all seemed to be quartering the territory, always watching for lemmings; they would not make more than a mile or two in a half hour; when one would come to the edge of the Churchill River, it would at once drop down to an elevation of not more than 50 feet above the water and fly directly to the other side; at other times, while hunting, they would be several hundred feet high. The entire migration seemed to move up the coast in a narrow strip, not more than a quarter to half a mile wide; they generally traveled separately, but most of them would usually be within sight of those ahead of them."—A. C. B.]

Courtship.—The courtship of this hawk appears to be performed in the air, as is the case with the marsh hawk. I have on several occasions seen two roughlegs in the spring soar upward close together, emitting their characteristic notes, a combination of musical whistles and hisses, and I have thought that this was part, perhaps the principal part, of their courtship. Henniger and Jones (1909) speak of these birds "circling high and playing with the wind in mating season."

Nesting.—The rough-legged hawk in its nesting habits is governed by its surroundings. On the bleak and treeless shores of Labrador this bird nests on the higher shelves of the cliffs and preferably on the tops of the cliffs. In forested regions, however, the nest is built in trees near their tops. But even in these forested regions a steep river bank of shelving rock often tempts the bird to place the nests here. Henniger and Jones (1909) state that the birds sometimes nest in "hollow trees, in crevices of rocks, in holes of river banks and in buildings."

Lucien M. Turner in his unpublished notes made early in the eighties in Labrador and Ungava says:

The rough-legged hawk is one of the most abundant of the birds of prey in Labrador and Ungava. It arrives about the last week in May at Fort Chimo and remains until the first week in October. Immediately on its arrival a locality is selected for a nest, as mating has evidently occurred before it appears in the vicinity of Fort Chimo; often the same place is resorted to where the same pair have reared their young for many seasons.

All the nests discovered by me were invariably placed on a ledge or projection of a high bluff. Strangely enough, should there be several ledges, apparently suitable in all respects, on the same bluff, the one nearest the top is selected. I suspected this to be done in order to allow the birds to have a greater view of the surrounding country for purposes of searching for food or to look for danger.

The nest is composed of sticks of various sizes together with a few grass or weed stalks placed irregularly crosswise. The particular location of the nest modifies the amount of nest material. A flat rock usually has but sufficient of these materials to prevent the eggs from rolling about. Where the place slopes, the nest is usually higher in front, often with nothing at the rear except the side of the cliff. In locations where the nest has been used for several years the amount of material accumulated is astonishingly large. Some nests are increased considerably each year, and other nests appear to have been only rearranged. The depression containing the eggs is quite shallow, and, in some instances, nearly flat.

The accumulations around the nest, such as refuse of food, is also surprising in quantity and, decomposing, forms a soil in which grow most luxuriantly grasses and other plants, thus marking the spot that might otherwise have been overlooked.

Alfred M. Bailey (1926) describes a nest on a cliff in Alaska that consisted of "a jumble of sticks cemented together by excrement." Roderick MacFarlane (1908) relates that 70 nests of this species were found in the Anderson River region. "About fifty-five of them were built in the crotches of the tallest trees, not far from the top, and at a height of from twenty to thirty feet from the ground. They were composed of small sticks and twigs, and comfortably lined with hay, moss, down and feathers. The remaining fifteen were placed near the edge of steep cliffs of shelving rock, or on the face of deep ravines and other declivitous river banks, and in make they were somewhat similar to the foregoing."

W. G. Sheldon (1912) relates of the closely allied European form, *B. lagopus lagopus*, that the nests he found in Lapland contained in the grass lining "fresh green shoots of pine and *Vaccinium*." Abel Chapman (1885) found a nest in Lapland on June 5 which was "a mass of dead sticks about two feet thick, with a layer of solid ice about six inches thick immediately under the new grass lining on which the three eggs were lying."

[All the nests found by Mr. Farley, at Churchill, "were on rocks except two that were placed on the tops of broken-off stub spruces;

these trees were in deep snowdrifts at some distance from the rocky shore. Some of the rock nests were built on big blocks of solid rock that lie plentifully all along the coast. The nests were frail affairs, although they had apparently been used for many years. I do not think that any of them were larger than the average crow's nest."]

Eggs.—[AUTHOR'S NOTE: The rough-legged hawk lays large sets, often as many as four or five and sometimes six, but oftener three or four and rarely only two. Major Bendire (1892) gives the following very good description of them:

Some are ovate, many short ovate, and others rounded ovate. The ground color in the more recently collected specimens is a pale greenish white, which appears to fade out in time, leaving the egg a dull dingy white. The shell is close grained and strong. There is an endless variety in the markings, both in regard to size and amount, in different specimens. In some they are fairly regular in shape as well as size, in others exactly the reverse. In some they are well defined, evenly colored throughout; in others quite clouded and of different tints. A few specimens are streaked and the markings run longitudinally from end to end. The spots and blotches consist of various shades of brown, the predominating tints being burnt umber and claret brown, and among these are mixed lighter shades of ochraceous, clay, fawn color, and écru-drab. Quite a number of specimens show also handsome shell markings of a rich heliotrope purple and pale lavender, mixed in and partly overlaid with darker tints. In many eggs the blotches are large and irregular in outline, and usually heaviest on the large end, but in no case do they hide the ground color. Others are regularly and sparingly marked over the entire egg, with fine dots of different shades of brown and lavender, giving the egg a flea-bitten appearance. While some eggs are but slightly marked, none are entirely unspotted.

The measurements of 50 eggs average 56.6 by 44.9 millimeters; the eggs showing the four extremes measure **62** by 46.5, 59 by 48.5, and **42.5 by 38** millimeters.]

Young.—Incubation is performed by both sexes, and its duration is generally stated as 4 weeks or 28 days (Burns, 1915). Lucien M. Turner in his unpublished notes stated that in the region of Fort Chimo, Ungava, "the young are hatched by July 20, a week earlier or later according to circumstances, and are able to fly by the first week in September. The young appear to be able to take care of themselves as soon as they leave the nest."

Plumages.—[AUTHOR'S NOTE: The downy young, when small, is well covered with long, thick, white down, tinged with "pale olive-buff" on the head and with "vinaceous-buff" on the back. A larger downy young is largely "smoke gray" on the upper parts and whiter below. The down also covers the front and sides of the tarsus.

The only large nestling I have seen is a young bird acquiring the juvenal plumage of the dark phase; it was taken in Alaska in July and is nearly fully grown and nearly fully feathered, but there is a large patch of white down on the upper breast and some on the belly and head. The new plumage varies in color from "warm sepia" to "Natal brown", edged or tinged on the head, breast, scapulars, and wing coverts with "Mikado brown." Immature birds in this phase are much like the adults, dark sooty brown above and below, except that the feathers of the head, breast, tibiae, and bend of the wing are edged or tipped with "tawny"; some white shows through on the under parts, and the tail is distinctly barred with gray on the outer webs and with white on the inner webs.

Immature plumages in the light phase are somewhat confusing, as there is much individual variation, but there seems to be a second-year, or subadult, plumage. The juvenal, or first-year, plumage is variegated or patterned above with "clove brown", "hair brown", and "smoke gray", with basal white showing through it; the under parts are "cream-buff" or buffy white at first, fading to white, streaked on the throat with "clove brown", heavily spotted or patterned on the breast with "Natal brown" or "snuff brown", heavily barred on the belly and tibiae with "bone brown" or "warm sepia" and with some "tawny" on the tibiae; the tail is basally white and tipped with grayish white; it has a broad, subterminal black band and three to five narrow black bars, becoming broken inwardly, on a white or grayish ground.

What looks like a second-year plumage, and is often regarded as the adult plumage, is similar to that of the first year, but is lighter on the head and neck, more buffy white and less dusky; the upper parts are dark sepia with "tawny" edges; the under parts, including the legs, are pale buff or buffy white, heavily streaked, spotted, or barred with "bister"; the large abdominal patches of "bister", so prominent in the adult, are only partially developed; the barring on the white half of the tail is reduced to median spots, and the dusky half of the tail shows indistinct grayish bars.

In the fully adult plumage the head and neck are white or creamy white, streaked with dusky bars, less heavily than in previous plumages, the white predominating; the breast is less heavily streaked or spotted with "bister" on a creamy-white ground; the abdominal patches of "bister" are larger and of a purer color; the tibiae are creamy, or buffy, white and nearly immaculate; the tarsi are the same color and quite immaculate; the inner two-thirds of the tail is white and the outer third is dusky, usually with no conspicuous barring in either zone.

Adults evidently have a complete annual molt between April and November. As this is mainly accomplished while the birds are on their breeding grounds, molting birds are scarce in collections.]

Food.—As before remarked, the rough-legged hawk is highly beneficial to man in its feeding habits, as it preys on harmful rodents

and insects. Seldom or never does it take birds. Dr. A. K. Fisher
(1893) gives a table showing the results of examinations of 49 stomachs of this bird. Forty contained mice—nearly all meadow mice—
and two contained rabbits, one a gopher and one a weasel. One
contained a lizard and 70 insects, and four were empty. Junius
Henderson (1927) quotes various observers and their stomach examinations and finds no record of bird remains. Field mice, so destructive to young orchards, were by far the most abundant. In one case
the stomach was "filled with grasshoppers", and the latter pests are
eagerly devoured by this hawk. In the North, lemmings constitute
the chief of its diet.

E. S. Cameron (1907) says that in Custer and Dawson Counties,
Mont., prairie dogs are the favorite food of this hawk, which, however, "is becoming very scarce from traps and poison put out for
wolves." Aretas A. Saunders (1911) found that in Gallatin County,
Mont., the roughleg feeds largely on pocket gophers. Huey (1924)
found in the stomach of a bird taken in California an adult female
pocket gopher and six grasshoppers. M. P. Skinner writes from
the Yellowstone National Park that "the food consists mostly of
mice and carrion" and adds ground squirrels to the list of rodents.
I have found the fur and bones of brown rats in the pellets of this
bird at Ipswich, Mass.

W. A. Smith, writing from Lyndonville, N. Y., communicates
the following interesting note about the specimens of this hawk received by his son, a taxidermist: "In each case a careful examination
of the stomachs revealed nothing but field mice, so it would seem
that they are a very beneficial bird to the farmer. However, one
which we received alive and only slightly wounded in one wing
has been kept alive for several months and will eat sparrows and
starlings greedily, as well as dead chickens, hens, or any animal.
Mice and small birds are devoured nearly whole and the bones,
feathers, and fur disgorged in the form of pellets." This last observation contradicts the erroneous statement that hawks and owls
pluck their bird victims so thoroughly before eating that the absence
of feathers in the pellets does not exclude birds from their dietary.

Although rodents constitute the chief and generally the only food
of this hawk, yet the prejudice against hawks is so great, and the
belief is so general that the larger the hawk the more damage it
does to poultry, game, and other birds, that it is difficult to persuade
the average gunner, farmer, or gamekeeper that the rough-legged
hawk is a friend of the agriculturalist and sportsman and not his
enemy. In April 1914, when I was staying at the heath-hen reservation at Marthas Vineyard, a rough-legged hawk was shot by the
English gamekeeper in charge, who stated his belief that the bird
had been feeding on the heath-hen chicks and probably on the sit-

ting adults. I opened the stomach in his presence and found it stuffed to its utmost capacity with the fur and bones of numerous field mice, and entirely destitute of bird bones or feathers, yet I very much doubt if this object lesson would prevent the gamekeeper from killing the next rough-legged hawk, if he could do so. Prejudices like this are difficult to eradicate.

It is generally believed that the roughleg captures and eats ducks crippled by gunners, and this was reported for Utah Lake by Henry W. Henshaw (1875), although he found the remains of mice only in the stomachs of 11 roughlegs captured in this region. Dr. Fisher (1893) quotes the above and adds: "The examination of such a considerable number of specimens from a locality in which multitudes of ducks occur, and the finding of nothing but the remains of mice is quite conclusive evidence that the former is not their favorite food. Recently Mr. Henshaw informed the writer that the above statement relative to this hawk feeding on water fowl was based on reports of gunners, which he now believes to be incorrect." [McAtee (1935) reports one pied-billed grebe, one ruddy duck, and two smaller birds identified in 99 stomachs examined, "but it seems probable that the first two mentioned were crippled or dead when found by the hawk."]

Kenneth Racey (1922), writing of this hawk in Washington State, says: "One was seen to rise from the ground and on going to the spot a dead Mallard was found with the breast eaten away. The Mallard had evidently been killed the day before by some hunter, as the feathers were covered with frost, but the breast had been freshly eaten." A roughleg shot on the prairie was "very fat, and its stomach contained the breast of a Mallard duck."

Forbush (1927) records that he has "seen a statement that remains of the western meadowlark have been found in its stomach." Turner says in his notes of the Ungava region: "I have never seen any feathers about the nests indicating that birds had been used as food for the young, except at a nest just back of the station at Davis Inlet a young *Dendragapus canadensis* of about two days of age was found lying near the side of a nest containing three young." H. J. Pearson (1898) found a headless young snow bunting beside a nest of young of the European form.

While records of rodent food of this hawk abound in literature, actual reports of the use of birds as food are exceedingly rare. I have collected all I could find. Dead birds as well as dead mammals and fish and even carrion are eaten by this hawk. Dr. Fisher (1893) quotes Maynard to the effect that they "feed upon fish and the dead animals cast up by the sea", and he quotes Vernon Bailey's account of their eating the skinned carcasses of muskrats.

Behavior.—As a rule the rough-legged hawk is an unsuspicious bird and can be more easily approached than most hawks. Particularly is this the case when the observer is mounted or in a carriage or automobile. Turner says in his Ungava notes: "At no time did I observe anything of a fierceness exhibited by these birds either when wounded or when their nest was approached. The male can seldom be secured near the nest, while the female is sometimes heedless of distance although rarely approaching very near." Their mild disposition is shown even by adults that have been tamed soon after being taken into captivity.

In nesting they maintain certain territorial rights; thus H. B. Bigelow (1902) found them very common in the cliffs of northern Labrador and "different pairs of hawks seemed to hold different tracts of country from which they drove all intruders." Charles A. Gianini (1917) found in Alaska that "evidently there is mutual respect between them and the bald eagle, for I have seen their nests on cliffs in close proximity to each other." They fly about their nest in great concern, scream loudly when a man approaches, so it is generally very easy to find the nest.

The flight of the rough-legged hawk, although generally slow and leisurely, is graceful and indicative of skill and power. In soaring, the wings and tail are spread to their full extent; the first half-dozen primaries are spread out separately like fingers and curve upward at their tops. On motionless wings, if the air currents are favorable, this bird may often be seen soaring high over the land rising higher and higher until it becomes a mere speck in the sky.

In searching for mice they often fly slowly, alternately flapping and sailing, close to the ground or even 50 yards up in the air. They often quarter the ground like marsh hawks or harriers, frequenting open fields and pastures and marshy places. I once saw one swooping down over the frozen surface of a pond and closely skimming it. Either the bird mistook the ice for water and was looking for swimming rodents or surface fish or else it was merely indulging in play. It is not uncommon to see this hawk skimming close to the surface of water, and one I watched in February at Ipswich flew from the region of the dunes over the sea, swooping down for a moment close to the waves at the bar. The vicinity to water seems always to attract this species.

They frequently hang in one place by rapid vibration of the wings, turn the head from side to side in looking down, and often drop their long-feathered tarsi preparatory to pouncing on the prey, only to draw them up behind when they change their mind. At other times, when luck is propitious, they partially close their wings and drop like a plummet. On one such occasion at Ipswich, Mass., in winter I saw the hawk fly off with a large mouse to the salt marsh, where it devoured its prey on an ice cake. Another time a fine roughleg pounced successfully on a cotton-tail rabbit and bore it off.

The favorable up-currents of air on the brow of a steep hill or cliff enable them to hang suspended in the air as motionless as a kite. Gravitation takes the place of the kite-string, and by skillful disposition of the plane of the wings to the up-current the bird remains motionless if the wind is steady. When the wind is irregular and flawy, the bird swings about more or less just as a kite acts under similar circumstances. When the wind drops for a moment the bird moves with rapid wing beats. This use of the up-currents over hill or cliff is a familiar habit of the roughleg, and I have frequently watched this habit both in Massachusetts and in Labrador. Another method for securing food is also resorted to, in which the bird sits on a rock or stub and watches for its prey.

Perched, the rough-legged hawk sits very erect, preferring dead trees or poles to living trees. At Ipswich I have frequently seen them perched on windmills. One bird that I watched seemed to have a special liking for this kind of perch, for on one occasion it visited and perched on three in succession. E. A. Kitchin (1918) records the following case of a bird collected from a telegraph pole by J. Hooper Bowles at Tacoma, Wash., on October 20, 1917: "The bird sat lengthwise of the cross-bar, on the sunny side of the pole, with wings half drooping. This odd attitude was observed by Mr. Bowles for a minute or more before collecting; when the hawk was brought to hand, he found the wings and tail soaking wet, which probably accounted for the strange position on the bar. A freshly eaten field mouse, found in its stomach, may have been caught swimming across one of the many channels of the flats, and the hawk had probably been obliged to take a partial dip to secure its prey."

Writing of this hawk, Major Bendire (1892) says that in autumn in the Harney Valley, Oreg., he had "often seen a dozen or two in a few hours' ride, usually standing singly on a little hillock in the open prairie, or perched upon a sage bush watching for prey"—small rodents and grasshoppers and occasionally rabbits. Dr. Fisher (1893) says: "The rough-leg is one of the most nocturnal of our hawks, and may be seen in the fading twilight watching from some low perch, or beating with measured, noiseless flight, over its hunting ground." Several times in January 1931 I saw a roughleg about sunset enter a grove of pines, apparently to roost there for the night. In so doing it disturbed a hundred or more crows that had been using this grove as a roosting place, and they flew about violently cawing and finally left the hawk in undisturbed possession. Crows frequently pursue roughlegs, sometimes darting at them from above and cawing loudly, but as a rule the hawk does not appear to notice them.

Once I saw a herring gull soar in close proximity for a few seconds to a soaring roughleg, which paid no attention to it.

On a January day at Ipswich I watched a roughleg perched in the top of a hickory tree near my house, a bird that seemed determined to clean its feathers thoroughly and rid them of all insect pests. From 8 : 30 to 10 : 30 a. m. it preened itself without intermission.

Francis H. Allen contributes the following notes:

One alighted in a willow on Great Neck, Ipswich, Mass., on November 11, 1929, in the middle of the tree, not on the top or in a commanding situation, and assumed an almost horizontal position. This seemed to be for resting, as the bird did not appear to be watching for prey.

On February 23, 1931, at Ipswich, I put up a bird in the dark phase at a distance of about 20 yards on a bushy hillside. It rose in the air and hung suspended against the northwest wind, adjusting itself by turning its partly closed tail, rotating it on a longitudinal axis, gaining considerable altitude in this way, and gradually moving off toward the southwest into the sun, where I finally lost sight of it.

Voice.—There is considerable variation in the notes of this hawk, which are often very loud. Those heard in spring where two birds circle around together, which I think are in the nature of a courtship song, I have described as a whistling, at first soft and musical and somewhat plaintive, the last part a hissing suggestive of the whistle of the red-tailed hawk but lacking the sound of escaping steam. Lucien M. Turner in his Labrador notes says: "The Eskimo apply the name *kin wi yuk* (in imitation of its note) to this species. The residents of Labrador term it the 'squalling hawk' from the noise it makes when the bird is alarmed."

Harrison F. Lewis (1927) relates of southern Labrador that— "during an afternoon spent on July 5 in the very rough country north of Bradore Bay, where suitable cliffs 50 to 100 feet high abound, I saw at least six pairs of these hawks and was almost never without one or two following me about and uttering loud protests." They "screeched loudly at me while I remained in their chosen locality." At Devils Mountain, near the Natashquan River about 70 miles above its mouth, I once watched a fine dark roughleg sail around the cliffs, where it was greeted by a salute of whistles from its young or its mate in some unseen aerie. Henniger and Jones (1909) described the voice thus: "Kle Kle Kle Kle Kle—Ree hee."

Sheldon (1912), writing of the European form, speaks of its cries as "mewing", and states that the mew of the female was in a "distinctly higher key than that of the male."

Field marks.—The large size of this hawk, with the broad but long wings and comparatively short tail—*Buteo* characteristics—at once attracts attention. Seen from above, as it wheels in flight, the white rump and base of the tail are sometimes very prominent, while seen from below the black patches near the wrist joints in the white of the lower surface of the wings, the black tips to the primaries and secondaries, and a broad black bar across the upper belly are all good field marks. The heavily feathered tarsi may often be made out as the bird drops its feet preparatory to a swoop. The white rump as well as the habit of quartering the ground might suggest a marsh hawk, but the smaller size of the marsh hawk, its slenderer form, narrower wings, and longer tail in proportion make the distinction an easy one.

To recognize the roughleg in the field one must take into consideration the great variations in its plumage. In the extreme dark phase it may appear as black as a crow both above and below, and it then lacks the white rump and other field marks. There is, however, generally some white to be seen in the wing feathers from below and some barring in the tail and, except in the extreme dark phase, a little white on the rump. In the light phase I have seen a bird with the white extending in the tail to within a short distance of the tip so as to give the effect of a white tail with a black terminal band. The dark band across the upper belly in the light phase, such a good field mark, is sometimes entirely lacking. In my notes I find records of 25 different roughlegs seen in the field in eastern Massachusetts where I recorded the plumage phases. Eight of these were in the dark phase, 12 in the light phase, and 5 intermediate (C. W. Townsend, 1920).

Fall.—The migration from the breeding grounds in the north depends on the snow fall. The earlier it comes, the earlier the birds migrate, and as the snow advances the hawks keep ahead of it, so that they can obtain their rodent food. I have seen this hawk at Ipswich as early as October 12, but the last days of the month are its usual time of arrival. Widmann (1907) gives November 1 for their arrival in Missouri. Singly or by twos or threes, the birds sometimes migrate in great numbers together. Fleming (1907) recorded an immense flight of rough-legged hawks in October 1895 at Toronto; "from the 26th to the 29th the birds were taken in dozens; I must have had over fifty brought to me in that time." Mr. Fleming, in a recent letter, states that besides these well over 50 more were mounted by taxidermists. "How many were killed at Toronto it is impossible to say, but it was in the days of flight shooting of hawks. They drifted westerly along the ridge that rises behind the city, and it was the custom for gunners to wait for them there. The 1895 flight was the greatest we have any knowledge of. and it was followed by a lesser one in 1896."

Enemies.—What these hawks may suffer from external parasites is well shown in the following report by T. T. and E. B. McCabe

(1928b). They shot a roughleg and found that the right tarsus and foot had been lost, possibly in a pole trap, some time previous.

The curious and pathetic point was that the head and neck, that is, all such parts as could not be reached by the bill, were literally swarming with lice, sometimes to the extent of dozens to the square centimeter. These had devoured all the softer, concealed parts of the head and neck feathers, so that while the rest of the body, which was quite free from vermin, was so densely coated with white under-plumage that it was very difficult to reveal even the principal inter-tract spaces, the bare skin of the infested areas was merely shingled over by the tips of the contour feathers. * * * The hawk had been able to strike its prey with one foot, but was being literally tormented to death, and deprived of its protection against the bitter cold, by the tragic circumstance of being unable to scratch its head!

Lucien M. Turner wrote that where the nest of this hawk is easily accessible, it "was often a matter of wonder to me how they escaped the ravages of foxes and other prowlers." Eternal watchfulness on the part of the hawks must be necessary, and few foxes, I imagine, would stand a determined onslaught by these birds.

The chief enemy of the rough-legged hawk is man. If the farmer and the gunner could be brought to look on this hawk in the same way that the ornithologist views him as a beneficent agent in the balance of nature, rodent pests would be much diminished and the nature lover would more frequently enjoy the spectacle of this splendid bird quartering the fields and soaring aloft. Owing to the generally unsuspicious character of the roughleg, its leisurely flight, and its large size, the average gunner is often enabled to shoot it, without the exercise of much skill, and at times of migration in numbers the slaughter is sometimes appalling, as has been stated above in the Toronto flight shooting.

William Brewster (1925) describes this ignoble and disastrous "sport" of shooting rough-legged hawks. He says the Connecticut River Valley "used to be one of its principal routes of migration through Massachusetts", and continues:

At Northampton, in the latter state, lived two gunners fond of shooting Hawks and very expert at it, who sometimes killed as many as twenty Roughlegs in the course of a single day. They began to hunt them systematically in 1879, and continued to bag them numerously up to 1887 or 1888, but were forced to discontinue the unworthy if exciting sport about 1890, because then and thereafter there were very few if any of the birds to be found in the neighborhood of Northampton, almost all having been apparently slain or driven to seek other haunts. The gunners commonly hunted them in an open buggy or "stone boat," drawn by a well-trained horse over smooth, grassy, interval lands bordering on the River, and shot at them mostly on wing as they flew from the tops of tall, isolated trees, chiefly elms, in which they were accustomed to perch. When approached in this manner they seldom left the tree, until the horse was stopped within gunshot of it. If he kept on past it they were unlikely to fly at all. Yet it was impossible for a man to get near them on foot in such open ground. All this was demonstrated to me on March 17, 1881, when I was driven out from Northampton in a buggy, to be shown how the thing was done. Upwards of twenty rough-legged hawks were seen that day, but because of the nervousness of our gun-shy horse only four were killed.

Other ways in which man cruelly destroys this hawk are by means of steel traps set on the tops of poles where the bird is in the habit of alighting and by means of poison bait put out for wolves and ground squirrels if not for the bird itself.

DISTRIBUTION

Range.—North America, south to California, Texas, Louisiana, and North Carolina. The Siberian race (*pallidus*) occurs in Alaska.

Breeding range.—The rough-legged hawk breeds **north** to Alaska (Noatak River and Fort Yukon); Yukon Territory (Herschel Island); northern Mackenzie (Anderson River, Franklin Bay, Horton River, Kogaryuak River, and Port Epsworth); probably northern Keewatin (Cape Fullerton); and northern Labrador (Cape Chidley). **East** to Labrador (Cape Chidley, Port Burwell, Okak, Kaipokok Bay, Tessiujak, Ailik, Davis Inlet, and Hamilton Inlet); and eastern Quebec (probably Chateau Bay, Bradore Bay, and Harrington). **South** to Quebec (Harrington); Ungava (Seal Lake); probably Manitoba (Norway House); and southern British Columbia (Quesnel). **West** to British Columbia (Quesnel); and Alaska (Kodiak Island, Popof Island, Amak Island, Herendeen Bay, Bethel, Igiak Bay, St. Michael, Golovin Bay, Kigluaik Mountains, and Noatak River).

Winter range.—In winter the roughleg is found **north** to British Columbia (Sumas and the Okanagan Valley); casually central Alberta (Glenevis); casually southern Saskatchewan (Eastend); North Dakota (Charlson and Argusville); Minnesota (Elk River, Fort Snelling, and Lanesboro); Wisconsin (Madison and Milton); Michigan (Kalamazoo, Ann Arbor, and Detroit); southern Ontario (Toronto and casually Ottawa); southern Quebec (Montreal and Quebec); New Hampshire (Jefferson); and Maine (Norway). **East** to Maine (Norway); Massachusetts (Boston and Edgartown); Long Island (Montauk); New Jersey (Morristown and Princeton); southeastern Pennsylvania (Philadelphia); casually Maryland (Baltimore and Sandy Spring); and casually the District of Columbia (Washington). **South** to casually the District of Columbia (Washington); Pennsylvania (Warren and Erie); Ohio (Medina, Columbus, and Greenville); Indiana (Richmond, Indianapolis, and Bicknell); Illinois (Rantoul and Canton); Missouri (St. Louis, Mount Carmel, and Concordia); Oklahoma (Norman and Tyrone); casually New Mexico (Zuni, Tularosa, and Rio Mimbres); rarely southern Arizona (Tucson and Fort Whipple); and southern California (rarely Santee). **West** to California (rarely Santee, Los Banos, San Jose, Hayward,

Newhope, and Dransfields) ; Oregon (Klamath Lake, Fort Klamath, and casually Seaside); Washington (Walla Walla and casually Tacoma); and southwestern British Columbia (Sumas).

Migration.—In common with many other hawks, the roughlegs sometimes migrate in large loosely organized flocks. Magee (1922) reports that "the line of greatest hawk migration between the eastern portion of the Upper Peninsula of Michigan and Canada is at Whitefish Point." According to Fleming an immense flight occurred in the latter part of October 1895 in the vicinity of Toronto, Ontario, and the birds continued to pass in decreasing numbers until December 5.

Spring migration.—Early dates of arrival in the North are: Manitoba—Aweme, March 11; Portage la Prairie, March 15; Treesbank, March 16; and Oak Point, April 7. Saskatchewan—Indian Head, April 1; and Dinsmore, April 3. Mackenzie—Fort Resolution, April 26; and Fort Simpson, April 28. Alberta—Carvel, March 24; Belvedere, March 25; Glenevis, March 26; Camrose, April 1; Mile 157, between Lac la Biche and Fort McMurray, April 16; and Sedgewick, April 20. Alaska—Bethel, April 15; and Beaver Mountains, April 24.

Late dates of spring departure from the winter quarters are: District of Columbia—Washington, March 25. Maryland—Sandy Spring, March 17; and Baltimore, April 15. Pennsylvania—Warren, April 15; Darby Creek, April 20; Darling, April 22; and Doylestown, April 27. New Jersey—Elizabeth, March 26; Morristown, March 28; New Brunswick, April 10; and Englewood, April 12. New York—Montauk, April 8; Locustgrove, April 15; Howard, April 18; and Ithaca, April 23. Connecticut—Hartford, April 9; and New Haven, April 20. Massachusetts—Northampton, April 5; Greenfield, April 13; and Harvard, April 24. Vermont—Clarendon, April 26. Maine—Lewiston, April 9; Dover-Foxcroft, April 10; and Auburn, April 12. Missouri—Columbia, March 14. Illinois—Ohio, April 11; Milford, April 15; Port Byron, April 20; and Glen Ellyn, April 23. Indiana—La Porte County, April 4; Bicknell, April 8; and Anderson, April 10. Ohio—Painesville, April 26; Ellsworth Station, April 28; and Oberlin, April 29. Michigan—Detroit, April 26; Blaney, April 29; Greenville, May 4; and Munuscong State Park, May 29. Ontario—Southmag, May 9; London, May 14; and Ottawa, May 21. Iowa—Keokuk, April 6; Brooklyn, April 10; Emmetsburg, April 18; and Ashton, May 1. Wisconsin—Unity, April 8; Madison, April 16; and Menomonie, April 19. Minnesota—White Earth, April 4; Minneapolis, April 20; Parkers Prairie, April 24; and Cass Lake, May 4. Kansas—Manhattan, March 30; Lawrence, April 2; and Ottawa, April 14. South

Dakota—Beresford, March 17; Huron, April 6; and Dell Rapids, May 2. North Dakota—Argusville, April 13; Grafton, April 23; and Jamestown, April 23. Manitoba—Shoal Lake, April 24; and Aweme, April 30. Colorado—Boulder, March 31; and Denver, April 21. Utah—Salt Lake County, April 17. Montana—Fort Custer, April 11; and Gallatin Valley, April 17. Alberta—Glenevis, May 27. California—Fresno, March 17; and Stockton, April 1. Oregon—Willows, April 2; Klamath Lake, April 12; and Multnomah County, April 18; Washington—Yakima County, April 3; and Pullman, April 3.

Fall migration.—Late dates of fall departure in the North are: Alaska—Fairbanks, November 11. Alberta—Fort McMurray, October 17; Boiler Rapid, October 22; Belvedere, October 23; and Glenevis, November 16. Mackenzie—Fort Smith, October 4; and five miles north of Fort Simpson, October 16. Saskatchewan—Eastend, October 27. Manitoba—Winnipeg, October 21; Treesbank, November 3; and Aweme, November 5.

Early dates of arrival in the fall are: Washington—Tacoma, October 20; Pullman, October 23; and Shoalwater Bay, October 31. Oregon—Seaside, October 13; Cold Spring Bird Reserve, October 23; and Netarts Bay, October 25. Alberta—Glenevis, August 9. Montana—Gallatin Valley, October 9. Wyoming—Yellowstone Park, September 17. Colorado—Boulder, October 15. New Mexico—Colfax County, September 30; and Pueblo Bonito, October 24. Manitoba—Alexander, August 23; and Aweme, September 10. North Dakota—Argusville, September 29; Jamestown, September 30; and Charlson, October 23. South Dakota—Sioux Falls, October 30; Forestburg, November 1; Lake Poinsett, November 2. Kansas—Topeka, September 30; Osawatomie, October 9; and Lawrence, October 20. Oklahoma—Norman, October 18. Minnesota—Lake County, August 29; Minneapolis, September 10; and Parkers Prairie, September 15. Wisconsin—Meridean, October 4; Madison, October 16; and Ashland, October 19. Iowa—Emmetsburg, September 9; Tabor, September 21; and Hudson, September 26. Ontario—Toronto, August 25; Point Pelee, August 25; and Toronto, September 18. Michigan—Ann Arbor, September 26; Sault Ste. Marie, October 14; and Blaney, October 17. Ohio—Columbus, October 21; Plainsville, October 30; and Huron, November 11. Indiana—Bicknell, November 6. Illinois—Port Byron, September 11; Catlin, September 21; and Glen Ellyn, October 12. Missouri—Columbia, November 27. Maine—Bangor, October 28. New Hampshire—Jefferson, September 12. Vermont—Wells River, October 16. Massachusetts—Harvard, September 30; and Northampton, November 1. Connecticut—East Hartford, October 11; and Preston, November 5. New York—Canandaigua, September 29; Rhinebeck, October 12; Howard, October

12; and New York City, October 31. New Jersey—Demarest, December 6; and Leonia, December 17. Pennsylvania—Jeffersonville, September 20; Wernersville, September 28; and Lima, October 20.

Casual records.—According to Reid (1884) a specimen of this species was taken in Bermuda and preserved in the Bartram collection. Other casual records are winter occurrences south of the normal range. Some of these are merely sight records and therefore are not entirely satisfactory. One was reported by Wayne as seen at Capers Island, S. C., on January 18, 1927; one was reported as seen at Mandeville, La., in February or March 1897, and another (thought to be this species) at West Baton Rouge, La., on April 6 and 7, 1903. Other writers on Louisiana birds refer to it as "not uncommon winter resident", "occasional winter visitor", and "rather rare winter visitor", but without citing any records. There are several published reports of sight records for Texas, among them being: Fredericksburg, January 15 to 29, 1894; Somerset, November 24, 1924; and Electra, March 9, 1921; but the only specimen record appears to be one in the Sennett collection, taken on the Aransas River on January 5, 1887. Rives (1890) says it is a rare winter visitor at Frenchcreek, W. Va., but gives no additional details.

Egg dates.—Alaska and Arctic Canada: 32 records, May 18 to July 13; 16 records, May 30 to June 20.

Labrador: 19 records, May 2 to June 23; 10 records, June 4 to 10.

BUTEO REGALIS (Gray)

FERRUGINOUS ROUGHLEG

HABITS

This latest name, *regalis*, is a very appropriate one for this splendid hawk, the largest, most powerful, and grandest of our Buteos, a truly regal bird. One who knows it in life cannot help being impressed with its close relationship to the golden eagle, which is not much more than a glorified *Buteo*. Both species have feathered tarsi, both build huge nests on cliffs or trees, and both lay eggs that are very similar except in size; the food habits, flight, behavior, and voice of the two are much alike.

The ferruginous roughleg is a bird of the western plains, the wide open spaces. It is equally at home on the grassy prairies, where it nests in the timber belts along the streams, or in the barren, treeless plains or badlands, where it is content to build its nest on some convenient cliff, butte, or cutbank. Its chief requirement seems to be a good supply of small rodents on which it feeds. It was well named the "California squirrel hawk", as it was known to prey largely on the ubiquitous ground squirrels that have become such a pest in several western States. The control of these pests by poison has resulted

in great mortality among many other forms of wild life and has
provoked bitter discussion. How much better it would have been
to encourage these and other useful hawks to do their good work!
But unfortunately the ignorant prejudice against all hawks has reduced this useful species to the verge of extinction, and allowed the
ground squirrels to increase.

Nesting.—My acquaintance with this magnificent hawk began in
North Dakota in 1901. On our first day there, May 30, we found
two nests in the heavy timber around Stump Lake. The first nest
was about 40 feet from the ground in the top of a tall swamp oak;
the hawk left the nest as we approached, uttering her harsh notes of
protest and sailing in majestic circles, as she mounted higher and
higher until a mere speck in the sky. It was a large nest, made of
heavy sticks, cow dung, and other rubbish and lined with grass and
strips of inner bark; it held five young hawks fully a week old.
The other nest was similar in construction but was only 20 feet up
in a leaning swamp oak on the edge of a little valley; it contained
three nearly fresh eggs. Both nests were in commanding situations
where the birds could have a good outlook. Two more nests were
found on June 4. One of these, containing two fresh eggs, was 30
feet up in a swamp oak, towering conspicuously above a strip of
timber along the lake shore. The bird in full melanistic plumage
was seen to leave the nest at short range. The nest was made of
large sticks and lined with dead flags, strips of the same, and a few
sprigs of green leaves; it measured 24 inches in diameter and 12
inches in height; the inner cavity was 9 inches wide and 4 inches
deep. The other nest, found that day, was 45 to 50 feet up in the
top of an elm (pl. 75).

During my two seasons in southwestern Saskatchewan we found
seven nests in 1905 and only three in 1906, illustrating the prejudice
of farmers and ranchmen against even this most useful hawk. These
nests were all in trees, willows, cottonwoods, and poplars, but at much
lower elevations; three of them were only about 10 feet from the
ground and the highest was only 30 feet up. A typical large nest
measured 36 inches in diameter and 24 inches high. Between May
30 and June 28 the nests contained young, most of which were
hatched before June 1. Nearly half of the adults seen were in the
melanistic phase, and in two cases we found a light bird mated with
a dark one.

In the regions where I have found this hawk breeding all the nests
I have seen have been in trees, and I believe it prefers to nest in trees
where these are available. It selects the largest trees it can find with
no special preference for any one species. I believe, however, that
its territory is selected on account of the food supply rather than by

the existence of suitable nesting sites. As it finds what it wants on the open grassy plains, arid badlands, sagebrush plains, and even deserts, it often builds its nest on hillsides, cutbanks, buttes, cliffs, or rocky pinnacles. I have never seen such a nest, but more than half of the nests for which I have the data before me were in such situations.

E. S. Rolfe (1896) found, in the Devils Lake region of North Dakota, "a somewhat carelessly constructed nest of the usual materials on the straw-covered roof of an abandoned stable" and another "on top of an old straw stack in the midst of a stubblefield", both of which were not far from more characteristic nesting sites. He says, however:

But the distinctive nest of this species, in this region at least, is placed on the ground on the summit, or well up the sides, of a hill that is crowned with stone and boulders, or along the verge of some stiff gorge through which a coulee finds its way. If on the summit of the hill, it is invariably enclosed and held in place by boulders, and if up the side of the hill, a jutting boulder forming a natural shelf is selected to stay the bulky, loosely-constructed nest in place. The nest material is uniformly sticks varying in size from that of a twig to one an inch or more in diameter and of all lengths suitable, well intertwined together, often, with one or more bleached buffalo bones. The lining is of turf, bunches of dried grass with roots adhering, well dried "cow chips" and the like, and the whole forms a structure suggesting that of the eagle as usually depicted in old-time illustrations, and, aside from its exposure to attack by small animals, somewhat superior to the average tree nest.

E. S. Cameron (1914) describes and illustrates some picturesque nests of this type in Montana. Some of these nests are evidently occupied for many years in succession. P. A. Taverner (1919) writes:

One built upon a salient buttress of a cliff had increased with annual additions until it formed a mass of material twelve or fifteen feet high. The lower masses of the nest were rotten and merged into the original clay foundation whilst it grew fresher towards the top until the final layer was of this year's construction—mostly sage-brush roots. In a little hollow adjacent to such a nest we found an accumulation of over a bushel of dried bones, and scraps of gophers that had been devoured by successive generations of young Rough-legs.

P. M. Silloway (1903) found nests of these hawks in Montana as high as 55 feet in pine trees; and Stanley G. Jewett (1926) records five nests found by him in Oregon at heights varying 6 to 9 feet in junipers. J. H. Bowles has sent me the following notes from eastern Washington:

There are two distinct types of nesting sites, one on the ground and the other in the little stunted juniper trees that grow scattered about on the sandhills. The ground nests are sometimes built on the ledge of a cliff, usually very easy of access, but oftener on an outcropping of rock on the side of a steep canyon where the collector can walk directly to them. Ground nests are seldom large, sometimes being simply the remains of a very old nest with only a few chunks of dry horse or cow dung added. The tree nests are very different, some of them being immense accumulations as large as very large eagle nests and taking up almost the entire tree. The magpie (*Pica pica hudsonia*) frequently builds in the same nest with the hawk, that of the magpie being usually underneath. In one example that we found the hawk had three eggs and the magpie seven, on April 28, 1928. The nests are built of sticks as large as the bird can carry, bleached cattle bones, and rubbish of all kinds. The lining consists mostly of sage bark, with practically always large dried chunks of horse or cow dung. Where dry dung can not be obtained, large dead roots are used. This very peculiar habit seems as typical with this hawk as cast snakeskin is with the crested flycatcher (*Myiarchus crinitus*).

The actions of the sitting bird vary greatly with individuals. Usually they will flush at some distance if they believe they are seen, but this is not always the case. One very striking example was about 9 feet up in a small locust tree on a deserted ranch and beside the road. We drove up to it, and the sitting bird raised herself up and looked at us, then settled back on her nest again. One of us then started to climb the tree, when the bird stood up again flapping her wings, working her claws, and opening her beak in a most formidable manner. She evidently had no intention of leaving, so we tried forcing her off with long sticks, being careful that we did not injure her. This was difficult because she kept falling back onto the nest, but we finally managed to force her up into the air so that a high wind that was blowing swept her away. We had only time hurriedly to take the eggs when she was back again. Oddly enough there were only two eggs, very small and considerably incubated.

Eggs.—The ferruginous roughleg lays ordinarily three or four eggs, sometimes only two, more rarely five, and as many as six have been recorded. They are ovate or elliptical-ovate in shape, and the shell is smooth or finely granulated. The ground color is white, creamy white, or pale bluish white. They are usually very handsome eggs, boldly marked with large blotches or spots of rich browns, from "burnt umber" or "chocolate" to "amber brown" or "tawny"; some are marked with lighter browns, "clay color" or "cinnamon-buff", or with shades of "vinaceous fawn color" under the browns; there are often underlying spots of "ecru-drab" or "Quaker drab." Some eggs are more evenly and some very sparingly spotted with the same colors; and some are nearly or quite immaculate. The measurements of 53 eggs average 61.2 by 48 millimeters; the eggs showing the four extremes measure **67.5** by 49, 62.7 by **51.3,** and 56 by **45.6** millimeters.

Young.—Incubation is said to last for about 28 days and to be shared by both sexes. Mr. Cameron (1914) says that the young remain in the nest until they are about two months old and fully fledged. Regarding the first flight from the nest of two young birds, he writes:

They were noticed to be very much on the alert, and Mr. Felton, desiring to obtain a photograph, crawled cautiously from above to within five yards of the nest. As he raised his Kodak both hawks took alarm, and boldly launched themselves from the eyrie in the direction of the creek below. After holding a straight course for about a quarter mile the fledglings seemed to lose heart; they circled right and left, and, again meeting, returned together to the cliff.

So far their graceful flight had been marked with almost adult ease; but both betrayed inexperience when trying to alight and capsized awkwardly upon the ledge.

The young are watched and cared for by their parents long after they have left the nest, guarded and fed, or taught to hunt for themselves.

Plumages.—The downy young ferruginous roughleg is covered with short, white, woolly down, but long and silky on the crown and tinged with gray on the crown, wings, and rump. The succeeding down, preceding the acquisition of plumage, is long, thick, and pure white. The flight feathers are the first to appear, when the young bird is less than half grown, followed closely by the plumage of the back, scapulars, and wing coverts, and then by that of the breast.

In fresh juvenal plumage the upper parts are from "clove brown" to "bister", edged on the head and back with "ochraceous-buff", on the scapulars with "tawny", and on the wing coverts with "russet"; the under parts are white, heavily suffused with "warm buff" on the upper breast and tibiae and suffused elsewhere with paler buff; there are narrow black shaft streaks on the sides of the breast, large blackish or dark sepia spots or patterns, edged with "tawny", on the flanks, and small scattered spots on the tibiae and tarsi; the tail is basally white, but largely "wood brown" and "mouse gray", the inner webs mainly white, and with about four indistinct dusky bars. This plumage is worn throughout the first year with no change except by wear and fading, all the buff and most of the rufous tints disappearing by wear or fading out to white.

The second-year plumage is acquired by a complete molt, probably prolonged during spring, summer, and fall. In this the upper parts are much like those of the adult, with very broad edgings of "cinnamon-rufous" or "Sanford's brown", broadest on the scapulars and darkest on the wing coverts; the under parts are much like those of the juvenal, except that the brown tibiae are acquired, varying from "tawny" to "russet", either heavily and thickly barred with black or more faintly with dusky; the belly and flanks are also more or less tinged, especially on the flanks, with "tawny" or "hazel" and more or less irregularly barred with black or dusky; the darker-colored birds may be the young of one of the dark phases; the tail has no barring and is largely whitish, mottled or clouded with gray, or extensively washed with "tawny" or "orange-cinnamon" on the outer webs.

Mr. Cameron (1914) says that the young bird requires four or five years to attain its fully adult plumage, but I should say that at the end of the second year the young bird molts into a plumage that is practically adult, although from then on the under parts continue to become more extensively white, nearly immaculate in the oldest birds, except for the brown tibiae and barred flanks; the tail becomes progressively whiter and finally pure white, except for faint gray or tawny clouding on the outer webs; and the upper parts become paler, with more white in very old birds.

The above descriptions apply to birds in the light phase. Dark-phase birds are not especially rare and are often found mated with light phase birds. Nearly half of the birds we saw in Saskatchewan were in melanistic plumage. Two young birds were taken from a nest and reared in captivity, one of which developed into a melanistic bird and one into the light phase. A brood of four young, taken from a nest in North Dakota in 1902 by Dr. Louis B. Bishop, developed into four dark juvenals.

In the extreme melanistic phase the entire plumage, except the tail, varies from "bister" to "bone brown", with faint "tawny" spots scattered over the belly, flanks, and upper and under tail coverts; the tail is mainly "neutral gray", mottled with white on the inner webs. A modification of this, which might be called an erythristic phase, is similar, except that the under parts are largely "hazel" or "burnt sienna", with more or less restricted dark centers on the feathers; the wing coverts and feathers of the upper back are edged with the same colors; and the upper tail coverts are mainly reddish brown.

Adults apparently have a complete annual molt from August to November.

Food.—The ferruginous roughleg is a highly beneficial hawk and should be encouraged as a great destroyer of injurious rodents. Mr. Cameron (1914) says that in eastern Montana this hawk "feeds chiefly upon prairie dogs and meadow mice." It eats snakes, but he thinks it never takes frogs. He has seen these hawks hunting in pairs and attacking jack rabbits; one which they killed, but could not carry away, weighed about 8 pounds. He writes:

On May 28, 1893, my wife and I witnessed the capture of a prairie dog by two of these hawks, and one of them was proceeding to devour it as we rode up. The methods of the crafty coyote and the Ferruginous Rough-leg are identical in "dog-towns." Both wait patiently, the hawk also on the ground, for a prairie dog to amble afield from its burrow, and thereupon make a dash, the first terrestrial, the latter aerial, to intercept it. A prairie dog always endeavors to gain its own burrow when danger threatens, and is marvellously quick to reach it, but if cut off from home, the beast becomes so bewildered that it neglects the nearer intermediate holes. When two coyotes, or two hawks hunt together, the fate of the intended victim is sealed, but with one assailant only, it has an even chance.

His opinion that this hawk never attacks poultry is confirmed by W. P. Sullivan, who for 16 years has protected several pairs that breed on his ranch; "they are constantly flying around the buildings, yet no chickens have ever been molested." He quotes Mr. Sullivan as follows:

I have watched the hawks often through glasses in our alfalfa field after the first crop has been taken off. The pocket gophers get pretty busy tunnelling, and pushing all the loose, damp earth up in piles on the surface. The hawks fly slowly over the field until they discover a fresh pile of damp earth. Here they will alight softly, and wait for the gopher to push close to the surface. They will then spread their wings, and, rising a few feet in the air, come down stiff-legged into the loose earth when the gopher is transfixed and brought out. I have seen them eat the gopher where caught and at other times carry it away.

During the nesting season some few birds are killed as food for the young; "until the nestlings were about two weeks old their food consisted partly of meadow-larks." In frequent visits to one nest there were seen in all nine prairie dogs, one cottontail rabbit, two bull snakes, and some remains of sharp-tailed grouse and meadow-larks. The contents of another nest were similar, including also the remains of a young magpie; the grouse remains were also probably from young birds. He gives an interesting account of an attempt by one of these hawks to carry off a cat; the hawk had risen to a height of about 25 feet with the surprised cat in its talons, when the cat opened hostilities with its claws and was promptly dropped, not much the worse for its experience.

Illustrating the value of this hawk to the wheat farmer, P. A. Taverner (1926) writes:

A conservative estimate of the requirements of a family of these large Hawks is surprising in its total. Two adults, from spring arrival to the birth of young, three months, consume not less than a gopher a day, 90 in all. After the young are out, four in the brood, and for two months at least, the family requirement can not average less than three gophers a day, or 180. Thereafter for one month, the six practically adult, though four are still growing, probably will require one gopher each day, or 180 more. A single gopher, under favourable circumstances, destroys at least one bushel of wheat. Supposing that one-tenth of this can be charged against the average gopher, we still have thirty-five bushels of wheat as the value of this one family of large Hawks for a single season. This can be translated into dollars and cents by multiplying the current price of wheat, and makes a sum that is well worth considering.

Nearly all observers agree that the food of the ferruginous roughleg consists almost exclusively of mammals, ranging in size from jack rabbits to meadow mice; as many as six or seven of the latter have been found in a single stomach. Snakes, lizards, and skunks are taken occasionally, as well as grasshoppers, crickets, and beetles.

Behavior.—The ferruginous roughleg appears sluggish at times, as it sits quietly on some low tree or fence post or even on the ground watching for its prey. At such times it is not particularly wary and can be approached within gun range if it thinks it is not observed. A rider on horseback or in an automobile has a better chance to approach the bird than a man on foot. A bird standing on the ground on a smooth, level prairie can sometimes be secured by driving an automobile at it at high speed; the surprised hawk must rise slowly, and the man with a gun is almost under it before the hawk can get away. When launching into the air off a perch or when rising from the ground, this big heavy hawk flaps heavily and awkwardly; but when well under way it soars gracefully and easily, mounting in great circles far up into the sky. I once saw nine of these great hawks in the air at one time, soaring majestically at varying heights over a burnt prairie near Quill Lake, Saskatchewan; these birds were probably hunting in true *Buteo* style. It is often more active in its hunting motions, beating over the open country a few feet above the ground, much after the manner of the American roughleg, but flying more swiftly and ready to pounce on any unlucky mammal it may surprise. Again it may stand patiently above the burrow of a prairie dog or ground squirrel, waiting to seize the unsuspecting animal if it shows its head.

In its behavior toward other bird species it seems to be a peaceful and harmless neighbor, except in the few cases where it needs food for its tender young. Magpies often build their nests in the same tree with it, and other birds seem to have very little fear of it. But it can defend its young with spirit, if necessary, as evidenced by Mr. Cameron's (1914) account of one that attacked a great horned owl and drove it away from the vicinity of its nest.

Voice.—The alarm notes heard when the nest is disturbed are recorded in my notes as a loud *kree-a*, or *ke-a-ah*, or again as a harsh *kaah, kaah*, like one of the notes of the herring gull.

Field marks.—The adult in the light phase is easily recognized from below by its practically white tail, unmarked by bars, and by its nearly all white under parts, broken only by the dark V formed by the brown tibiae converging to a point where the feet come together under the tail. As seen from above the head is whiter than most hawks, the back and shoulders more rufous, the tail whitish, and there is a light area in the widely extended primaries. This last character appears in all plumages. In immature plumages the under parts and tail are whiter than in other hawks. In the dark phase the plain light-colored tail and the light space in the primaries are distinctive.

Winter.—The ferruginous roughleg withdraws largely from the northern portions of its range in autumn to spend the winter in the milder portions of the United States. A young bird banded at Rosebud, Alberta, on July 7, 1930, was shot at Trinidad, Colo., on September 26, 1930; and another, banded at the same place the next day, was killed at Seagraves, Tex., on December 9, 1930. This hawk is fairly common in California in winter. While driving over the interior valleys during March, we occasionally saw one of these hawks standing on the ground on some grassy plain or cultivated field, probably hunting ground squirrels. The white breast shows up very conspicuously in such situations, but it may help to obliterate the outline of the hawk, as seen against the sky by its humble prey.

DISTRIBUTION

Range.—Southwestern Canada, Western United States, and northern Mexico.

Breeding range.—The ferruginous roughleg breeds **north** to Washington (Chelan); southern Alberta (Pigeon Lake, Red Deer River, Little Sandhill Creek, and Medicine Hat); Saskatchewan (Crane Lake, Quill Lake, Estlin, and Touchwood Hills); southern Manitoba (Margaret and Treesbank); and northeastern North Dakota (probably Grafton). **East** to eastern North Dakota (probably Grafton, Stump Lake, and Lake Washington); western South Dakota (probably Rapid City); northwestern Nebraska (probably Harrison); northeastern Colorado (Avalo, Sterling, and River Bend); probably western Kansas (Hays and Pratt); probably the panhandle of Texas (Potter and Armstrong Counties); and southwestern New Mexico (Fair View). **South** to southern New Mexico (Fair View); Arizona (20 miles southeast of Flagstaff); and central California (Cosumne River). **West** to California (Cosumne River and Jess Valley); Oregon (Boardman); and Washington (Chelan).

Winter range.—The ferruginous roughleg may sometimes winter throughout its entire breeding range. For example, a specimen was taken at Spokane, Wash., on January 22, 1926, while it has been observed at Edmonton, Alberta, and at Eastend, Saskatchewan, during this season.

Generally, however, the winter range extends **north** to Oregon (Fort Dallas); casually eastern Montana (Custer, Dawson, and Fergus Counties); and southern South Dakota (the Black Hills, Forestburg, and Vermillion). **East** to southeastern South Dakota (Vermillion); eastern Colorado (Boulder and Wray); western Kansas (Ellis); New Mexico (Engle and San Andreas Mountains); Texas (Somerset, Port Lavaca, and Corpus Christi); and Hidalgo (Real del Monte). **South** to Hidalgo (Real del Monte); and southern Lower California (Sierra de la Laguna). **West** to Lower California (Sierra de la Laguna, Santo Domingo, and Ensenada); California (Brawley, Corona, Paicines, Santa Cruz Mountains, casually Cotati, Marysville, and Red Bluff); and Oregon (Fort Dallas).

Spring migration.—Early dates of spring arrival, at points north of the regular winter range, are: North Dakota—Antler, March 24;

Charlson, March 25; Harrisburg, March 31; and Larimore, April 2.
Manitoba—Treesbank, March 15; Aweme, April 5; and Margaret,
April 5. Saskatchewan—Eastend, March 17; Skull Creek, March
18; Johnston Lake, March 23; Indian Head, March 28; and Ravens-
crag, April 3. Alberta—Nanton, March 14.

Fall migration.—Late dates of autumn departure from the north-
ern parts of the breeding range are: Alberta—Brooks, October 2.
Saskatchewan—Eastend, October 11; and Indian Head, October
29. Manitoba—Margaret, October 4; Treesbank, October 25; and
Aweme, November 1. North Dakota—Grafton, October 3; and
Charlson, October 19.

Casual records.—At least twice the ferruginous roughleg has been
detected in southern British Columbia, and eventually it may be
found breeding in that Province. On April 28, 1922, one was seen
at Osoyoos, while another was observed in the same region on May
22, 1922.

Other casual occurrences have been chiefly at points east of the
normal range. Among these are: Nebraska, one at Grand Island in
the winter of 1881, one found dead at Neligh, December 25, 1899,
one was taken at Warsaw in October 1917, and a specimen was
obtained at Ponca, on February 11, 1919; Wisconsin, a pair were
taken at Lake Koshkonong on November 10, 1893, and another in
the same area in October 1894; Iowa, a specimen was collected at
Ottumwa on November 4, 1914, and one was shot at Browns Lake
in September 1917; Illinois, one was reported to have been taken in
this State near the Mississippi River in 1876; and Indiana, a specimen was caught in a steel trap near Richmond on April 12, 1917,
while a second specimen was taken in the same locality on November
13, 1930.

Egg dates.—Canada: 33 records, April 26 to July 3; 16 records,
May 2 to 16.

Oregon and Washington to Dakotas: 94 records, March 24 to June
16; 48 records, April 16 to May 10.

California to Colorado and Texas: 23 records, February 28 to May
24; 12 records, April 23 to May 14.

AQUILA CHRYSAËTOS CANADENSIS (Linnaeus)

GOLDEN EAGLE

HABITS

This magnificent eagle has long been named the King of Birds,
and it well deserves the title. It is majestic in flight, regal in appearance, dignified in manner, and crowned with a shower of golden
hackles about its royal head. When falconry flourished in Europe
the golden eagle was flown only by kings. Its hunting is like that

of the noble falcons, clean, spirited, and dashing. It is a far nobler bird in every way than the bald eagle and might well have been chosen as our national emblem. But then the golden eagle is not a strictly American bird, as the bald eagle is.

The golden eagle, as a species, is widely distributed throughout the Northern Hemisphere; seven races have been described from various regions in Europe and Asia besides our North American form, which is a large, dark race. Our race was once more widely distributed than it is now. At the present time it is very rare as a breeding bird anywhere east of the Rocky Mountains. I have two birds in my collection that were taken from a nest on Waldens Ridge, in the Cumberland Mountains, Tenn., in 1902; they were raised in captivity for over a year before they died and were given to me. At the present time these eagles are probably more abundant in the wilder portions of southern California than anywhere else in this country, but even there they have decreased decidedly within the past few years. Their decrease is mainly due to the indiscriminate use of poisoned baits and to shooting and trapping by cowboys, ranchers, and hunters, with the erroneous notion that they do more harm than good. With the decrease in the number of eagles we may look for an increase in the number of ground squirrels.

Courtship.—The courtship of the golden eagle is much like that of the Buteos, to which it is closely related. It consists mainly of spectacular flight maneuvers, spiral sailings in ever-rising circles, in which the birds frequently come close together and then drift apart; as they pass they almost touch. Occasionally one will start a series of nose dives on half-closed wings, swooping up again between dives and giving vent to his joy in musical cries. This form of nuptial play is indulged in by both sexes and is kept up, more or less, all through the nesting season. Perhaps it is only a form of joyful exercise. The birds are apparently mated for life, and if one is killed the survivor immediately seeks a new mate.

Nesting.—My personal experience with golden eagles' nests is limited to seven nests found in Arizona and five in southern California, from all of which I collected only one egg. The Arizona nests were shown to me by my late lamented friend, Frank C. Willard, who, after many years of experience with them, knew where to find several pairs of these fine birds. Our first nest was a disappointment, as we found it occupied by a pair of western redtails. We had driven over the divide in the Mule Mountains, from Bisbee, to visit this long-established nest, which was located near the top of a high, rocky cliff, rising abruptly from a valley; but when we reached the top of the cliff, we saw the hawk fly off the nest (pl. 81).

The following day, April 5, 1922, we visited two nests near Tombstone. One was on a small ledge on the face of a bulging, rocky cliff on the steep side of a mountain; it was about 75 feet from the bottom of the cliff and 25 feet from the top, having a fine outlook over the valley far below. This was also occupied by redtails, but, as it was a fairly easy climb with the aid of ropes, I went down to it and secured the two hawk's eggs. It was a huge nest of large sticks, roots, and stems of yuccas and was lined with strips of yucca and other soft fibers.

The other pair of eagles had two nests, which they used in alternate years. One was in an easily accessible place on a low pinnacle of rock, but it was not in use. The alternate nest was on the farther side of a steep little mountain, which we reached by climbing up a steep slope to the rocky summit; here the ridge dropped off suddenly in rocky cliffs and steep slopes. At the brink of the cliff we could see no nest, but by rolling rocks over the edge we started the old eagle off her nest only about 12 feet below us. It was a difficult nest to reach from above on account of the overhanging cliff, but I found a place where I could climb to a ledge below it and come up to the nest on the ropes. It was located on an outlying spur of a high rocky cliff, about 125 feet up from the base. It was a large, old nest, 4 by 5 feet in diameter, made of large sticks, stems, and roots of yucca and other coarse materials; it was lined with grasses, weeds, strips of inner bark, and other soft fiber. Its contents were rather interesting, a small downy young, only a few days old, a very rotten egg, which burst in my hand, and the remains, mostly the hindquarters, of 12 rabbits. The eagle had flown off in silence and did not show herself again, even while I was sitting in the nest and admiring the view.

Two other nests were found in the Dragoon Mountains, one in the Huachucas and one in the Catalinas. All were similarly located in commanding positions on rocky cliffs, where the birds could look out over a wide expanse of open country. We found no tree nests in Arizona, where big trees are scarce, except in the canyons.

In southern California it was different. Here, in 1929, with the aid of Wright M. Pierce and E. L. Sumner, Jr., I saw five nests in a variety of situations, two in trees and three on cliffs. The cliff nests were very similar to those found in Arizona and similarly located on rocky cliffs in low mountain ranges or on rough, steep, rocky hills. To reach the nest from which I secured my only egg, we had a long, tough climb up to the head of a winding canyon among some rough, rocky hills on the Mojave Desert. Here we saw the nest on a high cliff above a steep, rocky slope; it was only about 20 feet up from the base of the cliff and about 30 feet down from the top. The old eagle flew off when she saw us coming and circled way off in the distance. We climbed to the foot of the cliff and halfway up to the nest, but only with ropes let down from above could

I negotiate the remaining few feet. The nest occupied the whole of a small shelf on a nearly perpendicular cliff. The nest measured about 4 feet in height and about 5 feet in width; it was a mass of large and small sticks, brush, and weeds and was profusely lined with dry and green sprigs of a stringy weed, which is very common here, and a few bits of down. It held one handsome egg on March 11 (pl. 81).

In Los Angeles County on February 28 we flushed an eagle off a tree nest, where she probably had eggs; we did not climb to it, since Mr. Sumner was planning to make a study of the young later on, as he had done previously. The nest was 65 feet from the ground in the largest of a small group of sycamores in a hollow among low grassy hills. The eagle flew off when we were 100 yards away and did not return (pl. 80).

The other tree nest, from which the eagle had been seen to fly on two previous occasions, was visited on March 8. It was about 60 feet up in a big eucalyptus and well hidden in the thick foliage. The tree stood in an open field among the foothills of a rocky range in Los Angeles County. There were no eagles about, and the nest had apparently been robbed. A short distance away, in a small clump of eucalyptus trees, was another old nest, probably an alternate site.

Much has been published on the nesting habits of the golden eagle in California, as the eggs are handsome and high priced and consequently very popular among collectors; I have seen many large series in California collections. A large majority of the nests seem to be placed in trees, mainly in various oaks, sycamores, redwoods, and pines. The heights from the ground vary from 20, or even 10, feet in low oaks up to 75 or 96 feet in tall pines or redwoods. The nest is made of large sticks, some over 2 inches in diameter, firmly interwoven, smaller sticks, twigs, brush, roots, grass, leaves, pieces of sacking, and other bulky rubbish; the lining is of softer materials, grasses, weeds, dead and green leaves, soft mosses, and lichens. Green grass, or green leaves, often attached to the twigs, are added from time to time, especially after the young are hatched. Milton S. Ray says in his notes: "The lining frequently varies with the particular pair of birds and also with the locality. A nest I found at a high altitude on a lofty and barren mountain side was merely lined with coarse roots. One in an oak-wooded canyon was lined with eucalyptus leaves, although no such trees were visible for miles around. Another nest was beautifully draped, hung, and lined with gray-green oak moss. So thickly was it covered with moss that it was very difficult to discern from a distance. Nests found in the humid coast belt in the great redwood forests were much more warmly lined; a typical nest was very thickly lined with rabbit fur also some moss and eagle down." New nests are sometimes quite small, 2½ to 3 feet in diameter and 18 inches high, but as they are added to from year to year they become quite bulky, 5 or 6 feet in diameter and 4 or 5 feet high.

In San Diego County a majority of the nests are on cliffs. While I was visiting James B. Dixon, at Escondido, he showed me a beautiful series of eggs that he had taken in that vicinity and pointed out some of the localities on rough, rocky mountains, where he had found the eagles nesting for many years. He (1911) says that each pair of birds has its own nesting and hunting range, from which others are driven out; but they have a peculiar habit of stealing materials from their neighbor's nest, which often results in a fight "over their stealings, diving and circling in the air and sometimes clashing together and falling thus several feet before breaking away from each other." He says further: "I have never yet found a nest that did not have some dagger leaves in it, and in some places the birds must have carried them for some distance. In other instances, pepper and eucalyptus leaves were used profusely in lining and were carried several miles as there were neither of these trees growing close by. The odor from either of these leaves is distasteful to bugs and lice of all kinds, and I think this the reason they took such pains to secure it when there was plenty of other nesting material close by."

Wilson C. Hanna (1930) has made some interesting observations on nest building activities, which begin in January in southern California. He says of one bird:

This bird would work pretty fast at nest building, as the following record indicates: 4:16 p. m., bird observed going to nest with stick in beak; 4:17, left nest; 4:19, returned to nest; 4:19½, left nest; 4:22, sailed by nest but did not go to it; 4:23, returned from the south with such a large piece of brush that it was hard to manage; 4:23½, left nest; 4:24½, returned to nest, descending from high above it; 4:26, left nest; 4:27, returned with stick; 4:29, left nest; 4:30, returned to nest from the north with stick; 4:31, left nest; 4:33, returned from the south over the nest and descended to it from the north; 4:37, left nest; 4:47, sailed over nest and then on out of sight in the distance.

Mr. Ray writes to me: "While engaged in nest building the eagles are seldom in evidence as they sail along close to the ground. On one occasion Rose Carolyn Ray noted a bird curving low over a hilltop and then beneath a huge oak, where, after rising straight up to a lofty bough, it placed the material it was carrying for the repair of the old nest which it later occupied. In leaving the tree the bird departed in the same unobtrusive manner in which it came."

If the first set of eggs is taken from an eagle's nest, the bird will often, but not always, lay a second set about a month later, sometimes in the same nest and sometimes in an alternate nest. The same nest may be used for many years in succession, but oftener the birds build two or more nests and use them alternately. Joseph R. Slevin (1929) has published an interesting history of seven California pairs, which illustrates the territorial habit. The sparrow hawk and the western kingbird have both been known to nest in the lower parts of golden eagles' nests.

Some of the older cliff nests are very large, as they last for many years in a secure and sheltered position, until the lower parts are quite thoroughly rotted. Bendire (1892) mentions one that was 7 feet high and 6 feet wide. F. C. Willard (1916a) tells of one that was "six feet one way by eight the other. Dried cactus leaves comprised most of it, but there were some sticks in the base of it." He writes further: "On one occasion I was interested in watching one collecting sticks for its nest. It would alight in the top of a half dead juniper tree, walk clumsily out on a dead branch and break off a stick with its beak. It carried this stick in its beak as far as I could see it, passing close by me enroute to its nest. I watched it make several trips, using a powerful glass to assure myself that it really carried the sticks in its beak and not in its talons. A short time thereafter I watched another eagle carrying dried 'nigger-head' leaves in its talons. It was using them as lining."

E. S. Cameron (1905) described and studied an eagle's nest in Montana that "was situated near the top of a scoriaceous rock in the badlands, a crimson pillar which crowned a high butte sloping abruptly to deep washouts. The upper part of this column consisted of easily detachable pink layers, called laterite by geologists, but scoriae of every color strewed the base which rested on red ochre clay reminiscent of a painter's palette. Placed in a hollow niche of the wall face the eyrie was entirely enclosed and sheltered on three sides by a dome of rock. On the fourth, and open side, the enormous sunken nest greatly overlapped the seemingly inadequate ledge, which served as a support, and thereby secured the safety of the eggs and young."

A totally different Montana nest was in a tall pine about halfway up a steep hillside. He (1908b) says:

The eyrie, which consists of an immense pile of pine sticks, rests upon, and is built around, a number of green boughs, while a dead projecting branch near the center forms a convenient perch for the parent eagles. As would naturally be expected in the present case, the vertical height of the nest greatly exceeds the diameter, and its width is much inferior to the nest upon the rock previously described. Nevertheless, as seen from below, it conveys an impression of strength, which is not belied when it is reached, for a six foot man can sit in it with ease. On May 11, the whole external circumference of the nest rim was interwoven with an ornamental binding of green pine tops.

Roderick MacFarlane (1908) found this eagle breeding nearly up to the Arctic coast; he writes: "From various points along the valley

of the Anderson River to its outlet in Liverpool Bay, and from near
the mouth of the Wilmot Horton River in Franklin Bay, an aggre-
gate of twelve nests of the golden eagle was procured in course of
the breeding seasons from 1861 to 1865, inclusive. Ten of them
were constructed on the side face, and within twenty or thirty feet
of the summit, of steep and difficult of access earth and shaly ravine
banks; and in the other two instances the nests were built near the
top of tall spruce pines."

Eggs.—The usual set of golden eagle's eggs is two; full sets of one
are common, sets of three rather rare, and at least one set of four has
been taken (Ray, 1928). Mr. Hanna (1930) writes: "Nests with complete sets of eggs that I have personally examined in southern California have had only one egg in 35 percent of the cases, two eggs in
60 percent and three eggs in 5 percent." The shape varies from
short-ovate to oval, or rarely to elliptical-oval; the shell is thick and
from finely to coarsely granulated. The ground color varies from
dull white to "cream-buff" or pinkish white. The variations in types
and colors of markings are endless, but series of eggs from the same
female usually run true to type. They are generally more or less
evenly marked with small blotches, spots, or fine dots, but often the
markings are unevenly distributed or concentrated at one end, and
some are evenly sprinkled with minute dots. The eggs are often
sparingly or faintly marked, or even nearly or quite immaculate.
The usual colors of the markings are "bay", "amber-brown", "hazel",
"tawny", "Mikado brown", "clay color", "vinaceous fawn color",
and various shades of "ecru-drab" or "Quaker drab." Some very
pretty eggs have large blotches or washes of the drabs overlaid with
browns. The measurements of 59 eggs in the United States National
Museum average 74.5 by 58 millimeters; the eggs showing the four
extremes measure 85.7 by 64.3, 67.5 by 53, and 70.7 by 49.4 millimeters.
An egg in the collection of C. S. Sharp measures 89 by 66.6 millimeters, the largest egg of which I have any record.

Young.—The period of incubation of the golden eagle has been
variously reported as from 28 to 35 days; the latter figure seems to
be based on the most accurate observation and is probably the most
nearly correct. Most observers agree that the male does not assist
the female in incubation, but he feeds his mate on the nest and helps
to care for the young by bringing in food, which his mate feeds to the
young, and by brooding the young occasionally himself. The incubating bird is easily frightened from the nest, although on rare
occasions she has been known to remain on the nest until the climber
has been near enough to touch her. She usually flies away in silence
and disappears entirely, or remains at a distance. Only once have I
ever seen an eagle return to her young while we were watching at a

long distance, and then only for a few seconds. I can find no au-
thenic record of an eagle attacking an intruder at her nest. If disturbed during the early stages of incubation, she may desert the
eggs but never the young, although she seems quite indifferent to
their welfare. Mr. Ray says in his notes: "Many birds resent any
interference with their nests and will frequently desert them whether
they are in the course of construction, completed, or even containing
partial or full sets of eggs. In some cases they have apparently
shown their extreme disfavor by casting the eggs out of the nest;
while I have never actually seen eagles engaged in taking such dras-
tic measures, on a number of occasions I have found eggs on the
ground just below the nests where an almost inaccessible situation
made it difficult to see how they could have been disturbed by any
outside agency."

Several British ornithologists have, at the cost of much effort,
personal discomfort, and risk, spent considerable time studying and
photographing the home life of the golden eagle. I would recom-
mend reading the published reports of H. B. Macpherson (1911),
H. A. Gilbert and Arthur Brook (1925), Duncan MacDonald (1926),
and Seton Gordon (1927). Much of what follows is taken from
their writings and from the observations of E. S. Cameron (1905 and
1908b) in Montana. I regret that space will not permit more
elaborate quotations from these interesting accounts.

E. L. Sumner, Jr., has sent some very full notes on the growth of a
brood of young eagles in California, which he measured and weighed
once a week from the time they hatched until they left the nest. The
loss in weight of the eggs prior to hatching is interesting; on February 27 the three eggs weighed 143, 143.4, and 133.7 grams; on March 20
the first egg had just hatched, and the other two eggs had shrunk in
weight to 128.6 and 126.2 grams. The newly hatched chick weighed
105 grams. A week later all three had hatched, and the chicks weighed
357.3, 232.3, and 98.2 grams, showing that they probably hatched at
intervals of two or three days. On April 3 the youngest and smallest
chick had disappeared and the other two had increased to 1,022.7 and
584.7, the older chick being then two weeks old, and ten times as heavy
as when hatched. From that time on both birds increased steadily in
weight, along slightly divergent lines; on May 8, when seven weeks
old, they weighed 3,851.7 and 2,801.7 grams. During the next week,
they both dropped off over 400 grams in weight, but regained this and
more during the following week, so that on May 22, when nine weeks
old, they weighed 4,061.7 and 2,981.7. This, compared with 4,169.4,
the weight of an adult male, emphasizes the lightness of the smaller
bird, probably a male. This was the last weighing, as the birds left
the nest during the following week.

Mr. Sumner noted that when first hatched the chick was unable to distinguish objects but could chirp incessantly. At the end of a week it could see well, move its head about, and bite at things. When two weeks old it could crawl and soon learned to rear up; but even when seven weeks old it could barely maintain its balance when placed on a limb.

Mr. Sumner's eaglets left the nest when between 9 and 10 weeks old; this was a tree nest and the eaglets had been often disturbed. Mr. Gordon gives the time in one case (1915) as 9 weeks and in another case (1927) as 11 weeks, saying: "The eyrie takes at least six weeks in the building or the repairing, and eagles continue to bring fresh fir branches and bunches of heather to the eyrie until the last fortnight before the nest is vacated—that is, until the eaglets are about nine weeks old."

The eaglet that Mr. Macpherson (1911) watched on a cliff nest began leaving the nest and wandering about on the ledge when a little over nine weeks old, but did not fly from the eyrie until about two weeks later. He noted that the young eaglet, while still in the downy stage, "was fed with great regularity twice a day—at daybreak and about 5 P. M." The food, mainly grouse and hares, is brought to the nest by both parents, but principally by the male. The game, at this age of the young, is stripped of fur or feathers and usually disemboweled before it is brought to the nest. The female does practically all the feeding, swallowing the intestines herself and picking out tidbits from the liver or other dainty morsels to feed to the young. All uneaten portions of the food are carried away. Later on, when the plumage is growing and the young eaglet is strong enough to tear up his own food, the game is left entire, he is taught to feed himself, and the remnants of the food are not so carefully removed. As the time draws near for him to leave the nest he is encouraged to exercise his legs and wings by placing the food beyond his reach on the ledge.

Of the eaglet's behavior Mr. Macpherson (1911) writes:

After his feast the Eaglet walked round the edge of the nest and began to play. He behaved exactly like a child thrown upon its own resources for amusement and compelled to fall back upon any handy article as a toy. Small pieces of heather in this case served his purpose, and he appeared to enjoy lifting them from the ground and throwing them down again. He also picked pieces of moss from the rocks and only desisted from this occupation after having completely stripped the walls of the eyrie. * * * He next began to make his toilet, carefully removing all the loose down, which was now freely coming away. This was accomplished with the aid of his beak, and, the task completed to his satisfaction, he lay down and went to sleep.

It often happens that one of a pair of eagle's eggs proves to be infertile. But oftener one of the eaglets disappears; the smaller and weaker bird may not be able to secure his share of the food and thus may weaken and die from exposure. The larger one, usually the female, often attacks and may kill her little brother. Seton Gordon (1927) twice witnessed spirited fights, of one of which he writes:

Twenty minutes after the parent had left the family, Cain commenced a very determined and entirely unprovoked attack upon her brother. She tore from his unfortunate person great billfuls of white down and even tiny feathers. Abel in desperation ran to the far side of the eyrie and lay there, quite still and very sullen. Cain thereupon stood up, flapped her downy wings, and uttered several wild and piercing yells of victory. There was an extraordinary and quite unearthly quality in these calls which deeply impressed itself upon my mind. Great billfuls of her brother's down adhered to her bill, and she had much trouble in ridding herself of the fruits of her easily gained victory.

Mr. Cameron (1905) says of the food of the young in Montana that "the nest always contained either sharp-tailed grouse, jack-rabbits, cotton-tails, mountain rats, meadowlarks or snakes", but no carrion. He says that the eagles catch a number of rattlesnakes. "According to eye-witnesses they feint several times at the snake to make it uncoil and seize it just behind the head with one foot, while gripping it further back with the other. The snake is then taken to a tree or rock and the head torn off, which according to one observer is immediately devoured, before the body is deposited in the eyrie."

Mr. Sumner, in California, found numerous ground squirrels and the remains of a cottontail, a crow, a meadowlark, and a gopher snake in the nest.

Young eagles remain in the vicinity of their nest for a long time after they leave it. They are probably at least three months old before they gain the full power of flight. They are partially fed by their parents at first and are watched and guarded by them until they learn to hunt for themselves, probably until early in fall. Dr. Loye Miller (1918) published the following account, as given to him by one of his students:

Last summer while my father and I were extracting honey at the apiary about a mile southeast of Thacher School, Ojai, California, we noticed a golden eagle teaching its young one to fly. It was about ten o'clock. The mother started from the nest in the crags, and roughly handling the young one, she allowed him to drop, I should say, about ninety feet, then she would swoop down under him, wings spread, and he would alight on her back. She would soar to the top of the range with him and repeat the process. One time she waited perhaps fifteen minutes between flights. I should say the farthest she let him fall was 150 feet.

My father and I watched this, spellbound, for over an hour. I do not know whether the young one gained confidence by this method or not. A few days later father and I rode to the cliff and out on Overhanging Rock. The eagle's nest was empty. (Miss F. E. Shuman.)

Plumages.—During the nest life of the eaglet the plumages may be roughly divided into three stages—four weeks in a pure downy stage, four weeks during which the plumage is growing, and three weeks in a nearly feathered stage. When first hatched it is completely and thickly covered, except on the toes and back of the tarsus, with short, thick, dirty-white or yellowish-white down, overlaid on the upper parts with a scanty growth of long, grayish-tipped, hairlike down. This is replaced later by a longer, thicker, woollier, pure-white down. At an age of four weeks the wing quills are sprouting and beginning to burst their sheaths. During the next week the tail quills appear. At the end of eight weeks Mr. Sumner's larger bird had a wing spread of 62 inches, primaries 11 inches long, and tail quills 7 inches. Meantime the body plumage has been growing, beginning with the scapular and back plumage during the fifth week; this is soon followed by the wing coverts and then the feathers on the sides of the breast. By the end of the seventh week the upper parts are fully feathered and the under parts largely so, but the head and neck are still downy and there is much down on the breast, flanks, and legs. At 10 weeks the juvenal plumage is practically complete, and the eaglet is ready to fly.

In fresh juvenal plumage the young eagle is considerably darker than the adult; the crown and hackles are darker and duller, not so golden; the upper parts vary from "blackish brown," or nearly black, to "clove brown", with a purplish sheen; the under parts are only a shade browner, with a purplish bloom on the breast; the basal third of the back feathers and the basal half of the breast feathers are pure white; a narrow white tip on the tail soon wears away, leaving a broad terminal band of brownish black, covering about one-quarter of the central rectrices and graduated up to one-half of the outer feathers; the rest of the tail is white, washed with gray on the outer webs and more or less spotted with black above the dark band; the remiges are black, with considerable white near the bases of the inner primaries and all the secondaries; the tarsi are dull white.

The juvenal plumage is worn for one year without change except by wear and fading. From that time on progressive changes take place through annual complete molts, toward maturity. The molts are mainly accomplished between April and July but may extend from March to October. The fully adult plumage is not complete until the bird is four years old or more. Meantime the white in the wings gradually disappears; the basal white in the body feathers grows less until there is little or none in the adult; the white in the tail decreases at each molt, becoming purer white, until the adult tail shows no white, but is more or less indistinctly and irregularly barred or spotted with very dark gray or brown; the feathers of the upper breast and the tibiae are edged with "ochraceous-tawny" or "tawny-olive" and the tarsi are pale brown or "tawny-olive."

Food.—The golden eagle is such a large and powerful bird that it can attack and kill many large mammals and birds, and it shows great courage in attacking animals larger than itself, many of which are capable of inflicting severe injury on the brave bird. The list of mammals recorded includes deer and their fawns, antelopes, lambs of mountain sheep, goats and their kids, domestic calves, lambs, dogs, cats, young pigs, foxes, hares, rabbits, ground and arboreal squirrels, raccoons, prairie dogs, woodchucks, marmots, spermophiles, porcupines, opossums, skunks, weasels, martens, pocket gophers, rats, mice, and moles. The list of birds is not so long, but it includes great blue heron, turkeys, geese, ducks, goshawk, red-tailed hawk and short-eared owl (both twice recorded), sage-hen and other grouse, ptarmigan, quails, band-tailed pigeon, crow, domestic poultry, curlews, plovers, kingfisher, meadowlarks, and thrushes. Birds, particularly the smaller species, are taken mainly during the nesting season as tender food for the young. But at all seasons mammals seem to be preferred. Eagles kill many snakes and an occasional tortoise; they often feed on carrion when live game is scarce.

The stomach contents of 30 golden eagles reported by Howard Kay Gloyd (1925) show their preference for mammals during the fall and winter months; 11 had eaten cottontail rabbits, 7 had taken jack rabbits, 9 prairie dogs, and 1 each had eaten a woodchuck, a ground squirrel, a short-eared owl, an opossum, a fox squirrel, and a red-tailed hawk.

There are numerous, apparently authentic, reports of these eagles killing large mammals. F. C. Willard (1916b) reports the killing of a four-point white-tailed deer in Arizona.

The deer had been pounced upon by one or more eagles as it floundered in the deep snow, and its back was fearfully lacerated by the talons. After it had succumbed, the carcass was dragged down-hill over one hundred yards until it lodged against a large boulder. Three eagles were feeding on it when first discovered by some prospectors. * * *

Recently two cowboys in the employ of Mr. Lutley came upon three eagles feeding upon the body of a calf about seven months old. * * * The back of this calf gave every evidence that it had been killed by the eagles.

C. F. Morrison (1889) reports that a golden eagle in Montana "had captured and killed a good sized Black-tail Deer, and was shot while sitting on its body." Mrs. Seton Gordon (1927), while watching a nest, saw "a wonderful sight. The cock eagle alighted, exhausted, at the eyrie with a roe-deer calf held in one great foot! The powerful bird arrived from below, and was only just able to raise himself to the nest with his large burden." A few days later there "were two more roe calves and the skeleton of the first" in the nest. Aiken and Warren (1914) write:

The Golden Eagle is reported to be one of the worst enemies of the mountain sheep, killing many of their lambs. A Mr. Waldron told Aiken that many years ago when driving on the plains with several others he saw an eagle of this species attack and kill an antelope. The bird pursued a bunch of the animals, singling out one, and when close enough struck it on the back with its talons, and while clinging there and tearing with claws and beak it at the same time beat its prey's sides with its wings. The men drove close enough to shoot the eagle, and found the antelope to be dead with its back badly torn by the bird. Aiken was also told that an eagle was seen to pounce upon a two-year-old calf near Hartsel but was driven away before any harm was done. Rather large prey for the bird to tackle.

M. P. Skinner's notes give a somewhat different impression, for he says: "I have made particular inquiries whether these eagles have ever been seen to kill mountain sheep lambs, but not one of our rangers had ever done so. In carrying on my inquiries outside the Park, I heard from one correspondent, previously unknown to me, that he had seen an attack wherein two golden eagles seemed to try to knock the lamb off its cliff, or at least to scare it so that it would fall. This inquiry extended to many parts of the United States and to some localities in Canada."

Mr. Cameron (1908b) states that R. L. Anderson came upon "three Golden Eagles which were devouring an adult buck antelope" in mid-winter in Montana. He continues:

Despite the bitterly cold weather, the antelope was warm and limber when found, as it had only been quite recently killed. The eagles had torn a large hole in its back with their terrible talons and were feeding on the kidneys and entrails. Mr. Anderson at once investigated the scene of the struggle and could easily read the gruesome details on the deep, crusted snow. The eagles had obviously stampeded a bunch of antelope, and then cut out a victim by a combined attack. Leaving the herd, the latter endeavored to escape down a small right hand draw, but after covering about a hundred yards was beaten back by the eagles. It then crossed a ridge on which the main antelope trail ran at right angles to its own and, hard pressed by its assailants, struggled down a narrow left hand draw to the place where it succumbed. Altogether the antelope could barely have covered three hundred yards after the first attack by the eagles. The victim, which had evidently offered a gallant resistance, seems to have made a stand in three places, chiefly where found, but also at points along the trail. The crimson stained snow and thickly strewn hair, added to the well defined wing prints of the flapping and dragging eagles, sufficiently revealed this prairie tragedy. One or more of the birds must have clung tenaciously to their quarry's back and from the deep wounds thus inflicted "the blood had spurted out as when a cow's horns are sawn off."

He also has much to say about the destruction of prairie dogs by these eagles. One of his pairs that lived near prairie-dog towns always had one or two of these animals in their nest. He says:

Now the destruction of prairie dogs is of the greatest benefit to the settlers, as in this locality (Knowlton) they have increased to an alarming extent. On some ranches the rodents play havoc with the crops and "dog towns" have encroached upon miles of good grazing land, reducing it to a desert. It is only necessary to read the forcible paper on "The Prairie Dog of The Great Plains" by Dr. C. Hart Merriam, quickly to realize what an unmitigated pest this animal becomes, and how rapidly its towns spread. As quoted therein, Professor W. W. Cooke computes that "32 prairie dogs consume as much grass as one sheep, and 256 prairie dogs as much as one cow."

Throughout the month of April, and for two days in May, allowing an average of three prairie dogs per diem, we get a total of 96 prairie dogs up to the time that the eaglets are hatched. Subsequently, until the young birds forage for themselves (about Aug. 1), if we allow only six of the rodents a day, the total is obtained of 540 prairie dogs for seventy-four days sustenance of four eagles. Thus we have a grand total of 636 prairie dogs during four months for one pair of eagles, which is probably well within the mark. An eagle intent on capturing a prairie dog floats leisurely above the "town" at a medium height on motionless wings. Preliminary inspection of the hunting ground is accomplished in wide circles or long sweeps, perhaps two or three miles each way, so as not to unduly alarm the game. Passing over at long intervals, the bird scans the dog town and judges of the prospect for a successful stoop. The "dogs" are of course immediately on the alert, but can only see their enemy for a short time on account of the high surrounding pine hills, and, indeed, most "dog towns" are too extensive for the denizens at one end to notice an eagle passing over at the other. Moreover, an unsuccessful eagle will keep on the wing for several hours, and it is almost certain that the hungry prairie dogs will relax their vigilance at last. When the eagle considers that a favorable chance has arrived it sinks lower, so as to bring the distance between itself and the animals to something like 75 or 100 yards. Should the latter still remain above ground, the royal bird suddenly folds its wings, and, with meteoric rush, falls head first towards the astounded prairie dogs. These scamper for their holes, but about three yards from the ground the eagle spreads its wings and, swiftly following the intended victim, darts out a cruel foot to grasp it. If the attack fails, as sometimes happens, the eagle mounts in a slow, reluctant manner which plainly reveals its disappointment.

This bold bird sometimes "catches a Tartar." Albert Lano (1922) had one brought to him that had attacked a porcupine; "it was literally covered underneath with quills. In fact there were a number of quills in the roof of its mouth. The body was much emaciated and many of the quills had penetrated deep into the flesh causing pus to form."

These eagles have often been known to attack foxes caught in traps, but the following spirited encounter, described by Mr. Gordon (1915) is unique:

The eagle was devouring the carcass of a blue hare when a fox sprang from the surrounding heather and seized the great bird by the wing. A well-contested struggle ensued in which the eagle made a desperate attempt to defend itself with its claws and succeeded in extricating itself from its enemy's grasp, but before it had time to escape Reynard seized it by the breast and seemed more determined than ever. The eagle made another attempt to overpower its antagonist by striking with its wings, but that would not compel the aggressor to quit its hold. At last the eagle succeeded in raising the fox from the ground, and for a few minutes Reynard was suspended by his own jaws between heaven and earth. Although now placed in an unfavorable position for fighting his courage did not forsake him, as he firmly kept his hold and seemed to make several attempts to bring the eagle down, but he soon found the strong wings of the eagle were capable of raising him, and that there was no way of escape unless the bird should alight somewhere. The eagle made a straight ascent and rose to a considerable height in the air.

After struggling for a time Reynard was obliged to quit his grasp, and descended much quicker than he had gone up. He was dashed to the earth, where he lay struggling in the agonies of death. The eagle made his escape, but appeared weak from exhaustion and loss of blood.

Hares, rabbits, and other smaller mammals are usually caught by chasing them in the open and pouncing on them, but Mr. Willard (1916a) witnessed another method: "In company with some friends one day, I watched a pair of these eagles hunting jack rabbits. They swooped down and drove the rabbit to cover under a mesquite bush. Then one alighted close by and began to walk toward the rabbit. He was so frightened he dashed from his shelter only to be snatched up by the other eagle which had been hovering close overhead."

Grouse, ptarmigan, and quail are also captured by swift pursuit in the air, as eagles are among the swiftest of fliers. Dr. A. K. Fisher (1893) quotes the following account by Robert Ridgway:

We were standing a few yards in the rear of a tent when our attention was arrested by a rushing noise, and upon looking up the slope of the mountain we saw flying down the wooded side with the rapidity of an arrow a Sage-Hen pursued by two Eagles. The Hen was about 20 yards in advance of her pursuers, exerting herself to the utmost to escape; her wings, from their rapid motion, being scarcely visible. The Eagles in hot pursuit (the larger of the two leading), followed every undulation of the fugitive's course, steadily lessening the distance between them and the object of their pursuit; their wings not moving, except when a slight inclination was necessary to enable them to follow a curve in the course of the fugitive. So intent were they in the chase that they passed within 20 yards of us. They had scarcely gone by, however, when the Sage Hen, wearied by her continued exertion, and hoping probably, to conceal herself among the bushes, dropped to the ground; but no sooner had she touched it than she was immediately snatched up by the foremost of her relentless pursuers, who, not stopping in its flight, bore the prize rapidly toward the rocky summits of the higher peaks, accompanied by its mate.

It can be seen from the foregoing quotations that the golden eagle is a very dangerous bird, a powerful influence for either good or evil according to the conditions in its habitat. Its natural and favorite food during most of the year consists of a long list of injurious rodents, which are prolific breeders. Where the eagles can keep these rodents in check, they are of great benefit to agriculture. But where they do much damage to domestic animals, the eagles may have to be controlled. Eagles kill some fawns and a great many grouse, but let us remember that all these wild creatures have existed for untold ages in apparent balance. Probably the eagle's victims include more of the weak and sickly individuals than of the strong and healthy ones, which greatly improves the strain and produces a healthier and more vigorous race by the survival of the fittest. We once found under a golden eagle's nest in California the dried remains of a wildcat.

Behavior.—The flight of the golden eagle is the embodiment of grace and power. To my mind it is more impressive than that of the bald eagle. The bald eagle is said to be swifter on the wing, but I doubt it. It is certainly inspiring to watch the spirited dash of this great bird in pursuit of its running or flying quarry. There are few swifter runners than the jack rabbit and few swifter flyers than the band-tailed pigeon, but this eagle is more than a match for either in an open chase. Mr. Gordon (1927) thinks "that the downward rush of the golden eagle is the swiftest thing, as it is the most magnificent thing, in the bird world." Its lofty soaring flight is equally grand, as it mounts in ascending spirals up into the clouds until lost to sight. Mr. Gordon (1915) again writes:

Then one day the north wind crossed the sea, and arrived at the eagle's home. And the eagle felt the cool arctic breeze and sailed out from his giant rocks which by now were burning hot in the fierce rays of the sun. With his pinions wide outstretched he leaned on the refreshing wind, which bore him strongly upward, without a single stroke of his wings to help him on his way. So he mounted higher and higher till he had risen far above his native hill-top, and was outlined, a mere speck, against the dark blue of the sky. Still upwards he sailed, and for sometime longer the watching stalker kept him in view, in the field of his glass. But at length he reached a point at which he was invisible, even by the aid of a telescope. From that point what a gorgeous panorama must have been laid out before his sight in the light of the summer sun. Even the highest tops were now far far below him, and the river in its windings down the great glen must have appeared as a thin silvery streak.

Lila M. Lofberg (1935) writes of her observations while watching a pair of golden eagles near their nesting site at Florence Lake, Calif.:

The most interesting thing that has occurred while I watched has been their aerial circus. Whether this occurs more than once a year I cannot say, but I have never seen it more than on one day during the season. A distant call first attracts my attention. This comes from a mere dot in the sky. The second bird then leaves its perch on the nesting ledge and soars in wide circles, upward. Before it can attain the height of its mate, the "dot" comes hurtling down with closed wings, at terrific speed. When not over a hundred feet from the ground and just as I am sure it will be dashed to pieces, out come the wings and this bird instantly goes into a series of daredevil stunts. It rolls, stands on its head or tail, or slides earthward sidewise, with extended wings. Between these it may perform flights that remind me of a skater cutting figures on the ice. When it has exhausted its repertoire it ends on a line with the nest. But instead of flying straight to it, the eagle makes three perfect loops in the air, coming out of the last within a couple of flaps (of the wings) of the ledge.

Meanwhile the one in the air has been forgotten entirely but soon the faint call reminds me to look upward to find that it, too, has become a dot. Upward starts the resting eagle. Down comes the distant one to go through the same routine. Always these flights end with those three loops that bring them onto the nesting ledge. For an hour or more they continue this exciting sport.

Then the one on the ledge fails to heed the call and remains until the other has alighted beside it. Then off they fly together toward Blaney Meadow, about five miles to the southeast of their home.

William Brewster (1925) witnessed a thrilling swoop of a golden eagle at a great blue heron:

Drifting, presently, over the place where the Heron had settled and evidently noticing the big bird for the first time, the Eagle checked his flight in the middle of a half-completed circle to poise for an instant on rapidly-vibrating wings, precisely as a Kingfisher will hover over a school of minnows. Then he swooped, apparently as straight and vertically as a heavy stone may fall, yet all the time *revolving like a spinning rifle bullet*, if more slowly, thereby showing us his (normally) upper and under parts alternately and making no less than four or five such turns before passing out of sight. Never before have I seen anything of the kind that seemed nearly so wonderful and impressive. As the great bird plunged headlong, from a height of at least one hundred yards, his wings, apparently set and almost closed, made a sound like that of a strong wind blowing through pine branches. His momentum must have been tremendous as he neared the earth. How it was finally checked and what else transpired behind the line of fallen trees I am, of course, unable to report. Without doubt the Eagle stooped at the Heron and quite as certainly failed to strike it down; for after an outburst of loud and prolonged squawking it rose above the trees and flew off at its very best pace, evidently badly frightened. Perhaps the Eagle had merely been amusing himself by bullying it, a diversion to which all strong-winged birds of prey are more or less inclined.

At another time he saw a young eagle attacked by an osprey, of which he writes:

After making the fruitless attempt to capture a Duck, he was assailed by an Osprey who kept darting down and striking at him from above, precisely as a Kingbird attacks Crows and other large birds. Every time the Osprey came within six or eight feet of him the Eagle would turn back downward and thrust up both feet with their talons extended, as if hoping to clutch his tormentor. This action was repeated at least half a dozen times, and performed so quickly that it was difficult to follow with the eye, although for a fraction of a second the upstretched legs and widespread talons showed distinctly enough against the sky.

Mr. Sumner has seen a young eagle pursued by a flock of avocets and driven away, one attacked by a blackbird, and one, which was standing on the ground, was attacked by a red-tailed hawk; the hawk—

which had been circling in the air, dove at him three times from a height of 300 to 400 feet. Each time the redtail dove the eagle jumped up from the ground and flung himself, while in the air, upside down so as to oppose his talons to those of the hawk. By and by the hawk stopped diving and began to circle again, the eagle staying where he was, but when the eagle got up and flew farther into the hawk's territory—flying leisurely—the redtail, although quarter of a mile or so from him, flapped his wings faster than I have ever seen a redtail flap, and was overhead in less than 30 seconds—like an airplane overtaking a freight train—and dove at him as before.

Mr. Skinner tells me these "eagles are much harassed by the ravens and crows"; he has often seen one "on the ground surrounded by a circle of ravens waiting for it to fly and the sport of mobbing to begin."

The prevailing belief that an eagle will attack anyone attempting to rob its nest is entirely erroneous. I can find no record of anyone being struck by an eagle at its nest, and only on very rare occasions has one been bold enough to even threaten the intruder. Evidently parental affection does not show itself in bravery, but hunger often makes the eagle bold and even savage; also a wounded eagle will show fight and even make an aggressive attack. Dr. E. W. Nelson (1887) says: "On one occasion a pair was disturbed by a friend of mine while they were feeding upon the remains of a hog in northern Illinois. As my friend approached the birds arose and swooped fiercely at him. Both birds were shot almost at the muzzle of the gun; the first fell dead almost at his feet; but this apparently seemed only to increase the rage of the survivor, which renewed the attack until it, too, was disabled."

Mr. Cameron (1907) relates the following story as told to him by a shepherd:

He narrated how from some distance away he saw an eagle stoop at one of the dogs, and hang above it as raptorial birds are wont to do when attacking ground game. The dog, not paralyzed like a hare, at the proximity of the great bird, ran towards its master, when the hovering and expectant eagle fixed one foot on each side of the collie's throat and endeavored to bear aloft the shrieking animal. The shepherd described how during the few minutes that he was running toward the struggling pair and trying, incidentally, to find a stick, the eagle made frantic efforts to carry away the dog, which seemed unable, when clutched in this manner, to make any attempt to free itself. According to the story, the bird was flying all the time, in any case flapping its wings, and, although prevented from rising by the weight of the quarry, it was able to drag the helpless dog to and fro. The eagle had, in fact, too good a hold for her own safety and was ignominiously killed by blows on the head with a stick.

An index to the food resources of the golden eagle in the mountains of northern British Columbia was afforded by a nest examined by Edward A. Preble. He says (MS. account):

The nest, built on a ledge overlooking the valley of the South Fork of Bear Creek, in the Babine Mountains, was found on August 3, 1913. It had just been vacated by the sole young bird that had been raised in it.

The nest was built in a vertical cleft or "chimney" in the cliff, and the site was plainly an old one, for the labors of successive years had reared a structure nearly 20 feet high. Access to the nest proper was somewhat of a problem, but by taking advantage of slight projections on the face of the cliff beside the nest I soon reached the top. Here was the usual depression, flattened out by the weeks of use by old and young. The chief interest, of course, centered in the remains left from the feasts that had contributed to the growth of the young eagle now about to begin its active life. I made no attempt to count the individuals represented by the remains, which, of course, included only those that chance had suffered to remain, but I was careful to identify all the species represented. Varying hares (*Lepus americanus*) and marmots formed the bulk. The latter were mainly the large hoary marmot (*Marmota caligata*), but one skull of the small relative of our eastern species (*M. m. ochracea*) was among the lot. Part of the skin and skeleton of a marten (*Martes americana*) proved somewhat of a surprise. These comprised the list of the mammals, the remains of smaller ones, if there had been any, not being in evidence. Among the birds whose relics had lodged in the structure was an adult goshawk, a genuine surprise, and a number of blue grouse (*Dendragapus o. richardsoni*). The cliffs on which the nest was placed held the homes of many bushy-tailed wood rats, and there were signs that they had occupied at one time the lower parts of the eagle's structure, but there was no evidence that any had been captured.

In my ascent of the nest I inadvertently disturbed a healthy colony of yellow jackets that had built their own home about halfway up the structure. Fortunately for the success of my deliberations aloft I was then unaware of this important circumstance; but, when nearly halfway down, I was met by an advance guard from the enraged colony. There was only one way out, and I made the remainder of the descent in record time, glad to escape with a moderate number of stings.

This episode over, I turned my attention to the young bird, which still occupied his perch on the verge of the broad ledge where I had first seen him. While I was at the nest he had uttered at intervals a querulous rattling call, evidently an appeal to his parents, one of which had been seen once or twice at a distance. At my approach he regarded me with a reserved indifference. He was evidently a male and fully grown, and his dark lustrous juvenal plumage was in perfect condition. Although I believed he had not yet flown, I concluded that he was able to take off. Under my judicious but firm encouragement he launched into the air, and after a few somewhat clumsy but effective attempts to master the art of balancing he soared and flapped off down the valley, finally perching awkwardly on the summit of a spruce nearly half a mile below. On our way out of the mountains about ten days later the eagles were still in the vicinity of the aerie.

There are many old tales of eagles carrying off young children, but most of them are pure fabrications by sensational reporters. An eagle, if pressed for food, might carry off a small baby that had been left in the open unprotected, but such an opportunity must occur very rarely. Stories of babies being found in eagles' nests, practically unharmed, are purely imaginary, as eagles are well known to kill their prey at once. Mr. Forbush (1927) has investigated a case, which seemed to him authentic; an eagle attacked a little girl, nine years old, and cut and bruised her arm quite badly before it was beaten off. It is doubtful if an eagle could lift anything heavier than a very small baby. Mr. Cameron (1905) says: "Personally I have never known an eagle to carry anything heavier than a seven pound jack-rabbit and would think eighteen pounds (the extreme weight of a jack-rabbit or a Scotch brown hare), to be the extent of the largest eagle's capacity. It follows, therefore, that the lambs taken are very small."

The weights of the fawns and the fox, referred to above, were not definitely known, but they probably did not exceed 18 pounds and may have weighed much less. An eagle in rising from level ground must use its feet to spring into the air; therefore, if one or both feet are needed to hold its prey, it is handicapped accordingly. From an easy take-off on a steep slope it could probably lift its own weight, 14 to 16 pounds, or perhaps more.

Mr. Ray's notes contain the following interesting items:

In one instance while at a lofty nest, just as I was about to examine the set of two eggs it contained, the massive bird, entirely unaware of my presence, came sailing in and lit upon the edge of the nest but a few feet away. It was, for me, an anxious moment. However, with a loud call and by waving my hat in the air, the badly frightened eagle immediately took flight.

On another occasion, while at an elevation of 7,500 feet in the high Sierras, I was crawling on the ground and just emerging from a dense thicket of buck brush after a fruitless search for a fox sparrow's nest, when I noticed a great shadow growing larger and larger on the ground at my feet. Now, as I stood up, I perceived just above my head a great golden eagle with pendant legs and outstretched claws. Quickly seizing a nearby stick and waving it above my head, I just narrowly missed striking the bird, which, apparently greatly surprised at my action, quickly sailed away. It was evident that the bird had mistaken me for some mammal as I emerged from the brush.

Voice.—The golden eagle is mainly a silent bird. It usually leaves its nest in silence and does not fly around and scream, as so many of the hawks often do. I have no record in my notes of ever having heard it. Bendire (1892) says: "The usual call note is a shrill 'keé,-keē,-keé,' uttered in a high tone; it is often heard in the early spring before nidification commences. Another note, not so frequently used—one of alarm—is 'kiah-kiah,' repeated a number of times."

Dawson (1923) writes: "In case of invasion, the king of birds can only lurk anxiously in the offing and give vent to his anxieties by a peculiar screaking, known throughout literature as a 'scream,' *cheop' cheop', tsyewk' tsyewk'*—slowly. This is a rather pathetic and quite inadequate sound, if intimidation be intended. Indeed, on occasion, it sounds more like the meditations of a young 'broiler' than it does like a master cry."

Field marks.—The adult golden eagle is a large dark-colored bird, appearing almost black in certain lights, with no white showing anywhere. In favorable lights at short range the golden hackles on the nape may show, but one must be very near to see the feathered tarsi. In the immature bird the white base of the tail is conspicuous, and also the white spaces in the wing formed by the white bases of the secondaries and inner primaries.

When the bird is soaring it holds the broad wings horizontally and not at an upward angle, as does the turkey vulture. The wing beats of the golden eagle are quicker and more vigorous than those of the bald eagle.

Winter.—The golden eagle is practically resident all the year round throughout most of its range, though many of the more northern birds are forced to drift southward during severe winters and wander about in search of a food supply. Deep snows and periodic scarcity of game make it hard for them to get a living. Lucien M. Turner says in his notes that it is very rare in northern Labrador and Ungava at any season, but occasionally a specimen "may be seen during the moderate periods occurring in winter." A. D. Henderson (1920) shot a golden eagle in northern Alberta in January 1907 and says: "The number of Eagles in the country that winter both Golden and Bald-headed, the Golden Eagle predominating, would hardly be believed unless actually witnessed. Every little muskeg had one or two and some four or five of the great birds perched on stubs or soaring overhead, all living on the rabbits which were present in thousands. Eagles and Ravens were a great nuisance to the trappers that winter, destroying many fine skins. The following winter you could walk many miles without crossing a rabbit track and the birds and animals of prey had departed."

DISTRIBUTION

Range.—The golden eagle is a circumpolar species of the Holarctic regions, ranging from northern Alaska, Ungava, the British Isles, Lapland, northern Russia, Siberia, and Kamchatka south to Lower California, central Mexico, North Carolina, northern Africa, Asia Minor, Persia, southern Tibet (Himalaya Mountains), Korea, and Japan. This outline represents the combined ranges of several subspecies, only one of which (*A. c. canadensis*) is found in North America.

The range in North America extends **north** to Alaska (Solomon, Fort Yukon, probably rarely near Point Barrow, and possibly Demarcation Point); Mackenzie (Fort Good Hope, Franklin Bay, Horton River, and McTavish Bay); northern Saskatchewan (Stone River); Manitoba (Hell Gate Gorge); and probably rarely the Ungava Peninsula (Fort Chimo). **East** to probably rarely Ungava (Fort Chimo); probably rarely Labrador (Lake Michikamau); Quebec (Anticosti Island); possibly Prince Edward Island; Nova Scotia (Colchester Island); formerly Maine (Sandy River Mountain); formerly New Hampshire (North Conway); formerly New York (West Point); formerly southeastern Pennsylvania (Lancaster County); and western North Carolina (Weaverville). **South** to North Carolina (Weaverville); eastern Tennessee (Mount Le Conte and Chilhowee Mountain); Texas (probably Kerrville, Rocksprings, probably Comstock, probably the Chisos Mountains, and Terlinqua); Nuevo Leon (Monterrey); Durango (Cuidad); probably Chihuahua (San Luis Springs); probably Sonora (San Bernardino River); and northern Lower California (San Quentin).

West to Lower California (San Quentin, San Fernando, and La Grulla); California (San Diego, Escondido, Santa Barbara, Santa Lucia Peak, Sur, Salinas, Watsonville, Santa Cruz, Nicasio, and probably Mount St. Helena); Oregon (Swan Lake, Fort Klamath, probably near Tillamook, and probably Astoria); Washington (Mount Rainier, Cascade Mountains, and Wrights Peak); British Columbia (Ducks, Bear Creek, and Nine-mile-mountain); and Alaska (probably Bethel, St. Michael, and Solomon).

It should not be understood that the golden eagle breeds throughout the vast area above outlined. The range is, in fact, discontinuous, as this species is chiefly a bird of the mountains, particularly of the western ranges. In the eastern part of the country it is known to nest only in the mountainous sections while breeding records are almost or entirely lacking from the Great Plains. It occurs regularly in Virginia and West Virginia in the mountains, though in recent years only in migration, so far as known.

Migration.—Eagles are sometimes found long distances from their breeding areas, while others will brave the winter in high latitudes. The species is known to winter casually north almost to the northern limits of its breeding range, and it is not known south of the southern parts of this range. Nevertheless, most of the northern breeding birds do withdraw southward during the winter season. Its movements appear to be wanderings rather than true migratory flights. Such travels are apparently influenced largely by the available food supply, and to some extent eagles will follow the migrations of other birds when these are being hunted for food, in consequence of which it is a fairly regular winter visitor in many large regions of its general breeding range where it is not known to nest.

Failure of the hares or other food is probably the cause of the appearance of relatively large numbers of eagles even in recent years. The low eastern ranges of the Appalachians in eastern Pennsylvania and New Jersey are thus still frequented by the golden eagle in migration, especially in autumn. Thus during 1935 (Broun, 1936) a total of 66 individuals were observed at Hawk Mountain Sanctuary near Drehersville, Pa.

Golden eagles have been noted to arrive on their breeding grounds in the North as follows: Yukon—Forty Mile, April 5; Alaska—St. Michael (probably winters casually), March 10; Kigluaik Mountains, March 27; Coal Harbor, April 3; and Mount McKinley, April 8.

They have been observed to leave in autumn from Alaska (Mount McKinley) on September 21, and Yukon (Plateau Mountain) on September 23.

Casual records.—Occasionally in winter golden eagles are detected in the Southeastern United States. There are several records for this section of the country, among which are the following: Five specimens were taken near De Funiak Springs, Fla., one in January 1896, one each on January 17 and 31, 1908, one on November 1, 1909, and one on February 3, 1910. One was killed at Walnut Hill near Tallahassee, Fla., on January 23, 1925, two were reported as seen at close range near Fort Drum, Brevard County, Fla., on November 4, 1888, and another was reported as having been killed in the "Alaqua section" in January 1928. One was poisoned by a sheep herder near Florala, Ala., on January 17, 1908; a second specimen for this State was obtained near Florence in March 1911; a third was taken at Preston in November 1921; and a fourth was taken December 1, 1923, near Prattville. Louisiana records are open to question, but a specimen was alleged to have been killed near Jackson about 1904, while the State Museum in New Orleans contains another presumed to be of local collection.

Egg dates.—Arctic America: 5 records, May 27 to June 29.

California to Texas: 272 records, February 9 to May 18; 136 records, February 26 to March 24.

HALIAEETUS ALBICILLA (Linnaeus)

GRAY SEA EAGLE

CONTRIBUTED BY FRANCIS CHARLES ROBERT JOURDAIN

HABITS

The gray sea eagle is included in the American list on the ground that it is resident on the west coast of Greenland, breeding up to latitude 70° N., and has also occurred as a casual on the Aleutian Islands (Unalaska, October 5, 1899) and has been recorded from Cumberland Sound (American Harbor, October 1877). It has also occurred accidentally off the coast of Massachusetts (off Nantucket Light Ship, November 14, 1914). These particulars are quoted from the fourth edition of the A. O. U. check-list.

Owing to the scarcity of material available, Dr. Ernst Hartert in his work on the Palearctic fauna treated this species binomially but pointed out that the Greenland race might have to be separated and gave the synonymy on page 1178. Although it was known that exceptionally large specimens had been obtained in Greenland, the matter remained undecided until the publication in 1931 of the third (posthumous) volume of E. Lehn Schiøler's great work on the birds of Denmark. In this fine volume the results of the study of Schiøler's great collection of skins from Greenland were for the first time rendered accessible to students.

Brehm's separation of the giant Greenland race is shown to be justified, although it can hardly be regarded as proved that Greenland birds wander south even to the North German islands, as he states. It seems more probable that winter visitors to North Ger-

many are birds of Scandinavian origin.   The figures in table 1 are
taken from Schiøler's work and show that on size alone the West
Greenland sea eagle (*Haliaeetus albicilla groenlandicus* Brehm)
deserves subspecific rank:

TABLE 1.—*Average measurements (in millimeters) of 31 young and 26 adult
skins of* Haliaeetus albicilla groenlandicus *and 12 young and 4 adult of* H. a.
albicilla, *from Scandinavia and Denmark.* (*After Schiøler, 1931*)

Measurement	groenlandicus				albicilla			
	Young females	Young males	Adult females	Adult males	Young females	Young males	Adult females	Adult males
Wing	696.9	660.2	682.2	637.1	670	637	652	615
Culmen	57	55.5	61.4	55.2	54	51	56.7	55
Middle toe	101.8	103.2	105.5	101.8	96.2	95.2		93

Although individual variation in this species is considerable, there
is little doubt that when sufficient material for comparison is available,
some further subdivision will be necessary.   Of the American records the Cumberland Sound birds probably belonged to the Greenland race; the Nantucket bird was immature and probably was also
of this form, though it is within the bounds of possibility that it was
a straggler from Iceland, where *H. a. albicilla* breeds.   The Unalaska
specimen, however, must belong to the form that breeds in northeastern Asia.   According to A. H. Clark (1910) and Dr. H. C.
Oberholser (1919) this race is so much smaller than the typical form
that it deserves recognition, and the latter writer suggests that
Hume's name, *H. a. brooksi*, can be used for it, though applied to a
winter bird in North India.

Spring.—In the case of such a widely distributed species, ranging
as it does over practically the whole of the Palearctic region, the
habits must necessarily vary according to the locality.   Thus, even
in the case of the Greenland sea eagle, the birds that breed in the
north are perforce migratory, moving southward on the approach
of winter, when their hunting grounds are frozen over, while in
southern Greenland, where the warm current keeps the coast more or
less open, they are sedentary, working their way northward along the
coast in spring.   Except in the far north, the adults remain all the
year round in the neighborhood of their nests.   Immature birds that
winter on the Schleswig coast generally leave about the end of
February for the north.

Courtship.—There seems to be little doubt that this eagle pairs
for life, as the same birds may be seen in one district for years.
When one of the pair is killed, the survivor, if it be the breeding season,
obtains a fresh partner within a few days.   Dr. H. L. Saxby (1874)

relates that in the Shetlands, while a nest was being built, the female bird was shot and immediately afterward the male disappeared, but returned in the course of a week with a new mate. The latter was also killed, but after an absence of about 10 days the male again returned with another female and succeeded in rearing a brood.

Little has been recorded of the actual courtship, though fights between rival males have been witnessed early in spring. Robert Gray (1861) mentions one case where no fewer than six birds were seen soaring in a group together. Two of them, probably males, attacked each other and fought viciously, the other four soaring leisurely round them and uttering their yelping notes. The fight continued till the two birds reached the ground, when one was found to be so injured that it was unable to rise and was killed by a shepherd who had been an interested spectator. Rasmus Müller (1906) also states that in Greenland he has occasionally seen fights between two of these birds, which have been carried on in the air for some time, but ended by their coming to the ground.

Nesting.—The difference in nesting sites is extraordinary and varies according to the locality. In Greenland the nest is always on a ledge of rock not far from the water, generally in one of the numerous fiords and within reach of a salmon river. Some nests are placed in situations difficult of access, others are comparatively easy to reach. The nest is an untidy heap of sticks and branches picked up from the shore, together with grass and seaweed, as well as bones and other remains of prey. In Iceland and northern Scandinavia the sites are very similar, but some of the Norwegian nests are to be found in the very middle of a huge colony of sea birds of various species on the precipitous sides of some small island. In countries like Denmark (formerly) and North Germany, the nests are almost invariably in big trees such as pines and oaks. These nests are built entirely by the birds themselves and are frequently occupied year after year, so that in time they become very large. In the marshes of the lower Danube, where the species is still quite common, the nests are generally in trees, sometimes at great heights, but also at times in quite small trees, though difficult to reach as the nest overhangs all round and the loose materials give no handhold. There are, however, exceptions, and I have seen a nest on an almost flat sandbank, only a foot or two above the water level. In the Lake of Antioch also the nests are to be found among the reedy shallows and formerly a similar site was occupied on Lake Menzala in Lower Egypt. In Iraq the nest has been found on the cliffs bordering the great rivers.

In some cases a single pair may have two or even three alternative sites, which are occupied in turn.

Eggs.—Normally two, but sometimes three in number, while very old birds may produce only one. They are usually white, not infrequently nest stained, but genuine markings, if they occur at all, must be of very rare occurrence. Some supposed instances are certainly due to faulty identification where other large raptorial birds are breeding. There is little or no gloss; some eggs are covered with shallow grooves; others show distinct but shallow pores. In size there is much variation. Schiøler (1931) records the measurements of 29 eggs of *H. a. groenlandicus* as averaging 75.3 by 58.3 millimeters. Maxima, 82.3 by 63.8; minima, 67.4 by 55.3 and 70 by 55 millimeters. Twenty-one eggs measured by the writer together with 11 by Dr. Rey of the same race average rather larger, 77.54 by 58.09: Maxima, 84.4 by 60.7 and 76.4 by 61.7; minima, 72.4 by 56.7 and 75.6 by 53.4 millimeters.

Eggs from Scotland are also large: Average of 52 (measured by the writer), 75.73 by 58.68 millimeters. Central European eggs are smaller: 21 Danish eggs measure 74.5 by 56.9, according to Schiøler (1931), while 12 eggs from Asia Minor average only 71.8 by 56.2 millimeters (writer's measurements). Dr. Rey gives the average weight of German eggs as 13.325 grams.

The eggs are laid at intervals of a day or two, and incubation apparently begins with the first egg laid, for there is generally a decided difference in size between the young. The greater part of the incubation is done by the female, but apparently the male takes some part. Siewert (1928) flushed only the hen from the eggs and many females have been shot from the nest. Saxby (1874), however, believed that the male was sometimes on the eggs, and in confinement he has been seen to take his turn. The period is still somewhat uncertain; Faber (1826) gave it as 35 days, but Krause (1926) says 40 to 42 and Hortling (1929) about 45 days.

Young.—The newly hatched young are covered with creamy or grayish buff down at first, and as they differ in size considerably, the smaller is generally bullied and not infrequently dies. In nests on open ledges or on the ground, the smaller bird can back away more readily when attacked and stands a better chance of survival than in a tree nest. The food is chiefly brought to the nest by the male and distributed to the young by the female. The growth of the young is a long process; Heinroth has given the weights of nestlings at different periods showing the rate of increase from the early stages to 42 days old—from 86 to 4,090 grams! The stay in the nest is at least six to seven weeks, so that the whole period from the laying of the first egg to the flight of the young lasts certainly not less than three months, and naturally only one brood is reared in the season.

H. H. Slater (1901) in Iceland once watched an old sea eagle giving instruction to its young. First it was shown how to fly in large even circles above the water; then several stoops were made at imaginary fish below. The old bird checked itself by spreading its wings long before reaching the water and once appeared to stoop when not above the water at all. Finally both birds went off together.

Plumages.—This subject has been fully treated not only by Witherby (1924, vol. 2, p. 172) but also by Naumann (vol. 5, p. 162) and Schiøler (1931, vol. 3, pp. 64 and 79). Probably the wholly white tail is not assumed till the fourth year. The bare, scutellated tarsus alone serves to distinguish it at a glance from the golden eagle.

Food.—This also varies according to locality. Greenland birds subsist largely on fish, especially salmon, and Brünnich's murres (*Uria l. lomvia*). Other birds recorded include eider duck, mallard, fulmar, and ptarmigan, while young seals are also occasionally taken. In the Shetlands and Hebrides lambs and even occasionally sheep, especially when in difficulties, were attacked; rabbits and domestic poultry were also taken and carrion freely eaten, as well as several species of gulls and auks. In central Europe many coarse fish are taken, and the diet includes young roedeer, lambs, hares, foxes, hamsters, rats, moles, and mice, as well as many species of birds (crows, owls, grebes, ducks [many species], coot, moor-hen, pheasant, lapwing, oystercatcher, curlew, and poultry). In Rumania I have seen a hedgehog at a nest with young, and on the South Russian Steppes it is said to feed largely on small mammals, such as spermophiles and *Spalax.*

Behavior.—At rest the sea eagle is a heavy, lumpish bird. On the low-lying shores of the Black Sea where trees are scarce, it may be seen sitting humped up on the mud flats, with the head sunk among the shoulder feathers, looking more like a stump or accumulation of rubbish around a stake than a bird.

The decrepit-looking hermit invites the attention of the hooded crows, which slyly approach, one bird distracting attention in front while the other from behind tweaks a tail feather of the great bird and hastily flaps out of reach as a huge wing is outspread and used to aim a blow at the aggressor. Once on the wing the whole appearance of the bird is at once altered. The broad wings, with each primary standing out by itself, and the wedge-shaped tail, pure white and transparent-looking in the adult, form an impressive picture as with slow flaps the great bird rises and soars in circles overhead. From time to time it may utter its cry, *gak-gak-gak-gak,* four times repeated with outstretched neck and widely opened beak, but for such a large bird, the notes are not striking, and there is another distress call, a high-pitched querulous chatter.

Enemies.—Although at times attacked and driven away from their breeding places by the larger falcons, the sea eagle stands in no real danger from them. In Rumania I have seen a Saker falcon (*Falco cherrug*) dash off its nest and strike a sea eagle full on the back, sending feathers flying, and in Scotland the peregrine (*F. peregrinus*) has been seen to stoop at it and drive it off. A fight has been recorded from Greenland with the snowy owl. In the case of a nest on the ground in the south a prowling fox might take small young, and on one occasion we found young dead in the nest, apparently killed by ants, but as a rule the only enemy of the sea eagle is man, whose chief weapons are the gun and poison. On one estate in West Ross alone, a single keeper shot 52 in 12 years, besides taking many eggs and young.

Fall.—The young soon disappear from the neighborhood of the nest, presumably driven away, and wander erratically southward. Evidently the mortality must be heavy, and practically all the specimens secured on migration are immature, the adults remaining (except in the high northern regions) in the neighborhood of their breeding place. Favorite haunts of these young birds are shallow lakes and morasses in southeastern Europe, where there are great assemblages of wildfowl, such as various species of wild ducks and coots.

The coots have their own special method of defense, by assembling in close order on the surface and throwing up with their powerful feet jets of water, which keep their assailants at a respectful distance.

DISTRIBUTION

Breeding range.—The gray sea eagle breeds on the west coast of Greenland from Cape Farewell to Disco Bay; on the east coast it is a rare straggler. It has bred at least once at Cumberland Sound on the west side of Davis Strait (Kumlien, 1879). Greenland birds have been described as a large race (Schiøler, 1931). *Haliaeetus a. albicilla* breeds in Iceland; formerly in western Ireland, Scotland, and the Faeroes; Norway; Sweden (about 20 pairs); Finland; Russia; Novaya Zemlya; Germany (chiefly Mecklenburg, Brandenburg, Pommern, and East Prussia); formerly in Denmark; Poland; the Baltic Republics; Corsica; Sardinia; Hungary; Yugoslavia; Albania; Rumania; Bulgaria; Greece; Asia Minor; formerly in Egypt; Syria; Iraq; North Persia; Turkestan; and in Siberia east to Mongolia. It is perhaps represented by a smaller race in northeastern Asia (Manchuria, Anadyr, and Kamchatka), on the Commander Islands, and in Japan.

Fall migration.—Juveniles from northern Europe drift southward over Helgoland and the Baltic, occasionally reaching the east coast of Great Britain. Young reared on the southern shores of the Baltic range south to the Mediterranean through Europe to the Canaries and North Africa (Algeria, Tunisia, and Egypt), and in Asia south to Persia, Baluchistan, and the northwestern provinces of India (Punjab, Northwest Provinces, and Sind). Also noted on passage in Foochow, South China, in September.

Casual records.—East Greenland, Angmagsalik, July 22, 1913; Massachusetts, off Nantucket Light Ship, November 14, 1914; Spain, Coto Doñana, December 28, 1898; and Aleutian Islands, Unalaska, October 5, 1899.

Egg dates.—Greenland: 14 records, March 31 to June 28; 3 records, April 15 to 29; and 9 records, May 1 to 26.

Iceland: Usually in April or early in May.

Northern Norway: From May 1 onward.

Scotland: March 15 to April 28; mostly between April 10 and 22.

Asia Minor: 9 records, January 20 to February 16.

Iraq: 3 records, January 28 to February 3.

HALIAEETUS LEUCOCEPHALUS LEUCOCEPHALUS (Linnaeus)

SOUTHERN BALD EAGLE

HABITS

For reasons that will be more fully explained under the northern race, I think the above name should be restricted to the bald eagles of the Southern States. In my opinion the breeding range of the southern bald eagle should not be considered to extend very far north of South Carolina, the Gulf States, and perhaps southern California.

On June 20, 1782, our forefathers adopted as our national emblem the bald eagle, or the "American eagle" as it was called, a fine *looking* bird, but one hardly worthy of the distinction. Its carrion-feeding habits, its timid and cowardly behavior, and its predatory attacks on the smaller and weaker osprey hardly inspire respect and certainly do not exemplify the best in American character. The golden eagle is a far nobler bird, but it is not strictly American. The wild turkey was suggested, but such a vain and pompous fowl would have been a worse choice. Eagles have always been looked upon as emblems of power and valor, so our national bird may still be admired by those who are not familiar with its habits. Its soaring flight, with its pure-white head and tail glistening in the sunlight, is really inspiring; and it adds grandeur to the scene as it sits in a dignified pose on some dead tree, its white head clearly visible against the dark green of the forest background.

Courtship.—I find practically nothing in print on this subject, but C. J. Pennock says in his notes: "During late September and through October may be said to be their mating season in Florida.

At this period they are to be seen flying over the marshes and open water, two or three in rapid chasing flights." They are probably mated for life, but if one of a pair is killed the survivor promptly secures a new mate, and occasionally the new mate is a bird in immature plumage. Almost always both birds of a breeding pair are white-headed adults.

I have seen an immature bird mated with an adult, and several other observers have reported it, but all seem to agree that it seldom occurs. I have never heard of a mated pair in which both birds were immature. Donald J. Nicholson, who has examined 125 eagles' nests, tells me that only once has he found an immature eagle mated with an adult.

Nesting.—My experience with the nesting habits of the bald eagle has been mainly in Florida, where this great bird is widely distributed, very common for a large bird, and so seldom disturbed by man that it nests, with confidence in its safety, often close to human habitations. I saw two occupied nests on golf courses, where players were passing daily almost under the nesting trees. And several nests were within sight of or even close to houses, or in open parks near much-traveled roads. During the winter of 1924–25, with the help of Oscar E. Baynard, we visited 18 eagles' nests in Pinellas County. These were located mainly near the shores of various bays or inlets, and all were in large longleaf pines, though two of the nesting trees were dead. The nests were placed 35 to 63 feet above the ground, about half of them being between 50 and 60 feet up. A typical nest was found on an island near Pass-a-grille on November 18, and the eagle flushed from the nest, but I did not climb to it until November 27, when it contained two eggs about one-quarter incubated. It was 40 feet up in a large pine in an open grove of longleaf pines; it rested on several branches and was made of large sticks and rubbish, with a lot of green and dry pine needles and Spanish moss in the flat top; in the center was a pretty little hollow, 20 inches in diameter and 4 or 5 inches deep, lined with the soft gray moss and small pine needles, in which the eggs were partially buried. It was a large nest, 7 feet high and 7½ by 5½ feet across the flat top. There was considerable white down scattered over the top of the nest. This pair of eagles laid a second set of eggs in the same nest later in the winter; I climbed to it on February 14 and found two eggs in it; I left them to hatch, as I wanted to photograph the young, but the eagles deserted the nest and the eggs never hatched (pl. 86).

I have seen three eagles' nests on the Florida Keys, the only nests I have ever seen in Florida that were not in pine trees. These were on the larger keys, where there was a heavy growth of large black mangroves, and the nests were in the main crotches of these trees at heights ranging from 30 to 40 feet; they were the usual large nests, 5 or 6 feet in height and about the same in diameter; one that I examined was lined with straw and grasses.

An interesting nest that I climbed to on November 26, 1911, near Mount Pleasant, S. C., was 45 feet up among the main branches of a longleaf pine; it was made of large pine sticks, cornstalks, sedges, and grasses and was deeply lined in the center, up to the level top, with soft grasses, Spanish moss, and feathers. No eggs were visible, but I found them deeply buried under fully 2 inches of the soft lining, completely concealed; the eggs had evidently been covered by the eagle when she left the nest.

J. R. Pemberton showed me a picturesque nest on Catalina Island, Calif., on February 22, 1929. The north end of the island terminates in a long, narrow cape, with steep, sloping sides leading up to a knife-edged, rocky ridge, 400 to 500 feet above the sea. On the top of a pinnacle of rock on the crest of this ridge was the eagle's nest. It was a laborious, but not a dangerous, climb to reach it, but it was well worth while. It was a shallow nest on the flat summit of the rock, about 6 feet long by 3 feet wide; it was made of dead sticks from the bushes that grew on the lower slopes and was profusely lined with grasses and decorated with a little white down (pl. 90). We found another old nest on San Nicholas Island, a great pile of sticks, 8 to 10 feet high, on a little shelf on an overhanging cliff.

These eagles are still fairly common on some of the other Santa Barbara Islands, nesting on rocky cliffs. W. Leon Dawson (1923) writes:

The nest, which is an immense pile of sticks, lined with fine twigs and grass, and other soft substances, is usually placed on some lesser promontory or a sharp, inaccessible ridge near the ocean. The historic pile figured on page 1713 measured twelve feet by six on top, the larger diameter being along the crest of the ridge; and contained no less than two wagonloads of accumulated materials. Another, from which the M. C. O. took two heavily incubated eggs on the 20th of March, 1919, was built up on a slanting ridge, so that the lower or seaward face was fourteen feet in depth, although the top of the nest was only four feet by six.

There are probably more bald eagles nesting in Florida than in any other State in the United States, and they are quite thickly concentrated in certain favorable localities. Donald J. Nicholson, who has had many years of experience with them, has sent me some voluminous notes on these birds. Pinellas County on the west coast and Brevard County on the east coast seem to be the centers of abundance. Mr. Nicholson mentions an area 3½ miles long and three-fourths of a mile wide, in which were seven occupied nests, three of them within a 1-mile circle, in Brevard County. The nesting season, he says, is quite prolonged, beginning sometimes in October, but usually not until November or later, and lasting all through winter and spring, even into June. There are two good reasons to account for such early nesting: First, it is desirable that the eaglets, which grow very slowly, have time to develop their protecting plumage before hot weather comes on early in spring; the hot sun might prove disastrous for the tender downy young, unless they were constantly brooded by their parents. Second, it is easier for the eagles to secure the large amount of food required by the eaglets during winter, when coots and other water fowl are abundant.

Mr. Nicholson mentions only three kinds of trees used in Florida, pines, cypresses, and mangroves, with a decided preference for pines. He says the height from the ground varies from 20 to over 100 feet but is usually between 45 and 70 feet. Oscar E. Baynard, who has climbed to between 250 and 300 nests, has found them as high as 140 feet.

Walter J. Hoxie (1888) watched a pair of eagles building a new nest, using some of the material from an old nest. The female did most of the building, and the male helped by bringing material. He says:

Having at last a foundation of about a foot thick, and four or five feet wide, as near as I could estimate, they proceeded to remove the material from the old partially repaired nest for the completion of the new one. The male bird worked fairly well at this task, and during the last day made at least three trips to one of the female. She apparently took great pains in the interior arrangements of her new home, frequently pulling out a quantity of trash upon the edge of the nest, and, after working around a while inside, tumbling it back again, shaking it up with a great rustling of wings and scratching of feet, which sent showers of little twigs and dirt upon the watcher below.

It is well known that, in Florida, great horned owls habitually use unoccupied eagles' nests, but a record of both species using the same nest simultaneously is unique. J. Warren Jacobs (1908) describes the finding of a huge nest in Florida that measured 15 feet in height and 8 feet in thickness. An eagle was incubating a set of eggs on the top of the great pile, and an owl flew "from a rude cavity in the side of the eagle's nest, in which she had formed a nest and deposited two eggs" 4 feet from the bottom of the pile. Mr. Nicholson once found an eagle incubating a great horned owl's egg.

In other parts of its range the bald eagle has been known to choose a variety of nesting sites. In the Middle Atlantic States nests have been found in oaks, chestnuts, pines, gums, and other trees. Bendire (1892) quotes Capt. B. F. Goss on two nests that he found on the ground on islands in Neuces Bay, Texas. Of one he says: "It consisted simply of a few sticks laid on the bare ground, not enough to make a single tier even, and these were covered with bones, feathers,

and fish scales, and the ground in the immediate vicinity was littered
with the remnants of their food and the excrement of the young."
The other was a massive structure at least 6 feet high and 5 feet in
diameter; he saw it fully 2 miles away.

Robert Ridgway (1877) had a nest shown to him in a very unusual
situation on an island in Pyramid Lake, Nev.: "This nest was placed
inside an oven-like cave about half-way up the side of the perpendicular rocks which formed this portion of the shore. The entrance
was about fifteen feet from the top of the rock, and the same distance
from the water, so it was inaccessible by any means then at command;
but it could be plainly seen by looking through a crevice in the top of
the rock. This nest was a huge bed of coarse sticks laid on the floor
of the cave, and scattered about were the bones of numerous animals
which were carried as food to the young."

I saw a nest in Texas about 50 feet up in a big live oak. Other
nests have been found there in pecans and in mesquites 10 or 15 feet
high (Lloyd, 1887). On Santa Margarita Island, Lower California,
Walter E. Bryant (1890) found a nest in a giant cactus.

Mr. Baynard told me that some pairs of eagles do not breed every
year; they may repair the nest and remain in the vicinity all through
the season without laying any eggs. This was true of one pair that
he and I watched. If the first set of eggs is taken, the eagle often
will lay a second set after an interval of four weeks or more. Mr.
Baynard says this happens in about half the cases, according to his
experience. In the one such case that I noted the interval was about
two months, and the second set was laid in the same nest. But often
another nest is used.

Eggs.—Two eggs almost invariably make up a full set for the bald
eagle, sometimes only one, and rarely three; in two or three cases four
eggs have been found in a nest, but these may have been the product
of two females. The eggs vary in shape from rounded-ovate to ovate,
the former predominating. The shell is rough or coarsely granulated.
The color is dull white or pale bluish white and unmarked, though
often nest stained. Very rarely an egg shows a few slight traces of
pale brown or buff markings. The measurements of 50 eggs from
Florida average 70.5 by 54.2 millimeters; the eggs showing the four
extremes measure 78.8 by 56.2, 71.1 by 57.6, and 58.1 by 47 millimeters.
The eggs are ridiculously small for so large a bird. (Compare the
relative sizes of the eggs of the ruddy duck, the sandpipers, or the
hummingbirds.) Consequently the little eaglet requires a long time
to develop.

Young.—The period of incubation is about 35 days, according to the
most careful observers, though it has been otherwise estimated. Both
parents assist in incubation and in the care of the young. Mr.

Nicholson tells me that at every nest he has visited after dark he has
found both birds at the nest, one incubating or brooding and one
perched near it. In one instance the incubating bird remained on the
nest until the climber nearly reached it. Usually an eagle will leave
its nest as soon as an intruder is seen approaching it, but occasionally
one will sit closely until the tree is rapped. The food of the young
seems to be about the same as that of the adult, to be referred to later.
The behavior and development of the young will be discussed under
the northern race, on which more information is available.

Plumages.—When first hatched the downy young eaglet is completely covered with long, thick, silky down, longest on the head; it
is "smoke gray" on the back, paler gray on the head and under parts,
and nearly white on the throat. When the young bird is about three
weeks old this light gray or whitish down is pushed out and replaced
by short, woolly, thick down of a dark, sooty-gray color, "hair brown"
to "drab." The plumage begins to appear on the body and wings,
scattered brownish-black feathers showing on the scapulars, back,
and sides of the breast, when about five or six weeks old; at this age
the wing quills are breaking their sheaths. At the age of seven or
eight weeks the eaglet is fairly well feathered, with only a little down
showing between the feather tracks, and the flight feathers are fully
half grown.

In fresh juvenal plumage the young eagle is uniformly dark colored "bone brown" to "clove brown" above and below; the flight
feathers are nearly black, but there is usually a slight sprinkling of
grayish white in the tail. This plumage is worn throughout the first
year without much change, except by wear and fading, the under
parts fading to "hair brown." After the first annual molt, the next
summer, the plumage becomes paler and much mixed with white in
very variable amounts. Individual feathers on the back, scapulars,
and breast are more or less extensively white, those of the breast and
belly being largely white in some specimens. I am not sure whether
this is a second or third year plumage, or both; if the latter, the third
year is whiter than the second. The tail is more extensively mottled
with white than in the first year, and the feathers of the crown and
occiput are broadly tipped with pale buff. After the next annual
molt the plumage of the body becomes darker, much like that of the
adult, but lightly tipped with white below and mottled with white
on the rump and upper tail coverts; the latter and the tail are now
quite extensively white; the head is mixed with white above, about
half white and half brown, and nearly clear, dirty white below. This
is probably the third year plumage. At the next annual molt, early
in the fourth year, the bird assumes a plumage that is practically
adult, with a pure-white head and tail; but usually remaining signs

of immaturity are seen, such as a few brown feathers in the head and some dusky mottling near the tip of the tail. The length of time required to assume the fully adult plumage does not seem to have been positively determined, and it may take longer than I have estimated. Adults and immature birds have one complete annual molt, which is very gradual, and prolonged through spring, summer, and fall. The flight feathers are molted mainly during July, August, and September.

Food.—The large amount of food found in the nests of bald eagles containing young indicates that the eaglets, even when small, are fed on much the same food that the adults eat, or that the adults devour much of the food that is brought to the nest, or perhaps both. Mr. Pennock (MS.) found in a nest with two very young eaglets, "certainly not over a few days old", an entire black duck, a headless black duck, and a headless mullet that had weighed 1½ to 2 pounds. In another nest he found a partly eaten lesser scaup duck, an entire horned grebe, and three other grebes more or less mutilated. Mr. Nicholson says (MS.) that the amount of food found in the nests is astonishing, and often much of it has not been touched. He lists rabbits, mostly marsh rabbits, other undetermined mammals, turtles, coots, Florida ducks, lesser scaup ducks, pied-billed grebes, little blue herons, snowy egrets, terns, killdeers, catfish (by far the most frequent species found and some up to 15 pounds in weight), black bass, sergeantfish, crevallé, pompano, and other fish. Under one nest he found between 40 and 60 skulls of mammals, about the size of rabbits. He has never found snakes in an eagle's nest, nor has he ever seen wool or bones of lambs, even in the heart of the sheep country. There is no doubt, however, that bald eagles do occasionally carry off lambs, as several good observers have seen them do it, and the bones have been found in and under their nests. Probably many of these were picked up dead, but sheep herders generally regard eagles as destructive enemies.

C. J. Maynard (1896) witnessed an attack by a bald eagle on a brood of young pigs; the old sow was defending them vigorously, but the eagle might have succeeded in securing one, if Mr. Maynard had not interfered. Dr. H. C. Oberholser (1906) gives many interesting details regarding the food of American eagles and says:

At favorable opportunities this eagle preys upon fawns, and pressed by hunger will sometimes attack a full-grown deer, particularly if the latter be wounded. Remains of a mule deer (*Odocoileus canus*) were found by Dr. E. A. Mearns in the stomach of one from the Mogollon Mountains, Arizona. Mr. E. W. Nelson is authority for the statement that in northern Alaska it feeds at times on young reindeer (*Rangifer arcticus*). Even the wily fox sometimes meets its fate at the talons of this powerful bird, as is shown by Mr. Vernon Bailey's report that at Provo, Utah, a farmer found a gray fox (*Urocyon scotti*), evidently just killed, which a pair of eagles was busy eating. Opossums (*Didelphis*) and raccoons

(*Procyon lotor*) are sometimes captured, but the nocturnal habits of these animals probably account for their not being more frequently obtained. Mr. Thomas McIlwraith mentions that an eagle shot on Hamilton Bay, Ontario, had the bleached skull of a weasel hanging firmly fastened by the teeth into the skin of its throat, a grewsome relic of a former desperate struggle.

Among the smaller mammals rabbits are often eaten, occasionally prairie dogs are taken, and, where they are plentiful, tree squirrels and ground squirrels, or spermophiles, form a large part of the food of the bald eagle. Domestic dogs have been attacked and such small fry as rats and mice are sometimes taken to the nest. Eagles, like most hawks and owls, cast up in the form of large pellets the indigestible portions of their food, such as bones, fur, and feathers.

In certain places, particularly in winter, bald eagles live largely on waterfowl, mainly geese, brant, ducks, and coots. This eagle is perfectly capable of catching a duck on the wing and frequently does so; but oftener the duck is pounced upon in the water or forced to dive again and again until it becomes exhausted and is easily captured; frequently two eagles join in the chase, which gives the poor victim a slim chance to escape. I have seen two eagles chasing a black duck in the air until it was forced down into the water. Ducks killed by sportsmen are often picked up by eagles. In Florida coots (*Fulica*) are very abundant in winter and furnish a favorite food supply for the eagles. Dr. W. L. Ralph (Bendire, 1892) says that many are caught on the wing; he found the remains of 13 in one nest. The interesting account, in a letter from John W. Baker to Charles F. Batchelder (1881), well illustrates the eagle's method of attack and the coot's attempt at escape. The eagle came daily and alighted in the top of a tree near the river where large numbers of coots were feeding.

At the first sight of the Eagle the Coots all huddled together, remaining so during his rest, swimming about aimlessly and casting uneasy glances up in the direction of their enemy. The moment the Eagle lifted himself from his perch, the Coots seemed to press towards a common centre until they were packed so closely together that they had the appearance of a large black mantle upon the water; they remained in this position until the Eagle made his first swoop, when they arose as one bird, making a great noise with their wings, and disturbance with their feet which continued to touch the water for the first fifty or one hundred feet of their flight. This seemed to disconcert the Eagle who would rise in the air only to renew his attack with great vigor.

These maneuvers were kept up, the Eagle repeating his attack with marvelous rapidity, until, in the excitement and hurry of flight, three or four Coots got separated from the main body; this circumstance the Eagle was quick to discover and take advantage of; it was now easy work to single out his victim, but usually long and hard to finally secure it. I have never seen him leave the field of battle, however, without a trophy of his prowess, though I have seen him so baffled in his first attempt to separate the birds, that he was compelled to seek his tree again to rest.

On one occasion, after separating his bird from the flock, he spent some minutes in its capture—the Coot eluding him by diving; this frequent rebuff seemed to provoke the Eagle to such an extent that he finally followed it under the water—remaining some seconds—so long, indeed, that I thought him drowned; he finally appeared, however, with the bird in his talons, but so weak and exhausted that he could scarcely raise himself above the water, and for the first thirty or forty yards of his flight his wings broke the surface of the water; very slowly he made his way to the nearest tree, where he alighted, on the lowest limb, to recover his spent strength.

William Brewster (1880) says that on the Virginia coast—

Geese and Brant form their favorite food, and the address displayed in their capture is very remarkable. The poor victim has apparently not the slightest chance for escape. The Eagle's flight, ordinarily slow and somewhat heavy, becomes, in the excitement of pursuit, exceedingly swift and graceful, and the fugitive is quickly overtaken. When close upon its quarry the Eagle suddenly sweeps beneath it, and, turning back downward, thrusts its powerful talons into its breast. A Brant or Duck is carried off bodily to the nearest marsh or sand-bar, but a Canada Goose is too heavy to be thus easily disposed of. The two great birds fall together to the water beneath, where the Eagle literally tows his prize along the surface until the shore is reached. In this way one has been known to drag a large Goose for nearly half a mile.

W. W. Worthington wrote to Major Bendire (1892) as follows:

The other day I noticed a Bald Eagle hovering over the sound, much the same as the Fish Hawk does when about to strike a fish. Suddenly he plunged down and grappled with what I supposed to be a large fish, but was unable to raise it from the water, and after struggling awhile he lay with wings extended and apparently exhausted. After resting a minute or two he again raised himself out of the water and I saw he had some large black object in the grasp of one of his talons, which he succeeded in towing along the top of the water toward the shore a short distance, and then letting go his hold. He was then joined by two other Eagles and by taking turn they soon succeeded in getting it to the shore. Investigation proved it to be a large Florida Cormorant, on which they were about to regale themselves.

During most of the year fish of various kinds furnish the eagle's main food supply. Many are picked up dead on the beaches or along the shores of lakes and streams, as these eagles are good scavengers. The osprey is systematically robbed, as nearly every observer or writer has noted. The eagle, from some favorable perch, watches for the return to its nest of this industrious fisherman, heavily laden with its prey. As the eagle starts in pursuit, the osprey mounts into the air in an endeavor to escape, but the eagle is too swift and too powerful for him, and the weaker bird is eventually forced to drop his prize, which his pursuer often dives down and catches before it falls to the ground. Sometimes the struggle is quite prolonged, but rarely does the osprey escape. Sometimes the eagle fails to catch the falling fish and it may be lost to both birds. Occasionally two eagles join in the chase, when the osprey soon gives up. Mr. Nicholson says in his notes: "I heard the angry cries of an osprey and, looking up, saw a bald eagle chasing the bird. The eagle flew over it making several quick dives, which were easily dodged by the osprey. But before we realized it, the eagle made one quick dive, turning upside down with talons outstretched, and took the fish from the grasp of the osprey. The eagle sailed away with the spoils, as if nothing had occurred. The osprey turned silently, with no pretense of fight, and flew down the river."

But where there are no ospreys to rob the eagle has to do its own fishing. Dr. Oberholser (1906) writes:

Sometimes from its perch on the summit of a dead tree it launches downward and, falling like a stone, seizes its prey; sometimes it hunts on the wing, much like an osprey, and when a fish is perceived poises by rapid wing-beats, finally dropping into the water even from a great height, and not infrequently becoming almost completely submerged; then, again, it varies this last method by flying leisurely along near the surface of the water. Audubon mentions that along Perkiomen Creek near Philadelphia, Pa., he saw it on several occasions wading in the shallows and striking at the small fish with its bill; and other observers elsewhere have noted a similar habit. It has been seen scrambling over the ice of a pond, trying to reach the fish below; and Mr. W. L. Dawson, in his "Birds of Ohio", says that at the Licking Reservoir, Ohio, it is reported in winter to watch near the air holes in the ice for the fish that come from time to time to seek the surface. Mr. J. G. Cooper has seen it catch a flying fish in the air, and the amazing celerity necessary for the performance of such an exploit may readily be imagined.

Again he writes:

The bald eagle does not disdain carrion, and in some parts of the arid West it lives at times to a considerable extent on the cattle and smaller domestic animals that fall victims to drought or other catastrophe. * * * Wilson tells that on one occasion when many thousands of tree squirrels were drowned in attempting to cross the Ohio River not far from Wheeling, W. Va., and a great number drifted to the shore, a bald eagle for several successive days regaled itself on them. Carrion was found in the stomachs of two eagles examined by Dr. A. K. Fisher. Mr. Horace A. Kline has seen this bird along the Wakulla River in Florida feeding on the carcass of an ox, again that of a sheep. * * * Sometimes it drives away the gathered vultures or dogs from their repast and keeps them at a respectful distance until its hunger is satisfied. Furthermore it does not hesitate even to pursue the vultures and compel them to disgorge, when if it fail to catch the coveted morsels before they reach the ground it alights and devours them. Audubon relates that on one occasion he saw it kill a vulture that for some reason was unable completely to disgorge.

Stories of eagles carrying off babies or small children are probably greatly exaggerated or imaginary, but Wilson (1832) relates the following: "A woman, who happened to be weeding in the garden, had set her child down near, to amuse itself while she was at work; when a sudden and extraordinary rushing sound, and a scream from her child, alarmed her, and starting up, she beheld the infant thrown down, and dragged some feet, and a large Bald Eagle bearing off a fragment of its frock, which being the only part seized, and giving way, providentially saved the life of the infant."

Apparently eagles do not attack the larger and more formidable birds, such as Ward's herons, American egrets, or sandhill cranes. Mr. Nicholson tells me that he has never found the bones or feathers of these birds in the eagles' nests and that on three occasions he has found the cranes nesting within plain sight of occupied eagles' nests and within 100 or 200 yards.

Behavior.—The flight of the bald eagle is powerful and impressive, but not so graceful or inspiring as that of the golden eagle. Its ordinary traveling flight appears heavy and labored, as it moves steadily along with slow beats of its great wings, but it is really much swifter than it seems, as is often the case with large birds. But in pursuit of its prey it develops marvelous speed, which the swiftest wildfowl can seldom escape. It often sails along on a level course on widespread wings for a considerable distance; again it soars in great circles to an immense height, from which it sometimes makes a thrilling dive at terrific speed on half-closed wings.

About its nest the bald eagle is an arrant coward, leaving the nest as the intruder approaches, flying about at a safe distance and squealing, or perching on a distant tree to watch proceedings. I have never had one even come within gunshot range when I was near the nest. Mr. Nicholson, in all his experience, has never had an eagle even threaten to attack him, except on two occasions, both by the same pair. In one case he was attacked by both birds, swooping alternately within 6 or 8 feet of him. Bendire (1892) mentions three cases where the eagles have attacked men attempting to rob the nests, but in no case was the man actually struck. The fierceness of eagles has been greatly exaggerated. They are really mild-tempered birds and often make gentle and devoted pets, when raised in captivity. They are easily raised, if not taken from the nest when too young; but they require an astonishing amount of food.

Voice.—The voice of the bald eagle seems to me to be ridiculously weak and insignificant, more of a squeal than a scream, quite unbecoming a bird of its size and strength. Dr. Ralph (Bendire, 1892) says: "The cry of the male is a loud and clear 'cac-cac-cac,' quite different from that of the female, so much so that I could always recognize the sex of the bird by it; the call of the latter is more harsh and often broken." Ralph Hoffmann (1927) says: "The cry of the Eagle, heard oftenest near its nest, is a high-pitched very metallic *kweek kuk kuk, kweek-a-kuk-kuk* with the quality of an unoiled castor."

Field marks.—An adult bald eagle is unmistakable, with its pure-white head and tail and its dark brown body; the head is conspicuous at a great distance, when the bird is perched on a tree, especially against a dark background. The juvenal first-year bird is uniformly dark colored and is easily confused with the golden eagle; but it lacks the golden hackles on the neck and head, and the young golden eagle has more white in the basal half of the tail than the first year bald eagle. Older bald eagles show more or less white on the breast and belly, which the golden eagle never shows. Both species show more or less white in the immature tail, but the bald does not have such a distinct dark band as the golden. I have noticed also that the head and neck of the bald eagle are stretched out much longer in flight than in any of the other hawks or eagles, except the caracara.

DISTRIBUTION

Range.—North America and northeastern Siberia; casual in Bermuda, accidental in Sweden.

Breeding range.—The bald eagle breeds **north** to northeastern Siberia (Bering Island); Alaska (Noatak River); Mackenzie (junction of the Peel and Mackenzie Rivers, Fort Anderson, McTavish Bay, and Artillery Lake); Manitoba (Fort Churchill); and probably Ungava (Ungava Bay). **East** to probably Ungava (Ungava Bay); southeastern Quebec (Wolf Bay and Anticosti Island); Newfoundland (Placentia Bay); Nova Scotia (probably Baddeck, Pictou, Grand Lake, Tangier, and Halifax); Maine (Deer Isle and Bath); Connecticut (formerly); New Jersey (Redbank, Sea Isle City, and Cape May); Virginia (Kilmarnock, Cobbs Island, Newport News, and probably Dismal Swamp); North Carolina (Cape Hatteras and Fort Macon); South Carolina (Waverly Mills, Cedar Island, Mount Pleasant, and Frogmore); Georgia (Savannah, Blackbeard Island, Darien, and St. Marys); and Florida (Allendale, Lake Monroe, Titusville, Merritts Island, probably Micco, probably Lake Worth, Miami, Cape Sable, and Key West). **South** to Florida (Key West, Tampa, Tarpon Springs, Whitfield, and Pensacola); Alabama (Perdido Bay); Mississippi (probably Biloxi); Louisiana (New Orleans, Avery Island, Mermenton, and Black Bayou); Texas (Belleville, Corpus Christi, and San Angelo); Arizona (Salt River Bird Reservation and Fort Whipple); and Lower California (San Francisco Island, Espiritu Santos Island, and Santa Margarita Island).[1] **West** to Lower California (Santa Margarita Island, Todos Santos Island, Laguna Hanson, and Guadalupe Canyon); California (San Clemente Island, Santa Catalina Island, Santa Rosa Island, Santa Barbara Island, Tulare Lake, Santa Clara County, Sacramento, Eagle Lake, and Tule Lake); Oregon (Fort Klamath, Bandon, Elkton, and probably near Astoria); Washington (Olympic Mountains, Quillayute Needles, and Neah Bay);

---

[1] The bald eagle also has been reported south to "central Mexico", and this statement has been repeated by many authors. A careful search of the literature and the files of the Biological Survey has failed to disclose any authentic records for this part of the continent. Nevertheless, it is to be expected that the species may occasionally breed on the coast of Tamaulipas and possibly rarely at interior points.—F. C. L.

British Columbia (probably near Comox, and Port Simpson); Alaska (Forrester Island, Craig, Sitka, Hawkins Island, Bethel, Unalaska Island, and Tanaga Island); and northeastern Siberia (Bering Island). In addition to breeding on Bering Island, the bald eagle is known to occur with fair regularity on the Arctic coast of Siberia (Nizhni-Kolymsk); the Commander Islands, and Kamchatka. It is reported to nest on the Kamchatka Peninsula (Kariaga), but this has not yet been verified.

The range above outlined is for the entire species, which has, however, been separated into two rather poorly defined subspecies. These intergrade extensively along the line of contact. The southern bald eagle (*H. l. leucocephalus*) is apparently confined to the Lower Austral Zone in South Carolina, Florida, the Gulf States, and Texas. The northern bald eagle (*H. l. alascanus* [=*washingtoniensis*]) occupies the rest of the range north to the Arctic regions.

Winter range.—The bald eagle is generally a resident species but probably retires southward in winter from the extreme northern parts of its range. It is known to winter north to Alaska (Craig, Captains Harbor, and Sitka); central Alberta (Mundare and Stony Plain); central Saskatchewan (Johnston Lake and East End); northern Minnesota (Elk River); Wisconsin (New London); Michigan (Sault Ste. Marie, Benzonia, Hillsdale, and Detroit); southern Ontario (Listowel and Toronto); and Quebec (Lac Tremblante and Godbout).

Casual records.—According to Reid (1884) the bald eagle has been recorded four times on Bermuda. No additional specimens have been noted since this report. One killed in Sweden about 1850 appears to be the only authentic record for Europe. This specimen apparently was still extant about 1880.

Egg dates.—Alaska and Arctic America: 62 records, March 24 to June 24; 31 records, May 7 to 14.

Maine to Michigan: 6 records, April 1 to 21.

New Jersey to Virginia: 75 records, February 2 to May 27; 38 records, February 27 to March 9.

Georgia and Florida to Texas: 62 records, October 30 to February 26; 31 records, December 8 to January 27.

Oregon to Mexico: 40 records, February 18 to April 1; 20 records, March 2 to 11.

HALIAEETUS LEUCOCEPHALUS ALASCANUS Townsend

NORTHERN BALD EAGLE

HABITS

I fully agree with Peters (1931) that Audubon's name, *washingtoniensis*, should be applied to this large, northern race, as it long antedates Townsend's (1897) *alascanus* and as Audubon's type, taken in Kentucky, was evidently a very large bird, apparently larger than any specimen we have from Alaska. Peters also has the distribution of the two races more nearly correct than that in the 1931 A. O. U. check-list. A glance at table 2, giving the wing measurements of 27 males and 26 females from various parts of North America, will show that there is a gradual decrease in size from Alaska to Florida. As it seems logical that the two races should be separated at a point midway between the two extremes, the line should be drawn somewhat south of North Carolina and the name *H. leucocephalus leucocephalus* should be restricted to birds of the Lower Austral Zone. Thus its range would correspond approximately with those of several other southern races. I have not seen enough material from southern California to form a definite opinion, but what little I have seen seems to indicate that the breeding birds of even Lower California are referable to the northern race (see egg measurements).

The largest two birds from Alaska measure exactly the same as the largest two from Massachusetts, all four immature birds. It is hardly likely that these Massachusetts birds came from Alaska, as bald eagles usually do not migrate far from their breeding grounds; they were probably reared in Maine, where bald eagles breed commonly, as this species is very rare as a breeding bird north of New England.

TABLE 2.—*Average wing measurements (in inches) of 27 males and 26 females of* Haliaeetus leucocephalus

Locality	Males		Females	
	Number	Wing average	Number	Wing average
Alaska	10	24. 07	6	25. 54
New England and New York	6	23. 33	10	25. 40
North Carolina	2	23. 50	5	24. 60
Georgia and Florida	9	20. 83	5	22. 65

When I visited Alaska in 1911, bald eagles were very common and conspicuous all along the coasts of southern Alaska and on some of the Aleutian Islands. While navigating the beautiful inside passages, from British Columbia northward, we noted that these fine birds were prominent features in the landscape; where the mountainous shores were heavily forested almost down to the water's edge, their snow-white heads were conspicuous at a long distance in sharp contrast against the dark-green background; and some of them were almost constantly in sight. About Unalaska they were especially abundant and not at all shy, frequently flying within easy gunshot range. They were especially bold about their nesting places, or near their favorite lookout points on the hilltops, where feathers and droppings indicated that they habitually used the same spot for a perch; at one such spot, on the crest of a steep, rocky hill, I surprised a large eagle that sailed back and forth several times, within a few feet of my head, squealing vigorously all the time, as if I were intruding on its home. They must have been very abundant on Atka Island at one time, for Austin H. Clark (1910) says that an Indian shot 175 there one winter "to prevent their making depredations on the young of a colony of blue foxes." But the situation has changed materially since the Alaska bounty law went into effect in 1917. During the first 10 years under this law it was reported that bounties were paid on 41,812 eagles. Since that time it has been estimated that the number has increased to over 50,000 and perhaps 70,000. At this rate of destruction the day may not be far distant when this splendid bird will be less often seen as a picturesque feature in the scenery along the inside passages of southern Alaska. Maj. Allan Brooks (1922) takes a more hopeful view:

It is impossible for anyone whose only acquaintance with Eagles is in the east to have any idea of their numbers on this portion of the Pacific coast; except the Raven, in many localities it is the commonest bird and I have often seen forty or more together.

As by far the greater portion of this region is totally uninhabited there is absolutely no chance of their numbers being seriously depleted by any system of destruction induced by a small bounty. The long winding inlets and channels which cut up the shore line of the whole of this region, together with the maze of islands more than doubles the total shore line, and affords a tremendous area (most of which is complete solitude) for the home of countless Bald Eagles.

Nesting.—In the Aleutian Islands we found these eagles breeding on Unalaska, Atka, Kiska, and Tanaga Islands; probably they breed on most of the other suitable islands. As there are no trees on any of these islands the nests were all placed on rocky cliffs or on pinnacles of rock; some were easily reached but some were inaccessible. All the nests contained young, half grown or more in July.

On the south side of the Alaska Peninsula, Charles A. Gianini (1917) found bald eagles nesting "on the cliffs overlooking the bay and further inland as well, but always near water." Another "nest was a mere depression in the heavy grass situated on the top of a butte inland and overlooking Big River."

On the coasts and islands of southern Alaska and British Columbia the eagles nest in large trees. Joseph Dixon (1909) says that "out of 25 nests observed, only two were in dead trees. The birds rarely build at the extreme end of a point of timber, but go back in the woods for fifty yards or so in order that the nest may be sheltered from the gales that rage at times." A nest that he found on Admiralty Island "was situated in the highest branches of a broken top spruce tree, 116 feet from the ground"; it measured "six feet four inches, by six feet eleven inches over-all, and the out-side depth was four feet. The nest cavity was lined with duck feathers, dry moss and grasses. It measured sixteen inches in diam-eter and was four inches deep." Of another huge nest that he found on Hawkins Island, Prince William Sound, he writes:

This nest was located in a large hemlock tree sixty-two feet from the ground. This was an immense pile of wood even for an eagle's nest. These are the actual measurements taken with a steel tape; outside diameter, eight by ten feet; depth, four feet; nest cavity, twelve by twelve inches; depth four inches. The nest was firmly supported by an eight-inch forked limb; but the lower portion of the nest was fast moldering away, and a green currant vine had become firmly anchored in the rotting wood and twined its graceful green tendrils around one side of the nest. The nest was practically level across the upper surface, which was carpeted with moss. The nest cavity was lined with gull feathers and fine dry moss. I stretched out across the narrowest diameter of the nest but my arms and legs extended were not visible from below. This nest must support at least a ton of snow during the winter, so I had no hesitancy in venturing out upon it.

Edward A. Preble (MS. account) says:

In the Mackenzie Valley, northern Canada, the bald eagle is generally dis-tributed but is nowhere really common. Here it usually nests in tall trees, as did those recorded by MacFarlane (1891) and observed by him on Lockhart and Anderson Rivers in the late 1860's. In my own experience I found them even fairly common only in the mountainous country just south of McTavish Bay, Great Bear Lake, late in August 1903. On a high cliff on the shores of Lake Hardisty a nest was observed on August 18, and near it lingered a pair of old birds, evidently still attending their young. To the northward of this point the birds were observed practically every day from August 22 to 27, and here several aeries, all on high cliffs on the low mountain chain that our canoe route pene-trated, indicated the section most favored by bald eagles in all the vast region covered by me during several summers' explorations. From all the Mackenzie region the bald eagle must absent itself from November to March.

Samuel F. Rathbun tells me that bald eagles are rather common along the coast and near some of the remote lakes in Washington State, where he knows of several nesting sites. Of one nest he says:

This structure was a very large affair and no doubt had been in use off and on over a period of years. It was placed at a height of 130 feet, in a large black cottonwood having a diameter of nearly 6 feet at its base. The tree grew on rather swampy ground, and other trees of the same kind were scattered about with some mixed growth, but as a whole the section was quite open. This pair of birds, after having been robbed laid a second set of eggs and raised a brood in another nest. He mentions another nest within a few miles of this locality that was at a height of between 160 and 180 feet, in a fir tree that was about 8 feet in diameter, measured at a man's height.

M. P. Skinner's notes from Yellowstone Park refer to two nests in the tops of lodgepole pines, one of which was occupied for four years in succession. Nests in Ontario have been recorded in chestnut, syca-more, elm, poplar, oak, and hemlock trees; one in a poplar was as low as 20 feet.

The very elaborate studies conducted by Dr. Francis H. Herrick on
the home life of the American eagle, and his numerous papers on
the subject, have given us a very complete picture of the nesting
activities of these great birds. His elaborate preparations, and the
great amount of time and effort devoted to this work, in spite of many
discouragements, can be appreciated only by reading these excellent
papers. Space will permit only a few extracts from them here, which
I think should be included under the northern race.

The "great nest" (pl. 92) at Vermilion, Ohio, one of several on
which his observations were made, has a history covering 35 years;
and for more than 80 years eagles have nested in that vicinity, during
which time six nests are known to have been occupied. The "great
nest" was built not later than 1890 and was added to and occupied
every year thereafter until it was blown down in a March storm in
1925. This nest, when measured in 1922, was 12 feet high and 8½ feet
across its top; the upper rim was 81 feet from the ground in the dead
top of a shellbark hickory. Dr. Herrick (1924b) says that the favorite
trees in that vicinity are the sycamore and the shellbark hickory, but
the elm is sometimes used, and he found one in an ash and one in a pin
oak. Of the structure of the nest he writes:

A nest of the first year consists of a great mass of sticks, gathered mainly
from the ground, borne to the nest-site in one or both talons, by either bird, and
laid individually with aid of the bill; as this mass of faggots grows, greater
attention is paid to the periphery, where the coarser materials are more carefully
and more effectively interlaid and adjusted; the center and interstices are filled
with dead weeds, cornstalks and stubble, with incidentally considerable earth
introduced with pieces of sod and with weeds. It is no wonder that with the
growth of years the core of such a structure comes to form a sodden mass of
vegetable mold. The largest sticks which I have taken from different nests were
a yard long and two inches thick, but many which I saw in a nest at Kelley's
Island this summer appeared to have a length of over six feet.

He describes the process of nest building as follows (1932):

In mild seasons the Vermilion eagles begin to rebuild or refit their old eyrie in
the first days of February, or, as we might say, they build a "new nest" atop of
the old, for the building impulses are purely instinctive, and the eagles' eyrie is
virtually a composite affair, being made up of the consolidated increments of as
many years as it has seen service. In winters severe enough to cut off their
usual sources of food, and to prolong their absence from their customary haunts,
the seasonal building activity may be delayed until the first of March, but with
both birds working this labor can be performed in a few hours or days.

After from ten to twelve weeks of daily use the top of the eyrie is apt
to be trodden flat, its surrounding sticks scattered and its straw bedding
ground to powder. The old eagles in each following year build a new rampart
of sticks, about a foot high, and fill up the intervening area with a thick layer
of dead grass or straw. This building fever is apt to recur with diminishing
force during the first weeks after the young are hatched, and their ardor
gradually wanes until it is finally satisfied by bringing only an occasional stick,
a wisp of dry grass, or a spray of oak leaves or of pine. Whole stalks of

field corn, and often still bearing their yellow ears, were commonly a late addition, and all the more noticeable when draped over the sides of the nest. A farmer who was working in his field at the back of the tower said that on February first of that year an eagle came down within two rods of where he was standing, seized a stalk of his corn and bore it away; and a number of years ago an eagle was seen at Vermilion by one of my students making for its nest with twenty-five or more feet of rope dangling from its talons.

The bald eagle probably nested at one time over much of New England, but there are no recent authentic records of its nesting in the three southern States though it probably still breeds sparingly in the wilder portions of Maine, New Hampshire, and Vermont. Eagles are often seen in spring in southern Massachusetts, especially on Cape Cod, and occasionally at other seasons. Many rumors have come to me of eagles' nests, and I have spent much time in investigating such reports, but always without success. I suppose that the eagles we see on Cape Cod come from the coast of Maine or from inland points in northern New England, as the distance is not great for so strong a flier.

The only New England nests I have seen were on the coast of Maine. Between April 20 and 24, 1900, Owen Durfee and I visited five nests in this region. In the heavily wooded portion of Arrowsic Island, near the mouth of the Kennebec River, we found two; one, about 60 feet up in a tall white pine, was evidently an old nest, but the other, to which I climbed, might have been occupied later, as we had seen eagles in the vicinity and Mr. Durfee had taken an egg from it the previous year. It was near the top of a large white pine on the side of a hill, from which the eagles could have a fine view; it was about 50 feet from the ground and was about 6 feet high and 5 feet broad. On the following day, April 21, we were guided to an occupied nest a few miles back of Phippsburg, Maine. This was in a large white pine, about 70 feet tall and 26 inches in diameter at a height of 5 feet, that stood in a large open space where most of the large trees had been cut off. The male eagle flew fom the nest tree when we were about 100 yards away, but the female did not leave the nest until we rapped the tree; both birds circled about at a safe distance, screaming or whistling weakly, but soon flew away and were seen or heard only occasionally in the distance. The nest rested on two large horizontal branches against the trunk, and its flat top was 52 feet above the ground. It was made entirely of large sticks, many of them an inch and a half thick; the nest was evidently an old one for the material in the lower part of it was well rotted; it measured 6 feet high and 6 feet wide; the center of the nest was well lined with dried grasses to a depth of 2 or 3 inches, making a circular cavity about 17 inches across and hollowed about 5 or 6 inches below the outer rim of the nest; on the top of the nest were a few sticks with usnea on them, a sprig of green white pine, and numerous bits of white down. The two eggs which it contained were one-half to two-thirds incubated (pl. 91).

On an island in Jericho Bay, Maine, on April 24, we found two more nests. One was in a dense virgin forest of spruce, fir, and hemlock; it was a huge mass of sticks built on the broken-off top of a dead spruce and only about 30 feet from the ground; it was well surrounded by taller live trees, admirably concealed from view and in no sense a lookout point. It was apparently unoccupied. The other nest was about 40 feet up in the very top of a dead yellow birch, only 10 inches thick near the base, in an exposed situation near the shore and visible at a long distance. The nest was fully 8 feet high and impossible to reach into without risking one's weight on some very rotten limbs; the tree was very shaky, and our spurs would not hold in the rotten wood. There was some white down on the nest, and we saw an eagle in the distance, so it was probably occupied.

The Rev. J. H. Langille (1884) writes of a very unusual nest as follows: "On the bank of Niagara River * * * was a farm which had not been occupied for several years, and which was some miles distant from the nearest residence. A missing board from the end of the barn giving access to a large quantity of straw in the mow, the Eagles had arranged a nest there, which contained young when discovered by the owner of the property."

Eggs.—The eggs of the northern bald eagle are similar in every way to those of the southern bird, except for a gradual increase in average size northward. The measurements of 50 eggs from Alaska and Arctic Canada, typical of this race, average 74.4 by 57.1 millimeters; the eggs showing the four extremes measure 84.3 by 58.9, 79.4 by 63.4, 69.6 by 54.6, and 70.2 by 53.1 millimeters. Four eggs from Iowa average 74.8 by 59.6 millimeters, larger than the Alaska average. The average of 7 eggs from Maine is 76.5 by 56.3, and the largest egg measures 81.3 by 57.7 millimeters. The measurements of 35 eggs from Pennsylvania to Virginia average 73.8 by 56.8, 4 eggs from South Carolina 70.9 by 55.4, 4 eggs from Texas 70 by 54.9, and 50 eggs from Florida 70.5 by 54.2 millimeters. A comparison of these figures with the average measurements of birds given elsewhere adds strength to the theory that the eagles of the northern half of the United States should be referred to the northern race. Strangely enough, the measurements of 16 eggs from southern California and northern Lower California average 75.3 by 57 millimeters, fully as large as the Alaska average, indicating that the northern race ranges far south on the Pacific coast.

Young.—Dr. Herrick's careful and prolonged studies of eagles' nests have added greatly to our knowledge of the home life of these great birds. Much of what follows has been taken from his published papers (1924a, b, c, and d, 1929, 1932, and 1933). He gives the period of incubation as 34 to 35 days under normal conditions, though interrupted incubation may require a somewhat longer time. Both sexes share in the duties of incubation and care of the young; of which he (1929) writes:

In conducting the shifts a rather definite formula was observed. The sitting bird would give a sharp chitter when wishing to be relieved; the mate, if within hearing, came to the eyrie, moved up close, and the exchange was quickly made. If the eggs were left for only the shortest time, they were carefully covered with a great quantity of grass, stubble, and other convenient nest material, and the scrupulous covering and uncovering process would sometimes last from five to ten minutes. * * *

The eagle is the greatest home-keeper of his class. His eyrie is his castle, which, as we have seen, he will at times defend against all comers. In it his eaglets spend the first ten weeks of their life—from mid-April until early July, upon the southern shore of Lake Erie—and it is the occasional rendezvous, lookout point, and dining table for the elder pair for the remainder of the year.

In his final paper (1933) he writes:

Many times I have been impressed by the behavior of the mother eagle when rain or hail descended upon her down-clad young. As I approached the woods one mid-May morning the female eagle was on the nest, and whether because of seeing me or not, she presently withdrew to a tree-perch. Then, just as I entered the grove a brisk shower started, and the eagle at once returned to her young ones. Frightened at my ascent of the tower, she was off again, but, as the shower continued, returned in a few moments after I had entered the tent. She stood facing the wind and rain, with half-open wings, and afforded good shelter for the month-old eaglets huddled beneath her. In a few minutes this shower passed, and as the sun broke out she went back to her perching tree and spread her drooping wings to dry, in precisely that attitude assumed in times of great heat and humidity. Now, a quarter of an hour had hardly passed before the clouds again closed in and darkened above us; another downpour was under way, and the faithful mother sped back to her charges, and there she remained fending them with her stalwart body until this final shower was over. * * * Branches of pine and other green vegetation were always brought to the Vermilion nests both early and late in the season, and leaves were occasionally eaten by both adult and young eagles, as proved by their castings, but what significance this may have, if any, has not been ascertained.

He says elsewhere (1924c): "In 1923, if our estimate of the incubation period is correct, Eaglet No. 1 spent seventy-two and Eaglet No. 2 seventy-four days, in this case continuously, in the eyrie. Allowing then from 10 to 11 weeks for the life of the young Eagles in the nest, about one-half of this period, or five weeks, is passed in the white and gray down stages and the other half in the juvenal dress."

Although often two, and sometimes three, eaglets are hatched, the larger number is seldom raised to maturity, and often only one eaglet lives to grow up. The young hatch at intervals of a few days and the first one hatched, often the female, is larger and stronger than the other. The larger eaglet often abuses the smaller one and gets more than its share of the food, until the poor little one succumbs and dies of weakness and exposure. Dr. Herrick (1932) writes:

Two eaglets were hatched in that season on about April 24 and 28, and the younger bird was handicapped not only on account of its lesser age, but from the tempestuous weather and the shower of abuse it daily received from its older companion. The mother eagle constantly disregarded the needs of its puny infant, but bestowed every attention on her more vociferous offspring. Thus, on May 18, when the eagle brought in a large fish, the older nestling got 76 pieces, but the younger only 2, and a bad drubbing from his nest-mate in the bargain. On the following day rain and hail beat so relentlessly on the great nest that this much abused eaglet, then hardly able to crawl beneath the sheltering wings of its mother, finally succumbed and was trampled into the great mass of withered grass that lined its bed. It should be noticed that this harsh treatment of the younger bird had often occurred when the parent was away and when there was no contest over the food.

Both parents bring food to the nest and both assist in feeding the young. Dr. Herrick (1929) describes the process as follows:

The female eagle has been brooding her callow young, which are now in white down and about two weeks old. She deliberately rises, walks over to the carcass of a large fish, stands on it and begins tearing off small pieces of the flesh and passing them to the three eaglets, which line up before her.

Twenty minutes later the male drops on the eyrie and immediately joins his mate in the work of satisfying the appetites of their hungry brood. The old eagles bend to their task and pass up bits of food at the rate of about five to the minute. At least the passes are at this rate, but the proffered food is not always taken. It may indeed go the rounds, to be eaten finally by one of the old birds. [Pl. 93.]

When the eaglets are older and strong enough to tear up their own food, they are taught to do so. A family feast, presided over by the mother eagle, who has just arrived with a fish, is thus described by the same observer (1929):

Her young, all aquiver with excitement, continue to crouch and squeal, with their wings half spread, but they seldom venture to advance. The old bird now seizes her quarry, which appears to be a lake catfish of about four pounds in weight, and with one foot drags it to the center of the nest.

Standing on it there, she begins ripping it up without further ceremony. With swift thrusts of her bill she detaches large pieces of the white flesh and, taking a glance around at each upward stroke, swallows them in rapid succession. Then to the nearest bird, which by this time has edged up to its parent, she passes several pieces from bill to bill, and goes to work again on her own account.

When eaglet number two has been served in the same fashion, she moves a few steps away; whereupon number one seizes the carcass and, spreading over it, claims it as his own. Squealing, with head down, but for some moments without touching a morsel, he warns all intruders away. Mean-

while the other eaglet, drawing nearer, with head extended, watches the feeding bird and seldom venturing to interfere, patiently awaits its turn.

He relates (1924c) another instance as follows:

After a repast of a quarter of an hour the first Eaglet gave way to the
other bird which laid hold of the prey with one talon, dragged it aside and
set to work; not feeling satisfied, however, the first bird went after the
chicken again, but was immediately warned off. For two minutes they stood,
with wings raised, facing each other, like fighting cockerels, until the bird
which had taken first chance by an adroit thrust snatched the chicken with
one talon and, dragging it to the opposite side of the nest, began treading
it with both feet; after each hasty mouthful it glanced around to watch its
nest-mate. The robbed bird stood still, as if dazed, for some moments, and
after having flapped a few times settled down to watch for another opening;
with lowered head it moved very slowly towards the feeding bird, following
its every movement intently, and now an interesting thing happened: the
Eaglet that was feeding tore out pieces of the flesh and intestines and thrice
offered them to Eaglet number two who received them in bill and deposited
them at his feet without swallowing a morsel. He was not to be thus beguiled,
however; watching his chance, he seized the whole carcass and having deposited
it beside the proffered pieces went to feeding in earnest.

With the growth of the first plumage, when about a month old,
the eaglet spends much time preening its new feathers and gradually
disposing of its old gray down.

At this stage preening was the order of the day and for a week or more
the young "bird o' freedom" presented a most ragged and disreputable appearance. When thus actively engaged, and with the eyes often closed, the light
down was sent flying to the breeze; gray fluffy sprigs of their natal covering
were clinging to all parts of the nest, to neighboring trees, and when the wind
was right at a later time, some of it even floated into our tent. A pair of
House Sparrows, which were then nesting in the side of the eyrie, were most
diligent in collecting this treasured down, and in early June one would see
these little vagabonds steal up to the edge of the nest, snatch a few coveted
sprigs and hurry back to their retreat.

With the increase in size and strength comes an increase in activity,
with more time devoted to play and exercise in preparation for
flight. Activities begin by walking or jumping about the nest, which
soon becomes trodden quite flat, picking up and playing with sticks,
learning to grasp objects in the talons, and stretching and flapping
their growing wings. With tail raised and head lowered the eaglet
backs up to the edge of the nest and shoots its liquid excreta clear of
the nest to form a "whitewashed" circle on the ground below. Later
on the flight exercises begin in earnest, of which Dr. Herrick (1924c)
writes:

After a while a simple routine is established—raising the wings until they
seem to touch over the back, taking a few strokes and jumping; the flapping
gradually comes to take their feet above the floor of the eyrie and at eight
weeks of age they may be able to rise two feet or more in the air; this ability
attained, they are liable to go higher and higher and in a fairly stiff breeze,
which helps to sustain if not to stimulate them, they begin to soar and hover.

In 1922 we said "good-bye" to the Eaglets more than once before knowing the long practise they required to produce that perfect coordination of muscles and nerves which was necessary for confidence in the air. During the last week of regular eyrie life in that year they would sometimes rise to a height of fifteen feet, and soar for a full minute, going even beyond the confines of the nest and always with talons down to facilitate landing upon their return.

At last the day comes for the eaglets to leave the nest. Sometimes they do so voluntarily; but in some cases it seems necessary to use persuasion. In Dr. Herrick's (1924c) "first season with the Eagles the young seemed disinclined to leave their eyrie and were finally starved out and lured away." After two days of scanty feeding and two days of fasting, "as the old Eagle with the fish was circling just above the nest the Eaglet was jumping with legs rigid and flapping frantically; suddenly it leaped into the air, and for a second seemed to hang, as if poised over the eyrie; at that moment the circling Eagle began to scream, and swooping down at the hovering and now screaming youngster passed him within six feet; a minute later the Eaglet, still holding to the air, drifted fifteen feet or more beyond the margin of the nest; with vigorous wing-beats it began to move eastward, following the mother bird with the fish and made a full mile in its first independent flight; it finally landed in the branches of a tree on the edge of a strip of woods and doubtless was there allowed to feed on the tantalizing fish."

For some time after they leave the nest, probably all through their first summer, the young eagles associate with their parents in the home territory and frequently return to the nest or other favorite perches. But they are eventually driven out to earn their own living and seek new territory. They are never allowed to establish a breeding station near their parental home.

Food.—Eagles feed their young on much the same food as they eat themselves, with perhaps a somewhat larger proportion of chickens, other birds, and small mammals. As the bulk of the food of adults consists of fish, so it does of the young. Dr. Herrick (1924c) says that in 1922 fish made up 70 percent of the food fed to the young, and in 1923 fish constituted 96 percent of their food. Among the fish fed to the young were carp, pike, catfish, and sheepshead. Chickens, broiler size, were brought to the nest only about 12 times during the two seasons, and once a bird that looked like a killdeer. Crows, grebes, muskrats, rabbits, squirrels, and rats have been found in the nests. In one nest, which was destroyed, were 14 muskrat traps with the bones of the rats attached.

Probably most of the fish taken are dead or dying fish, picked up along the shores or floating on the surface of lakes, ponds, or streams. But eagles are perfectly capable of catching live fish, as referred to elsewhere. On Cape Cod, Mass., large numbers of herring, or alewives, run up the rivers and small streams in spring to spawn in the lakes and ponds. After the spawning season is over the shores of many ponds are lined with the dead bodies of herring. Here the eagles gather at that season to feed on this plentiful food supply. William Brewster (1925) writes:

During the continuance of spring freshets, Suckers and Pickerel, dead or dying, are washed ashore more or less numerously and eaten greedily by Eagles, even when in putrid condition. Later in the season these and other fish of goodly size are often snatched up while basking in the sun or swimming at or very near the surface of the water. For whenever it suits his needs or whim the Eagle will catch living and vigorous fish quite as adroitly as can any Osprey, although pursuing the sport in a somewhat different way. Thus he commonly swoops at the fish from a tree on shore, along a comparatively slight downward incline, or perhaps somewhat more abruptly, after hovering for a moment over the water at a height no greater than fifteen or twenty feet. In either case he is likely to capture such prey without wetting more than his feet and legs and never, I believe, will completely immerse himself to secure it as the Osprey does habitually, because accustomed to descend directly from greater heights, with much more impetus.

On the coasts of British Columbia and Alaska the eagles gather in enormous numbers to feast on the great schools of salmon and herring that are running up the rivers to spawn. Many are caught alive at the mouths of the streams or in the rapids, but eagles are too lazy to catch living fish when they can gorge themselves on the countless numbers of dead ones that line the streams after the spawning season. Bears, gulls, and ravens join in this feast, which lasts only during spring and summer. Joseph Dixon (1909) writes:

By the first of May the eagles are on the lookout for schools of herring that usually make their appearance about this time. One afternoon I noticed a commotion out in the bay where a flock of loons were fishing, then an eagle left a nearby perch, swoopt down, struck a fish in the water and returned to his perch where he gave a shrill scream. At the sound, eagles began to come from all directions to the spot where he had secured his fish, and within five minutes there were more than twenty eagles assembled. Only the first ones secured fish, as the fish which had evidently been driven to the surface of the water by the loons, went down again; the eagles returned to their perches to begin another vigil and soon all was quiet again.

Major Brooks (1922) says:

When fish are easy to capture as during the salmon and herring runs these undoubtedly comprise the bulk of the Eagles' food, but at other times fish are scarce and beyond the Eagles' ability to capture them; crabs may form a good portion of their diet in the summer but during most of the winter and up to June or even July Ducks and other waterfowl form the bulk of their food. These are taken in the water, usually after a long chase the victim being picked up as he comes to the surface. The only chance a Duck has is to get on the wing at all costs. I have repeatedly seen such clumsy risers as Goldeneyes and Scoters get away by shooting to the surface right under the Eagle's tail and instantly taking wing before he can swing around. I have never seen one of these Eagles catch a bird on the wing, although they sometimes make a determined effort even after such strong fliers as Geese, Brant and Pintail. * * *

The grouse of these islands—the Sooty Grouse—have a hard time. As soon as the broods are hatched they are led out by the mother bird to sun themselves on the sand dunes among the small spruces along the shoreline. During this season this strip is carefully covered by low-flying Eagles which quarter the ground just as a Marsh-hawk covers a marsh, except that the Eagle flies a little higher and usually on motionless wings. The result was an almost complete extermination of the Grouse, broods of one or two chickens only were seen in a few places, and twice single chicks without any parent.

I. J. Van Kammen (1916) writes: "At Unalaska there was found at different times around several eagle eyries the feathered remains of nearly every species common to those parts indicating that seemingly a bird diet is as desirable as one of fish. Among the water birds found were puffins, auklets, murres, murrelets, guillemots, ducks, and several species of waders, while among the song bird victims were Alaskan Longspurs, Aleutian Rosy Finches, Western Savannah Sparrows, Shumigan Fox Sparrows, and a sub-species of the Song Sparrow."

To sum up, the eagle's bill of fare is most varied, especially during the seasons when fish are not easily obtained. It includes all kinds of waterfowl, grebes, loons, gulls, any of the Alcidae, cormorants, coots, all kinds of ducks and geese, grouse, ptarmigan, and even the smaller land birds. Many kinds of small mammals, as mentioned above, are taken; fox farmers complain that eagles kill many young and even adult foxes; even the porcupine has been attacked, with disastrous results for the eagle; hunters complain that eagles kill young fawns and sometimes older deer. As eagles do not disdain carrion they may often be seen in company with ravens feeding on the carcasses of any animals they can find.

Behavior.—Eagles, like many other birds of prey, spend much of their time sitting immovable on some favorite perch, where they will remain for hours, unless disturbed, moved by hunger, or stimulated to action by the sight of game. Nothing, however, within their range of vision escapes their notice. Dr. Herrick (1924c) watched one, standing guard within sight of its nest; "for three and one half hours he had not apparently moved and had not been seen to lift even a foot."

Eagles are generally not gregarious, but they often gather in flocks about their fishing grounds in southern Alaska. J. S. Dixon (1909) "once saw more than 15 eagles sitting in a single spruce tree waiting for a school of herring, and at a distance it appeared like a magnolia tree in blossom because only the white heads were discernible." Alfred M. Bailey (1927) "saw a flock containing at least 300 March 10 at Klawack, where herring were schooling. Twenty-one birds were counted in one tree." Such sights are unknown in other parts of the range of the bald eagle. Mr. Brewster (1925) once counted 25 in a single day at Lake Umbagog, Maine, and Henry Beston writes to me that in October 1932 he saw an unusual flight of bald eagles over the waters of Damariscotta Lake in Maine; it was a pleasant, warm day with a light northwest wind, favorable for migration. Some 30 of the birds were performing their aerial evolutions, and "the effect was rather that of a kind of swarming. They were so high that the white heads and tails of the mature birds could be identified with certainty in only three or four cases. After remaining in view for about four minutes the whole gathering vanished, and my impression was that they disappeared aerially in a generally southeasterly direction." Several other local observers noticed the flight and said that they had never seen one like it.

Eagles have been seen on several occasions to alight on water, float about for several minutes as lightly as a gull, probably in pursuit of fish, and then arise from the surface with no great difficulty. Occasionally one may fasten its claws on a fish that is too big for it to lift, which results in a struggle that is unpleasant or even dangerous for the eagle. But the eagle is a powerful bird and can probably lift an object of its own weight; one has been known to carry a lamb over a distance of 5 miles.

I have referred to the bald eagle elsewhere as an arrant coward, and so I have always found it; but Dr. Herrick (1929) says that it "will sometimes put up a stiff fight in the defense of its nest, or when hard pressed on the ground." He cites an instance where an eagle, caught in a trap, put up such a vigorous fight that its captors were unable to release it and had to kill it; and "to the end it was fiercely defiant"; though repeatedly beaten down, "in an instant he was on his feet again, as indomitable as ever", and his courage was "persistent to the last." I heard of another eagle, in a similar predicament, that was very docile; it allowed itself to be freed from the trap, without any show of hostility, and then quietly flew away. Only in rare instances have men been attacked at the nest. Major Bendire (1892) mentions a nest at which one of the eagles always threatened him, swooping down at him, "sometimes as close as 20 feet."

Enemies.—Eagles have no serious enemies except man. Most of the feathered foes that attack them are usually regarded with dignified indifference, as if they were only annoying pests. It is a well-known habit of the eagle to attack and rob the osprey, but few people have seen the tables turned. Once, while watching the graceful evolutions of an eagle and an osprey sailing about away up in the sky, as I thought in play, I was surprised to see the osprey swoop downward and almost strike the eagle; the eagle quickly turned over, back downward, and presented his claws, which sent the osprey scaling off in a hurry. This maneuver was repeated several times. It was too late in the season, August 16, for the osprey to have young in its nest. A somewhat different method is thus described by Freeman F. Burr (1912):

The Eagle had just forced the Osprey to drop a fish, but had failed to catch it as it fell. The smaller bird then withdrew to a point about fifty feet above, and suddenly swooping down, attempted to strike the Eagle on the back. Just as it looked certain that the broad back must receive the full force of the stroke, up went one great wing, with an agility and a skill that would have done credit to a practised boxer, and the Osprey was tossed aside with apparently almost no effort. This was repeated several times; when the Osprey, evidently discouraged, gave up the unequal fight and winged away toward the far side of the lake. Immediately the Eagle dropped to the water, and picking up the fish made off with it.

Eagles are often attacked by crows, just as these black rascals will attack any large bird of prey; and occasionally the crow pays the extreme penalty for its audacity. Mr. Brewster (1925) relates the following surprising incident: "An immature Bald Eagle perched on a stub on B Point was harassed for several moments by a Crow of whose noisy and threatening demonstrations it took little apparent notice at first; but when the Crow alighted on its back about between the shoulders and began pecking at its head the Eagle spread its wings and swooped down a steep incline to plunge headlong into the Lake where it almost completely immersed itself, thereby escaping for the moment from its tormentor who, however, did not let go his hold until just as the water was reached."

The eagle despises the crow, but does not fear it; ordinarily it treats the crow with indifference, but when it has eggs in its nest any approaching crows are promptly driven away. Hawks also are not tolerated near an eagle's nest containing young. Almost any small bird will fearlessly attack an eagle or any other predatory bird that comes too near its nest. The fiery little kingbird will even invade the eagle's territory to attack it, even alighting on a perch above the eagle's nest and darting down at it. Dr. Herrick (1932) relates the following incident: "The mother eagle had but just dropped a fish on the eyrie, and taken a favorite perch 100 feet from our tower and from a Blue Gray Gnatcatcher's nest that was affixed to the lofty branch of an elm just below the tent. The eagle was beset by this pair of indignant gnatcatchers, which buzzed about her like so many angry wasps. I could see one of the eagle's wings drop, as she started to relax, but there was no peace for the tired bird and after ducking her head time and again at the thrusts of her pigmy assailants, she left this perch and went to one farther away in the forest."

Voice.—Mr. Brewster (1925) describes the eagle's notes very well, as follows:

The commonest and most characteristic utterance of the Bald Eagle is singularly out of keeping with the bird's imposing size and not undignified bearing. Weak in volume and trivial in expression it consists of seven or eight notes given rather quickly, but haltingly and with apparent difficulty, as if their author were choking or gasping for breath. It cannot fitly be called a scream, but is rather a snickering laugh expressive of imbecile derision, rather than anything else. My notes render it thus—*Ki-ki-ki-ki-ki-ki-ker.* I am not sure that this outcry is ever made by Eagles less than a year old. Younger ones frequently utter a shrill, querulous squealing *pee, pee-e, pee-e* having a rising inflection and suggestive of hunger unappeased and insatiable.

Dr. Herrick (1933) writes:

Notwithstanding the many days and weeks spent with these eagles I have only once or twice seen them to good advantage when making their famous ear-splitting screams. This once happened when I was taking motion pictures of the female on her tree-perch, one hundred feet away, and the scream was occasioned, I think, by a distant glimpse which she got of her mate, who was at that time recreant to his domestic duties. Bending down somewhat, the head is gradually elevated until at the climax of the scream it is directed to the zenith and nearly or quite touches the back.

Economic status.—Throughout most of its range in central and eastern North America the northern bald eagle is too thinly distributed to be of any great importance economically. It destroys many fish, but mainly those of the least food value; by far the greater part of these are dead fish, which would only pollute the waters. Domestic poultry is seldom disturbed; Dr. Herrick counted only 13 chickens in two years of study. The number of game birds and other small birds destroyed is insignificant. Most of the small mammals on which it feeds are more or less injurious. But on the coasts of British Columbia and Alaska, where eagles are enormously abundant, the case is very different. Vast numbers of eagles have been killed under the bounty system, which has caused much concern among bird protectionists and much controversy over the justification for such slaughter. The salmon fisheries claim that the eagles injure their business seriously by devouring enormous numbers of salmon, but they forget that eagles are too lazy to catch live fish when they can pick up dead ones and that probably the bulk of their food consists of dead and dying salmon and herring that have finished spawning. Here too considerable damage is done to wildfowl, ducks and geese, and other game birds. Eagles undoubtedly kill some lambs of mountain sheep, kids of mountain goats, and young fawns, but there is little, if any, evidence that this damage is extensive, especially as eagles are scarce in the interior. Where eagles are sufficiently abundant and are known to be doing serious damage to salmon fisheries, fur-farming activities, or other human interests they should be reduced in numbers. There is no danger of their extermination in the vast uninhabited regions of Alaska. Elsewhere we can afford to protect such a picturesque feature as our national emblem.

Winter.—Throughout much of its range the northern bald eagle is permanently resident. But from the extreme northern portions in the interior, when the lakes and rivers are frozen and the ground is covered deeply with snow, it is difficult or impossible for the eagles to find food; they must then retire to the seacoast or to a milder climate where they can find open water. Eagles are found all winter on the coast of Maine. On the Hudson River, north of New York City, they are often seen floating down the river on ice cakes. Dr. Herrick (1924c) says of his Ohio eagles: "In ordinary seasons, according to Mr. and Mrs. [Otto] Buehring, they are away only from six to eight weeks, or from mid-November to mid-January; but in the season of 1921–22, which was one of the mildest on record, they were missed for barely a fortnight in the latter part of December. In the winter of 1922–23, which continued rather mild until January, both birds remained in the neighborhood, and were even seen resting on the nest itself at the very end of December."

THALLASOAETUS PELAGICUS (Pallas)

STELLER'S SEA EAGLE

HABITS

The claim of this magnificent eagle to a place on our check-list is based on its accidental occurrence on the Pribilof Islands, Kodiak Island, and the Aleutian Islands, the first two records being substantiated by specimens. The Pribilof record is thus given by G. Dallas Hanna (1920) : "A bird of this species was shot and wounded on St. Paul Island, December 15, 1917, but fell into the sea. Five days later it was picked up on the beach in badly decomposed condition. Enough of the specimen could be saved however to enable the identification to be made in the National Museum."

A Kamchatkan sea eagle, as this bird was formerly called, was taken on Kodiak Island, Alaska, August 10, 1921. Charles H. Gilbert (1922), who was a member of the party, reports the circumstances as follows:

The bird was shot by a native who formed a member of our party, and who frequently hunts about Karluk Lake. He stated that he had observed this strange eagle on a number of previous occasions and had tried to capture it, and that this was the only bird of the kind he had ever seen. Bald Eagles were abundant in this locality.

My few notes state that the head was not white but was covered with feathers variegated much as in the Golden Eagle. The tail and leg feathers were white, as were the entire front edges of the wings. The feet and bill were extremely powerful, obviously larger than in the Bald Eagle, which we had for comparison. Both bill and feet were strikingly bright in color, of a deep golden yellow, which covered also the bare portion of the legs. The tail feathers were graduated.

Austin H. Clark (1910) saw a large eagle that was probably of this species near Unalaska on May 26, 1906. As it sailed almost directly over his head and very low down, and as he was familiar with the bird elsewhere, it hardly seems likely that he could have been mistaken in a bird of such distinctive field marks. We failed to find this eagle during the month we spent among the Aleutian Islands and no other observers seem to have recorded it from there.

Dr. Leonhard Stejneger (1885) says: "Pallas was very much mistaken in giving Bering Island as the true habitat of this bird. This mistake arose from his having misunderstood Steller's description of the bald eagle as referable to *Th. pelagicus.* The habitat is especially the mainland of Kamtschatka, where it is abundant, but also all the countries bordering the Okotsk Sea. On Bering Island it is only an occasional visitor, being chiefly an inland bird preferring the quiet rivers and lakes surrounded by dense forests."

We know very little about the habits of this great eagle beyond the account of it given by Pallas, based on Steller's notes, and translated by Cassin (1856) as follows:

This very large bird is frequent in the islands between Kamschatka and the American continent, especially in the islands noted for the unfortunate shipwreck and death of Bering. It appears very rarely in Kamschatka itself. In the highest rocks overhanging the sea, it constructs a nest of two ells in diameter, composed of twigs of fruit and other trees, gathered from a great distance, and strewed with grass in the centre, in which are one or two eggs, in form, magnitude and whiteness, very like those of a Swan. The young is hatched in the beginning of June, and has an entirely white woolly covering. While Steller was cautiously viewing such a nest from a precipice, the parents darted with such unforeseen impetuosity as nearly to throw him headlong; the female having been wounded, both flew away, nor did they return to the nest which was watched for two days. But, as if lamenting, they often sat on an opposite rock. It is a kind of bird, bold, very cunning, circumspect, observant, and of savage disposition. Steller saw a Fox (*Vulpes lagopodus*) It lives also on dead substances cast up by the sea, and various offscourings carried off by one and dashed upon the rocks, and afterwards torn in pieces. of the ocean.

The following account by Dr. Heinrich Bolau (1892) adds a little to our knowledge of this rare eagle:

Very little is known of the *Haliaetus pelagicus* in its free state. The Dorries Brothers, collectors who resided for many years in Amour, in Eastern Siberia, and during that time watched the animal world very closely, saw only four white-shouldered sea eagles among the many common sea eagles in the neighborhood of Vladivostock, and only two black Corean eagles; and never succeeded in shooting one of these rare birds. The Russian explorer Von Middendorff speaks of the sea eagle as being very cautious. Although he found many nests, he very seldom saw the birds; apparently they were on the high seas busily fishing. In August, so says our authority, the sea eagles were quite numerous on the south coast of the Ohotsk Sea, where they preferred to build their nests on the summits of the cliffs, which frequently project singly and like towers from the surface of this sea; and therefore their nests, were very inaccessible. Consequently, the eggs of our birds are unknown, nor is anything known in regard to the number and treatment of their young. About the middle of October these eagles move southward, flying high in the air. In the winter they go to Japan and the North of China, returning in summer to their breeding grounds in Kamchatka.

The Ainos raise the young as an article of trade, and the Giljaks sell the white tail feathers to the Japanese, who prize these feathers highly and are willing to pay high prices for them. The Japanese like them to use in window decoration.

In captivity the sea eagles are very quiet, generally keeping away from the other birds in the cage. Their food consists of fish and meat. Their sharp, penetrating cry is as powerful as their bodies, and, in their native land, can be heard above the noise of storm and surf.

Plumages.—Dr. Stejneger (1885) describes the natal down as "pure white all over." He also describes "a downy young, just assuming the first plumage, the feathers coming out copiously on head, upper neck, interscapular region, wings, breast, anal region, and tail. The predominant color of the feathers is a dull blackish brown, slightly lighter brownish on the middle of the feathers of the top of the head; the inner webs of the wing-coverts are lighter, some at the base whitish, secondaries and scapulars margined with lighter brown. The tail-feathers, of which only the tips are visible, are white near the end, mottled with blackish further up. The down covering the rest of the body is of a smoky brownish gray."

A young bird is described by Baird, Brewer, and Ridgway (1905) as follows: "Dark umber or blackish-brown, the feathers of the head and neck with lighter shaft-streaks; tertials (except at ends) and basal third, or more, of inner webs of tail feathers, white; tail-coverts much mixed with the same."

The bird shown in Cassin's (1856) plate is evidently an older, but still immature, bird. It has a wholly white, wedge-shaped tail, but all the rest of the plumage is brownish black. This eagle must be easily recognizable at a long distance in its strikingly marked adult plumage, white forehead, white wing coverts, white belly and tibiae, and cuneate white tail. Travelers in the Bering Sea region should keep a sharp lookout for it.

DISTRIBUTION

Range.—Northeastern Asia, casual in Alaska. Steller's sea eagle breeds in northeastern Siberia, Kamchatka, probably Sakhalin Island, and possibly occasionally on the Commander Islands. In winter it occurs south to Korea, Japan, the coasts of Amur and Ussuri, and the Riu-kiu Islands.

Casual records.—This eagle was recorded as seen, but not taken, on Unalaska Island on May 26, 1906. As specimen confirmation was

lacking, this record was properly not accepted as sufficient evidence to
warrant adding the species to the North American list. Subsequently, however, two specimens were taken, the first on December
15, 1917, on St. Paul Island of the Pribilof group, and the second
on August 10, 1921, on Kodiak Island.

PANDION HALIAËTUS CAROLINENSIS (Gmelin)

AMERICAN OSPREY

HABITS

The osprey, as a species, is widely distributed throughout the world
and has been divided into five subspecies. Our race breeds in North
America but wanders to Central and South America. It has a wide
range over most of this continent, but, as a breeding bird, it is rare
or widely scattered throughout most of this range. In a few favorable localities, mainly along the Atlantic coast, it is very abundant
and breeds in several more or less dense colonies. As it lives entirely
on fish, it naturally prefers to live in the vicinity of the seacoast or
near some large body of water, lake, or stream, where it can find an
abundance of its finny prey. Given this food supply, it makes little
difference to the osprey what its surroundings are. It is equally at
home near the shore of some remote wilderness lake, on timbered or
open islands along the coast, in the valleys of inland streams, in open
farming country, or even close to houses. In the last two localities
it is jealously protected and often encouraged to breed by placing
cart wheels or other supports for its nest in trees or on poles. In
the region where I am most familiar with it, it has become a common
dooryard bird, almost a domestic pet, and consequently very tame.

The history of the status of the osprey in Massachusetts is rather
interesting, as illustrating how little some of the early writers on
local ornithology knew about the birds of the State outside of the
limited regions with which they were familiar. Dr. J. A. Allen
(1869) wrote: "It seems at first a little strange that this noble bird
should not be found breeding anywhere on the Massachusetts coast.
* * * The present puny second forest-growth affords it no suitable breeding places, and this is no doubt the reason of its being now
but a transient visitor here." This remarkable statement shows
lamentable ignorance of the nesting requirements of the osprey and
a lack of acquaintance with the forests of Bristol County, which in
those days were far from "puny." This error was repeated by Minot
(1877) and Stearns (1883).

It was my old field companion, Frederic H. Carpenter (1887) who
first called our attention to the large breeding colonies of these fine
birds in southern Massachusetts. It was he who first introduced me
to these interesting colonies, with which we have kept in close touch
ever since.

The changes in the distribution of the nesting birds during the past 50 years, in the area covered by our observations, are also interesting. When our records began, in 1882, there were over 80 occupied nests in the rather limited area that we hunted, on foot, in Rehoboth, Seekonk, and Swansea in Massachusetts and in Warren and Barrington in Rhode Island. As time went on, we enlarged our field and discovered a number of outlying nests in neighboring towns, both north and south of the region named. The northernmost nest, north of Taunton, was 18 miles from the nearest salt water; and some of the Rehoboth nests were 12 miles inland. The interesting point is that these inland nests have been gradually disappearing, until now not one of the 81 nests recorded in 1882 is in existence. The ospreys are now all concentrated near the shores of Mount Hope and Narragansett Bays and their tributaries. What caused this wholesale evacuation is a mystery. Considerable egg collecting was done in certain parts of the area, but no more than, if as much as, in the area where the birds still breed. There are just as many suitable trees as ever, and many perfectly sound nesting trees have been abandoned. There may be fewer fish in the inland ponds and streams, though there has been no noticeable increase in pollution. The only answer seems to be that the birds have decreased in numbers, from some unknown cause, and the remaining birds are concentrated where there is a better food supply and where they are more rigidly protected. In the area that we now cover, there are between 50 and 60 pairs of birds nesting, where there were at least twice that number 50 years ago. Formerly we could visit between 30 and 40 breeding pairs in a day on foot, but now our best recent record is 56 pairs seen with the help of an automobile.

Spring.—Throughout all the northern part of its range the osprey is migratory. In much of Florida and in the Gulf States the osprey is present all winter, but C. J. Pennock tells me that it is absent from northern Florida, Wakulla County, "from about the middle of November until early February." Arthur T. Wayne (1910) says that it is absent from South Carolina "from December until very late in February." Mr. Pennock's earliest date for Delaware is March 16. In southern New England, the ospreys appear with considerable regularity during the last week in March; my earliest date is March 15, but Forbush (1927) has a record for March 7. Usually only a few individuals are seen here in March, the main body arriving during the first week in April. The males are said to precede the females. Their migrations are probably influenced by the movements of the fish on which they prey; we usually see them at about the time that the alewives, or herring, are starting to run up the rivers. In other parts of the country, their arrival is equally subject to climatic and food conditions. M. P. Skinner's notes give the dates of arrival in Yellowstone National Park as ranging, over a period of seven years, from April 9 to 25.

Courtship.—I believe that ospreys are mated for life, as is the case with many other large birds. Dr. Harry C. Oberholser (1897) tells the following pathetic story, illustrating the constancy of a bereaved mate:

At a time when one of the birds, presumably the female, was on the nest, a bolt of lightning struck the tree, killing the bird and demolishing the nest. Strangely enough, the other osprey when returning only to find his home desolated, took up his station upon the top of one of the uninjured trees close at hand, and throughout the remainder of the summer was seen day after day, month after month, keeping his lonely vigil, apparently mourning the loss of his mate. * * * He remained until late in September, but at the time that the other ospreys departed he too disappeared. The next spring, however, found him again at his post, and throughout the whole summer he continued just as before ; but in the ensuing autumn, joining the company of his fellow ospreys in their journey to the southland, he departed, this time to return no more.

But such constancy is not the invariable rule. I have known of several cases where one of a pair has been shot and the survivor has secured a new mate. I also knew of a case where both of a pair were shot and a new pair appropriated the nest.

As soon as the ospreys arrive on their breeding grounds they inspect the old nest and begin repairing it. One bird, probably the female, stands on the nest, and receives and arranges the material brought in by her mate; it is interesting to see these great birds flying home with a long string of seaweed or cornstalks trailing out behind. Mrs. Irene G. Wheelock (1904) says that the osprey breaks off the dead twigs from a tree, sweeping down on them and seizing them in its feet.

All is activity in the colony, as new birds are arriving at intervals, looking up their old nests or seeking locations for new ones. As most of the birds are already mated, courtship is mainly a nuptial display, an expression of joy at their home coming, or an exhibition of exuberant spirits. It consists mainly of aerial gymnastics in which both sexes indulge, chasing each other in swift pursuit-flight, soaring, scaling, circling, dodging with rapid turnings or quick dashes downward, as they sweep, now low, now high, in wide circles. Several pairs are often seen in the air together, and sometimes trios, all screaming their notes of love or excitement.

Copulation is performed on the nest or on a branch of a tree; the male stands on the back of the female, balancing himself by waving his wings, and making connections for a few seconds.

William Brewster (1925) saw what was probably a male—

mount to an immense height above the Lake near Great Island, to drift slowly eastward over the forest, poising or hovering all the while on set or loosely-flapping wings, uttering almost ceaselessly a shrill, screaming *cree-cree-cree* wholly different from the ordinary musical outcry of his kind. All this was kept up fully fifteen minutes. Of course it represented the characteristic love-flight of the Osprey, often witnessed at the Lake in early spring, and not unlike that performed by several other species of Hawks found in New England. Finally a female Osprey appeared, swinging around and around in wide circles a thousand feet below the other bird. He, however, continued to hover, flutter, and scream at his former level.

Nesting.—What was once, probably, the largest known breeding colony of ospreys formerly existed on Plum Island at the eastern end of Long Island, N. Y. Charles Slover Allen (1892) gives a very interesting account of this colony, which had been protected for many years by the former owners, the Jerome family. When Mr. Allen first visited this island in 1879, Mr. Jerome "claimed that fully two thousand nightly roosted on the island, and that over five hundred nests had been built there." But Mr. Allen "finally reduced these numbers one half." In 1885, this island "was sold to a syndicate who planned the construction of large hotels and cottages; since then all has completely changed."

Probably most of the ospreys from Plum Island moved over to Gardiners Island, only a few miles distant, which now holds the largest breeding colony of which we have any record. The size of this colony has been variously estimated, but I doubt if any accurate census has ever been taken. Good descriptions of this colony have been written by Dr. Frank M. Chapman (1908), who estimated the number of nests as 150 to 200; by Clinton G. Abbott (1911), who estimated 200 nests; and by Capt. C. W. R. Knight (1932), who thought the number exceeded 300. Gardiners Island is about 7 miles long and 3 miles wide and contains about 3,000 acres.

Our scattered colony, in southern Massachusetts and eastern Rhode Island, could now be covered by a circle 8 miles in diameter, and contains about 60 occupied nests, possibly a few more; it formerly covered more than twice this area and contained much more than twice this number of nests.

Bendire (1892) mentions a colony on Seven Mile Beach in southern New Jersey, in which "several hundred pairs have nested every season." In other parts of the country the colonies are usually smaller, or more scattered. Mr. Abbott (1911) found a colony of 30 nests at Great Lake, N. C., in 1909.

Donald J. Nicholson tells me that in 1910 there were at least 75 occupied ospreys' nests in the cypresses around Lake Istokpoga in Florida; and in Volusia County "possibly hundreds of their nests can be found in the cypress swamps near Maytown, 30 being visible from a lofty cypress." William G. Fargo writes to me of a colony of 12 or more nests that he found near Old Tampa Bay, Fla., "of which at least nine were within an area of about 100 acres."

The region with which I am most familiar, southern Massachusetts and eastern Rhode Island, is largely an open farming country, with considerable heavily wooded territory scattered through it, with numerous streams and small lakes and with many large residential estates near the shores of the salt-water bays. Formerly many ospreys nested in the wooded sections, far from human habitations. The nests were usually placed in the largest trees they could find, tall solitary white pines, or large oaks on the edges of the woods, and generally not far from some lake or stream, where the fishing was good. Comparatively few were more or less hidden within the woods, in almost any kind of large tree, or on the top of some large dead stub. These woodland nests have nearly all disappeared, except in a few large groves near the shores, where they are protected, but even these are decreasing.

The nests in the open farming country and on residential estates seem to be the most successful and to last the longest. Here the ospreys seem to have no special preference for any species of tree and are not at all particular as to its height. Security and a good food supply seem to be all that they require. Sycamores, locusts, and elms figure most prominently in my notes; but we have also found nests in various oaks, ashes, tupelos, maples, red cedars, wild cherries, willows, pines, and even apple trees. Many of the nests are in partially, or wholly, dead trees; although the birds sometimes build in a dead tree, I believe that in most cases the tree is killed by the weight of the nest or by the saline character of the nesting material and of the birds' food; I have known of many cases where the tree has died and fallen after the ospreys had built in it. Some of the occupied nests have been in trees standing in water; one such in a pond was so low that it could be looked into from a boat.

Many nests are built on poles near houses (pl. 97), a cart wheel or some other support having been attached to the top of it to hold the nest. Similar supports are often placed in trees by the landowners, who protect the ospreys and encourage them to nest near their houses or on their farms as picturesque features or because they are supposed to drive away other hawks. Ospreys often build on telegraph or telephone poles, where the cross arms and wires give good support, much to the annoyance of the linemen who have to remove the nests. Harry S. Hathaway (1905) says that "one pair in Bristol was so persistent in 'sticking' to the same pole after it had been pulled down that they built it up four times, and it was only after a 'ground' had been made by the wet mass in a rain, which set the pole and nest afire, that they deserted it."

A better location, recently adopted, is the steel framework of a high-tension-line tower. Nests are also built occasionally on buildings or on unused chimneys. Mr. Forbush (1927) tells a remarkable story of a pair of ospreys that had a nest on the chimney of a vacant house. A new family moved in, removed the nest, and shot one of the birds. The survivor secured a new mate and rebuilt the nest. By the time that the birds were at last driven away, after repeated attempts to rebuild the nest, the chimney was found to be completely filled with rubbish. I once saw a nest on an electric-light transformer to which a white flag was attached. Several nests have been on unused windmill towers. Mr. Hathaway (1905) says: "One of the most unique situations that has come to my attention is a nest in Portsmouth, built on an old windmill, which has had the 'floats' blown off, and the nest is so placed that, when the rudder turns, the sitting bird, on her nest, swings round and round with every breeze."

In our territory I have never seen a nest on the ground. The height above ground has varied from 10 or 15 feet, in cedars, locusts, or on poles, to 50 or 60 feet, in tall pines or elms; but most of the nests have been under 40 feet. The nests vary greatly in size; nests on artificial supports are usually very flat, from a few inches to a foot high, and they are not built up from year to year, as the tree nests are; the latter often increase to enormous size until they fall or break down the tree; the tallest one I have seen was built up to a height of 10 feet before the tree and all collapsed.

None of my earliest nests are still in existence. One nest that I first saw in 1891 was still occupied in 1935; it is in a locust, now dead. There are two others that I have recorded as occupied for 41 years, one in an elm and one in a locust. Another, still occupied in 1935, has been occupied for about 45 years, according to an interested neighbor; this is artificially supported in a locust close to a much-traveled road. A few other nests have lasted for 30 years or more, but most of them last for much shorter periods. Mr. Hathaway (1905) refers to a nest in this territory that has been used annually since "about 1780, and was until recently still occupied" by successive pairs of birds.

The foregoing remarks all refer to nesting habits in Massachusetts and Rhode Island. C. S. Allen (1892) describes some interesting nests on Plum Island, as follows:

The first Fish Hawk's nest shown to me by Mr. Jerome was fairly in his dooryard, close by his front gate, and only about fifty yards from his house. It was placed upon an old pile of fence rails, rotted to black mould in the center, but kept up by the yearly addition of fresh rails. Mr. Jerome said that to his knowledge this nest had been occupied every year for forty years. It likewise had been added to yearly until its bulk of sticks, sods, cow dung, decayed wood, seaweed, etc., would amount to at least three carloads, in addition to what had rotted and fallen to the ground. The nest was only seven or eight feet from the ground, so that by stepping on a projecting rail I could readily see the three beautiful spotted eggs within, which I promised not to disturb. Mr. Jerome could pass close to the pile of rails without the birds leaving the nest, while I could not get nearer than thirty or forty feet. * * *

Out on the sandy meadow to the southward were what at a distance appeared to be two gigantic mushrooms about seventy-five yards apart. A nearer approach disclosed the fact that they were cedar trees twenty feet high; the trunks were about one foot in diameter and without a limb for the first ten feet. The whole top of each tree was involved in a huge nest. These nests, Mr. Jerome said, had been occupied every year for forty years, each year the Hawks repairing them and adding to their bulk. These nests were so unusually large that they are worthy of description. Each nest involved the whole tree, even to the lowest branches. At the base loose sticks, six to twelve feet in length, were spread out so as to form a projecting platform ten to fifteen feet in diameter, forming complete protection from below. The base of the solid portion of the nest was about eight feet across, sloping up to the level top, which was about four feet across, and very firm and solid, and readily bearing my weight. The bulk of this nest was about equal to three cartloads. The central part of the nest consisted of a mass of sand and decayed matter from the old nests, much of which had fallen through to the ground. The base of the nest consisted of long sticks, oyster stakes, etc., loosely put together and extending beyond the longest limbs of the tree, making it over twelve feet in diameter. Each year for many years the nest had been repaired and built up with every kind of material that had been washed ashore or could be picked up in the fields. The center of the nest, nearly five feet high, was composed of clods and sand and the decayed remains of material added many years before.

Of the famous colony on Gardiners Island, Mr. Abbott (1911) writes:

Ospreys' nests on Gardiner's Island are placed in almost every conceivable situation. They are on trees by scores, both high up and low down; on rocks and boulders, whether on land or in the water; on sheds and buildings; on fences and walls; on piles of debris; on old stumps; on a floating wooden platform intended for the fishermen's use; on a channel buoy; on sand-bluffs; on pieces of wreckage, driftwood, and fish-boxes. The birds even attempted to build on the slender stakes supporting the fish-nets! In all of these varied nesting-sites, however, it will be noted that at least the suggestion of an eminence has probably first attracted the Osprey to the spot. Similarly, many of the ground nests are found to be very close to some prominent object—itself incapable of supporting the nest—such as a post, a notice-sign, a telegraph pole, or a pointed stone. The high, shelving beach, with its tempting piles of seaweed, probably appealed to some of the first ground-nesters as an "eminence," and their offspring have come back and chosen a similar nesting-site. At all events, in 1910 there was a succession of no less than twenty-two nests at intervals varying from eleven yards to three hundred yards along the beach, on the south-westerly side of Gardiner's Island. Some of the most recent additions to the beach-nesting colony had certainly quite lost any instinctive attraction for an "eminence"; their nests being a mere scattering of sticks in the edge of the marsh-grass—in location suggesting more the humble home of the Tern than the eyrie of the noble Osprey.

In the southern Atlantic and Gulf States the ospreys nest very commonly in living or dead cypresses, about the shores of lakes, along the banks of streams, or on the borders of swamps. Some of these old stubs, which resist decay for many years, even when standing in water, offer ideal nesting sites.

In Florida I have seen many nests in such locations, as well as in tall pine trees; most of the nests in pines range from 25 to 70 feet from the ground, but Mr. Nicholson tells me that he has seen them as high as 110 feet. In southern Florida, they often nest in low mangroves, 15 to 20 feet above the water, according to Mr. Pennock. Among the Florida Keys I once saw a nest built on the tops of some little low mangroves, with its base only a few inches above the water; we could look into it from a boat.

Mr. Skinner's notes mention some interesting nests in the Yellowstone National Park. One of the most famous nests is on "Eagle Nest Rock", a lofty pinnacle in Gardner Canyon; he says that this has been occupied—

each year since 1875 at least, but one spring I found as many as seven adult osprey in its immediate vicinity. * * * The nests at Eagle Nest Rock and in the Grand Canyon of the Yellowstone are on the tips of pinnacles of rock jutting out from the canyon slopes. As these are usually the only nests seen by visitors, they unconsciously form the opinion that all osprey nests are on rock pinnacles.

But this is not true, even for the majority of the Yellowstone osprey, for the original and most used sites are on the tips of dead trees, and on the tips of living lodgepole pine and spruce trees. In 1914, I estimated there were 25 nests in the Grand Canyon; adding in the Eagle Nest Rock site and all others there may be 30 rock pinnacle sites in Yellowstone National Park that are occupied by osprey nests. But there are twice as many tree sites as that around Yellowstone Lake alone.

He says he has seen the nests "floored with cedar bark. Often these osprey added a rim of green pine tips to their nest. In each case it looked like a large wreath of green laid on the nest floor, surrounding the eggs and sitting bird."

Bendire (1892) writes:

The most picturesque nesting site of the Osprey I ever saw was located in the midst of the American Falls of Snake River, Idaho. Right on the very brink of these, and about one-third of the way across, the seething volume of water, confined here between frowning walls of basalt, was cleft in twain by a rocky obstruction which had so far withstood the ever eroding currents, and this was capped with a slender and fairly tapering column of rock rising directly out of the swirling and foaming whirlpool below. On the top of this natural monument, whose apex appeared to me to be scarcely 2 feet wide, a pair of Ospreys had placed their nest and were rearing their young amidst the never ceasing roar of the falls directly below them.

About the inland lakes of California the favorite nesting sites seem to be the broken tops of dead pine trees, sometimes standing in or near the water and sometimes several miles from it; some of these are very lofty, 75 to 112 feet from the ground and often inaccessible. But on the coastal islands, where there are no large trees, the ospreys build their nests on pinnacles of rock or on out-lying rocks, where they are not easily reached. In Lower California they sometimes nest in the giant cacti, which offer firm support and discouragement to climbers; on the islands here they build ground nests on the higher beaches.

The enormous nests of the osprey are made mainly of large sticks, sometimes 4 feet long or longer and as large as a man's wrist, mixed with sods and almost anything that the birds can pick up. As they last for many years, with annual additions, the older material becomes thoroughly rotted, and the nests become heavy enough to break down any but the stoutest trees.

C. S. Allen (1892) records the following list of material that he personally observed in the nests on Plum Island:

Brushwood, barrel staves, barrel heads, and hoops; bunches of seaweed, long masses of kelp, mullein stalks and cornstalks; laths, shingles, small pieces of boards from boxes; parts of oars, a broken boat-hook, tiller of a boat, a small rudder, and parts of life preservers; large pieces of fish nets, cork, and cedar net floats, and pieces of rope, some of them twenty feet in length; charred wood, sticks from hay bales, and short, thick logs of wood; a toy boat, with one sail still attached; sponges, long strings of conch eggs, and eggs of sharks and dogfish; a small axe with broken handle, part of a hay rake, old brooms, an old plane, a feather-duster, a deck swab, a black-ing-brush, and a bootjack; a rubber boot, several old shoes, an old pair of trousers, a straw hat, and part of an oil skin "sou'wester"; a long fish line, with sinkers and hooks attached, wound on a board; old bottles, tin cans, oyster shells, and large periwinkle shells, one rag doll, shells and bright colored stones, a small fruit basket, part of an eel pot, a small worn out door mat; wings of ducks and gulls, sometimes with parts of the skeleton attached, and one fresh crow's wing, as already related. A strange feature was the frequent presence of *bleached bones* from the pasture, as the ribs and long bones of sheep and cattle, and especially *sheep skulls*. Nearly all the old nests had masses of dried cow dung, and large pieces of sod, with grass still growing.

Others have noted similar interesting collections of materials in the nests, but Chester C. Lamb (1927) found some of the most unique nests on Natividad Island, Lower California; he says: "All the nests examined were made partly of Black-vented Shearwater wings, and of one nest seen, all except a part of the foundation was entirely made of wings."

Eggs.—The osprey lays almost invariably three eggs, occasionally only two and more rarely four. T. E. McMullen's series of 100 sets contains 12 sets of four and none of two. Griffing Bancroft's series of 49 sets contains 4 sets of four and 11 sets of two. My experience has been that sets of four are less than 5 percent of the total. F. A. E. Starr tells me that he knows of a set of five, and Mr. Allen (1892) once found five in a nest on Plum Island. Reginald Heber Howe, Jr., (1895) reports a remarkable brood of seven young, only four of which survived. One of our pairs laid sets of four eggs for three years in succession and then laid a set of two. Another pair laid two sets of four, one in 1897 and one in 1902, with normal sets in the intervening years.

The eggs of the osprey are the handsomest of all the hawks' eggs; they show considerable variation, and the coloring is very rich; a selected series of them is a great addition to an egg collector's cabinet. I shall never forget my envious enthusiasm when a rival boy collector showed me the first fish hawk's eggs I had ever seen. Nor could I ever forget the peculiar pungent odor that clings to these eggs after many years in the cabinet, a fragrant reminder of many hard climbs.

The eggs are usually more elongated than other hawks' eggs, but they vary greatly in shape from ovate to short-rounded, elliptical, or elongate-ovate. The shell is fairly smooth and finely granulated. The ground color, which is often largely or wholly concealed, may be white, creamy white, pinkish white, "pale pinkish cinnamon", "fawn color", "light pinkish cinnamon", or "vinaceous-cinnamon". They are usually heavily blotched and spotted with dark rich browns, or bright reddish browns, "bone brown", "liver brown", "bay", "chestnut", "burnt sienna", or various shades of "brownish drab." Rarely they are marked with only the drabs, but often with both browns and drabs. The markings are sometimes concentrated at one end, or they form a ring, leaving much of the ground color exposed. Some are marked like duck hawks' eggs, or caracaras' eggs, and some like red-shouldered hawks' eggs. Very rarely an egg is nearly immaculate. The brighter colors fade with age.

The measurements of 312 eggs in Mr. McMullen's collection average 61 by 45.6 millimeters; the eggs showing the four extremes measure **68.3** by **50.4**, **55.2** by **45.5**, and 60 by **41.7** millimeters. In the Bancroft collection there is a longer egg, measuring **69.5** by 48, and a shorter egg, measuring **54.8** by **42.6** millimeters.

Young.—Incubation, which is apparently performed solely by the female, lasts for about 28 days; the male feeds the female at the nest while she is incubating, but she sometimes leaves the nest for exercises or to fish for herself. Only one brood is raised in a season, but, if the eggs are taken, a second set will usually, though not always, be laid within three or four weeks. The earliest date on which I have found newly hatched young in Massachusetts is May 25; the latest date for unhatched eggs is June 18. The young remain in the nest about eight weeks. I have seen the young leave the nest as early as July 26, but most of them do not leave until the first week in August or later.

At first the young are very weak and helpless, lying prone in the nest and hardly able to lift up their heads. At this early stage, I suppose, they are fed on semidigested, regurgitated food, "fish chowder." When ten days or two weeks old, they are able to sit up and move about some. At this age, they are fed on bits of raw fish, the male bringing in the fish, which is fed to the young by the female. Mr. Skinner, in his notes, describes a typical feeding scene very well, as follows:

I noted the male on the stub of a dead tree not far away, but with a fish. He had cut off the head and disposed of the entrails. A moment after, two or three screams sounded and the male dropped down on the nest with the fish. Then the female stood up quietly and the young birds immediately became interested. The father stood on the fish, that I judged had been about a pound in weight, and tore it up, giving it bit by bit to the mother, and occasionally a tiny piece directly to a young bird. It was the mother, however, that did most of the feeding to the youngsters. After receiving the fish in not larger than half-inch pieces, she 'chewed' them a bit, and then gave the nestlings some after reducing the size somewhat. The proceedings were very orderly, the young birds remaining quietly in their places and not moving toward the fish only a few inches from them. Apparently, the trout was completely devoured; as I did not see any other disposal made of the bones and skin, I assumed that they were eaten along with the flesh.

Young ospreys are fed at infrequent intervals. I have read that they are fed only twice, or possibly three times, a day, before 8.30 a. m., around noon, and after 4.30 p. m. I have never watched a nest all day, but I have seen them fed at various times during both forenoon and afternoon. I believe that there are no regular feeding times but that feeding depends on the size and number of the young and the size of the fish caught. A large fish might serve for more than one meal. The feeding time also depends on when the adult succeeds in catching the fish, which is fed as soon as it is caught.

During the first few weeks, the young are only scantily covered with down, which matches their surroundings in the nest; the nests are usually in open situations, exposed to the full glare of the midsummer sun; they therefore suffer greatly from the heat, panting with open mouths and with moisture dripping from their tongues. The mother appreciates this and spends much time on hot days standing over them with half-open wings to shield them from the sun. After the young are well feathered, during the last few weeks of nest life, this protection is no longer needed, and the young are left alone in the nest for long periods.

Young ospreys are well camouflaged with concealing coloration during the downy stage, so well, indeed, that they might easily be overlooked by an aerial enemy; they are also past masters in the art of "freezing", or feigning death. At a note of warning from their parents they lie flat in the nest with wings partly extended and neck stretched out on the floor of the nest, or hanging down among the outer sticks, and there they stay, absolutely motionless, until their mother gives them a note of assurance. They will even allow themselves to be handled without showing any signs of life, except for the motion of breathing or the winking of an eyelid. I have seen young ospreys hold this hiding pose for over an hour, even when partially fledged. Once, as I approached an osprey's nest in a grove, the old birds made a great outcry; and, as I came near enough to see the nest, no young were visible. I withdrew and concealed myself. After the old birds had flown away and all was quiet, I saw three half-grown young stand up in the nest and watch for their mother's return. She came at last, saw me, and gave the warning cry. The young immediately dropped down out of sight; and although I remained in the vicinity for over an hour, the young never showed themselves again.

Very different behavior was noted in another nest in the same grove, which held two large young, fully fledged and nearly ready to fly. These youngsters evidently had nothing to fear, for they stood up in the nest constantly, craning their necks to watch my movements, in spite of the warning cries of both parents, who seemed greatly concerned. Probably the hiding pose is of importance only while the young are small and subject to attack by flesh-eating birds and is no longer necessary after they are large enough to defend themselves.

There is evidently considerable mortality among young ospreys; I have repeatedly noted that nests, occupied by brooding birds in May, were empty and deserted long before the time for the young to have flown. I once found a half-grown young lying on the ground, under a nest I was watching, with a badly crushed skull and one claw torn out. There had been a severe thunder storm the night before, which may have caused the young bird to fall out of the nest, and the skull may have been crushed by striking a stone wall directly under the tree.

Mr. Abbott (1911) gives a good account of the behavior of young ospreys, as follows:

Not until they are well feathered have I ever heard them emit anything approaching Osprey-like sounds; I have then observed them imitate the cry of their parent overhead, in a charmingly babyish and amusing manner.

At this latter age they add to the death-feigning instinct of the earlier period, a most interesting habit, which we may term "looking fierce." If, as they lie flat in the nest, they are approached too closely or touched, the first sign of life is a bristling of the feathers on the back. If the intrusion be continued they rise suddenly in the nest and turn toward one with ruffled feathers and glaring eyes, which, coupled with a desire to bite when opportunity offers, is evidently calculated to scare the boldest of assailants. It does not take one long to discover, however, that this display of fierceness is mere show, and that even with its formidable bill the young bird is apparently incapable of inflicting a painful wound. The attitudes assumed by

young Ospreys during this "looking fierce" operation are often ludicrous in
the extreme. They will spread or trail their wings, lower their heads in wicked
fashion, raise their crests, and in general assume as formidable an aspect
as possible. Sometimes they exhibit the power of extending the feathers of
the throat and cheeks, forming a sort of mask. After standing for a few moments in this "terrifying" attitude, the strength of the young bird begins to
ebb and his muscles to relax; he will fall back on his "heels," and his head
will begin to droop forward. At this stage he will often be resting on "all
fours," so to speak, the "shoulders" of his wings acting as supports to the
fore-part of his body. They gradually give way, however, and the bird's bill
comes closer and closer to the nest, until at last he is once more in his original
prone and death-like position.

By the time the young are five or six weeks old they are strong
enough to stand up and feed themselves. The parent no longer tears
up the food and feeds them, as described above, but drops the fish
in the nest and flies away. The young then take turns feeding, standing over the fish, or on it, in a crouching attitude, with wings half
spread and drooping. They are well behaved; I have never seen
any evidence of quarreling; even with a small fish, they seem content to take turns. They are very neat in their sanitary habits; after
a meal each bird backs up to the edge of the nest and squirts its
excrement clear of the nest. As the wings develop faster than the
young bird's strength, they are allowed to droop, or are used as
additional supports. About two weeks before the young are ready to
fly, they begin their wing exercises, standing up and flapping the
wings vigorously for several minutes at a time. After a week or so of
this exercise, their wings are strong enough to lift them up a few
feet above the nest, and then to attempt short, uncertain flights to
nearby branches or to a perch above the nest. Finally, confidence in
the power of its wings, or the example set by its parents, prompts
the boldest of the young to make its first real flight away from the
nest, a supreme moment in its life. With surprising ease it sails or
flaps along, but it soon becomes tired and looks for a place to perch.
Its attempts to alight on a treetop are awkward and uncertain; it
has not learned to grasp a slender perch and finds it difficult to get
its balance with much flapping of wings and wiggling of tail. It
may be forced to alight on the ground to rest; I have often seen
one do this; and it can rise from the ground quite easily. Such
flights are short at first and the young always seem glad to return
to the firm flat top of the nest, which will be their headquarters,
bedroom, and dining room for several weeks yet.

Throughout the summer the young ospreys associate with their
parents, playing with them in flying exercises, following them to the
fishing grounds, and learning to fish for themselves. This latter they
seem to do instinctively, as Mr. Forbush (1927) says: "They require
no teaching, as individuals that have been brought up by hand and

have never seen their parents catch fish, will begin fishing for themselves as soon as they have fully mastered the intricate problem of flight. At first they have very little success. I have seen a young bird plunge into a river seven times in succession without securing a fish, but the bird did not appear to be in the least discouraged, for it continued to follow the river and scan its waters in search of a victim."

Plumages.—The young osprey, when first hatched, is entirely unlike the young of any other hawk. It is not naked, as has been said, but is completely covered with very short, soft down, protectively colored. The color is mainly in shades of "pale pinkish buff" tinged with "cinnamon" on the crown; the lores, auriculars, and a large spot on the occiput are "bone brown", and the sides of the neck are tinged with this color; the shoulders, back, wings, and rump, except for a wide buffy stripe down the center of the back, are "bone brown", with "wood brown" tips; the entire under parts are "pale pinkish buff" and unmarked. This down is worn with but little change except fading until the plumage appears. There is no secondary down, as in young eagles and many hawks. A larger young bird, about a foot long, probably about three weeks old, shows some slight changes; the lores and auriculars are darker, brownish black, the latter nearly clear black; short feathers have appeared on the hind neck, "cinnamon-buff" to "pinkish cinnamon" in color; small feathers, similarly colored, are appearing on the wings and in the broad central, dorsal stripe, which has now faded to dull white; the dark areas, in which the down is still short and thick, are now "bister" in color; and the central belly is "drab."

When about four weeks old, the plumage begins to appear, the black primaries showing first, then the black and white pattern of the head and the dusky, yellowish-tipped plumage of the mantle; when five weeks old the young bird begins to look like a real osprey. At this stage, in fresh juvenal plumage, the crown is buffy white, heavily streaked with brownish black; the hind neck is tinged with "cinnamon-buff"; the rest of the upper parts are "warm sepia" to "bister", the feathers broadly tipped with "cream-buff"; the tail is broadly tipped with "cinnamon-buff"; the throat and fore breast are washed with "cinnamon-buff", and the rest of the under parts are white.

This plumage is worn, with only slight and gradual changes, throughout the first year. The buff tips fade out to white and then wear away during the first winter. The head and neck become whiter. Some body molt begins late in fall or early in winter and is prolonged through spring and summer. The wings and tail may be molted in spring; but probably oftener in summer or fall. When the young bird is 18 months old, its plumage is practically adult.

Adults have a similar, prolonged molt, which may be in evidence during any month in the year.

The American osprey is supposed to have much less spotting on the breast than the European bird, but this character is none too well marked and none too constant in the series I have examined. The spotted breast, in the American bird, is said by some writers to be a character of the female, but in a series of 33 males and 22 females I find but little evidence of it. Among birds with white or nearly white breasts I find 13 males and 4 females; with lightly spotted breasts, there are 14 males and 9 females; and with heavily spotted breasts, I find 6 males and 9 females. A heavily spotted bird shows on each feather a large, concealed spot of "olive-brown" and a smaller, subterminal, triangular spot of "clay color" and "snuff brown", with a dusky shaft streak; there are all gradations from the above to a bird with only the shaft streaks. In the European bird, the throat and upper breast are pale brown, sometimes tinged with rusty, forming a broad pectoral band. Very few American birds even approach this condition; these may be the younger birds, for I believe that the breast becomes whiter with advancing age.

Food.—Fish is the almost exclusive food of the osprey, well named the fish hawk. The following species have been recorded in its food: Alewife or herring, bluefish, blowfish, bonito, bowfin, carp, catfish, eel, flounder, flying fish, goldfish, hornpout, menhaden, mullet, perch, pickerel, pike, salmon, shad, squiteague, sucker, sunfish, tomcod, trout, and whitefish; doubtless many others might be included. As the osprey is not a deep diver, it catches only such fish as swim on or near the surface, or in rather shallow water. Walter B. Savary writes to me of the following amusing incident: "This summer, while watching a fish hawk at his fishing, I saw him catch and lose four blowfish (*Spheroides maculatus* Nichols), the fish escaping each time by inflating itself until the hawk's talons lost their hold. The bird was near at hand and I, through my field glasses, could see the fish as he blew himself up, and, when he fell, lie on the surface until he could deflate. The bird never got above ten feet from the water before the fish got loose."

Audubon (1840) says that the fish hawk catches flying fish while they are swimming near the surface but does not attempt to catch them in the air. I was much surprised one day to see an osprey flying over at short range with a small flounder in its talons; the hawk's claws were embedded in the back of the fish, whose white belly and twitching tail were clearly seen. I marveled at the bird's ability to dive deeply enough to capture a fish on the bottom, until I remembered that these small flounders often swim into shallow water; but I still marvel at the keen vision needed to locate a fish that matches the bottom so closely. The osprey is a clean sportsman and prefers to catch living fish, but it is not above picking up a dead fish if it is still fresh; but it is not a carrion feeder like the bald eagle and will not touch a tainted fish.

It has been stated repeatedly by good authorities, on apparently reliable evidence, that the osprey sometimes tackles a fish too big for it to lift, is unable to release its grip, is dragged under water, and is drowned. The evidence is too convincing and there is too much of it to dispute the fact, as the drowned osprey has been found on several occasions, sometimes still attached to the dead fish. Mr. Nicholson tells me that he has seen an osprey dive beneath the surface and never appear again. But it seems to me incredible that such a skillful fisherman would be foolish enough to tackle a fish big enough and strong enough to drag under water so powerful a bird with such a broad expanse of wing. It seems still more inconceivable that a bird that can so easily drop a fish in the air or at its nest cannot release its claws under water, even to save its life. There must be some other explanation for what has occurred; possibly the large, horny scales on the back of a sturgeon might entrap the claws of an osprey, if the bird were rash enough to tackle it. An osprey has been known to break its wing in diving (Fisher, 1893); I once caught one on the ground with a broken humerus.

Most of us have seen the osprey's method of hunting, as it flies along at a moderate height above the water, scanning the surface for its prey, flapping or sailing, or stopping to hover above a likely spot. Its keen eyes can sometimes locate a fish when flying at a height of 100 or 200 feet, but oftener it hunts at 30 or 100 feet above the water. When a fish is sighted, it plunges downward with half-closed wings enters the water with a splash, sending the spray flying, striking the water breast first, with wings extended upward, and seizes the fish in its strong talons; usually it does not go much below the surface, but sometimes it disappears for an instant or shows only the tips of its wings above the water. If successful, as it usually is, it rises heavily from the water with the fish in its talons, shakes the water from its plumage, and flies away to its nest or favorite perch. But not every attempt is successful, and not every fish seen is in a position that will insure a successful dive, so the osprey may pass on or even check its plunge in midair; but it keeps on trying until its persistence is rewarded. Mr. Skinner says of the ospreys in Yellowstone National Park: "Not only do they plunge into lakes and quiet stretches of streams, but I have also seen them hunt the swollen and raging Gardiner River in flood."

The fish is invariably carried head first; probably it is usually caught that way, the approach from the rear being oftenest successful; but if caught otherwise it is turned about in the air. A small fish is carried in one claw, but a large one requires both, one claw in advance of the other. Just before the bird alights, the hinder claw is released to grasp the perch. I have seen an osprey bring to its nest a fish that must have weighed at least four pounds; I have read that it can carry one of 6, or even 8 pounds, though the latter seems unlikely.

Having secured its fish, the osprey flies with it to its nest, to some favorite perch, or to an unused nest, to eat it. It holds the fish down under one foot and is very deliberate about eating it; one that I watched waited half an hour before beginning to eat, and at the end of another half hour it had not finished; when I moved it flew away with much of the fish uneaten.

Another alighted with a fish on a half-built nest but did not start to eat it at once; when its mate came in and alighted beside it the first bird spread its wings and tail, crouching over the fish to guard it, until its mate flew away; this was before the eggs were laid and each bird had to fish for itself; eventually it flew away with the fish and half an hour later returned to the nest to eat it.

Mr. Abbott (1911) quotes Ernest H. Baynes, who had two young ospreys as pets, as follows:

They often began by picking out the eyes, perhaps because those organs were conspicuous and easily removed. They held their food in their claws, and usually before seizing any part of it, they would "finger" it, so to speak, with their bills, as though feeling for a good hold. They would tear off large pieces, jerk them backwards into the throat and swallow them. They ate every part of the fish except the harder bones. Tough pieces were removed by a steady upward pull, and the ends of bones were twisted off with a pivotal movement such as a man would use to draw a nail with a pair of pincers. Later, they ejected the bones and other indigestible particles in the form of pellets.

If the osprey ever takes any kind of food but fish it must be on very rare occasions; I can find very little positive evidence of it. It has been reported as eating young ducks, snakes, and frogs. Witherby (1924) says that it has been known to take chickens and that beetles have been found in its stomach. I once found a domestic pigeon in a nest, but this was probably brought in with other miscellaneous material so often found in the nests. Dr. Robert C. Murphy told Mr. Abbott a remarkable story of an osprey that was killed by a woman while it was raiding her henyard; I quote it, in part as follows: "The woman told me that on the afternoon of the previous day, which had been rainy, she had been disturbed by a commotion among her chickens, and on going into her yard, had found the Hawk with its talons sunk in a hen, and flapping violently in an attempt to fly off with its prey. She had killed the robber with a stick, and had freed the hen, which, however, died during the night. The Hawk which she gave me, was in a starved and emaciated condition, and was, of course, much bedraggled from lying out of doors in the rain."

Benjamin R. Warriner (1934) saw an osprey catch a turtle, of which he says: "When he rose from the water he carried a black object which I could not at first identify. Then I could see—a turtle some six or seven inches across. The Osprey fought desperately to hold his victim, but of course the turtle's bony covering prevented the bird's claws from penetrating below the surface. Suddenly he lost his hold and the turtle came down hard upon a big stump, and bounced off into the water."

Behavior.—The flight of the osprey is powerful and well sustained, swift and dashing at times, but oftener slow and heavy. It soars at times on motionless wings but ordinarily proceeds with deliberate flappings. In ordinary traveling flight the wing is somewhat flexed at the wrist joint; even when the bird is soaring the wings are not held so straight out as they are in the Buteos. This curvature of wing makes the osprey recognizable often at a long distance. When hovering over a fish, the wings are flapped rapidly, but the body is not held in the vertical position assumed by the kingfisher. In flight the feet are extended under the tail.

The behavior of ospreys about their nests is characteristic, quite uniform, and quite different from that of any other bird of prey. As soon as a stranger is seen approaching, the sitting bird, who is always on the lookout for trouble, rises up and begins her musical, whistling cackle; as the man draws near, her notes become shriller, increasing to loud, ear-piercing screams, intercepted by the short, sharp, impulsive *ick, ick, ick* notes, on a lower key and more metallic. She soon leaves the nest and circles about, screaming lustily; she is then joined by her mate and perhaps one or two of her neighbors. If the man climbs the tree, he must expect to be attacked, or at least threatened, although this does not always happen. The method of attack is to make vicious swoops at the intruder, but seldom striking him.

I have frequently been attacked and even struck on the head and shoulders, but have never even been scratched; a rush of wings, as the bird swerves uncomfortably close, is the usual experience. Many birds are content to fly about and scream at a safe distance, but they do not fly away and desert their nests, as other hawks do. Ospreys that nest near houses seem to recognize harmless friends and pay no attention to familiar human beings; but with strangers their behavior is very different.

Only once have I ever found an osprey "asleep at the switch" in daylight. We visited an old nest 60 feet up in a tall pine in some woods but saw no birds about it and concluded that it was unoccupied. After a while we heard an osprey whistling in the vicinity; we returned to the big pine and rapped it; much to our surprise the bird flew off the nest. That was the only time I have ever had to rap a tree to flush an osprey, except at night when they probably sleep on the nest. I believe that both birds spend the night in the nest tree; we have several times in complete darkness flushed both birds while climbing the tree.

In spite of their frequent plunges into the water, ospreys like to bathe occasionally. Pearson, Brimley, and Brimley (1919) say: "A curious habit of the Osprey is that of 'foot-washing.' From flapping in wide circles over the lake a bird may be seen suddenly to half close its wings and glide toward the water in a long, gentle sweep. When almost touching the surface, the feet are dropped to the full extent of the long legs and a horizontal flight of fifteen or twenty yards follows, while the feet drag in the water. The reason for this action is doubtless to cleanse the toes and claws of the fish-slime that must necessarily accumulate on them."

L. McCormick-Goodhart (1932) watched an osprey bathing on a sandy point and writes:

The bird stands in about six inches of water, and bathes in the same manner as other birds, by ducking himself under and then vigorously flapping his wings. On May 15, 1932, however, I witnessed what appeared to be a new method of bathing. When I observed the bird this time (through a 16x binocular) it was flying towards me, about six feet above the surface. It was observed suddenly to descend into the water, and then adopt a sort of vertical American-eagle attitude while flapping its wings two or three times before rising again. It then again flew along the water, keeping the same general direction, and repeated this form of immersion some five times, finally rising to a normal flight.

I once saw a different method of bathing. While driving past a small pond we saw an osprey perched on a low stake on the edge of the pond, and we stopped to watch him. He rose, circled around the pond once or twice, and then dove into the water head first, going entirely below the surface; he rose to the surface almost immediately and flopped along it for a few yards, flapping his wings vigorously and ducking his head under water occasionally. He then rose, shook the water from his plumage, circled the pond again, and flew back to his perch. This performance was repeated three or four times at intervals of only a few minutes. We did not see him preen his plumage. He was evidently not fishing but seemed to be enjoying his bath.

In its relations with other species the osprey is a peaceful, gentle, and harmless neighbor. Only such species as might harm its eggs or young are attacked or driven away. If unmolested it attends strictly to its own business, in which it is very industrious. A very large proportion of the osprey's nests that I have seen have contained one or more nests of the English sparrow or the starling among the lower sticks. Mr. Abbott (1911) says that "Purple Grackles especially, commonly build in convenient niches among the sticks even of the ground nests. Being naturally gregarious, they will congregate to the number of six or seven pairs in one Osprey's nest. * * * Ospreys are recorded to have admitted House Wrens and even Night Herons as basement tenants. On the beaches, Meadow Mice have found the nests to be convenient mounds under which to construct their multifarious run-ways."

Mr. Abbott (1911) found the ospreys nesting in the night-heron colony on Gardiners Island, apparently peacefully. But Mr. Allen (1892) gives a different impression of the behavior of the ospreys toward the herons, which the former may have regarded as enemies; he writes:

In the swamp near the Fish Hawks' nests was a colony of Night Herons, nesting in the smaller trees near the swamp. Almost daily a flock of Crows from Connecticut were accustomed to rob this heronry, covering the ground with the shells of the eggs they had eaten, and occasionally treating a few Fish Hawks' nests in the same way. The Fish Hawks seemed to unjustly accuse the Herons of this robbery, as the Herons were constantly persecuted by the Hawks. Whenever a Heron appeared he was instantly set upon by one or more of them, and the Herons would seek safety in the thick underbrush where the Hawks could not follow them. Herons were killed, however, almost daily by the Hawks.

Enemies.—In addition to the thoughtless gunner, who shoots every large bird that he can, and the greedy egg collector, who takes all the eggs he can get for exchange, the osprey has a number of natural enemies, most of which are more annoying than harmful. The bald eagle is undoubtedly its worst enemy. Its well-known habit of attacking the osprey, to rob it of its well-earned prey, has been described in my account of the southern bald eagle and many times by other writers, so I shall not repeat it here. In Florida, where the ospreys often nest well inland, the eagles lie in wait for them as they fly from their fishing grounds back to their nests. The osprey seldom escapes from these attacks, but a clever attempt at dodging, when pursued by two eagles, is thus described by Henry G. Vennor (1876):

On first hearing the shrill screams of its pursuers, the poor bird made desperate efforts by straight flight to reach the drowned wood-lands in which its nest and young were located; but long before it reached these its course was intercepted by one of the Eagles, while the other made repeated and fierce stoops at it from above. The Fish Hawk, however, still held on firmly to its prize, and made repeated attempts to baffle the onsets of the Eagle, in many of which it was successful. Before long both birds had risen to a great height—the one alternatingly surmounting the other; but we could still detect every now and then the gleam of the fish in the sunlight. Suddenly, the

Fish Hawk was seen to descend with great velocity towards the water, and we thought the poor bird had been struck, and perhaps mortally wounded. It, however, as suddenly checked its downward course, and the Eagle which had as quickly followed it, shot past and far below it; and now once more the pursued bird made straight for its nesting site, but again was intercepted by the other Eagle, which made desperate by the protractedness of the chase, struck fiercely at it with piercing screams. Baffled on every side, wearied and blinded with the repeated buffettings of the Eagles, the Fish Hawk, with a scream of rage, let go its prize, which fell head long towards the water.

The osprey will drive away an eagle or any other bird of prey from its nest and young, but I once saw one attack an eagle several miles away from any nest and long after its young were on the wing. Several times it swooped down, as both birds were circling high in the air, almost striking the eagle; the latter turned on its back each time and presented its talons, which sent the osprey scaling off in a hurry. Edward Fuller (1891), quoting "a gentleman who witnessed a scene of this kind", describes a joint attack on an eagle by a colony of ospreys, as follows:

They seemed to have formed a sort of colony for mutual protection, and the moment their foe, the Eagle, made his appearance among them, the cry of alarm was raised, and the vigilant colonists, hurrying from all quarters, attacked the robber without hesitation, and always succeeded in driving him away.

There was always a desperate battle first before the savage monarch could be routed, and I have seen them gathered about him in such numbers, whirling and tumbling amidst a chaos of floating feathers through the air, that it was impossible for a time to distinguish which was the Eagle, until having got enough of it against such fearful odds, he would fain turn tail, and with most undignified acceleration of flight would dart toward the covert of the heavy forest to hide his baffled royalty, and shake off his pertinacious foes amidst the boughs.

Dr. Theodore Gill (1901) quotes an interesting account of the persecution of ospreys by man-o'-war-birds, as observed by I. Lancaster in southern Florida. The ospreys seemed to be in mortal terror of these pirates, who not only made them drop their fish by merely threatening an attack, but sent them, screaming, back to the land in hurried flight. The reason for the ospreys' dread of these black-winged rascals is told in his thrilling account of an attack, by a number of man-o'-war-birds, on an osprey that they had robbed. The poor bird was chased about in the air and all his frantic attempts to escape were headed off, until he became so exhausted that he dropped into the water. Even there his tormentors continued their attack forcing him under the water, until he was finally killed.

Crows are always on the lookout for unguarded eggs and have been known to puncture and suck ospreys' eggs; consequently the ospreys always drive them away from their nesting grounds. Owen Durfee describes such an instance in his notes; he was "interested, in watching a fish hawk flying along near its nest, to see a crow fly up and chase the fish hawk, diving down on its back. Finally the fish hawk grew tired of this and made a savage swoop at the crow. Apparently the crow did not heed this warning, for he once more rose and dove at the fish hawk. Then the latter bird really made things lively for a few moments for the crow, attacking him from all directions, and finally driving him off into the woods, but never appearing to really strike him, the crow dodging in fine shape."

Large herons have been suspected of eating the very small young, and hence are potential enemies. Mr. Nicholson tells me that "the osprey is extremely pugnacious towards both the black and the turkey vultures, which may venture near its nest, and drives them away with great fierceness and display of anger. Possibly the blacks have been guilty of going to the nests and stealing choice morsels or killing the young. I twice found great holes in osprey eggs that were drained of their contents, which I took to be the work of fish crows."

Almost any small bird will drive away from the vicinity of its nest any predatory bird. The kingbird is one of the most aggressive defenders of its home territory and drives the osprey ignominiously away. Blackbirds, grackles, and starlings attack the osprey singly or mob him in flocks, swarming around him until he is glad to beat a retreat. I have seen even a barn swallow chasing one. Unless pestered with overwhelming numbers, the osprey pays but little heed to these small tormentors. Mr. Abbott (1911) says that on Gardiners Island the ospreys that have their nests near a colony of common terns "are being continually harried by the Terns. I have seen an Osprey driven from her nest by a Tern three or four times within a quarter of an hour."

Voice.—The osprey is a noisy bird about its nest, and its vocal efforts are most interesting and quite characteristic. One reason for its popularity among farmers is that it is a good "watch dog", always alert and sure to give the alarm with its loud cries as soon as a stranger approaches. The alarm note begins with a loud, rich, musical whistle, *cheeap, cheeap*, many times repeated; as excitement increases, this is lengthened into a much shriller angry scream of great intensity, interrupted with, or ending in, a sharp, metallic *ick, ick, ick*, a harsh rasping note on a lower key. When undisturbed it has a weak note, which reminds me of the "cheeping" note of a young chicken. There is also a soft note of greeting, as the bird returns to its mate or young, sounding like *chirrup*. Mr. Abbott (1911) has described the alarm notes very well, as follows: "The commonest note is a shrill whistle, with a rising inflection: *Whew, whew, whew, whew, whew, whew, whew*. This is the sound usually heard during migration; and when the bird is only slightly alarmed. When she becomes thoroughly aroused, it will be: *Chick, chick, chick, cheek, cheek, ch-cheek, ch-cheek, cheereek, chezeek, chezeek*, gradually increasing to a frenzy of excitement at the last. Another cry sounds like: *Tseep, tseep, tseep-whick, whick, whick-ick-ick-ck-ck*, dying away in a mere hiccough. And there are endless variations quite incapable of syllabification."

Dr. Frank M. Chapman (1908) calls the food call "a high, rapidly uttered *tweet-tweet-tweet"*; probably it is also a note of assurance to the young that danger has passed. William Brewster (1925) says that "in calm summer weather their musical whistled calls, not unlike those of the Purple Martin, but much louder, fall pleasantly on the ear at frequent intervals, coming from far and near over the shining Lake." Dr. Winsor M. Tyler contributes the following impressions:

The note most frequently heard from the osprey, as it moves northward in migration, pausing to fish in the lakes, ponds, and larger rivers which it meets on its journey, is a rather shrill squeal—not full-voiced, but with a slight hissing quality. The note sounds petulant, and calls to mind a smaller bird than this great hawk with its fine spread of wings.

About its nest the osprey is a noisy bird. It flies off when approached, giving a long series of notes, somewhat whistled in quality, but often harsh and rasping. These notes are uttered with the beak open throughout the series and suggest the cry of the yellowlegs. They may be written *ku-ku-ku,* or *ke-ke-ke,* and are delivered at the rate of the flicker's shouted *wik-wik-wik,* and nearly on the same pitch, although the voice may slide upward a little way, or, at the end, downward to a considerable degree. A modification of this note, shortened to two rapid syllables, the first sharply accented, *ke-oo,* bears a decided resemblance to the call of the evening grosbeak.

The bird on its breeding ground has also a dull chatter and the squeal mentioned above, the inflection often varying, the pitch sometimes sweeping upward, and sometimes dropping a little. All these notes give an impression of querulousness rather than of hostility, and seem inadequate to express the emotions of so large a bird.

During the autumnal migration, and during winter, as I have seen the osprey in Florida, it is for the most part silent.

Field marks.—The bend in the wing and the manner of flight, described above, will serve to distinguish the osprey from other hawks at a great distance. The white breast can be seen almost as far away, and the distinctive head markings are a good field mark at short range.

Fall.—Mr. Hathaway (1905), referring to southern New England, says: "The return movement to their southern habitat commences in August, and probably their place is taken by birds from breeding places in the more northern states which pass on, or linger as the weather may be favorable or not, as late as October or early November. A large majority of the birds which are shot in the fall by hunters are young birds of the year, leaving us to conclude that the adults are the first to depart, while the young follow as they grow stronger and are able to stand the long flight to warmer climes."

Dr. Charles W. Townsend (1905) records the heaviest flight through Essex County, Mass., during the last week in September and says that it "generally precedes a heavy flight of ducks." Mr. Skinner's dates for Yellowstone National Park are about the same; his latest date is October 7.

Throughout the southern portions of its range the osprey is permanently resident; its numbers are greatly increased in winter with migrants from farther north.

DISTRIBUTION

Range.—As a species the osprey is nearly cosmopolitan, and the form of the Western Hemisphere (*carolinensis*) ranges north to Alaska and Labrador (accidental in Greenland) and south casually to Paraguay, Argentina, and Chile.

A tropical form, *Pandion haliaëtus ridgwayi* Maynard, has been described. It is resident in the Bahama Islands, the coasts of Yucatan, and in British Honduras.

Breeding range.—The breeding range extends **north** to Alaska (Kowak Delta, Fort Yukon, and Kandik River); Mackenzie (Fort Rae and Bear Lake River); northern Alberta (Poplar Point); northern Saskatchewan (Ile a la Crosse, Knee Lake, and Churchill River); northern Manitoba (Grass River, probably Churchill, and probably York Factory); northern Ontario (Poplar River and Moose Factory); Quebec (Lake Mistassini, Godbout, and Anticosti Island); and Labrador (Northwest River and White Bear River). **East** to Labrador (White Bear River and Wolf Bay); Newfoundland (St. George Bay); Nova Scotia (Sydney, Antigonish, and Halifax); Maine (North Haven and Jericho Bay); southeastern New Hampshire (Manchester); Massachusetts (Swansea, Wareham, and Falmouth); New York (Plum Island and Gardiners Island); New Jersey (Red Bank, Pennsville, and Cape May); Virginia (Chincoteague Island, Hog Island, and Newport News); North Carolina (Orton Lake); South Carolina (Charleston); Georgia (Savannah, Blackbeard Island, Cumberland Island, and St. Marys); and Florida (St. Augustine, New Smyrna, Wekiva River, Lake Istokpoga, and Florida Keys.) **South** to Florida (Florida Keys, Bocagrande, Marquesas Keys, St. Marks, Alaqua Bayou, and Pensacola); Alabama (Perdido Bay, Orange Beach, and Spring Hill); Louisiana (New Orleans and Bayou Sara); probably rarely Texas (Refugio County and Corpus Christi); and Lower California (Tres Marias Islands and Cape San Lucas). **West** to Lower California (Cape San Lucas, Santa Margarita Island, Natividad Island, Cerros Island, San Benito Island, San Geronimo Island, San Martin Island, and Todos Santos Island); California (San Diego, San Clemente Island, San Nicolas Island, Clearlake, Garberville,

Humboldt Bay, and Requa); Oregon (Ochoco River); Washington
(Bellingham and Simiahoo); British Columbia (Chilliwack, Court-
enay, Friendly Cove, probably Massett, and Atlin); and Alaska
(Sitka, Tocotna, North Fork, Nulato, and Kowak Delta).

Winter range.—In winter the osprey is found **north** to California
(Farallon Islands); Arizona (Salt River); Texas (Eagle Pass and
Rockport); Louisiana (State Game Preserve and New Orleans);
Mississippi (Natchez); and Florida (St. Marks and Fruit Cove).
East to Florida (Fruit Cove, Daytona, Ponce de Leon Inlet, Orlando,
Royal Palm Hammock, and Alligator Lake); Haiti (Seven Brothers
Islands and Monte Christi); Puerto Rico (Mameyes and Vieques
Island); the Lesser Antilles (St. Vincent, the Grenadines, Carriacou,
Grenada, and Trinidad); probably eastern Venezuela (Waini River
and Barima River); probably British Guiana (Abary River); west-
ern Brazil (Caicara); and Paraguay (Rio Negro). **South** to
Paraguay (Rio Negro); rarely Argentina (Tucuman); and rarely
Chile (Reloncavi Bay). **West** to rarely Chile (Reloncavi Bay);
northwestern Peru (Santa Luzia); Ecuador (Chone); Colombia
(Bonda); Panama (Farfan); Costa Rica (Guacimo); Nicaragua
(Escondido River); Oaxaca (Ventosa Bay); Federal District of
Mexico (Lake Chalco); Lower California (Natividad Island and
Cerros Island); and California (San Diego, Santa Cruz Island, and
Farallon Islands).

Ridgway (1874, p. 324) reported the osprey as resident at Mount
Carmel, Ill., but it is believed that was an error. A specimen reported as seen at Morristown, N. J., on December 25, 1918, probably
was misidentified.

Spring migration.—Early dates of arrival in the spring are:
Georgia—Savannah, February 24; St. Marys, February 24; Darien,
March 10; and Athens, April 9. South Carolina—Charleston, Feb-
ruary 14; Frogmore, March 3; and Columbia, April 13. North Caro-
lina—Raleigh, March 5; and Walke, March 21. Virginia—Newport
News, February 17; Hicks Wharf, March 15; and Camp Eustis,
March 19. District of Columbia—Washington, March 19. Mary-
land—Baltimore, March 11; Rock Hall, March 14; Chestertown,
March 16; and Cambridge, March 17. Pennsylvania—Wayne, Feb-
ruary 10; Germantown, February 18; Westtown, March 17; and Lan-
caster, March 30. New Jersey—Cape May, March 13; Mount Bethel,
March 14; New Brunswick, March 19; and Asbury Park, March 25.
New York—Shelter Island, March 18; Gardiners Island, March 20;
and Rochester, March 31. Connecticut—New Haven, March 24;
Portland, March 25; and East Hartford, April 1. Massachusetts—
Taunton, March 16; Dartmouth, March 29; and Boston, March 30.
Vermont—St. Johnsbury, March 19; Wells River, March 28; and

Rutland, April 19. New Hampshire—Concord, April 8; Peterborough, April 10; and Durham, April 13. Maine—Orono, March 27; Eagle Island, April 5; and Pittsfield, April 6. Quebec—Montreal, April 10; East Sherbrooke, April 14; and Quebec City, April 29. New Brunswick—Scotch Lake, April 8; St. John, April 11; and St. Andrews, April 14. Nova Scotia—Bridgetown, April 18; and Pictou, April 12. Prince Edward Island—Alberton, April 24; and North River, May 1. Newfoundland—Raleigh, May 10. Arkansas—Pike County, April 2; and Amity, April 11. Tennessee—Nashville, April 14. Kentucky—Lexington, April 17; and Pine Mountain, April 28. Missouri—Warrensburg, April 8; Auburn, April 20; and St. Louis, April 23. Illinois—Shawneetown, April 1; Alton, April 4; and Elgin, April 12. Indiana—Terre Haute, March 22; Indianapolis, March 25; and Rockville, April 1 (there also is an unusually large number of observations for Indiana early in March, and some as early as February 20, Bloomington, and February 11, La Fontaine). Ohio—Columbus, March 28; Barnesville, April 1; Youngstown, April 3; and Oberlin, April 5. Michigan—Ann Arbor, March 10; Sault Ste. Marie, March 27; and Pontiac, April 8. Ontario—Ottawa, April 10; London, April 11; and Toronto, April 16. Iowa—La Porte City, March 15; Sioux City, March 19; and Davenport, March 28. Wisconsin—Milwaukee, March 10; Foxlake, March 21; and Madison, April 1. Minnesota—Fort Snelling, April 11; Elk River, April 13; and Lanesboro, April 15. Kansas—Lawrence, April 14; and Ellis, April 16. Nebraska—Neligh, April 1; and Lincoln, April 23. South Dakota—McCook Lake, April 19. North Dakota—Jamestown, April 23; and Snyder Lake, April 24. Manitoba—Margaret, April 21; and Aweme, May 2. Saskatchewan—Indian Head, April 8; and Crooked Lake, May 3. Mackenzie—Fort Resolution, May 10. Colorado—Loveland, April 8; Durango, April 10; and Denver, April 25. Wyoming—Yellowstone National Park, April 9; Laramie, April 23; and Jackson Hole, May 2. Idaho—Minidoka Bird Refuge, March 1; and Rathdrum, April 1. Montana—Columbia Falls, April 16; and Bitterroot Valley, April 17. Alberta—Pine Lake, April 26; and Banff, April 29. Oregon—Mercer, April 1 (may be a rare and local resident). Washington—Cashmere, March 27; Puyallup, March 31; Chelan, April 17; and Tacoma, April 18. British Columbia—Okanagan Landing, March 27; Mirror Lake, April 2; Courtenay, April 3; Chilliwack, April 6; Hastings, April 10; and Masset, April 24. Alaska—Nulato, May 8; and Kowak River, June 3.

Fall migration.—Late dates of fall departure are: Alaska—Bethel, September 7; and Kowak River, September 20. British Columbia—Chilliwack, October 11; Mirror Lake, October 20; and Okanagan Landing, October 30. Washington—Yakima, November 15. Alberta—Belvedere, September 17. Montana—Rockhill, October 2.

Idaho—Priest River, October 6. Wyoming—Yellowstone National Park, October 4. Colorado—Yuma, September 29. Mackenzie—Great Slave Lake, September 20; and Fort Good Hope, October 2. Saskatchewan—Indian Head, September 10. Manitoba—Treesbank, October 3; and Aweme, October 12. South Dakota—Grand River Agency, October 7; and Vermillion, October 15. Nebraska—Badger, October 1; and Jackson Lake, October 9. Kansas—Lawrence, October 10. Oklahoma—Norman, November 25. Minnesota—Lanesboro, October 21. Wisconsin—Elkhorn, October 1; Racine, October 3; and Burlington, November 10. Iowa—Osage, October 13; Davenport, November 1; and Keokuk, November 12. Ontario—Point Pelee, October 14; Ottawa, October 17; and Port Dover, October 19. Michigan—Sault Ste. Marie, October 19; South Lyon, October 21; and Detroit, November 18. Ohio—Upper Sandusky, October 12; Austinburg, October 29; and Columbus, October 30. Indiana—New Harmony, October 28 (once observed on December 6); and Crawfordsville, November 21. Illinois—Lake Forest, September 26; Horseshoe Lake, September 29; and Springfield, October 10. Missouri—Marionville, September 13; Monteer, September 22; and Iberia, October 15. Kentucky—Bardstown, October 18. Tennessee—Nashville, October 7. Prince Edward Island—Alberton, September 25. Nova Scotia—Sable Island, September 30; and Pictou, October 13. New Brunswick—Scotch Lake, November 7. Quebec—Montreal, September 17; and Quebec City, September 26. Maine—Phillips, October 9; Lewiston, October 13; Avon, October 26; and Owls Head Light Station, October 28. New Hampshire—Concord, September 21. Vermont—Woodstock, October 19; and Wells River, November 8. Massachusetts—Harvard, October 18; Danvers, October 24; Marthas Vineyard, October 30; and Boston, November 8. Connecticut—Hadlyme, October 28; Meriden, October 29; and Danbury, November 13. New York, Ithaca, October 25; Howard, November 12; and Fire Island Light, November 25. New Jersey—Elizabeth, October 13; Sandy Hook, October 16; Hackettstown, November 2; and Morristown, December 25. Pennsylvania—Renovo, November 7; Doylestown, November 9; Berwyn, November 13; and York, December 15. Maryland—Chestertown, October 23; and Baltimore, November 8. District of Columbia—Washington, November 30. Virginia—Backbay, November 20. North Carolina—Raleigh, October 5; and Knotts Island, November 17. South Carolina—Charleston, December 6. Georgia—Atlanta, October 20.

More definite information concerning the fall migration of ospreys on the Atlantic coast is afforded by a consideration of the available banding data. Five birds banded as fledglings at Avalon, N. J., one in June and four in August, were all recovered in September and

October, four in the former month and one in the latter. Three
of the September recoveries were in West Virginia and one was in
North Carolina. The October bird was taken at McIntyre, Fla.
Another, banded at Gardiners Island, N. Y., in July, was caught
in a steel trap the following September at Hancock, Md.; one
banded in July at Milford, Del., was taken at Sitlington, Va., in
September; and another banded in July at Gardiners Island, N. Y.,
was recaptured at Shell, S. C., also in September, which is clearly
the principal month of the southward movement on the Atlantic
coast. While the West Virginia records seem to indicate a migration
route in the interior, H. H. Bailey (1913, p. 281) has recorded one
that came on board a vessel about 100 miles east of Cape Hatteras,
N. C., on October 16, 1911. The banding files contain the records
of two more recoveries in this same general area. One, banded
at Orient, Long Island, N. Y., on June 30, 1928, was retaken 60
miles offshore on September 20, 1928; the other, also banded at
Orient on July 23, 1933, was shot 73 miles at sea off Cape Hatteras on
October 9, 1933.

The distance traveled in migration by ospreys is indicated by
two banding records. One, banded at Slaughter Beach, Del., on
April 25, 1932, was retaken on the Milk River, Jamaica, British
West Indies, on March 10, 1933; while the other, banded at the
same place on April 26, 1934, was recovered at El Mojan-Estado,
Zulia, Venezuela, on June 28, 1935.

Casual records.—According to Macoun (1903, p. 261) "a single
specimen was obtained at Godhaven, Greenland, by Mr. E. Whymper
and sent to the Museum at Copenhagen."

Egg dates.—Quebec: 35 records, May 24 to June 28; 18 records,
May 28 to June 8.

New York and New England: 48 records, April 25 to June 18;
24 records, May 6 to 18.

Delaware and New Jersey: 513 records, April 24 to June 16; 257
records, May 3 to 25.

Maryland and Virginia: 90 records, March 10 to May 30; 45
records, April 29 to May 9.

Florida: 19 records, December 4 to April 28; 10 records, March
13 to April 10.

California: 15 records, March 14 to May 30; 8 records, April 2
to May 3.

Lower California and Mexico: 29 records, January 22 to April 16;
15 records, February 20 to April 8.

LITERATURE CITED

ABBOTT, CLINTON GILBERT.
   1911. The home-life of the osprey.
AIKEN, CHARLES EDWARD HOWARD, and WARREN, EDWARD ROYAL.
   1914. The birds of El Paso County, Colorado. II. Colorado College Publ.,
      sci. ser., vol. 13, pp. 497–603.
ALLEN, CHARLES SLOVER.
   1892. Breeding habits of the fish hawk on Plum Island, New York. The
      Auk, vol. 9, pp. 313–321.
ALLEN, JOEL ASAPH.
   1869. Notes on some of the rarer birds of Massachusetts. Amer. Nat., vol.
      3, pp. 505, 568–585, 631–648.
AMERICAN ORNITHOLOGISTS' UNION.
   1910. Check-list of North American birds. Ed. 3.
   1931. Check-list of North American birds. Ed. 4.
ANDERSON, RUDOLPH MARTIN.
   1897. Nesting habits of Krider's hawk. The Museum, vol. 3, pp. 188–190.
ANTHONY, ALFRED WEBSTER.
   1893. Birds of San Pedro Martir, Lower California. Zoe, vol. 4, pp.
      228–247.
AUDUBON, JOHN JAMES.
   1835. Ornithological biography. Vol. 2.
   1840. The birds of America. Vol. 1.
AUSTIN, OLIVER LUTHER, JR.
   1932. The birds of Newfoundland Labrador. Mem. Nuttall Orn. Club,
      no. 7.
BAILEY, ALFRED MARSHALL.
   1926. A report on the birds of northwestern Alaska and regions adjacent
      to Bering Strait. Part VI. The Condor, vol. 28, pp. 121–126.
   1927. Notes on the birds of southeastern Alaska. The Auk, vol. 44, pp.
      184–205.
BAILEY, BERT HEALD.
   1917. Description of a new subspecies of the broad-winged hawk. The
      Auk, vol. 34, pp. 73–75.
BAILEY, FLORENCE AUGUSTA MERRIAM.
   1896. Notes on some of the birds of southern California. The Auk, vol. 13,
      pp. 115–124.
   1902. Handbook of birds of the Western United States. Ed. 1.
   1903. The Harris hawk on his nesting ground. The Condor, vol. 5, pp.
      66–68.
   1915. A family of North Dakota marsh hawks. Bird-Lore, vol. 17, pp.
      431–438.
   1916. Meeting spring half way. The Condor, vol. 18, pp. 183–190.
   1928. Birds of New Mexico.
BAILEY, HAROLD HARRIS.
   1913. Birds at sea. The Auk, vol. 30, pp. 281–282.
   1925. The birds of Florida.
BAILEY, VERNON.
   1902. Notes *in* Florence Merriam Bailey's "Handbook of Birds of the
      Western United States."

BAIRD, SPENCER FULLERTON; BREWER, THOMAS MAYO; and RIDGWAY, ROBERT.
    1905. The land birds of North America.
BANGS, OUTRAM.
    1901. On an apparently unnamed race of *Buteo borealis*. Proc. New
        England Zool. Club, vol. 2, pp. 67–69.
    1920. A new red-shouldered hawk from the Florida Keys. Proc. New
        England Zool. Club, vol. 7, p. 35.
BANGS, OUTRAM, and PENARD, THOMAS EDWARD.
    1920. Two new American hawks. Proc. New England Zool. Club, vol. 7,
        pp. 45–47.
BARLOW, CHESTER.
    1895. In the haunts of the white-tailed kite. The Oologist, vol. 12, pp. 97–
        101.
    1897. Some notes on the nesting habits of the white-tailed kite. The Auk,
        vol. 14, pp. 14–21.
BARROWS, WALTER BRADFORD.
    1912. Michigan bird life.
BARTSCH, PAUL.
    1897. A few notes on the avifauna Columbiana. The Auk, vol. 14, p. 326.
BATCHELDER, CHARLES FOSTER.
    1881. The bald eagle (*Haliaëtus leucocephalus*) as a hunter. Bull. Nuttall
        Orn. Club, vol. 6, pp. 58–60.
BAXTER, LUCY V.
    1906. A tragedy. Bird-Lore, vol. 8, p. 68.
BAYNARD, OSCAR EDWARD.
    1909. Notes from Florida on *Catharista urubu*. The Oologist, vol. 26, pp.
        191–193.
    1910. Additional notes on the breeding of *Catharista urubu*. The Oologist,
        vol. 27, p. 106.
    1913. Breeding birds of Alachua County, Florida. The Auk, vol. 30, pp.
        240–247.
BELDING, LYMAN.
    1890. Land birds of the Pacific district.
BENDIRE, CHARLES EMIL.
    1892. Life histories of North American birds. U. S. Nat. Mus. Spec. Bull. 1.
BENNERS, GEORGE B.
    1887. A collecting trip in Texas. Ornithologist and Oologist, vol. 12, pp.
        65–69.
    1889. Nesting of the swallow-tailed kite in Texas. Ornithologist and
        Oologist, vol. 14, pp. 83–85.
BIGELOW, HENRY BRYANT.
    1902. Birds of the northeastern coast of Labrador. The Auk, vol. 19, pp.
        24–31.
BISHOP, LOUIS BENNETT.
    1901. Rachitis in young red-shouldered hawks. The Auk, vol. 18, pp. 192–
        193.
    1912. An apparently unrecognized race of the red-shouldered hawk. The
        Auk, vol. 29, pp. 232–233.
BLINCOE, BENEDICT JOSEPH.
    1922. Unusual flight of black vultures in Nelson County, Kentucky. The
        Auk, vol. 39, p. 416.
BOLAU, HEINRICH.
    1892. Sea eagles in the Hamburg Zoological Garden. Sci. Amer., vol. 67,
        no. 8, p. 121, Aug. 20.

BRANDT, HERBERT WILLIAM.
1924. The nesting of the short-tailed hawk. The Auk, vol. 41, pp. 59–64.
BREWSTER, WILLIAM.
1880. Prowess of the bald eagle (*Haliaëtus leucocephalus*). Bull. Nuttall
Orn. Club, vol. 5, pp. 57–58.
1883. On a collection of birds lately made by Mr. F. Stephens in Arizona.
Bull. Nuttall Orn. Club, vol. 8, pp. 21–36.
1906. The birds of the Cambridge region of Massachusetts.
1925. The birds of the Lake Umbagog region of Maine. Bull. Mus. Comp.
Zool., vol. 66, pt. 2, pp. 211–402.
BROOKS, ALLAN CYRIL.
1917. Birds of the Chilliwack district, B. C. The Auk, vol. 34, pp. 28–50.
1922. Notes on the abundance and habits of the bald eagle in British Columbia. The Auk, vol. 39, pp. 556–559.
1927. Notes on Swarth's report on a collection of birds and mammals from
the Atlin region. The Condor, vol. 29, pp. 112–114.
1928. Should we protect the marsh hawk? Amer. Game, vol. 17, pp. 88, 91.
BROUN, MAURICE.
1935. The hawk migration during the fall of 1934, along the Kittatinny
Ridge in Pennsylvania. The Auk, vol. 52, pp. 233–248.
1936. Hawk Mountain Sanctuary. Nature Mag., vol. 27, pp. 367–368.
BROWN, NATHAN CLIFFORD.
1878. A list of birds observed at Coosada, central Alabama. Bull. Nuttall
Orn. Club, vol. 3, pp. 168–174.
1879. A list of birds observed at Coosada, central Alabama. Bull. Nuttall
Orn. Club, vol. 4, pp. 7–13.
BRYANT, HAROLD CHILD.
1921. Red-bellied hawk eats caterpillars. The Condor, vol. 23, p. 65.
BRYANT, HENRY.
1861. A list of birds seen at the Bahamas, from Jan. 20th to May 14th,
1859. Proc. Boston Soc. Nat. Hist., vol. 7, p. 102.
BRYANT, WALTER (PIERC) E.
1888. Birds and eggs from the Farallon Islands. Proc. California Acad.
Sci., ser. 2, vol. 1, pp. 25–50.
1890. A catalogue of the birds of Lower California, Mexico. Proc. California Acad. Sci., ser. 2, vol. 2, pp. 237–320.
BUNKER, CHARLES DEAN.
1917. American goshawks in Kansas. The Auk, vol. 34, pp. 87–88.
BURNS, FRANKLIN LORENZO.
1906. A vulture pie. Wilson Bull., vol. 18, p. 25.
1911. A monograph of the broad-winged hawk (*Buteo platypterus*).
Wilson Bull., vol. 23, pp. 139–320.
1915. Comparative periods of deposition and incubation of some North
American birds. Wilson Bull., vol. 27, pp. 275–286.
BURR, FREEMAN FOSTER.
1912. Note on the bald eagle and osprey. The Auk, vol. 29, p. 393.
BURROWS, D. B.
1917. White-tailed hawk. The Oologist, vol. 34, pp. 78–81.
BURTCH, VERDI.
1911. A nest of the red-tailed hawk. The Oologist, vol. 28, p. 50.
1927. Near cannibalism in a *Buteo*. The Auk, vol. 44, pp. 248–249.
CAMERON, EWEN SOMERLED.
1905. Nesting of the golden eagle in Montana. The Auk, vol. 22, pp.
158–167.

CAMERON, EWEN SOMERLED—Continued.

1907. The birds of Custer and Dawson Counties, Montana. The Auk, vol.
24, pp. 241–270.

1908a. Changes of plumage in *Buteo swainsoni*. The Auk, vol. 25, pp.
468–471.

1908b. Observations of the golden eagle in Montana. The Auk, vol. 25,
pp. 251–268.

1913. Notes on Swainson's hawk (*Buteo swainsoni*) in Montana. The
Auk, vol. 30, pp. 167–176, 381–394.

1914. The ferruginous rough-leg, *Archibuteo ferrugineus* in Montana. The
Auk, vol. 31, pp. 159–167.

CARPENTER, FREDERIC HOWARD.

1887. The occurrence of the osprey in the fauna of Bristol County, Mass.
Publ. Bristol County Orn. Club, no. 2.

CARRIKER, MELBOURNE ARMSTRONG, JR.

1902. Notes on the nesting of some Sioux County birds. Proc. 3d Ann.
Meet. Nebraska Orn. Union, pp. 75–89.

CARROLL, JAMES JUDSON.

1900. Notes on the birds of Refugio County, Texas. The Auk, vol. 17, pp.
337–348.

CASSIN, JOHN.

1856. Illustrations of the birds of California, Texas, Oregon, British and
Russian America.

1860. Letter to the editor. The Ibis, vol. 2, pp. 103–104

CHAMBERS, WILLIE LEE.

1921. A flight of Harris hawks. The Condor, vol. 23, p. 65.

1924. Another flight of Harris hawks. The Condor, vol. 26, p. 75.

CHAPMAN, ABEL.

1885. A bird's nesting ramble in Lapland. The Ibis, ser. 5, vol. 3, pp. 158–
184.

CHAPMAN, FRANK MICHLER.

1891. On the birds observed near Corpus Christi, Texas, during parts of
March and April, 1891. Bull. Amer. Mus. Nat. Hist., vol. 3,
pp. 315–328.

1894. On the birds of the Island of Trinidad. Bull. Amer. Mus. Nat. Hist.,
vol. 6, pp. 1–86.

1908. Camps and cruises of an ornithologist.

1917. The distribution of bird life in Colombia.

1929. My tropical air castle: Nature studies in Panama.

1933. The migration of turkey buzzards, as observed at Barro Colorado
Island, Canal Zone. The Auk, vol. 50, pp. 30–34.

CLARK, AUSTIN HOBART.

1905. Birds of the southern Lesser Antilles. Proc. Boston Soc. Nat. Hist.,
vol. 32, pp. 203–312.

1910. The birds collected and observed during the cruise of the United
States Fisheries steamer *Albatross* in the North Pacific Ocean,
and in the Bering, Okhotsk, Japan, and Eastern Seas, from April
to December, 1906. Proc. U. S. Nat. Mus., vol. 38, pp. 25–74.

CLARK, JOSIAH HUNTOON.

1900. The giant cactus as a nesting place for the western red-tailed hawk.
The Oologist, vol. 17, p. 126.

COALE, HENRY KELSO.

1925. Habits of the marsh hawk. The Auk, vol. 42, p. 269.

COOPER, JAMES GRAHAM.
  1870. Geological survey of California. Ornithology. Vol. 1: Land birds.
  1890. A doomed bird. Zoe, vol. 1, pp. 248-249.
COUES, ELLIOTT.
  1874. Birds of the Northwest. U. S. Geol. Surv. Terr. Misc. Publ. 3.
  1878. Swallow-tailed kite in Dakota in winter. Bull. Nuttall Orn. Club,
    vol. 3, p. 147.
COURT, EDWARD JOSEPH.
  1924. Black vulture nesting in Maryland. The Auk, vol. 41, pp. 475-476.
CRAM, WILLIAM EVERETT.
  1899. Winter bird notes from southern New Hampshire. Bird-Lore, vol. 1,
    pp. 180-184.
CRIDDLE, NORMAN.
  1917. The red-tailed hawk in Manitoba. Ottawa Nat., vol. 31, pp. 74-76.
DALGLEISH, JOHN J.
  1880. List of occurrences of North American birds in Europe. Bull. Nuttall
    Orn. Club, vol. 5, pp. 65-74, 141-150, 210-221.
DANFORTH, STUART TAYLOR.
  1928. Birds observed in the vicinity of Santiago de Cuba. Wilson Bull.,
    vol. 39, pp. 178-182.
DARLINGTON, PHILIP JACKSON, JR.
  1930. Notes on the senses of vultures. The Auk, vol. 47, pp. 251-252.
DAWSON, WILLIAM LEON.
  1923. The birds of California. Vol. 3.
DEANE, RUTHVEN.
  1907. Unusual abundance of the American goshawk (*Accipiter atricapillus*).
    The Auk, vol. 24, pp. 182-186.
DIXON, JAMES BENJAMIN.
  1902. Early nesting of the western red-tail. The Condor, vol. 4, p. 46.
  1911. The golden eagle. The Oologist, vol. 28, pp. 126-129.
  1928. Life history of the red-bellied hawk. The Condor, vol. 30, pp. 228-236.
DIXON, JOSEPH SCATTERGOOD.
  1906. Land birds of San Onofre, California. The Condor, vol. 8, pp. 91-98.
  1909. A life history of the northern bald eagle. The Condor, vol. 11, pp.
    187-193.
DONAHUE, RALPH J.
  1923. The food of young hawks. The Oologist, vol. 40, p. 134.
DOOLITTLE, EDWARD ARTHUR.
  1919. A flight of broad-winged hawks and rough-legs in Lake Co., Ohio.
    The Auk, vol. 36, p. 568.
DuMONT, PHILIP ATKINSON.
  1935. An autumnal flight of broad-winged hawks in eastern Iowa. Iowa
    Bird Life, vol. 5, p. 5.
EATON, ELON HOWARD.
  1910. Birds of New York.
EIFRIG, CHARLES WILLIAM GUSTAVE.
  1907a. American goshawk (*Accipiter atricapillus*) versus man and barred
    owl. The Auk, vol. 24, pp. 437-438.
  1907b. The American goshawk near Ottawa. Ottawa Nat., vol. 21, pp.
    96-97.
ERRINGTON, PAUL LESTER.
  1930. Territory disputes of three pairs of nesting marsh hawks. Wilson
    Bull., vol. 42, pp. 237-239.
  1932. An encounter between a Cooper's hawk and a horned owl. Wilson
    Bull., vol. 44, p. 189.

FABER, FRIEDRICH.
    1826. Ueber das Leben der hochnordischen Vögel.
FARLEY, JOHN AUSTIN.
    1923. Breeding of the goshawk in Massachusetts.   The Auk, vol. 40, pp.
        532–533.
FERGUSON, ALFRED LUDLOW, and FERGUSON, HENRY LEE.
    1922. The fall migration of hawks as observed at Fishers Island, N. Y.
        The Auk, vol. 39, pp. 488–496.
FIGGINS, JESSE DADE.
    1923. The breeding birds of the vicinity of Black Bayou and Bird Island,
        Cameron Parish, Louisiana.   The Auk, vol. 40, pp. 666–677.
FINLEY, WILLIAM LOVELL.
    1905. Photographing the aerie of a western red-tail.   The Condor, vol. 7,
        pp. 3–7.
    1906. Life history of the California condor.   Part I.—Finding a condor's
        nest.   The Condor, vol. 8, pp. 135–142.
    1908. Life history of the California condor.   Part III.—Home life of the
        condors.   The Condor, vol. 10, pp. 59–65.
    1910. Life history of the California condor.   Part IV.—The young condor
        in captivity.   The Condor, vol. 12, pp. 5–11.
FISHER, ALBERT KENRICK.
    1893. The hawks and owls of the United States in their relation to agri-
        culture.   U. S. Dept. Agr., Div. Orn. and Mamm., Bull. 3.
FLANAGAN, JOHN HENRY.
    1901. The use of old nests.   The Oologist, vol. 18, pp. 12–13.
FLEMING, JAMES HENRY.
    1907. Birds of Toronto, Canada.   The Auk, vol. 24, pp. 71–89.
    1924. The California condor in Washington: Another version of an old
        record.   The Condor, vol. 26, pp. 111–112.
FORBES, HENRY STONE, and FORBES, HILDEGARDE BOUGHTON.
    1927. An autumn hawk flight.   The Auk, vol. 44, pp. 101–102.
FORBUSH, EDWARD HOWE.
    1927. Birds of Massachusetts and other New England States.   Vol. 2.
FOWLER, FREDERICK HALL.
    1903. Stray notes from southern Arizona.   The Condor, vol. 5, pp. 68–71.
FULLER, EDWARD.
    1891. The American osprey.   The Oologist, vol. 8, pp. 145–147.
GANIER, ALBERT FRANKLYN.
    1902. The Mississippi kite (*Ictinia mississippiensis*).   The Osprey, vol. 6,
        pp. 85–90.
GARDNER, LEON LLOYD.
    1930. On the body temperature of nestling altricial birds.   The Auk, vol.
        47, pp. 367–379.
GIANINI, CHARLES ALFRED.
    1917. Some Alaska Peninsula bird notes.   The Auk, vol. 34, pp. 394–402.
GILBERT, CHARLES HENRY.
    1922. Kamchatka sea eagle at Kodiak, Alaska.   The Condor, vol. 34, p. 66.
GILBERT, HUMPHREY ADAM, and BROOK, ARTHUR.
    1925. The secrets of the golden eagle and of other rare birds.
GILL, THEODORE.
    1901. The osprey or fish hawk; its characteristics and habits.   The Osprey,
        vol. 5, pp. 11–12, 25–28, 40–42, 60–61, 73–76, 92–93, 105–106, 124–
        125, and 141.

GLOYD, HOWARD KAY.
  1925. Field studies of the diurnal Raptores of eastern and central Kansas.
      Wilson Bull., vol. 37, pp. 133–149.
GOELITZ, WALTER ADOLPH.
  1916. A strange nesting of the barred owl and red-shouldered hawk.
      Wilson Bull., vol. 28, pp. 105–106.
GOLSAN, LEWIS SAMUEL, and HOLT, ERNEST GOLSAN.
  1914. Birds of Autauga and Montgomery Counties, Alabama. The Auk,
      vol. 31, pp. 212–235.
GORDON, SETON PAUL.
  1915. Hill birds of Scotland.
  1927. Days with the golden eagle.
GOSS, NATHANIEL STICKNEY.
  1891. History of the birds of Kansas.
GRAY, ROBERT.
  1861. The birds of the West of Scotland, including the outer Hebrides.
GREEN, JAMES.
  1927. Turkey buzzards eat pumpkins. Bird-Lore, vol. 29, pp. 117–118.
GRINNELL, JOSEPH.
  1900. Birds of the Kotzebue Sound region. Pacific Coast Avifauna, no. 1.
  1917. An invasion of California by the eastern goshawk. The Condor, vol.
      19, pp. 70–71.
GRINNELL, JOSEPH, and STORER, TRACY IRWIN.
  1924. Animal life in the Yosemite.
GRISCOM, LUDLOW.
  1923. Birds of the New York City region.
  1932. The distribution of bird life in Guatemala.
GUTHRIE, JOSEPH EDWARD.
  1931. Red-shouldered hawks and chimney swifts. Iowa Bird Life, vol. 1,
      p. 35.
HAHN, WILLIAM, Jr.
  1927. Florida red-shouldered hawk. The Oologist, vol. 44, pp. 141–142.
HANNA, G. DALLAS.
  1920. Additions to the avifauna of the Pribilof Islands, Alaska, including
      four species new to North America. The Auk, vol. 37, pp. 248–254.
HANNA, WILSON CREAL.
  1930. Notes on the golden eagle in southern California. The Condor, vol. 32,
      pp. 121–123.
HANTZSCH, BERNHARD.
  1929. Contributions to the knowledge of the avifauna of northeastern
      Labrador. Can. Field-Nat., vol. 43, pp. 11–18. (Translated by
      M. B. A. Anderson and R. M. Anderson.)
HARTERT, ERNST.
  1912–1921. Die Vögel der paläarktischen Fauna. Vol. 2.
HATHAWAY, HARRY SEDGWICK.
  1905. The American osprey. Roger Williams Park Mus. Bull. 10.
HEINROTH, OSKAR, and HEINROTH, MAGDALENA.
  1926–27. Die Vögel Mitteleuropas. Vol. 2.
HENDERSON, ARCHIBALD DOUGLAS.
  1920. Golden eagle and horned owl. The Oologist, vol. 37, p. 122.
  1924. Nesting habits of the American goshawk. Can. Field-Nat., vol. 38,
      pp. 8–9.
HENDERSON, JUNIUS.
  1927. The practical value of birds.

HENNIGER, WALTHER FRIEDRICH, and JONES, LYNDS.
  1909. The falcons of North America. Wilson Bull., vol. 16, pp. 205–218.
HENSHAW, HENRY WETHERBEE.
  1875. Report upon ornithological collections made in portions of Nevada,
        Utah, California, Colorado, New Mexico, and Arizona during the
        years 1871, 1872, 1873, and 1874. Wheeler's Rep. Geogr. and Geol.
        Expl. and Surv. West 100th Merid., vol. 5, chap. 3, pp. 131–507.
  1901. First impressions of Hawaiian birds. Bird-Lore, vol. 3, pp. 119–125,
        153–159.
HERRICK, FRANCIS HOBART.
  1924a. An eagle observatory. The Auk, vol. 41, pp. 89–105.
  1924b. Nests and nesting habits of the American eagle. The Auk, vol. 41,
        pp. 213–231.
  1924c. The daily life of the American eagle: Late phase. The Auk, vol.
        41, pp. 389–422, 517–541.
  1924d. Family life of the American eagle. Nature Mag., vol. 4, pp. 133–140.
  1929. The eagle in action. Nat. Geogr. Mag., vol. 55, pp. 635–660.
  1932. Daily life of the American eagle: Early phase. The Auk, vol. 49,
        pp. 307–323, 428–435.
  1933. Daily life of the American eagle: Early phase (concluded). The
        Auk, vol. 50, pp. 35–53.
HERRICK, HAROLD.
  1879. Notes on some birds of Chatham, N. J. Forest and Stream, vol. 12,
        p. 165.
HERSEY, FRANK SEYMOUR.
  1923. A nesting red-shouldered hawk's hearty meal. The Auk, vol. 40,
        p. 693.
HESS, ISAAC ELNORE.
  1910. One hundred breeding birds of an Illinois ten-mile radius. The Auk,
        vol. 27, pp. 19–32.
HOFFMANN, RALPH.
  1927. Birds of the Pacific States.
HOLT, ERNEST GOLSAN, and SUTTON, GEORGE MIKSCH.
  1926. Notes on birds observed in southern Florida. Ann. Carnegie Mus.,
        vol. 16, pp. 409–439.
HOOPES, BERNARD A.
  1873. Description of a new variety of Buteo. Proc. Acad. Nat. Sci. Phila-
        delphia, vol. 25, pp. 238–239.
HORTLING, IVAR.
  1929. Ornitologisk Handbok.
HOWE, REGINALD HEBER, Jr.
  1895. A large brood of ospreys. The Auk, vol. 12, p. 389.
HOWELL, ARTHUR HOLMES.
  1932. Florida bird life.
HOWES, PAUL GRISWOLD.
  1926. A turkey vulture's nest in the State of New York. Bird-Lore, vol.
        28, pp. 175–180.
HOXIE, WALTER JOHN.
  1886. Breeding habits of the black vulture. The Auk, vol. 3, pp. 245–247.
  1888. A bald eagle's nest. Ornithologist and Oologist, vol. 13, pp. 63–64.
HUDSON, WILLIAM HENRY.
  1920. Birds of La Plata.

HUEY, LAURENCE MARKHAM.
    1913. Nesting notes from San Diego County. The Condor, vol. 15, p. 228.
    1924. Notes from southern California. The Condor, voi. 26, pp. 74–75.
HUNTER, JOSEPH SLAYTON.
    1898. Hawk killed by rattlesnake. The Osprey, vol. 3, p. 46.
JACKSON, THOMAS HOOPES.
    1903. The turkey vulture and its young. Bird-Lore, vol. 5, pp. 184–187.
JACOBS, JOSEPH WARREN.
    1908. Bald eagle (*Haliaetus leucocephalus*) and great horned owl (*Bubo virginianus*) occupying the same nest. Wilson Bull., vol. 20, pp. 103–104.
JEWETT, STANLEY GORDON.
    1926. The ferruginous rough-leg nesting in Oregon. The Condor, vol. 28, pp. 245–246.
JOHNSON, CHARLES EUGENE.
    1925. Kingfisher and Cooper's hawk. The Auk, vol. 42, pp. 585–586.
JOURDAIN, FRANCIS CHARLES ROBERT.
    1925. The Dobrogea, bird life in the western Balta. Oologists' Rec., vol. 5, pp. 49–56.
JOUY, PIERRE LOUIS.
    1893. Notes on birds of central Mexico, with descriptions of forms believed to be new. Proc. U. S. Nat. Mus., vol. 16, pp. 771–791.
KEMPTON, RUSSELL MARSHALL.
    1927. Notes on the home life of the turkey vulture. Wilson Bull., vol. 39, pp. 142–145.
KENNARD, FREDERIC HEDGE.
    1894a. The habits and individualities of the red-shouldered hawk (*Buteo lineatus*) in the vicinity of Brookline, Mass. The Auk, vol. 11, pp. 197–210.
    1894b. The young of the red-shouldered hawk (*Buteo lineatus*). The Auk, vol. 11, pp. 270–280.
KERMODE, FRANCIS.
    1904. Catalogue of British Columbia birds.
KITCHIN, EDWARD ALEXANDER.
    1918. The rough-legged hawk in western Washington. The Condor, vol. 20, p. 91.
KNIGHT, CHARLES WILLIAM ROBERT.
    1932. Photographing the nest life of the osprey. Nat. Geogr. Mag., vol. 62, pp. 247–260.
KNIGHT, ORA WILLIS.
    1908. The birds of Maine.
KRAUSE, GEORG.
    1926. Oologia universalis palaearctica.
KRIDER, JOHN.
    1879. Forty years notes of a field ornithologist.
KUMLIEN, LUDWIG.
    1879. Contributions to the natural history of Arctic America, made in connection with the Howgate Polar Expedition, 1877–78. U. S. Nat. Mus. Bull. 15.
KUMLIEN, LUDWIG, and HOLLISTER, NED.
    1903. The birds of Wisconsin.
LACEY, HOWARD.
    1911. The birds of Kerrville, Texas, and vicinity. The Auk, vol. 28, pp. 200–219.

LAINE, W. H.
1928. Marsh hawk hatches prairie chicken. Can. Field-Nat., vol. 42, p. 47.
LAMB, CHESTER CONVERSE.
1927. The birds of Natividad Island, Lower California. The Condor, vol. 29, pp. 67–70.
LANG, HERBERT.
1924. *Ampullarius* and *Rostrhamus* at Georgetown, British Guiana. The Nautilus, vol. 37, pp. 73–77.
LANGILLE, JAMES HIBBERT.
1884. Our birds in their haunts.
LANO, ALBERT.
1922. Golden eagle (*Aquila chrysaetos*) and porcupine. The Auk, vol. 39, pp. 258–259.
LAW, JOHN EUGENE.
1919. Problem: Do birds mate for life? The Condor, vol. 21, pp. 26–27.
LAWRENCE, GEORGE NEWBOLD.
1874. The birds of western and northwestern Mexico, etc. Mem. Boston Soc. Nat. Hist., vol. 2, pp. 265–319.
1876. Birds of southwestern Mexico collected by Francis E. Sumichrast for the United States National Museum. U. S. Nat. Mus. Bull. 4.
LEWIS, HARRISON FLINT.
1927. Notes on the birds of the Labrador Peninsula in 1925 and 1926. The Auk, vol. 44, pp. 59–66.
LLOYD, WILLIAM.
1887. Birds of Tom Green and Concho Counties, Texas. The Auk, vol. 4, pp. 181–193.
LOFBERG, LILA McKINLEY.
1935. Random notes on raptors at Florence Lake, California. The Condor, vol. 37, pp. 171–173.
LOOMIS, LEVERETT MILLS.
1890. Summer birds of the mountain portions of Pickens County, South Carolina. The Auk, vol. 7, pp. 30–39.
LYLE, ROBERT BARTON.
1931. Black vulture nesting in Washington County, Tennessee. The Auk, vol. 48, p. 598.
LYON, JAMES A., Jr.
1893. Collecting black vulture's eggs. The Oologist, vol. 10, pp. 55–56.
MacDONALD, DUNCAN.
1926. Some notes on the golden eagle. Brit. Birds, vol. 19, pp. 218–225.
MacFARLANE, RODERICK ROSS.
1891. Notes on and list of birds and eggs collected in Arctic America, 1861–1866. Proc. U. S. Nat. Mus., vol. 14, pp. 416–466.
1908. List of birds and eggs observed and collected in the North-West Territories of Canada, between 1880 and 1894. *In* Charles Mair's "Through the Mackenzie Basin."
MACOUN, JOHN.
1903. Catalogue of Canadian birds. Part 2.
1909. Catalogue of Canadian birds. Ed. 2.
MACPHERSON, H. B.
1911. The home-life of a golden eagle. Ed. 3.
MAGEE, MICHAEL JARDEN.
1922. Hawk migration at Whitefish Point, Upper Peninsula of Michigan. The Auk, vol. 39, pp. 257–258.

MAILLIARD, JOSEPH.
1908. Cooper hawks attacking crows. The Condor, vol. 10, p. 129.
MAY, JOHN BICHARD.
1935. The hawks of North America.
MAYNARD, CHARLES JOHNSON.
1896. The birds of eastern North America. Ed. 2.
McATEE, WALDO LEE.
1935. Food habits of common hawks. U. S. Dept. Agr. Circ. 370.
McCABE, THOMAS TONKIN, and McCABE, ELINOR BOLLES.
1928a. Hawks and kingfisher. The Auk, vol. 45, p. 374.
1928b. Notes on certain injured birds. The Condor, vol. 30, pp. 190–191.
McCORMICK-GOODHART, LEANDER.
1932. Bathing behavior of the osprey. The Auk, vol. 49, pp. 463–464.
MEARNS, EDGAR ALEXANDER.
1886. Some birds of Arizona. The Auk, vol. 3, pp. 60–73.
MERRIAM, FLORENCE AUGUSTA. (See Bailey, Florence A. Merriam.)
MERRILL, JAMES CUSHING.
1879. Notes on the ornithology of southern Texas, being a list of birds ob-
        served in the vicinity of Fort Brown, Texas, from February, 1876,
        to June, 1878. Proc. U. S. Nat. Mus., vol. 1, pp. 118–173.
MICHAEL, CHARLES WILSON.
1921. Pileated woodpecker versus Cooper hawk. The Condor, vol. 23,
        p. 68.
MICHENER, HAROLD.
1930. Hawks unwelcome visitors at banding stations. The Condor, vol. 32,
        p. 212.
MILLER, JOHN PAUL.
1931. The red-tailed hawk (*Buteo borealis* (Gmelin)) in relation to the
        control of the Columbian ground squirrel (*Citellus c. columbianus*
        (Ord)). The Murrelet, vol. 12, pp. 46–49.
MILLER, LOYE HOLMES.
1918. First flights of a young golden eagle. The Condor, vol. 20, p. 212.
1926. The food of the white-tailed kite. The Condor, vol. 28, pp. 172–173.
1930. Further notes on the Harris hawk. The Condor, vol. 32, pp. 210–211.
MINOT, HENRY DAVIS.
1877. The land-birds and game-birds of New England.
MITCHELL, HORACE HEDLEY.
1924. Birds of Saskatchewan. Can. Field-Nat., vol. 38, pp. 101-118.
MORGAN, Mrs. A. B.
1915. The story of a red-tailed hawk—in two parts. Bird-Lore, vol. 17,
        pp. 8–11, 113–117.
MORRISON, CHARLES F.
1887. Field notes on some birds of Colorado. Ornithologist and Oologist,
        vol. 12, pp. 27–28.
1889. Golden eagle in Montana. Ornithologist and Oologist, vol. 14, p. 25.
MOUSLEY, HENRY.
1918. Further notes and observations on the birds of Hatley, Standstead
        County, Quebec, 1916–1917. The Auk, vol. 35, pp. 289–310.
MÜLLER, RASMUS.
1906. Vildet og Jagten i Sydgrönland.
MUNRO, JAMES ALEXANDER.
1919. Notes on some birds of the Okanagan Valley, British Columbia.
        The Auk, vol. 36, pp. 64–74.

MURRAY, JAMES JOSEPH.
    1928. The range of the black vulture. Bird-Lore, vol. 30, p. 116.
NAUMAN, E. D.
    1929. The raptor's mistake. Wilson Bull., vol. 41, p. 252.
NAUMANN, JOHANN FRIEDRICH.
    1905–?1909. Naturgeschichte der Vögel Mitteleuropas. 12 vols.
NELSON, EDWARD WILLIAM.
    1877a. Birds of northeastern Illinois. Bull. Essex Inst., vol. 8, pp. 90–155.
    1877b. Notes upon birds observed in southern Illinois, between July 17
            and September 4, 1875. Bull. Essex Inst., vol. 9, pp. 32–65.
    1887. Report upon natural history collections made in Alaska. U. S.
            Signal Service, Arctic ser., no. 3.
NICHOLSON, DONALD JOHN.
    1926. Nesting habits of the everglade kite in Florida. The Auk, vol. 43,
            pp. 62–67.
    1928a. A swallow-tailed kite in Monroe County. Florida Nat., vol. 8, pp.
            20–21.
    1928b. Habits of the black vulture in Florida. The Oologist, vol. 45, pp.
            21–24.
    1930. Habits of the Florida red-shouldered hawk. Wilson Bull., vol. 42, pp.
            32–35.
NOBLE, GEORGE KINGSLEY.
    1919. Notes on the avifauna of Newfoundland. Bull. Mus. Comp. Zool.,
            vol. 62, pp. 543–568.
NUTTALL, THOMAS.
    1832. A manual of the ornithology of the United States and of Canada.
OBERHOLSER, HARRY CHURCH.
    1896. A preliminary list of the birds of Wayne County, Ohio.
    1897. An osprey's vigil. The Osprey, vol. 2, p. 50.
    1906. The North American eagles and their economic relations. U. S.
            Biol. Surv. Bull. 27.
    1919. Notes on North American birds. VII. The Auk, vol. 36, pp. 81–85.
PEABODY, PUTNAM BURTON.
    1895. On the nesting of Krider's hawk (*Buteo borealis krideri*) in Min-
            nesota. The Auk, vol. 12, pp. 11–16.
PEARSON, HENRY JOHN.
    1898. Notes on the birds observed on Waigats. The Ibis, ser. 7, vol. 4,
            pp. 185–208.
PEARSON, THOMAS GILBERT.
    1919. Turkey vulture. Bird-Lore, vol. 21, pp. 319–322.
    1921. Bird life of southeastern Texas. The Auk, vol. 38, pp. 513–523.
PEARSON, THOMAS GILBERT; BRIMLEY, CLEMENT SAMUEL; and BRIMLEY, HERBERT
        HUTCHINSON.
    1919. Birds of North Carolina.
PENNOCK, CHARLES JOHN.
    1890. Note on the nesting of *Buteo brachyurus* at St. Marks, Florida. The
            Auk, vol. 7, pp. 56–57.
PETERS, JAMES LEE.
    1931. Check-list of birds of the world. Vol. 1.
PEYTON, LAURENCE GORHAM.
    1915. Nesting of the white-tailed kite at Sespe, Ventura County, California.
            The Condor, vol. 17, pp. 230–232.

PICKENS, ANDREW LEE.
    1927. Some unusual records for South Carolina. The Auk, vol. 44, pp.
        573–574.
PICKWELL, GAYLE BENJAMIN.
    1930. The white-tailed kite. The Condor, vol. 32, pp. 221–239.
    1932. Requiem for the white-tailed kites of Santa Clara Valley. The Con-
        dor, vol. 34, pp. 44–45.
PINDAR, LEONIDAS OTLEY.
    1925. Birds of Fulton County, Kentucky. Wilson Bull., vol. 37, pp. 77-78.
PRESTON, JUNIUS WALLACE.
    1886. Nesting of the swallow-tailed kite in Becker County, Minnesota.
        Ornithologist and Oologist, vol. 11, pp. 181–183.
    1888. Nesting habits of the broad-winged hawk. Ornithologist and
        Oologist, vol. 13, pp. 19–21.
RACEY, KENNETH.
    1922. Notes on the rough-legged hawk (Archibuteo lagopus sancti-johannis).
        The Murrelet, vol. 3, no. 2, p. 6.
RAWSON, CALVIN L. (J. M. W.).
    1882. Sharp shinned hawk. Ornithologist and Oologist, vol. 6, pp. 89–91.
RAY, MILTON SMITH.
    1926. The discovery of the nest and eggs of the western goshawk in Cali-
        fornia. The Condor, vol. 28, pp. 258–261.
    1928. A record set of eggs of the golden eagle. The Condor, vol. 30, p. 250.
REDINGTON, PAUL GOODWIN.
    1932. Policies of the Bureau of Biological Survey relative to the control
        of injurious birds. U. S. Dept. Agr. Misc. Publ. 145.
REID, PHILIP SAVILE GREY.
    1884. The birds of Bermuda. Part IV of Contributions to the natural
        history of the Bermudas. U. S. Nat. Mus. Bull. 25, pp. 163–279.
RHOADS, SAMUEL NICHOLSON.
    1893. The birds observed in British Columbia and Washington during
        spring and summer 1892. Proc. Acad. Nat. Sci. Philadelphia, 1893,
        pp. 21–65.
RIDGWAY, ROBERT.
    1874. The Lower Wabash Valley, considered in its relation to the faunal
        districts of the eastern region of North America; with a synopsis
        of its avian fauna. Proc. Boston Soc. Nat. Hist., vol. 16, pp.
        304–332.
    1877. Report on the United States geological exploration of the fortieth
        parallel. Vol. 4, pt. 3: Ornithology.
    1881a. A hawk new to the United States. Forest and Stream, vol. 16, p.
        206, Apr. 14.
    1881b. On a tropical American hawk to be added to the North American
        fauna. Bull. Nuttall Orn. Club, vol. 6, pp. 207–214.
    1884. Description of a new race of the red-shouldered hawk from Florida.
        Proc. U. S. Nat. Mus., vol. 7, pp. 514–515.
    1886a. Description of a melanistic specimen of Buteo latissimus (Wils.)
        Proc. U. S. Nat. Mus., vol. 9, pp. 248–249.
    1886b. A nomenclature of colors for naturalists.
    1912. Color standards and color nomenclature.

RILEY, JOSEPH HARVEY.

1902. Notes on the habits of the broad-winged hawk (*Buteo platypterus*) in the vicinity of Washington, D. C. The Osprey, vol. 6, pp. 21–23.

1908. Notes on the broad-winged hawks of the West Indies, with description of a new form. The Auk, vol. 25, pp. 268–276.

RIVES, WILLIAM CABELL.

1890. A catalogue of the birds of the Virginias. Proc. Newport Nat. Hist. Soc., 1890–91, Doc. 7.

ROBERTS, THOMAS SADLER.

1919. A review of the ornithology of Minnesota.

1932. The birds of Minnesota. Vol. 1.

RODDY, HARRY JUSTIN.

1888. Feeding habits of some young Raptores. The Auk, vol. 5, pp. 244–248.

ROLFE, EUGENE S.

1896. Nesting of the ferruginous rough-leg. The Osprey, vol. 1, pp. 8–10.

1897. Short-eared owl and marsh hawk. The Nidologist, vol. 4, pp. 39–41.

RUST, HENRY JUDSON.

1914. Some notes on the nesting of the sharp-shinned hawk. The Condor, vol. 16, pp. 14–24.

SALVIN, OSBERT.

1861. A list of species to be added to the ornithology of Central America. The Ibis, vol. 3, pp. 351–357.

SASS, HERBERT RAVENEL.

1930. Kings of winter. Good Housekeeping, vol. 90, no. 2, pp. 32–33, 202–216, February.

SAUNDERS, ARETAS ANDREWS.

1906. The habits of the black vulture in Nicaragua. Bird-Lore, vol. 8, pp. 165–167.

1911. A preliminary list of the birds of Gallatin County, Montana. The Auk, vol. 28, pp. 26–49.

1913. A study of the nesting of the marsh hawk. The Condor, vol. 15. pp. 99–104.

SAVAGE, WILLIAM.

1900. Cooper's hawk. Western Orn., vol. 5, pp. 6–8.

SAXBY, HENRY LINCKMYER.

1874. The birds of Shetland.

SCHIØLER, EILER LEHN.

1931. Danmarks Fugle. Vol. 3: Rovfugle.

SCLATER, PHILIP LUTLEY.

1910. Revised list of the birds of Jamaica. Reprinted from the Handbook of Jamaica for 1910.

SCOTT, WILLIAM EARL DODGE.

1886. On the avi-fauna of Pinal County, with remarks on some birds of Pima and Gila Counties, Arizona. The Auk, vol. 3, pp. 421–432.

1889. On the specific identity of *Buteo brachyurus* and *Buteo fuliginosus*, with additional records of their occurrence in Florida. The Auk, vol. 6, pp. 243–245.

1892. Observations on the birds of Jamaica, West. The Auk, vol. 9, pp. 120–129.

SENNETT, GEORGE BURRITT.

1878. Notes on the ornithology of the lower Rio Grande of Texas. Bull. U. S. Geol. and Geogr. Surv. Terr., vol. 4, pp. 1–66.

SENNETT, GEORGE BURRITT—Continued.

1879. Further notes on the ornithology of the lower Rio Grande of Texas, with observations made during the spring of 1878. Bull. U. S. Geol. and Geogr. Surv. Terr., vol. 5, pp. 371–440.

SETON, ERNEST THOMPSON.

1890. The birds of Manitoba. Proc. U. S. Nat. Mus., vol. 13, pp. 457–643.

1908. Bird records from Great Slave Lake region. The Auk, vol. 25, p. 68.

SHARP, CLARENCE SAUGER.

1902. Nesting of Swainson hawk. The Condor, vol. 4, pp. 116–118.

1906. Nesting of the red-bellied hawk. The Condor, vol. 8, pp. 144–148.

SHELDON, W. G.

1912. Notes on the breeding habits of some of our winter migrants in the Swedish provinces of Jemtland and Lapland. Brit. Birds, vol. 5, pp. 150–157.

SHELLEY, LEWIS ORMAN.

1930. Taming the marsh hawk. Bull. Audubon Soc. New Hampshire, vol. 9, pp. 37–40.

SHIELDS, ALEXANDER McMILLAN.

1895. Nesting of the California vulture. The Nidiologist, vol. 2, pp. 148–150.

SIEWERT, HORST.

1928. Der Seeadler. Journ. für Orn., Jahrg. 76, pp. 204–214.

SILLOWAY, PERLEY MILTON.

1903. The birds of Fergus County, Montana.

SIMMONS, GEORGE FINLAY.

1915. On the nesting of certain birds in Texas. The Auk, vol. 32, pp. 317–331.

1925. Birds of the Austin region.

SIMPSON, CHARLES TORREY.

1923. Out of doors in Florida.

SIMPSON, RALPH B.

1909a. American goshawk nesting in Pennsylvania. The Oologist, vol. 26, pp. 85–87.

1909b. More about the Pennsylvania goshawks. The Oologist, vol. 26, pp. 119–120.

1911. The sharp-shinned hawk. The Oologist, vol. 28, pp. 54–56.

SINGLEY, JOHN ALLEN.

1886. Nesting of the great horned owl in Texas. Ornithologist and Oologist, vol. 11, pp. 123–124.

SKINNER, MILTON PHILO.

1928. Kingfisher and sharp-shinned hawk. The Auk, vol. 45, pp. 100–101.

SLATER, HENRY HORROCKS.

1901. Manual of the birds of Iceland.

SLEVIN, JOSEPH RICHARD.

1929. A contribution to our knowledge of the nesting habits of the golden eagle. Proc. California Acad. Sci., ser. 4, vol. 18, pp. 45–71.

SMITH, AUSTIN PAUL.

1915. Birds of the Boston Mountains, Arkansas. The Condor, vol. 17, pp. 41–57.

SMYTH, ELLISON ADGER, Jr.

1912. Birds observed in Montgomery County, Virginia. The Auk, vol. 29, pp. 508–530.

STEARNS, WINFRID A.

1883. New England bird life. (Edited by Elliott Coues.)

STEJNEGER, LEONHARD.
    1885. Results of ornithological explorations in the Commander Islands and in Kamtschatka. U. S. Nat. Mus. Bull. 29.
STOCKARD, CHARLES RUPERT.
    1904. Nesting of the woodpeckers and the vultures in Mississippi. The Auk, vol. 21, pp. 463–471.
STODDARD, HERBERT LEE.
    1931. The bobwhite quail: Its habits, preservation and increase.
STONE, CLARENCE FREEDOM.
    1899. Nest building of Cooper's hawk. The Oologist, vol. 16, pp. 67–68.
STONE, WITMER.
    1922. Hawk flights at Cape May Point, N. J. The Auk, vol. 39, pp. 89–91.
SUMNER, EUSTACE LOWELL, Jr.
    1931. Some observations on bird behavior. The Condor, vol. 33, pp. 89–91.
SUTTON, GEORGE MIKSCH.
    1925. Notes on the nesting of the goshawk in Potter County, Pennsylvania. Wilson Bull., vol. 37, pp. 193–199.
    1927. The invasion of goshawks and snowy owls during the winter of 1926–1927. The Cardinal, vol. 2, pp. 35–41.
    1928. Notes on a collection of hawks from Schuylkill County, Pennsylvania. Wilson Bull., vol. 40, pp. 84–95.
SWARTH, HARRY SCHELWALD.
    1920. Birds of the Papago Saguaro National Monument and the neighboring region, Arizona.
    1922. Birds and mammals of the Stikine River region of northern British Columbia and southeastern Alaska. Univ. California Publ. Zool., vol. 24, pp. 125–314.
    1926. Report on a collection of birds and mammals from the Atlin region, northern British Columbia. Univ. California Publ. Zool., vol. 30, pp. 51–162.
TALBOT, DANIEL HECTOR.
    1882. The swallow-tailed kite in Dakota. Bull. Nuttall Orn. Club, vol. 7, p. 59.
TAVERNER, PERCY ALGERNON.
    1919. The birds of Red Deer River, Alberta. The Auk, vol. 36, pp. 1–21.
    1926. Birds of western Canada. Victoria Mem. Mus. Bull. 41.
    1927. A study of Buteo borealis, the red-tailed hawk, and its varieties in Canada. Victoria Mem. Mus. Bull. 48.
    1936. Taxonomic comments on red-tailed hawks. The Condor, vol. 38, pp. 66–71.
TAVERNER, PERCY ALGERNON, and SWALES, BRADSHAW HALL.
    1907. The birds of Point Pelee. Wilson Bull., no. 60, pp. 82–99.
TAYLOR, HENRY REED.
    1895. Collecting a condor's egg. The Nidiologist, vol. 2, p. 88.
TAYLOR, WILLIAM PENN, and SHAW, WILLIAM THOMAS.
    1927. Mammals and birds of Mount Rainier National Park.
THOMAS, EDWARD SINCLAIR.
    1928. Nesting of the black vulture in Hocking County, Ohio. Ohio State Mus. Sci. Bull., vol. 1, no. 1, pp. 29–35.
THOMAS, GERALD BAMBER.
    1908. The Mexican black hawk. The Condor, vol. 10, pp. 116–118.
THOMPSON, ERNEST EVAN. (See Seton, Ernest Thompson.)

TODD, WALTER EDMOND CLYDE, and WORTHINGTON, WILLIS WOODFORD.

1911. A contribution to the ornithology of the Bahama Islands. Ann. Carnegie Mus., vol. 7, pp. 388–464.

TOWNSEND, CHARLES HASKINS.

1887. Field notes on the mammals, birds, and reptiles of northern California. Proc. U. S. Nat. Mus., vol. 10, pp. 159–241.

1897. Descriptions of a new eagle from Alaska and a new squirrel from Lower California. Proc. Biol. Soc. Washington, vol. 11, pp. 145–146.

TOWNSEND, CHARLES WENDELL.

1905. The birds of Essex County, Massachusetts. Mem. Nuttall Orn. Club, no. 3.

1906. Notes on the birds of Cape Breton Island. The Auk, vol. 23, pp. 172–179.

1913. Some more Labrador notes. The Auk, vol. 30, pp. 1–10.

1920. Supplement to the birds of Essex County, Massachusetts.

1927. Notes on the courtship of the lesser scaup, Everglade kite, crow, and boat-tailed and great-tailed grackles. The Auk, vol. 44, pp. 549–554.

1930. Pursuit and capture by birds of prey. Bull. Essex County Orn. Club, 1930, pp. 55–61.

TOWNSEND, MANLEY BACON.

1914. Turkey vultures in northwestern Iowa. Bird-Lore, vol. 16, pp. 279–280.

TROWBRIDGE, CHARLES CHRISTOPHER.

1895. Hawk flights in Connecticut. The Auk, vol. 12, pp. 259–270.

TURNER, LUCIEN MCSHAN.

1886. Contributions to the natural history of Alaska. Part 5: Birds.

URNER, CHARLES ANDERSON.

1925. Notes on two ground-nesting birds of prey. The Auk, vol. 42, pp. 31–41.

VAN KAMMEN, I. J.

1916. Relative to the bald eagle in Alaska. The Oologist, vol. 33, pp. 156–158.

VENNOR, HENRY G.

1876. Our birds of prey, or the eagles, hawks, and owls.

WARREN, BENJAMIN HARRY.

1890. Report on the birds of Pennsylvania. Ed. 2.

WARRINER, BENJAMIN R.

1934. An osprey note. The Migrant, vol. 5, p. 43.

WAYNE, ARTHUR TREZEVANT.

1910. Birds of South Carolina. Contr. Charleston Mus., no. 1.

WETMORE, ALEXANDER.

1920. Observations on the habits of birds at Lake Burford, New Mexico. The Auk, vol. 37, pp. 393–412.

1926. Observations on the birds of Argentina, Paraguay, Uruguay, and Chile. U. S. Nat. Mus. Bull. 133.

1931. The avifauna of the Pleistocene of Florida. Smithsonian Misc. Coll., vol. 85, pp. 1–45.

WHEATON, JOHN MAYNARD.

1882. Report on the birds of Ohio. Geol. Surv. Ohio, vol. 4, pt. 1, Zoology, sect. 2, pp. 187–628.

WHEELOCK, IRENE GROSVENOR.

1904. Birds of California.

WIDMANN, OTTO.

1907. A preliminary catalog of the birds of Missouri.

WILLARD, FRANCIS COTTLE.

1916a. The golden eagle in Cochise County, Arizona. The Oologist, vol. 33, pp. 3–8.

1916b. Notes on the golden eagle in Arizona. The Condor, vol. 18, pp. 200–201.

WILLETT, GEORGE.

1927. Notes on the occurrence and distribution of some southeastern Alaskan birds. The Condor, vol. 29, pp. 58–60.

WILLIAMS, MERTON YARWOOD.

1921. Notes on the fauna of lower Pagwachuan, lower Kenogami, and lower Albany Rivers of Ontario. Can. Field-Nat., vol. 25, pp. 94–98.

WILLIAMSON, EDWARD BRUCE.

1913. Actions of nesting red-shouldered hawks. The Auk, vol. 30, pp. 582–583.

1915. Actions of the red-tailed hawk. The Auk, vol. 32, pp. 100–101.

WILSON, ALEXANDER.

1832. American ornithology.

WILSON, FRANK NORMAN.

1927. A citizen of the marshes. Bird-Lore, vol. 29, pp. 397–402.

WITHERBY, HARRY FORBES, et al.

1924. A practical handbook of British birds.

WOOD, CASEY ALBERT.

1922. Economic status of *Coragyps urubu* in British Guiana. The Auk, vol. 39, pp. 255–256.

WOOD, JOHN CLAIRE.

1906. Green leaves in nest. The Oologist, vol. 23, p. 5.

WOOD, NORMAN ASA.

1911. The results of the Mershon expedition to the Charity Islands, Lake Huron. Wilson Bull., vol. 23, pp. 78–112.

1932. Harlan's hawk. Wilson Bull., vol. 44, pp. 78–87.

WRIGHT, ALBERT HAZEN, and HARPER, FRANCIS.

1913. A biological reconnaissance of Okefinokee Swamp: The birds. The Auk, vol. 30, pp. 477–505.

Southern California.   © W. L. Finley and H. T. Bohlman.

PAIR OF CALIFORNIA CONDORS.

Ventura County, Calif., March 17, 1929.   A. C. Bent.

© S. B. and L. G. Peyton.

NEST SITE AND NEST OF CALIFORNIA CONDOR.

Southern California, March 23, 1906.   W. L. Finley and H. T. Bohlman.

Two days old.

Southern California, April 26, 1906.   W. L. Finley and H. T. Bohlman.

Thirty-five days old.

YOUNG CALIFORNIA CONDOR.

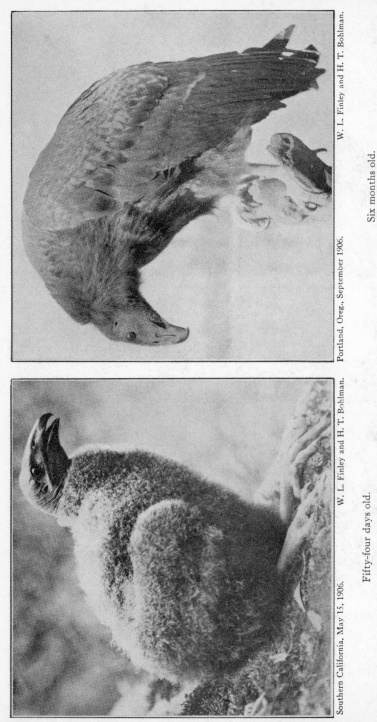

W. L. Finley and H. T. Bohlman.

Portland, Oreg., September 1906.

Six months old.

W. L. Finley and H. T. Bohlman.

Southern California, May 15, 1906.

Fifty-four days old.

YOUNG CALIFORNIA CONDOR.

Southern California, April 11, 1906.   © W. L. Finley and H. T. Bohlman.

Pair of adults caressing.

From the Big Sur, Calif., 1896.   F. H. Holmes.

CALIFORNIA CONDORS.

Klamath Indian Reservation, Oreg., May 20, 1926.   J. E. Patterson.

Under a mountain cliff.

Orleans County, N. Y., May 28, 1927.   W. A. Smith.

In a hollow log in woods.

NESTS OF TURKEY VULTURES.

Near Salem, N. J., April 23, 1930.   R. L. Coffin.

In a stall in a deserted barn.

Near Medford, N. J., May 17, 1929.   R. L. Coffin.

Double nest in a laurel thicket.

NESTS OF TURKEY VULTURES.

Catonsville, Md., May 21, 1933.   W. B. Tyrrell.

About 3 or 4 days old.

Catonsville, Md., June 4, 1933.   W. B. Tyrrell.

About 17 days old.

YOUNG TURKEY VULTURES.

Catonsville, Md., July 17, 1933.   W. B. Tyrrell.

About 60 days old.

Catonsville, Md., July 25, 1933.   W. B. Tyrrell.

About 68 days old.
YOUNG TURKEY VULTURES.

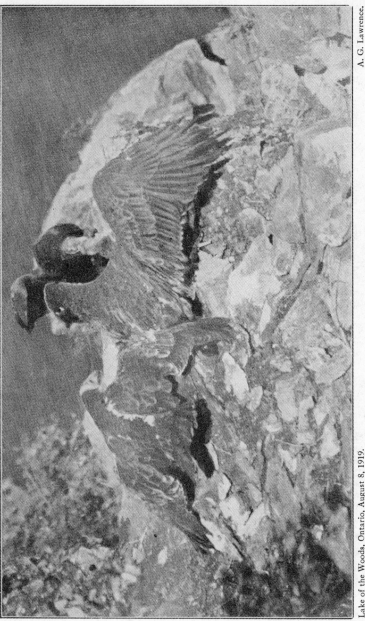

A. G. Lawrence.

Lake of the Woods, Ontario, August 8, 1919.   TURKEY VULTURE FEEDING FULLY GROWN YOUNG.

Duval County, Fla., February 25, 1933.   S. A. Grimes.

Leaving nest in a hollow cypress.

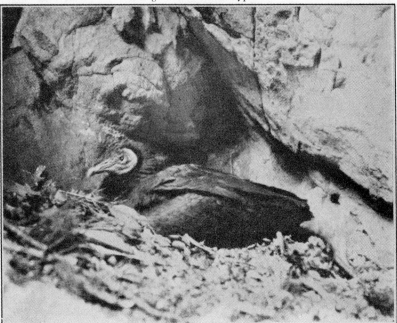

Hocking County, Ohio, April 1926.   E. S. Thomas.

On nest on a cliff.

BLACK VULTURES

Florida.   O. E. Baynard.

In a saw-palmetto patch.

NESTING OF BLACK VULTURE

Florida.

D. J. Nicholson.

Downy young.

Near Nashville, Tenn., 1920.

H. S. Vaughn.

Nest in a cliff.

BLACK VULTURE.

Baker County, Fla., October 18, 1925.   S. A. Grimes.

Pinellas County, Fla., April 21, 1925.   A. C. Bent.

BLACK-VULTURE ROOSTS.

A. A. Allen.

Taylor Creek, Fla., April 13, 1924.

NEST SITES OF SWALLOW-TAILED KITE.

Collier County, Fla., April 6, 1928.   D. J. Nicholson.

Taylor Creek, Fla., May 11, 1924.   A. A. Allen.

NESTS OF SWALLOW-TAILED KITES.

Ventura County, Calif., April 27, 1929.   A. C. Bent.

NEST SITE AND NEST OF WHITE-TAILED KITE.

Santa Clara County, Calif., June 1928.   G. B. Pickwell.

NEST SITES OF WHITE-TAILED KITE.

Santa Clara County, Calif., July 27, 1928.   G. B. Pickwell.

Nest-leaving age.

Southern California, 1927.   © W. M. Pierce.

Fledgling.

YOUNG WHITE-TAILED KITES.

G. B. Pickwell.

WHITE-TAILED KITE NEAR ITS NEST

Santa Clara County, Calif., June 1928.

Near Vicksburg, Miss., May 26, 1901. A. F. Ganier.

Near Vicksburg, Miss., May 25, 1902. A. F. Ganier.

NESTS OF MISSISSIPPI KITES.

Near Vicksburg, Miss., August.   Wing-tipped adult.   A. F. Ganier.

Medicine Lodge, Kans., June 1932.   Eggs in nest.   Walter Colvin.

MISSISSIPPI KITE.

Brevard County, Fla., May 1925.   D. J. Nicholson.

Young about to leave the nest.

Brevard County, Fla., April 19, 1931.   D. J. Nicholson.

Nest in sawgrass.

NESTS OF EVERGLADE KITES.

St. Johns Marsh, Fla., March 7, 1926.   W. A. Smith.

Downy young in nest.

Brevard County, Fla., March 15, 1927.   W. L. Dawson.

Nest and eggs.
EVERGLADE KITE.

W. L. Dawson.

EVERGLADE KITES.

Brevard County, Fla., March 14 and 17, 1927.

Nelson County, N. Dak., June 3, 1901.   A. C. Bent.

Lakeville, Mass., April 30, 1898.   A. C. Bent.

NESTS OF MARSH HAWKS.

Utah County, Utah, May 17, 1929.   R. G. Bee.

Fifteen days old.

Utah County, Utah, June 2, 1929.   R. G. Bee.

Thirty-one days old.

YOUNG MARSH HAWKS.

A. A. Allen.

Canoga, N. Y., June 26, 1914.   Male at nest.

MARSH HAWKS.

A. A. Allen.

Juvenal plumage.

Ithaca, N. Y.

Hennepin County, Minn., June 1929.   S. A. Grimes.

Female protesting.

St. Martins, Quebec, May 28, 1931.   W. J. Brown.

Large set of eggs.

MARSH HAWK.

F. N. Wilson.

FEMALE MARSH HAWK BROODING YOUNG.

Near Ann Arbor, Mich., June 11, 1927.

F. N. Wilson.

Near Ann Arbor, Mich., June 17, 1927.

FEMALE MARSH HAWK AND YOUNG.

Huachuca Mountains, Ariz., May 28, 1922.   A. C. Bent.

Taunton, Mass., June 17, 1900.   A. C. Bent.

NESTS OF SHARP-SHINNED HAWKS.

A. A. Allen.

Ithaca, N. Y., July 5, 1924.   Nestlings.

A. A. Allen.

Ithaca. N. Y.   Juvenal plumage.

YOUNG SHARP-SHINNED HAWKS.

Isle Jesus, Quebec, August 1930.   W. J. Brown.

Fledgling.

Caldwell, N. J., July 15, 1928.   R. T. Peterson.

Four weeks old.

YOUNG SHARP-SHINNED HAWKS

A. C. Bent.

Carver, Mass., May 22, 1904.

NEST SITES OF COOPER'S HAWKS.

F. C. Willard.

Huachuca Mountains, Ariz.

Taunton, Mass., May 7, 1920.   A. C. Bent.

Erie County, N. Y., June 7, 1928.   S. A. Grimes.

NESTS OF COOPER'S HAWKS.

Taunton, Mass., June 18, 1903.   A. C. Bent.

Ten days old.

Taunton, Mass., June 30, 1903.   A. C. Bent.

Twenty-two days old.

YOUNG COOPER'S HAWKS.

Pomona, Calif.   Twenty-eight days old.   W. M. Pierce.

Kent, Conn., August 1903.   A. C. Bent.

Full juvenal plumage.

YOUNG COOPER'S HAWKS.

A. C. Bent.

Fox Island River, Newfoundland, June 10, 1912.

NESTING SITES OF EASTERN GOSHAWKS.

J. R. Gillin.

Wayne County, Pa., April 24, 1935.

Wayne County, Pa., April 24, 1935.   J. R. Gillin.

Huntington, Mass., April 19, 1931.   A. A. Cross.

NESTS OF EASTERN GOSHAWKS.

A. M. Bailey and R. T. Niedrach, Courtesy Colorado Museum of Natural History.

EASTERN GOSHAWK.

Colorado, May 30, 1936.

Whartons Lake, Alberta, June 5, 1926.   R. H. Rauch.

Whartons Lake, Alberta, June 3, 1926.   R. H. Rauch.

NEST SITE AND YOUNG OF EASTERN GOSHAWK.

M. S. Ray.

Eldorado County, Calif., April 15, 1931.

M. S. Ray.

Eldorado County, Calif., May 24, 1926.

NEST SITES OF WESTERN GOSHAWKS.

A. C. Bent.

Carver, Mass., May 4, 1901.

A. C. Bent.

Rehoboth, Mass., May 5, 1928.

NEST SITES OF EASTERN RED-TAILED HAWKS.

Taunton, Mass., April 18, 1906.   A. C. Bent.

Typical nest.

Minneapolis, Minn., fall 1929.   S. A. Grimes.

First-year plumage.

EASTERN RED-TAILED HAWK.

Nashville, Tenn.

H. S. Vaughn.

ADULT EASTERN RED-TAILED HAWK.

Near Nashville, Tenn.   H. S. Vaughn.

One hundred feet above ground

NEST SITE OF EASTERN RED-TAILED HAWK.

A. C. Bent.

Stump Lake, N. Dak., June 1, 1901.

A. D. Patton.

Custer County, S. Dak., April 18, 1925.

Cochise County, Ariz., April 8, 1922.

A. C. Bent.

In a soapweed yucca.

W. L. Dawson.

In an ocotillo bush.

NEST SITES OF WESTERN RED-TAILED HAWKS.

Southern California, March 17, 1928.   E. L. Sumner, Jr.

Spring Lake, Oreg., May 20, 1929.   J. E. Patterson.

Nest in a juniper.

NESTING OF WESTERN RED-TAILED HAWKS.

Oregon, May 3, 1902.   W. L. Finley and H. T. Bohlman.

Thirteen days old.

Oregon, May 24, 1902.   W. L. Finley and H. T. Bohlman.

Thirty-four days old.

YOUNG WESTERN RED-TAILED HAWKS.

Bassinger, Fla., March 21, 1925.   A. C. Bent.

Glades County, Fla., February 15, 1930.   A. C. Bent.

NEST SITES OF FLORIDA RED-TAILED HAWKS.

Taunton, Mass., April 8, 1906.   A. C. Bent.

Seekonk, Mass., May 6, 1905.   A. C. Bent.

NESTING OF NORTHERN RED-SHOULDERED HAWKS.

A. D. Dubois.

NEST OF NORTHERN RED-SHOULDERED HAWK.

Tompkins County, N. Y., April 30, 1910.

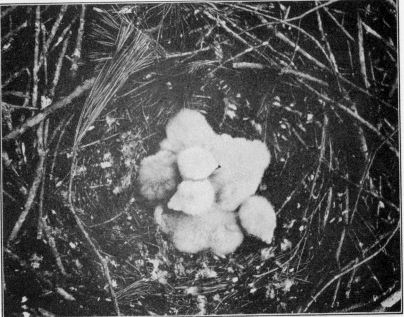

Raynham, Mass., May 15, 1902.

A. C. Bent.

Averaging seven days old.

Raynham, Mass., May 22, 1902.

A. C. Bent.

Same birds one week later.

YOUNG NORTHERN RED-SHOULDERED HAWKS.

Matteson, Ill., May 24, 1931. K. Abegg. Twenty-one days old.

Matteson, Ill., June 8, 1931. K. Abegg. Thirty-six days old.

YOUNG NORTHERN RED-SHOULDERED HAWKS.

Incubating.

Cook County, Ill., April 18, 1929.   A. M. Bailey.

Defiant.

NORTHERN RED-SHOULDERED HAWK.

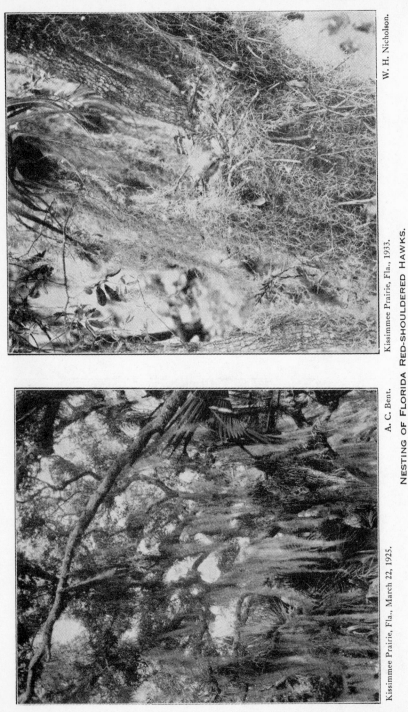

W. H. Nicholson.

Kissimmee Prairie, Fla., 1933.

A. C. Bent.

Kissimmee Prairie, Fla., March 22, 1925.

NESTING OF FLORIDA RED-SHOULDERED HAWKS.

San Diego County, Calif., April 9, 1929.   A. C. Bent.

NEST SITES OF RED-BELLIED HAWKS.

Riverside County, Calif., April 9, 1929.   W. M. Pierce.

E. L. Sumner, Jr.

Nest in above site.

NESTING OF RED-BELLIED HAWKS.

A. C. Bent.

NEST SITES OF ZONE-TAILED HAWKS.

Catalina Mountains, Ariz., April 18 and 19, 1922.

Point Isabel, Tex., May 14, 1924.   Herbert Friedmann.

Brownsville, Tex., June 1929.   A. H. Cordier.

YOUNG SENNETT'S WHITE-TAILED HAWKS.

Crane Lake, Saskatchewan, June 3, 1905.   A. C. Bent.

NESTING OF SWAINSON'S HAWK.

© W. M. Pierce.

Mojave Desert, Calif., June 18, 1916.

NESTING OF SWAINSON'S HAWK

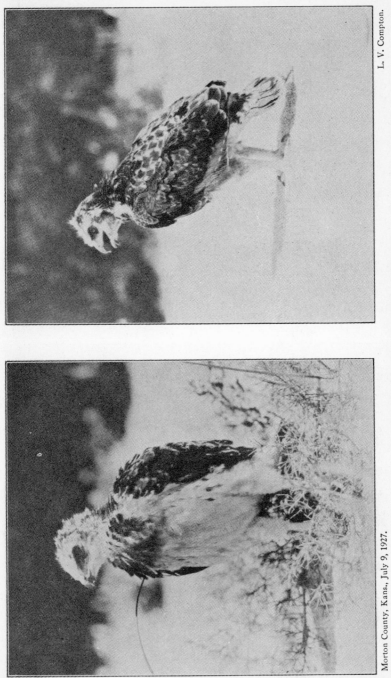

L. V. Compton.

Morton County, Kans., July 9, 1927.

FLEDGLING SWAINSON'S HAWK.

Grant County, Wash., May 25, 1932.   S. F. Rathbun.

Nest site.

W. L. Dawson.

Juvenal plumage.

SWAINSON'S HAWK.

S. F. Rathbun.

Grant County, Wash., May 25, 1931.

W. L. Dawson.

A desert nest.

NEST SITES OF SWAINSON'S HAWKS.

Carver, Mass., May 22, 1904.

A. C. Bent.

A. C. Bent.

Rehoboth, Mass., June 3, 1928.

NEST SITES OF BROAD-WINGED HAWKS.

Taunton, Mass., May 22, 1900. · · · · · · · · · · · · · · · A. C. Bent.

Kent, Conn., May 1908. · · · · · · · · · · · · · · · H. K. Job.

NESTS OF BROAD-WINGED HAWKS.

Pensacola, Fla., June 20, 1934.   C. Kingsbury.

Hennepin County, Minn., July 1929.   S. A. Grimes.

YOUNG BROAD-WINGED HAWKS.

Lake Istokpoga, Fla., March 30, 1923.   H. W. Brandt.

NEST OF SHORT-TAILED HAWK.

Pima County, Ariz., May 19, 1922.   A. C. Bent.

NEST SITE OF MEXICAN BLACK HAWK.

Pima County, Ariz., May 20, 1922.

A. C. Bent.

Pima County, Ariz.

F. C. Willard.

NEST SITES OF MEXICAN GOSHAWKS.

Pima County, Ariz., May 20, 1922.   A. C. Bent.

Pima County, Ariz.   F. C. Willard.

NESTING OF MEXICAN GOSHAWKS.

Churchill, Manitoba, June 26, 1933.   A. Marguerite Heydweiller.

Golovin Bay, Alaska, July 1921.   A. M. Bailey.

NESTS OF AMERICAN ROUGH-LEGGED HAWKS.

Stump Lake, N. Dak., June 4, 1901.   A. C. Bent.

Crane Lake, Saskatchewan, June 4, 1905.   A. C. Bent.

NEST SITES OF FERRUGINOUS ROUGHLEGS.

Walter Colvin.

EGGS OF FERRUGINOUS ROUGHLEG.

Paca County, Colo., May 10, 1921.

Russell Reid.

YOUNG FERRUGINOUS ROUGHLEGS.

Near Amidon, N. Dak., May 30, 1928.

Near Patricia, Alberta.   G. F. Sternberg.

NEST SITE OF FERRUGINOUS ROUGHLEG.

G. F. Sternberg.

Near Patricia, Sand Creek Badlands, Alberta.

YOUNG FERRUGINOUS ROUGHLEGS.

Los Angeles County, Calif., February 28, 1929.   A. C. Bent.

Huachuca Mountains, Ariz., May 15, 1922.   A. C. Bent.

NEST SITES OF GOLDEN EAGLES.

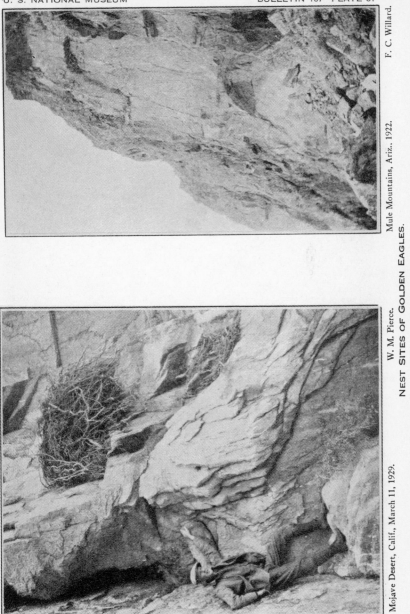

F. C. Willard.

Mule Mountains, Ariz., 1922.

W. M. Pierce.

NEST SITES OF GOLDEN EAGLES.

Mojave Desert, Calif., March 11, 1929.

Southern California, March 9, 1929.   E. L. Sumner, Jr.

Southern California, March 10, 1919.   W. M. Pierce.

EGGS OF GOLDEN EAGLES.

Southern California, March 25, 1928.   E. L. Sumner, Jr.

Just hatched.

Southern California, April 3, 1927.   W. M. Pierce.

Two weeks old.

YOUNG GOLDEN EAGLES.

Southern California, April 23, 1929.   E. L. Sumner, Jr.

Thirty-three days old.

Southern California, May 8, 1927.   W. M. Pierce.

Seven weeks old.

YOUNG GOLDEN EAGLES.

Southern California, May 15, 1927.   W. M. Pierce.

Eight weeks old.

Southern California, June 2, 1928.   E. L. Sumner, Jr.

Ten weeks old.

YOUNG GOLDEN EAGLES.

Pinellas County, Fla.　　　　　　　　　W. G. Fargo.

The author climbing.

Pinellas County, Fla., February 14, 1925.　　A. C. Bent.

Another nest.

NEST SITES OF SOUTHERN BALD EAGLES.

Pinellas County, Fla., February 14, 1925.   W. G. Fargo.

Duval County, Fla., December 11, 1930.   S. A. Grimes.

NESTS OF SOUTHERN BALD EAGLES.

Duval County, Fla., December 30, 1930.   S. A. Grimes.

About one week old.

Duval County, Fla., January 16, 1931.   S. A. Grimes.

About three weeks old.

YOUNG SOUTHERN BALD EAGLES.

Duval County, Fla., January 25, 1931.   S. A. Grimes.

About five weeks old.

St. Johns County, Fla., February 10, 1931.   S. A. Grimes.

About seven weeks old.

YOUNG SOUTHERN BALD EAGLES.

Santa Catalina Island, Calif., February 22, 1929.   A. C. Bent.

J. R. Pemberton climbing.

Santa Cruz Island, Calif., March 2, 1917.   M. C. Badger.

S. B. Peyton climbing.

NEST SITES OF BALD EAGLES.

A. C. Bent.

Bird's-eye view of nest.

NESTING OF NORTHERN BALD EAGLE.

Phippsburg, Maine, April 21, 1900.

Owen Durfee climbing.

Vermilion, Ohio.

F. H. Herrick.

NORTHERN BALD EAGLES ON THE "GREAT NEST."

F. H. Herrick.

NORTHERN BALD EAGLE.

Vermilion, Ohio.

F. H. Herrick.

NORTHERN BALD EAGLE.

Vermilion, Ohio.

F. H. Herrick.

NORTHERN BALD EAGLE FEEDING YOUNG.

Vermilion, Ohio.

Swansea, Mass., May 4, 1913.

A. C. Bent.

Dighton, Mass., May 4, 1913.

A. C. Bent.

NESTING SITES OF AMERICAN OSPREYS.

J. E. Patterson.

Tule Lake, Calif., April 1920.

H. S. Hathaway.

Swansea, Mass., October 31, 1897.

Lower California, 1926.   © W. M. Pierce.

Gardiners Island, N. Y., July 30, 1927.   L. W. Turrell.

NESTS OF AMERICAN OSPREYS.

Lyme, Conn., June 2, 1934.   A. C. Bent.

NESTS OF AMERICAN OSPREYS.

Duval County, Fla., April 1932.   S. A. Grimes.

Swansea, Mass., May 25, 1902.   A. C. Bent.

Young just hatched.

NESTS OF AMERICAN OSPREYS.

Duval County, Fla., April 29, 1931.   S. A. Grimes.

Largest about one week old.

Swansea, Mass., June 30, 1917.   A. C. Bent.

Nearly half grown.

YOUNG AMERICAN OSPREYS.

Warren, R. I., June 27, 1903.   A. C. Bent.

Adult alighting on nest.

Swansea, Mass., July 16, 1905.   A. C. Bent.

Young nearly grown.

AMERICAN OSPREYS.

INDEX

Abbott, C. G., on American osprey, 355, 358, 363, 368, 371, 373.
abbreviatus, Buteo albonotatus, 216.
Abegg, Klauss, viii.
Accipiter cooperi, 112.
   velox velox, 95.
Accipitriidae, 44.
Aiken, C. E. H., on Swainson's hawk, 222.
Aiken, C. E. H., and Warren, E. R., on golden eagle, 304.
alascanus, Haliaeetus leucocephalus, 333, 334.
albicaudatus, Buteo albicaudatus, 221.
albicaudatus albicaudatus, Buteo, 221.
albicaudatus colonus, Buteo, 221.
albicaudatus hypospodius, Buteo, 216.
albicilla, Haliaeetus, 315.
   Haliaeetus albicilla, 316, 320.
albicilla albicilla, Haliaeetus, 316, 320.
albicilla brooksi, Haliaeetus, 316.
albicilla groenlandicus, Haliaeetus, 316, 318.
albonotatus, Buteo, 212.
albonotatus abbreviatus, Buteo, 216.
Allen, C. S., on American osprey, 355, 357, 360, 371.
Allen, F. H., on American rough-legged hawk, 278.
Allen, J. A., on American osprey, 352.
alleni, Buteo lineatus, 197, 199, 208, 210.
American osprey, 352.
American rough-legged hawk, 269.
Anderson, R. L., 305.
Anderson, R. M., on Krider's hawk, 166.
Anderson, W. B., 199.
Angel, Hector, 11.
Animal Industry, U. S. Bureau of, 41.
Anthony, A. W., 206.
   on western red-tailed hawk, 172.
   on zone-tailed hawk, 213.
anthracina, Urubitinga anthracina, 259.
anthracina anthracina, Urubitinga, 259.
anthracina cancrivorus, Urubitinga, 264.
antillarum, Buteo platypterus, 251.
Aquila chrysaëtos canadensis, 293.
Astur atricapillus atricapillus, 125.
   atricapillus striatulus, 138, 139.
Asturina nitida, 268.
   plagiata plagiata, 264.
atratus, Coragyps atratus, 28.
atricapillus, Astur atricapillus, 125.
atricapillus striatulus, Astur, 138, 139.
Audubon, J. J., 147, 330.
   on American osprey, 366.
   on black vulture, 29, 35-37, 39.
   on broad-winged hawk, 246.

Audubon, J. J., on California condor, 8.
   on eastern goshawk, 131.
   on Harlan's hawk, 175, 177.
   on marsh hawk, 91.
   on Mississippi kite, 63, 67, 68.
   on North American white-tailed kite, 56, 62.
   on sharp-shinned hawk, 98.
   on swallow-tailed kite, 45, 49.
aura septentrionalis, Cathartes, 12.
Austin, O. L., Jr., on eastern red-tailed hawk, 165.
Avery, W. C., 118.
Bachman, John, 37, 39.
Badger, M. C., 57.
Bailey, A. M., on American rough-legged hawk, 271.
   on northern bald eagle, 345.
Bailey, B. H., on broad-winged hawk, 243.
Bailey, Mrs. Florence M., on California condor, 2.
   on Harris's hawk, 142.
   on marsh hawk, 88, 89, 91.
   on red-bellied hawk, 199.
   on Sennett's white-tailed hawk, 220.
   on turkey vulture, 19.
Bailey, H. H., on American osprey, 379.
   on short-tailed hawk, 257.
   on swallow-tailed kite, 45, 46.
Bailey, Vernon, 327.
   on American rough-legged hawk, 275.
   on Harris's hawk, 144, 145.
Baird, Brewer, and Ridgway, on California condor, 2, 7.
   on Steller's sea eagle, 351.
   on swallow-tailed kite, 50.
Baker, J. H., on everglade kite, 72.
Baker, J. W., on southern bald eagle, 328.
Bald eagle, northern, 333.
   southern 321.
Bancroft, Griffing, 360.
Bangs, Outram, on Florida red-tailed hawk, 178.
   on insular red-shouldered hawk, 208.
Bangs, Outram, and Penard, T. E., on North American white-tailed kite, 54.
Barlow, Chester, on North American white-tailed kite, 56, 58.
Barrows, W. B., on North American white-tailed kite, 63.
Batchelder, C. F., on southern bald eagle, 328.

399

O

MANUAL OF THE TREES OF NORTH AMERICA
by Charles Sprague Sargent

The greatest dendrologist America has ever produced was without doubt Charles Sprague Sargent, Professor of Arboriculture at Harvard and Director of the Arnold Arboretum in Boston until his death in 1927. His monumental "Manual of the Trees of North America," incorporating the results of 44 years of original research, is still unsurpassed as the most comprehensive and reliable volume on the subject. Almost every other book on American trees is selective, but this one assures you of identifying any native tree; it includes 185 genera and 717 species of trees (and many shrubs) found in the United States, Canada, and Alaska. 783 sharp, clear line drawings illustrate leaves, flowers, and fruit.

First, a 6-page synoptic key breaks trees down into 66 different families; then, an unusually useful 11-page analytical key to genera helps the beginner locate any tree readily by its leaf characteristics. Within the text over 100 further keys aid in identification. The body of the work is a species by species description of leaves, flowers, fruit, winterbuds, bark, wood, growth habits, etc., extraordinary in its fullness and wealth of exact, specific detail. Distinguishing features of this book are its extremely precise locations and distributions; flower and leaf descriptions that indicate immaturity variations; and a strong discussion of varieties and local variants.

Additional useful features are a glossary of technical terms; a system of letter keys classifying trees by regions; and a detailed index of both technical and common names (index, glossary, and introductory keys are printed in both volumes.) Students and teachers of botany and forestry, naturalists, conservationists, and all nature lovers will find this set an unmatched lifetime reference source. "Still the best work," Carl Rogers in "The Tree Book."

Unabridged and unaltered reprint of the 2nd enlarged 1926 edition. Synopsis of Families. Analytical Key of Genera. Glossary. Index. 783 illustrations, 1 map. Total of I + 89lpp. 5⅜ x 8. T277 Vol I Paperbound **$2.00**
T278 Vol II Paperbound **$2.00**
The set **$4.00**

STUDIES ON THE STRUCTURE AND DEVELOPMENT OF VERTEBRATES
by Edwin S. Goodrich

This monumental work by the greatest comparative anatomist of modern times was recognized as the definitive study of the field immediately upon publication. Its wealth of factual detail plus brilliant exposition and theory have made it an indispensable text and reference work for anatomists, medical students, morphologists, histologists, embryologists, students of evolution, zoologists, and every person interested in biology, no matter what his field of specialization.

The skeleton, fins and limbs, head, vascular, respiratory, excretory, genital and nervous systems, and the subdivision of the body-cavity, are covered from fish to the higher mammals. Among the features of the book is the account of the structure and evolutionary history of the ossicles of the ear, extensive cranial studies, a detailed and complete classification of the vertebrate phylum including the sub-orders, the treatment of the separate divisions of the coelom and the diaphragm, the concise and lucid coverage of embryology, and the illustrations (over 300 of them by the author) which are among the very finest ever presented in a single work. The entire study is characterized by the unity and meticulous attention to detail of one man thoroughly familiar with his subject.

"For many a day this will certainly be the standard text-book," JOURNAL OF ANATOMY. "The reviewer knows of no other book in English which covers so thoroughly the more modern work on many aspects of the subject," SCIENCE PROGRESS.

Enlarged by a 69-page biographical study of his life and work by A. C. Hardy. 754 figures. Bibliography of 1186 references. Index. 2 volumes; total 906pp. 5⅜ x 8. 2 vol. set.

S449, 450 Paperbound **$5.00**

TREES OF THE EASTERN AND CENTRAL UNITED STATES AND CANADA

by W. M. Harlow (Professor of Wood Technology, College of Forestry, State University of New York, Syracuse)

This is an extensively revised edition of the standard semipopular TREES OF THE EASTERN UNITED STATES AND CANADA. It covers the important native trees found in this area, and the more common escapes.

A concise introduction first provides you with all that you need to know for general identification of trees, and an 8-page easy-to-use synoptic key enables you to locate at a glance the group or genus of any tree you are likely to meet. Supplementary keys in the main text enable you to locate with ease the various species.

More than 140 different trees are described in detail, with information on general appearance, habit of growth, leaf forms, flowers, fruit, twig appearance, bark, and other features; habitat, distribution by states, commercial use, woodlore, and other information is also given.

Treatment throughout this fascinating book is semipopular; it is neither a slight sketch, nor a monograph for the specialist. It is a middle-range book, carefully written for the intelligent reader who is sincerely interested in accurate information about trees.

With this book you will be able to identify at sight almost any tree you are likely to encounter, not only by group, but also by species. You will be able to tell a sugar maple from a Norway maple, a black oak from a white oak, a horse-chestnut from a chestnut. You will know which trees have edible fruit, which are suitable for house plantings; what climate and drainage conditions trees need, and much other useful and interesting information.

More than 600 photographs and figures. xiii + 288pp. 4⅝ x 6½.

T395 Paperbound $1.35

FRUIT KEY AND TWIG KEY TO TREES AND SHRUBS

FRUIT KEY TO NORTHEASTERN TREES
TWIG KEY TO THE DECIDUOUS WOODY PLANTS OF EASTERN NORTH AMERICA

by William M. Harlow
(Professor of Wood Technology, College of Forestry, State University of New York, Syracuse)

Bound together for the first time in one volume, these handy, accurate, and easily used keys to fruit and twig identification are the only guides of their sort with photographs — over 350 of them, of nearly every twig and fruit described — making them especially valuable to the novice.

The fruit key (dealing with both deciduous trees and evergreens) begins with a concise introduction, explaining simply and lucidly the process of seeding, and identifying the various organs involved: the cones and flowers, and their component parts and variations. Next the various types of fruits are described — drupe, berry, pome, legume, follicle, capsule, achene, samara, nut — and fruiting habits, followed by a synoptic summary of fruit types.

The introduction to the twig key tells in plain language the process of growth, and its relation to twig morphology through leaf scars, branch scars, buds, etc. For the benefit of the unwary, poison-ivy, poison-oak, and poison-sumac are immediately and fully described.

Identification in both books is easy. There is a pair of alternative descriptions of each aspect of the specimens. Your choice of the fitting one leads you automatically to the next proper pair. At the end of the chain is the name of your specimen and, as a double check, a photograph. More than 120 different fruits and 160 different twigs are distinguished.

This exceptional work, widely used in university courses in botany, biology, forestry, etc., is a valuable tool and instructor to the naturalist, woodsman, or farmer, and to anyone who has wondered about the name of a leafless tree in winter, or been intrigued by an interestingly shaped fruit or seed.

Over 350 photographs, up to 3 times natural size. Bibliography, glossary, index of common and scientific names, in each key. Total of xvii + 126pp. 5⅝ x 8⅜. Two volumes bound as one.   T511 Paperbound $1.25